SOLUTIONS MANUAL
JAN WILLIAM SIMEK
California Polytechnic State University

Organic
Chemistry
Fourth Edition

L.J. WADE, JR.

Prentice Hall, Upper Saddle River, NJ 07458

Associate Editor: Mary Hornby
Acquisitions Editor: Matt Hart
Special Projects Manager: Barbara A. Murray
Supplement Cover Manager: Paul Gourhan
Supplement Cover Designer: PM Workshop Inc.
Manufacturing Buyer: Ben Smith

Printed in the United States of America

10 9 8 7 6 5 4 3 2 1

ISBN 0-13-974023-6

Prentice-Hall International (UK) Limited, *London*
Prentice-Hall of Australia Pty. Limited, *Sydney*
Prentice-Hall Canada, Inc., *London*
Prentice-Hall Hispanoamericana, S.A., *Mexico*
Prentice-Hall of India Private Limited, *New Delhi*
Prentice-Hall of Japan, Inc., *Tokyo*
Simon & Schuster Asia Pte. Ltd., *Singapore*
Editora Prentice-Hall do Brazil, Ltda., *Rio de Janeiro*

TABLE OF CONTENTS

Hints for Passing Organic Chemistry

Do you want to pass your course in organic chemistry? Here is my best advice, based on twenty-plus years of observing students learning organic chemistry:

Hint #1: *Do the problems*. It seems straightforward, but humans, including students, try to take the easy way out until they discover there is no short-cut. Unless you have a measured IQ above 200 and comfortably cruise in the top 1% of your class, *do the problems*. Usually your teacher (professor or teaching assistant) will recommend certain ones; try to do all those recommended. If you do half of them, you will be half-prepared at test time. (Do you want your surgeon coming to your appendectomy having practiced only *half* the procedure?) And when you do the problems, keep this Solutions Manual CLOSED. Avoid looking at *my* answer before you write *your* answer—your trying and struggling with the problem is the most valuable part of the problem. Discovery is a major part of learning. Remember that the primary goal of doing these problems is *not* just getting the right answer, but understanding the material well enough to get right answers to the questions you haven't seen yet.

Hint #2: *Keep up*. Getting behind in your work in a course that moves as quickly as this one is the Kiss of Death. For most students, organic chemistry is the most rigorous intellectual challenge they have faced so far in their studies. Some are taken by surprise at the diligence it requires. Don't think that you can study all of the material the couple of days before the exam—well, you can, but you won't pass. Study organic chemistry like a foreign language: try to do some every day so that the freshly-trained neurons stay sharp.

Hint #3: *Get help when you need it*. Use your teacher's office hours when you have difficulty. Many schools have tutoring centers (in which organic chemistry is a popular offering). Here's a secret: absolutely the best way to cement this material in your brain is to get together with a few of your fellow students and make up problems for each other, then correct and discuss them. When *you* write the problems, you will gain great insight into what this is all about.

Purpose of this Solutions Manual

So what is the point of this Solutions Manual? First, I can't do your studying for you. Second, since I am not leaning over your shoulder as you write your answers, I can't give you direct feedback on what you write and think—the print medium is limited in its usefulness. What I *can* do for you is:
1) provide correct answers; the publishers, Professor Wade, Dr. Meisenheimer (my reviewer), and I have gone to great lengths to assure that what I have written is correct, for we all understand how it can shake a student's confidence to discover that the answer book flubbed up; 2) provide a considerable degree of rigor; beyond the fundamental requirement of correctness, I have tried to flesh out these answers, being complete but succinct; 3) provide insight into how to solve a problem and into where the sticky intellectual points are. Insight is the toughest to accomplish, but over the years, I have come to understand where students have trouble, so I have tried to anticipate your questions and to add enough detail so that the concept, as well as the answer, is clear.

It is difficult for students to understand or acknowledge that their teachers are human (some are more human than others). Since I am human (despite what my students might report), I can and do make mistakes. If there are mistakes in this book, they are my sole responsibility, and I am sorry. If you find one, PLEASE let me know so that it can be corrected in future printings. Nip it in the bud.

What's New in this edition?

Better answers! Part of my goal in this edition has been to add more explanatory material to clarify how to arrive at the answer.

Better graphics! The print medium is very limited in its ability to convey three-dimensional structural information, a problem that has plagued organic chemists for over a century. I have added some graphics created in the software, Chem3D®, to try to show atoms in space where that information is a key part of the solution.

Appendix 2! Acidity and basicity are such critical properties of organic molecules that I wanted to give a summary of the effects of molecular structure on acidity and basicity.

Web stuff! See the next topic.

Some Web Stuff

Since the last edition of this Solutions Manual, the World Wide Web has become ubiquitous and a common tool for information access. I have a rudimentary web page where I intend to post any corrections that may arise. So if you're curious, try our department web site at **http://www.calpoly.edu/~chem/**, then follow the "Faculty" button and "Simek, Jan" button, and follow the links. My web page will be perpetually under construction, just like our lives.

Prentice-Hall maintains a web site dedicated to the Wade text: try **www.prenhall.com/wade**.

Acknowledgments

No project of this scope is ever done alone. These are team efforts, and there are several people who have assisted and facilitated in one fashion or another who deserve my thanks.

Professor L. G. Wade, Jr., your textbook author, is a remarkable person. He has gone to extraordinary lengths to make the textbook as clear, organized, informative and insightful as possible. He has solicited and followed my suggestions on his text, and his comments on my solutions have been perceptive and valuable. We agreed early on that our primary goal is to help the students learn a fascinating and challenging subject, and all of our efforts have been directed toward that goal. I have appreciated our collaboration.

My former student, current friend and colleague, and now professor, Dr. Kristen Meisenheimer, has reviewed every problem and every solution. Her precision, diligence, and sensitivity (and diplomacy!) have made this a much better supplement. I will be forever grateful for her enthusiasm, wisdom, and devotion.

The people at Prentice-Hall have made this project possible. Good books would not exist without their dedication, professionalism, and experience. Among the many people who contributed are: Lee Englander, who connected me with this project; Matthew Hart and John Challice, Chemistry Editors; and Mary Hornby, Associate Editor.

The entire manuscript was produced using *ChemDraw®*, the remarkable software for drawing chemical structures developed by CambridgeSoft Corp., Cambridge, MA. We, the users of sophisticated software like ChemDraw, are the beneficiaries of the intelligence and creativity of the people in the computer industry. We are fortunate that they are so smart.

Finally, I appreciate my friends who supported me throughout this project, most notably my good friend of over thirty years, Judy Lang. The students are too numerous to list, but it is for them that all this happens.

Jan William Simek
Department of Chemistry and Biochemistry
Cal Poly State University
San Luis Obispo, CA 93407
Internet: jsimek@calpoly.edu

DEDICATION

To my inspirational chemistry teachers:

Joe Plaskas, who made the batter;

Kurt Kaufman, who baked the cake;

Carl Djerassi, who put on the icing;

and to my parents:

Ervin J. and Imilda B. Simek,

who had the original concept.

SYMBOLS AND ABBREVIATIONS

Below is a list of symbols and abbreviations used in this Solutions Manual, consistent with those used in the textbook by Wade. (Do not expect all of these to make sense to you now. You will learn them throughout your study of organic chemistry.)

BONDS

———	a single bond
═══	a double bond
≡≡≡	a triple bond
▬▬◄	a bond in three dimensions, coming out of the paper toward the reader
▥▥▥	a bond in three dimensions, going behind the paper away from the reader
- - - - - -	a stretched bond, in the process of forming or breaking

ARROWS

⟶	in a reaction, shows direction from reactants to products
⇌	signifies equilibrium (not to be confused with resonance)
⟷	signifies resonance (not to be confused with equilibrium)
⤴⤵	shows direction of electron movement: the arrowhead with one barb shows movement of one electron; the arrowhead with two barbs shows movement of a pair of electrons
⊢⟶	shows polarity of a bond or molecule, the arrowhead signifying the more negative end of the dipole

SUBSTITUENT GROUPS

Me	a methyl group, CH_3
Et	an ethyl group, CH_2CH_3
Pr	a propyl group, a three carbon group (two possible arrangements)
Bu	a butyl group, a four carbon group (four possible arrangements)
R	the general abbreviation for an alkyl group (or any substituent group not under scrutiny)
Ph	a phenyl group, the name of a benzene ring as a substituent, represented:

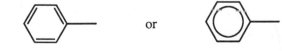

         or

Ar	the general name for an aromatic group

continued on next page

SUBSTITUENT GROUPS, continued

Ac   an acetyl group: $CH_3-\overset{\overset{\textstyle O}{\|}}{C}-$

Cy   a cyclohexyl group:

Ts   tosyl, or *p*-toluenesulfonyl group: CH_3-

Boc   a *t*-butoxycarbonyl group (amino acid and peptide chemistry): $(CH_3)_3C-O-\overset{\overset{\textstyle O}{\|}}{C}-$

Z, or   a carbobenzoxy (benzyloxycarbonyl) group (amino acid and peptide chemistry):
Cbz

$-CH_2-O-\overset{\overset{\textstyle O}{\|}}{C}-$

REAGENTS AND SOLVENTS

DCC   **di**cyclohexyl**c**arbodiimide $-N=C=N-$

DMSO   **d**i**m**ethyl**s**ulf**o**xide

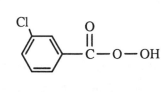

ether   diethyl ether, $CH_3CH_2OCH_2CH_3$

MCPBA   *meta*-**c**hloro**p**eroxy**b**enzoic **a**cid $\overset{\overset{\textstyle O}{\|}}{C}-O-OH$

MVK   **m**ethyl **v**inyl **k**etone

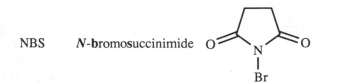

NBS   *N*-**b**romo**s**uccinimide

continued on next page

Symbols and Abbreviations, continued

REAGENTS AND SOLVENTS, continued

PCC pyridinium chlorochromate, $CrO_3 \cdot HCl \cdot N$

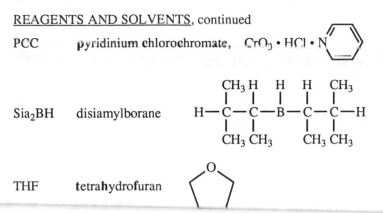

Sia$_2$BH disiamylborane

THF tetrahydrofuran

SPECTROSCOPY

IR	**i**nfrared spectroscopy
NMR	**n**uclear **m**agnetic **r**esonance spectroscopy
MS	**m**ass **s**pectrometry
UV	**u**ltra**v**iolet spectroscopy
ppm	**p**arts **p**er **m**illion, a unit used in NMR
Hz	hertz, cycles per second, a unit of frequency
MHz	megahertz, millions of cycles per second
TMS	tetramethylsilane, $(CH_3)_4Si$, the reference compound in NMR
s, d, t, q	**s**inglet, **d**oublet, **t**riplet, **q**uartet, referring to the number of peaks an NMR absorption gives
nm	nanometers, 10^{-9} meters (usually used as a unit of wavelength)
m/z	mass-to-charge ratio, in mass spectrometry
δ	in NMR, chemical shift value, measured in ppm
λ	wavelength
ν	frequency

OTHER

a, ax	axial (in chair forms of cyclohexane)
e, eq	equatorial (in chair forms of cyclohexane)
HOMO	**h**ighest **o**ccupied **m**olecular **o**rbital
LUMO	**l**owest **u**noccupied **m**olecular **o**rbital
NR	no reaction
o, m, p	*ortho, meta, para* (positions on an aromatic ring)
Δ	when written over an arrow: "heat"; when written before a letter: "change in"
δ^+, δ^-	partial positive charge, partial negative charge
hν	energy from electromagnetic radiation (light)
$[\alpha]_D$	specific rotation at the D line of sodium (589 nm)

1-1 Na $1s^2 2s^2 2p^6 3s^1$

Mg $1s^2 2s^2 2p^6 3s^2$

Al $1s^2 2s^2 2p^6 3s^2 3p_x{}^1$

Si $1s^2 2s^2 2p^6 3s^2 3p_x{}^1 3p_y{}^1$

P $1s^2 2s^2 2p^6 3s^2 3p_x{}^1 3p_y{}^1 3p_z{}^1$

S $1s^2 2s^2 2p^6 3s^2 3p_x{}^2 3p_y{}^1 3p_z{}^1$

Cl $1s^2 2s^2 2p^6 3s^2 3p_x{}^2 3p_y{}^2 3p_z{}^1$

Ar $1s^2 2s^2 2p^6 3s^2 3p_x{}^2 3p_y{}^2 3p_z{}^2$

1-2 In this book, lines between atomic symbols represent covalent bonds between those atoms. Nonbonding electrons are indicated with dots.

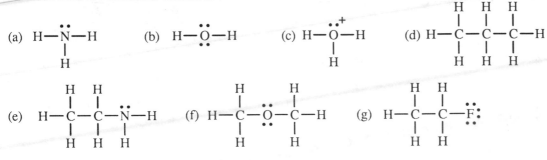

The compounds in (h) and (i) are unusual in that boron does not have an octet of electrons—normal for boron because it has only three valence electrons.

1-3

(a) :N≡N:

(b) H—C≡N:

(c) H—Ö—N̈=Ö:

(d) Ö=C=Ö

(e) H—C̈=N—H with H below C

(f) :O: double bonded to C, H—C—Ö—H

(g) H—C=C—C̈l: with H, H below

(h) H—N̈=N̈—H

(i) H—C=C—C—H with H's

(j) H—C=C=C—H with H's

(k) H—C≡C—C—H with H's

1

1-4

(a) :N≡N: (b) H—C≡N: (c) H—O̤—N=O   (d) O=C=O

(e) H—C=N—H   (f) H—C—O—H   (g) H—C=C—Cl: (h) H—N=N—H
        | | |
       H   H H

There are no unshared pairs in parts (i), (j), and (k).

1-5 The symbols "δ^+" and "δ^-" indicate bond polarity by showing partial charge. (In the arrow symbolism, the arrow should point to the partial negative charge.)

   δ^+ δ^- δ^+ δ^- δ^+ δ^- δ^+ δ^- δ^- δ^+
(a) C—Cl   (b) C—O   (c) C—N   (d) C—S   (e) C—B

   δ^+ δ^- δ^+ δ^- δ^- δ^+ δ^- δ^+ δ^+ δ^-
(f) N—Cl   (g) N—O   (h) N—S   (i) N—B   (j) B—Cl

1-6 Non-zero formal charges are shown beside the atoms.

(a), (b), (c):
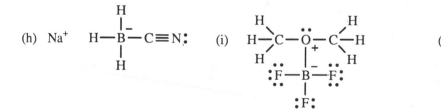

(d) Na⁺ :O—H   (e) H—C⁺—H   (f) H—C⁻—H   (g) Na⁺ H—B⁻—H
                              | |

(h), (i), (j):
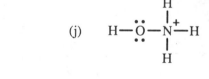

(k), (l):
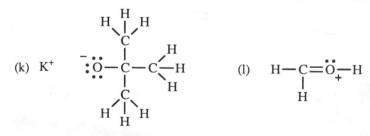

1-7 Resonance forms in which all atoms have full octets are the most significant contributors. In resonance forms, ALL ATOMS KEEP THEIR POSITIONS—ONLY ELECTRONS ARE SHOWN IN DIFFERENT POSITIONS.

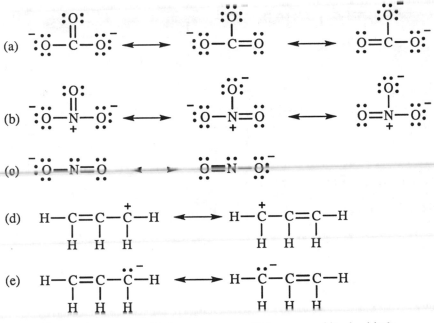

(a)

(b)

(c)

(d) H—C=C—C⁺—H ⟷ H—C⁺—C=C—H
         | | | | | |
         H H H   H H H

(e) H—C=C—C⁻—H ⟷ H—C⁻—C=C—H
         | | | | | |
         H H H   H H H

(f) Sulfur can have up to 12 electrons around it because of its d orbitals.

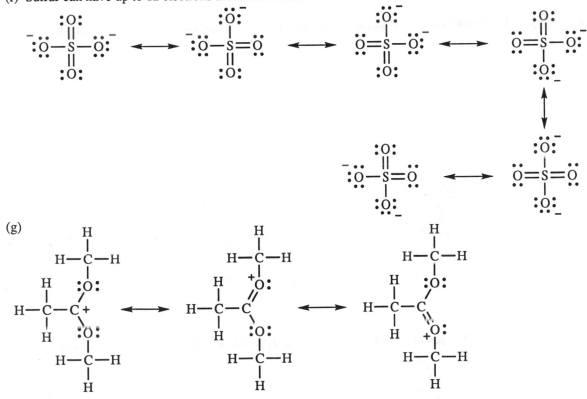

(g)

3

1-8 Major resonance contributors would have the lowest energy. The most important factors are: maximize full octets; maximize bonds; put negative charge on electronegative atoms; minimize charge separation.

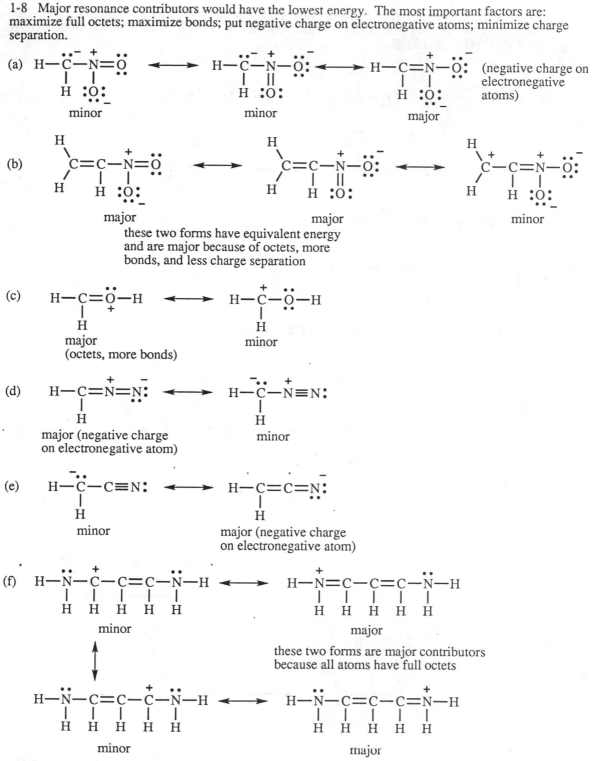

(a) (negative charge on electronegative atoms)

minor   minor   major

(b) major   major   minor

these two forms have equivalent energy and are major because of octets, more bonds, and less charge separation

(c) major (octets, more bonds)   minor

(d) major (negative charge on electronegative atom)   minor

(e) minor   major (negative charge on electronegative atom)

(f) minor   major

these two forms are major contributors because all atoms have full octets

minor   major

4

1-8 continued

(g)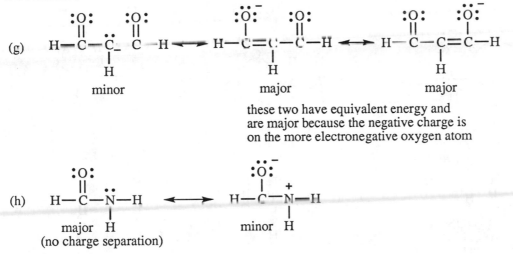

minor   major   major

these two have equivalent energy and
are major because the negative charge is
on the more electronegative oxygen atom

(h)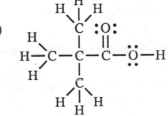

major H   minor H
(no charge separation)

1-9 Your Lewis structures may look different from these. As long as the atoms are connected in the same order and by the same type of bond, they are equivalent structures. For now, the exact placement of the atoms on the page is not significant. A Lewis structure is "complete" with unshared electron pairs shown.

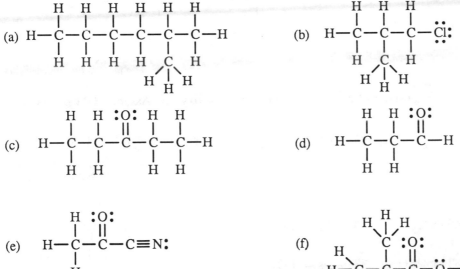

(g) (same as (c))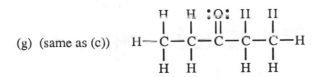

5

1-10

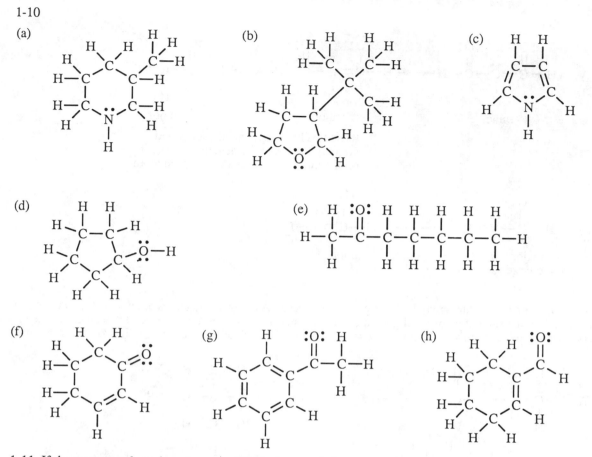

(a)

(b)

(c)

(d)

(e)

(f)

(g)

(h)

1-11 If the percent values do not sum to 100%, the remainder must be oxygen. Assume 100 g of sample; percents then translate directly to grams of each element.

There are usually many possible structures for a molecular formula. Yours may be different from the examples shown here.

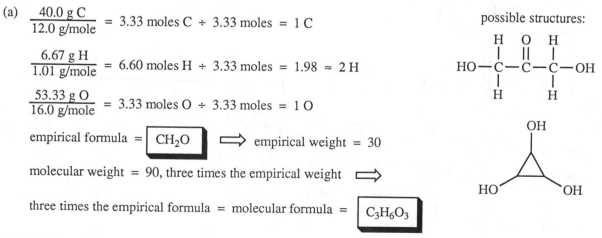

(a) $\dfrac{40.0 \text{ g C}}{12.0 \text{ g/mole}}$ = 3.33 moles C ÷ 3.33 moles = 1 C

possible structures:

$\dfrac{6.67 \text{ g H}}{1.01 \text{ g/mole}}$ = 6.60 moles H ÷ 3.33 moles = 1.98 ≈ 2 H

$\dfrac{53.33 \text{ g O}}{16.0 \text{ g/mole}}$ = 3.33 moles O ÷ 3.33 moles = 1 O

empirical formula = CH_2O ⟹ empirical weight = 30

molecular weight = 90, three times the empirical weight ⟹

three times the empirical formula = molecular formula = $C_3H_6O_3$

1-11 continued

(b) $\dfrac{32.0 \text{ g C}}{12.0 \text{ g/mole}}$ = 2.67 moles C ÷ 1.34 moles = 1.99 ≈ 2 C

possible structures:

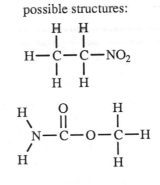

$\dfrac{6.67 \text{ g H}}{1.01 \text{ g/mole}}$ = 6.60 moles H ÷ 1.34 moles = 4.93 ≈ 5 H

$\dfrac{18.7 \text{ g N}}{14.0 \text{ g/mole}}$ = 1.34 moles N ÷ 1.34 moles = 1 N

$\dfrac{42.6 \text{ g O}}{16.0 \text{ g/mole}}$ = 2.66 moles O ÷ 1.34 moles = 1.99 ≈ 2 O

empirical formula = $\boxed{C_2H_5NO_2}$ ⟹ empirical weight = 75

molecular weight = 75, same as the empirical weight ⟹

empirical formula = molecular formula = $\boxed{C_2H_5NO_2}$

(c) $\dfrac{37.2 \text{ g C}}{12.0 \text{ g/mole}}$ = 3.10 moles C ÷ 1.55 moles = 2 C

$\dfrac{7.75 \text{ g H}}{1.01 \text{ g/mole}}$ = 7.67 moles H ÷ 1.55 moles = 4.95 ≈ 5 H

$\dfrac{55.0 \text{ g Cl}}{35.45 \text{ g/mole}}$ = 1.55 moles Cl ÷ 1.55 moles = 1 Cl

empirical formula = $\boxed{C_2H_5Cl}$ ⟹ empirical weight = 64.46

molecular weight = 64, same as the empirical weight ⟹

empirical formula = molecular formula = $\boxed{C_2H_5Cl}$

There is only one structure possible with this molecular formula:

$$\begin{array}{cc} \text{H} & \text{H} \\ | & | \\ \text{H}-\text{C}-\text{C}-\text{Cl} \\ | & | \\ \text{H} & \text{H} \end{array}$$

7

1-11 continued

(d) $\dfrac{38.4 \text{ g C}}{12.0 \text{ g/mole}}$ = 3.20 moles C ÷ 1.60 moles = 2 C

$\dfrac{4.80 \text{ g H}}{1.01 \text{ g/mole}}$ = 4.75 moles H ÷ 1.60 moles ≈ 2.97 ≈ 3 H

$\dfrac{56.8 \text{ g Cl}}{35.45 \text{ g/mole}}$ = 1.60 moles Cl ÷ 1.60 moles = 1 Cl

empirical formula = $\boxed{C_2H_3Cl}$ $\Longrightarrow$ empirical weight = 62.45

molecular weight = 125, twice the empirical weight $\Longrightarrow$

twice the empirical formula = molecular formula = $\boxed{C_4H_6Cl_2}$

possible structures:

1-12

(a) 5.00 g HBr x $\dfrac{1 \text{ mole HBr}}{80.9 \text{ g HBr}}$ = 0.0618 moles HBr

0.0618 moles HBr $\Longrightarrow$ 0.0618 moles H_3O^+ (100% dissociated)

$\dfrac{0.0618 \text{ moles } H_3O^+}{100 \text{ mL}}$ x $\dfrac{1000 \text{ mL}}{1 \text{ L}}$ = $\dfrac{0.618 \text{ moles } H_3O^+}{1 \text{ L solution}}$

pH = $-\log_{10}[H_3O^+]$ = $-\log_{10}(0.618)$ = $\boxed{0.209}$

(b) 2.00 g NaOH x $\dfrac{1 \text{ mole NaOH}}{40.0 \text{ g NaOH}}$ = 0.0500 moles NaOH

0.0500 moles NaOH $\Longrightarrow$ 0.0500 moles ⁻OH (100% dissociated)

$\dfrac{0.0500 \text{ moles } ^-OH}{50. \text{ mL}}$ x $\dfrac{1000 \text{ mL}}{1 \text{ L}}$ = $\dfrac{1.0 \text{ moles } ^-OH}{1 \text{ L solution}}$

$[H_3O^+]$ = $\dfrac{10^{-14}}{[^-OH]}$ = $\dfrac{10^{-14}}{1.0}$ = 1.0×10^{-14}

pH = $-\log_{10}[H_3O^+]$ = $-\log_{10}(1.0 \times 10^{-14})$ = $\boxed{14.0}$

1-13

(a) By definition, an acid is any species that can donate a proton. Ammonia has a proton bonded to nitrogen, so ammonia can be an acid (although a very weak one). A base is a proton acceptor, that is, it must have a pair of electrons to share with a proton; in theory, any atom with an unshared electron pair can be a base. The nitrogen in ammonia has an unshared electron pair so ammonia is basic. In water, ammonia is too weak an acid to give up its proton; instead, it acts as a base and pulls a proton from water to a small extent.

(b) water as an acid: $H_2O + NH_3 \rightleftharpoons {}^-OH + NH_4^+$

water as a base: $H_2O + HCl \rightleftharpoons H_3O^+ + Cl^-$

(c) methanol as an acid: $CH_3OH + NH_3 \rightleftharpoons CH_3O^- + NH_4^+$

methanol as a base: $CH_3OH + H_2SO_4 \rightleftharpoons CH_3OH_2^+ + HSO_4^-$

1-14

(a) HCOOH   + $^-$CN $\rightleftharpoons$ HCOO$^-$ +   HCN   FAVORS
     stronger   stronger   weaker   weaker   **PRODUCTS**
     acid   base   base   acid
     pK$_a$ 3.76   pK$_a$ 9.22

(b) CH$_3$COO$^-$ +   CH$_3$OH $\rightleftharpoons$ CH$_3$COOH   +   CH$_3$O$^-$ FAVORS
     weaker   weaker   stronger   stronger   **REACTANTS**
     base   acid   acid   base
                            pK$_a$ 15.5   pK$_a$ 4.74

(c) CH$_3$OH   +   NaNH$_2$ $\rightleftharpoons$ CH$_3$O$^-$ Na$^+$ +   NH$_3$ FAVORS
     stronger   stronger   weaker   weaker   **PRODUCTS**
     acid   base   base   acid
     pK$_a$ 15.5   pK$_a$ 33

(d) Na$^+$ $^-$OCH$_3$ +   HCN $\rightleftharpoons$ HOCH$_3$ +   NaCN   FAVORS
     stronger   stronger   weaker   weaker   **PRODUCTS**
     base   acid   acid   base
                                pK$_a$ 9.22   pK$_a$ 15.5

(e) HCl   +   H$_2$O $\rightleftharpoons$ H$_3$O$^+$ +   Cl$^-$ FAVORS
     stronger   stronger   weaker   weaker   **PRODUCTS**
     acid   base   acid   base

The first reaction in Table 1-5 shows the K$_{eq}$ for this reaction is 160, favoring products.

(f) H$_3$O$^+$ +   CH$_3$O$^-$ $\rightleftharpoons$ H$_2$O   +   CH$_3$OH   FAVORS
     stronger   stronger   weaker   weaker   **PRODUCTS**
     acid   base   base   acid

The seventh reaction in Table 1-5 shows the K$_{eq}$ for the *reverse* of this reaction is 3.2×10^{-16}.
Therefore, K$_{eq}$ for this reaction as written must be the inverse, or 3.1×10^{15}, strongly favoring products.

9

1-15

CH$_3$—C—$\ddot{\text{O}}$—H + H$^+$ $\rightleftharpoons$ CH$_3$—C—$\overset{+}{\underset{|}{\ddot{\text{O}}}}$—H
with :O: double bond above each C, H below the second

Protonation of the double-bonded oxygen gives three resonance forms (as shown in Solved Problem 1-5(c)); protonation of the single-bonded oxygen gives only one. In general, the more resonance forms a species has, the more stable it is, so the proton would bond to the oxygen that gives a more stable species, that is, the double-bonded oxygen.

1-16 In Solved Problem 1-4, the structures of ethanol and methylamine are shown to be similar to methanol and ammonia, respectively. We must infer that their acid-base properties are also similar.

(a) This problem can be viewed in two ways. 1) Quantitatively, the pK_a values determine the order of acidity. 2) Qualitatively, the stabilities of the conjugate bases determine the order of acidity (see Solved Problem 1-4 for structures): the conjugate base of acetic acid, acetate ion, is resonance-stabilized, so acetic acid is the most acidic; the conjugate base of ethanol has a negative charge on a very electronegative oxygen atom; the conjugate base of methylamine has a negative charge on a mildly electronegative nitrogen atom and is therefore the least stabilized, so methylamine is the least acidic.

$$\text{acetic acid} \quad > \quad \text{ethanol} \quad > \quad \text{methylamine}$$
$$\text{p}K_a\ 4.74 \qquad \text{p}K_a \approx 15.5 \qquad \text{p}K_a \approx 33$$
$$\text{strongest acid} \qquad\qquad\qquad\qquad \text{weakest acid}$$

(b) Ethoxide ion is the conjugate base of ethanol, so it must be a stronger base than ethanol; Solved Problem 1-4 indicates ethoxide is analogous to hydroxide in base strength. Methylamine has pK_b 3.36. The basicity of methylamine is between the basicity of ethoxide ion and ethanol.

$$\text{ethoxide ion} \quad > \quad \text{methylamine} \quad > \quad \text{ethanol}$$
$$\text{strongest base} \qquad\qquad\qquad\qquad \text{weakest base}$$

1-17 Curved arrows show electron movement, as described in text section 1-14.

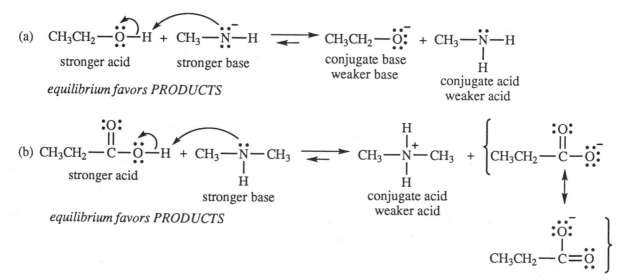

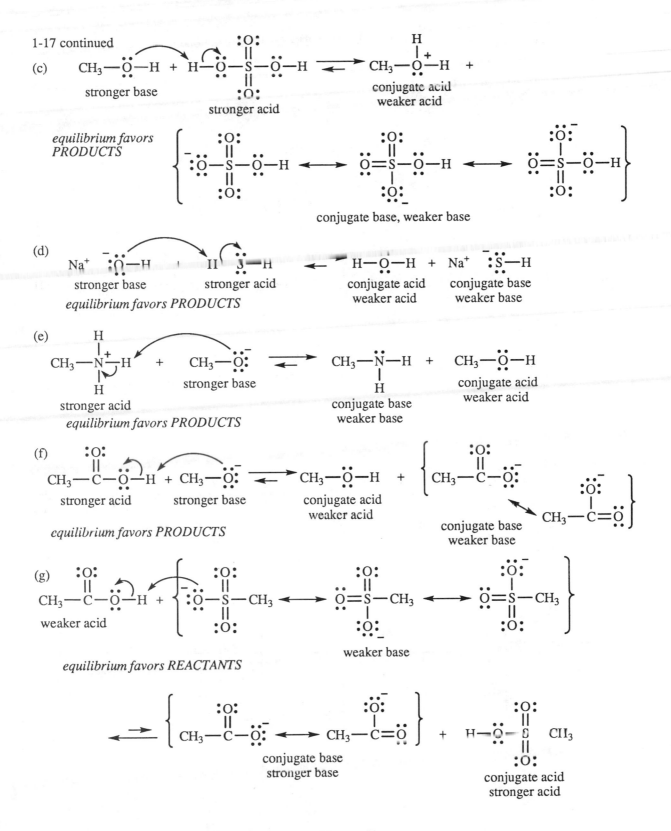

1-17 continued

(c) CH₃—O—H + H—O—S—O—H ⇌ CH₃—O—H +
    stronger base   stronger acid   conjugate acid
                                             weaker acid

equilibrium favors PRODUCTS

conjugate base, weaker base

(d) Na⁺ :O—H + H—S—H ⇌ H—O—H + Na⁺ :S—H
    stronger base   stronger acid   conjugate acid   conjugate base
                                      weaker acid   weaker base
    equilibrium favors PRODUCTS

(e) CH₃—N⁺—H + CH₃—O: ⇌ CH₃—N—H + CH₃—O—H
    stronger acid   stronger base   conjugate base   conjugate acid
                                      weaker base   weaker acid
    equilibrium favors PRODUCTS

(f) CH₃—C—O—H + CH₃—O: ⇌ CH₃—O—H +
    stronger acid   stronger base   conjugate acid
                                      weaker acid
    equilibrium favors PRODUCTS
                                      conjugate base
                                      weaker base

(g) CH₃—C—O—H + :O—S—CH₃ ⇌ O=S—CH₃ ⇌ O=S—CH₃
    weaker acid
                                      weaker base
    equilibrium favors REACTANTS

    ⇌ CH₃—C—O: ⇌ CH₃—C=O + H—O—S—CH₃
         conjugate base   conjugate acid
         stronger base   stronger acid

11

1-18 (a) and (b) are presented in the Solved Problem. The newly formed bond is shown in bold.

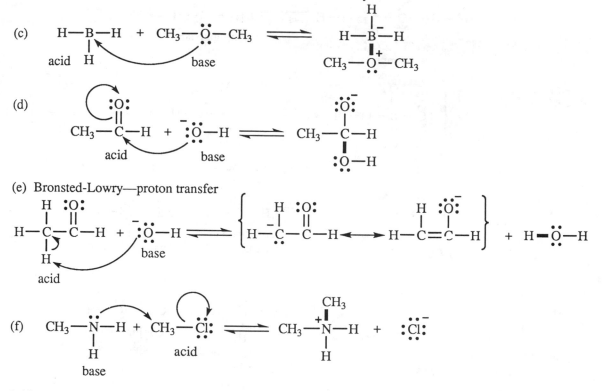

(e) Bronsted-Lowry—proton transfer

1-19

Learning organic chemistry is similar to learning a foreign language: new vocabulary, new grammar (reactions), some new concepts, and even a new alphabet (the symbolism of chemistry). This type of definition question is intended to help you review the vocabulary and concepts in each chapter. All of the definitions and examples are presented in the Glossary and in the chapter so this Solutions Manual will not repeat them. Use these questions to evaluate your comprehension and to guide your review of the important concepts in the chapter.

1-20 (a) CARBON!   (b) oxygen   (c) phosphorus   (d) chlorine

1-21

valence e⁻ → 1	2	3	4	5	6	7	8
H							He (2e⁻)
Li	Be	B	C	N	O	F	Ne
				P	S	Cl	
						Br	
						I	

1-22

(a) ionic   (b) covalent (H—O⁻) and ionic (Na⁺ ⁻OH)

(c) covalent, but the C—Li bond is strongly polarized

(d) covalent   (e) covalent (CH₃—O⁻) and ionic (Na⁺ ⁻OCH₃)

(f) covalent (HCO₂⁻) and ionic (HCO₂⁻ Na⁺)   (g) covalent

1-23

(a)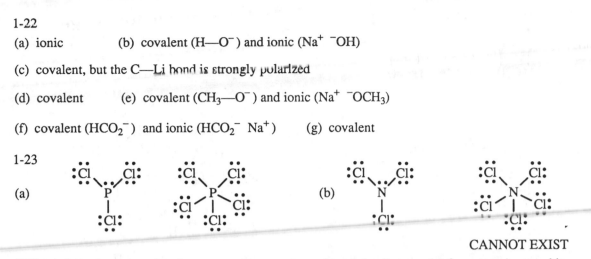

(b)

CANNOT EXIST

NCl₅ violates the octet rule; nitrogen can have no more than eight electrons (or four atoms) around it. Phosphorus, a third-row element, can have more than eight electrons because of d orbitals, so PCl₅ is a stable, isolable compound.

1-24 Your Lewis structures may look different from these. As long as the atoms are connected in the same order and by the same type of bond, they are equivalent structures. For now, the exact placement of the atoms on the page is not significant.

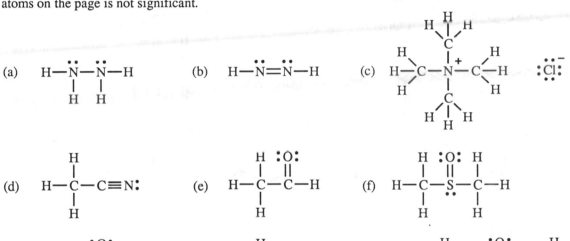

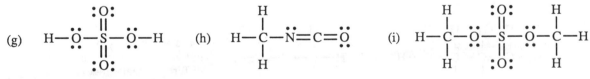

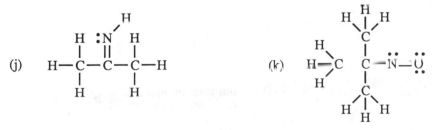

1-25

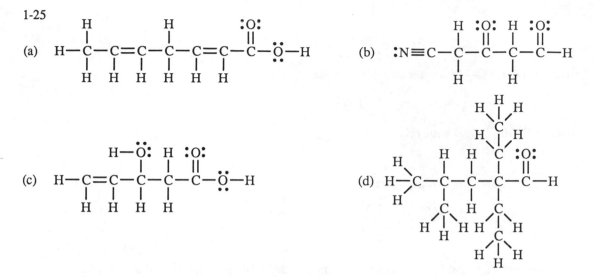

(a) H—C—C=C—C—C=C—C—O—H

(b) :N≡C—C—C—C—C—H

(c) H—C=C—C—C—C—O—H

(d) H—C—C—C—C—C—H

1-26 In each set below, the second structure is a more correct line formula. Since chemists are human (surprise!), they will take shortcuts where possible; the first structure in each pair uses a common abbreviation, either COOH or CHO. Make sure you understand that COOH does not stand for C—O—O—H. Likewise for CHO.

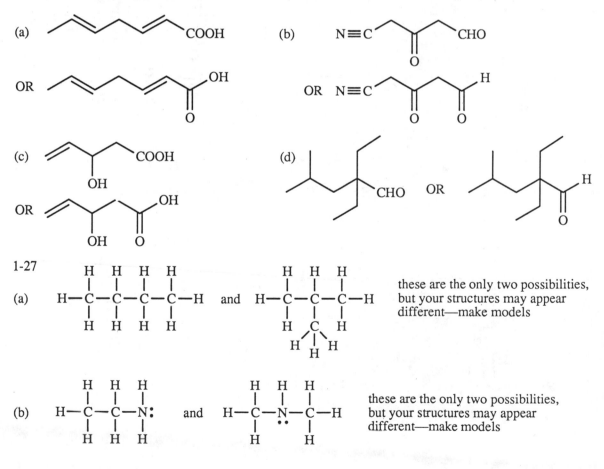

(a) ⎯⎯⎯⎯⎯COOH

OR ⎯⎯⎯⎯⎯OH

(b) N≡C⎯⎯⎯CHO

OR N≡C⎯⎯⎯H

(c) ⎯⎯⎯COOH OH

OR ⎯⎯⎯OH OH O

(d) ⎯⎯⎯CHO   OR   ⎯⎯⎯H

1-27

(a) H—C—C—C—C—H   and   H—C—C—C—H

these are the only two possibilities, but your structures may appear different—make models

(b) H—C—C—N: and   H—C—N—C—H

these are the only two possibilities, but your structures may appear different—make models

14

1-27 continued

(c) There are several other possibilities as well. Your answer may be correct even if it does not appear here. Check with others in your study group.

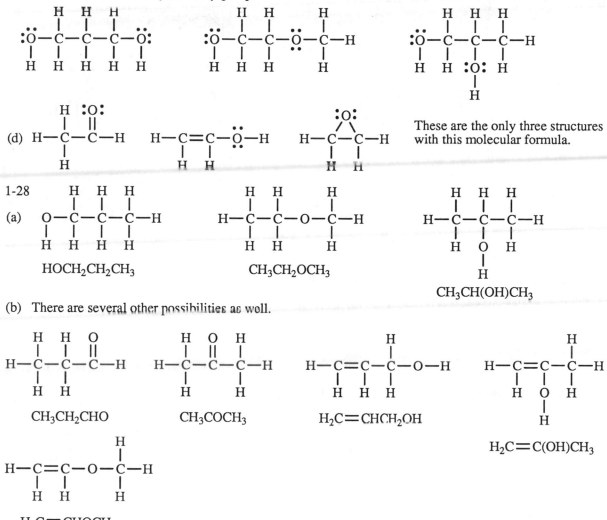

(d)

These are the only three structures with this molecular formula.

1-28

(a)

HOCH$_2$CH$_2$CH$_3$ CH$_3$CH$_2$OCH$_3$ CH$_3$CH(OH)CH$_3$

(b) There are several other possibilities as well.

CH$_3$CH$_2$CHO   CH$_3$COCH$_3$ H$_2$C=CHCH$_2$OH   H$_2$C=C(OH)CH$_3$

H$_2$C=CHOCH$_3$

1-29 General rule: molecular formulas of stable hydrocarbons must have an even number of hydrogens.

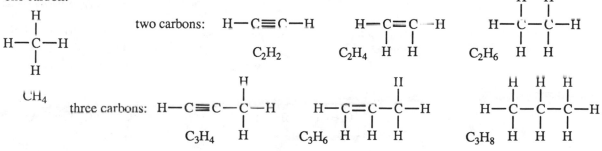

one carbon:

CH$_4$

two carbons: C$_2$H$_2$ C$_2$H$_4$ C$_2$H$_6$

three carbons: C$_3$H$_4$ C$_3$H$_6$ C$_3$H$_8$

1-30

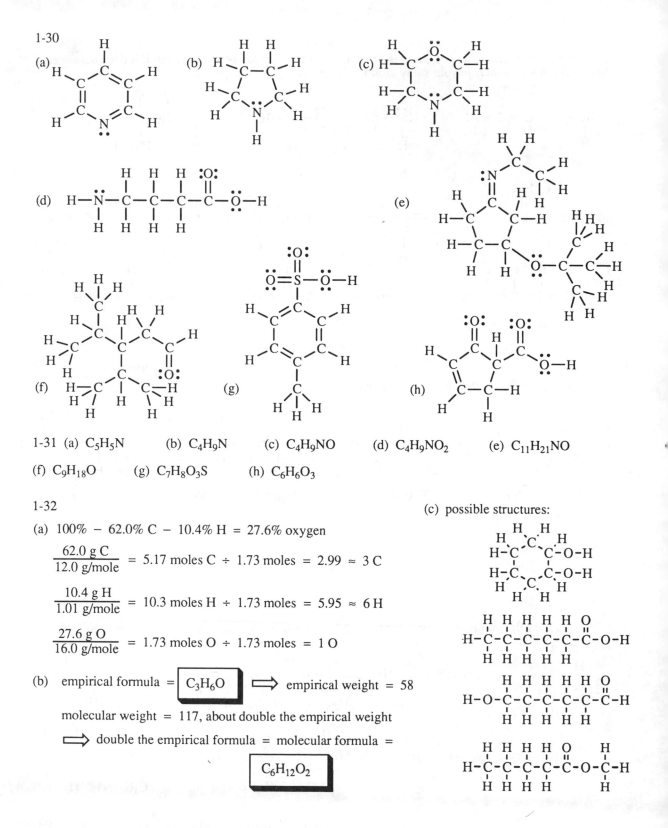

1-31 (a) C_5H_5N (b) C_4H_9N (c) C_4H_9NO (d) $C_4H_9NO_2$ (e) $C_{11}H_{21}NO$

(f) $C_9H_{18}O$ (g) $C_7H_8O_3S$ (h) $C_6H_6O_3$

1-32

(a) $100\% - 62.0\%$ C $- 10.4\%$ H $= 27.6\%$ oxygen

$$\frac{62.0 \text{ g C}}{12.0 \text{ g/mole}} = 5.17 \text{ moles C} \div 1.73 \text{ moles} = 2.99 \approx 3 \text{ C}$$

$$\frac{10.4 \text{ g H}}{1.01 \text{ g/mole}} = 10.3 \text{ moles H} \div 1.73 \text{ moles} = 5.95 \approx 6 \text{ H}$$

$$\frac{27.6 \text{ g O}}{16.0 \text{ g/mole}} = 1.73 \text{ moles O} \div 1.73 \text{ moles} = 1 \text{ O}$$

(b) empirical formula = $\boxed{C_3H_6O}$ $\Longrightarrow$ empirical weight = 58

molecular weight = 117, about double the empirical weight

$\Longrightarrow$ double the empirical formula = molecular formula =

$$\boxed{C_6H_{12}O_2}$$

(c) possible structures:

16

1-33 Non-zero formal charges are shown by the atoms.

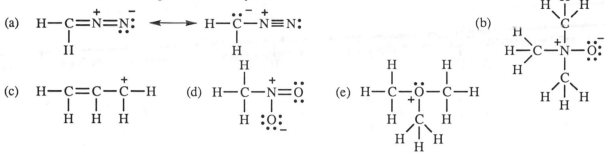

(a) H—C=N=N̈: ⟷ H—C̈—N≡N:
          | |
          H   H

(b)

(c) H—C=C—C⁺—H   (d) H—C—N⁺=Ö: (e) H—C—Ö⁺—C—H
        | | | | | | |
        H H H   H :Ö:⁻   H ⟍C⟋ H
                                              H | H
                                                 H

1-34 The symbols "δ⁺" and "δ⁻" indicate bond polarity by showing partial charge. Electronegativity differences greater than 0.5 are considered large.

(a) $\overset{\delta^+}{C}$—$\overset{\delta^-}{Cl}$ large (b) $\overset{\delta^-}{C}$—$\overset{\delta^+}{H}$ small (c) $\overset{\delta^-}{C}$—$\overset{\delta^+}{Li}$ large (d) $\overset{\delta^+}{C}$—$\overset{\delta^-}{N}$ small (e) $\overset{\delta^+}{C}$—$\overset{\delta^-}{O}$ large

(f) $\overset{\delta^-}{C}$—$\overset{\delta^+}{B}$ large (g) $\overset{\delta^-}{C}$—$\overset{\delta^+}{Mg}$ large (h) $\overset{\delta^-}{N}$—$\overset{\delta^+}{H}$ large (i) $\overset{\delta^-}{O}$—$\overset{\delta^+}{H}$ large (j) $\overset{\delta^+}{C}$—$\overset{\delta^-}{Br}$ small

1-35

(a) different compounds—a hydrogen atom has changed position

(b) resonance forms—only the position of electrons is different

(c) resonance forms—only the position of electrons is different

(d) resonance forms—only the position of electrons is different

(e) different compounds—a hydrogen atom has changed position

(f) resonance forms—only the position of electrons is different

(g) resonance forms—only the position of electrons is different

(h) different compounds—a hydrogen atom has changed position

(i) resonance forms—only the position of electrons is different

(j) resonance forms—only the position of electrons is different

1-36

(a)

H :Ö: H :Ö:⁻
| ‖ | |
H—C—C—C̈⁻—H   ⟷   H—C—C=C—H
| | | |
H   H   H   H

17

1-36 continued

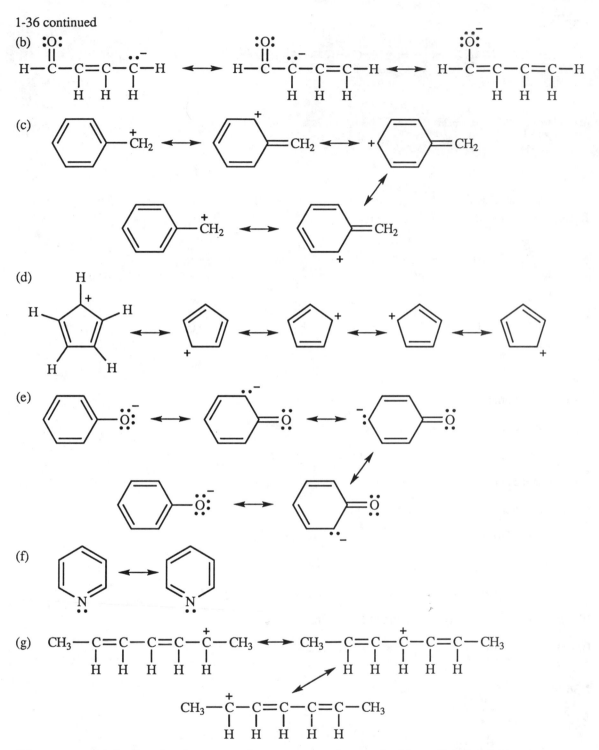

(b) ... ⟷ ... ⟷ ...

(c) ... ⟷ ... ⟷ ...
... ⟷ ...

(d) ... ⟷ ... ⟷ ... ⟷ ... ⟷ ...

(e) ... ⟷ ... ⟷ ...
... ⟷ ...

(f) ... ⟷ ...

(g) $CH_3-C=C-C=C-\overset{+}{C}-CH_3 \longleftrightarrow CH_3-C=C-\overset{+}{C}-C=C-CH_3$

$CH_3-\overset{+}{C}-C=C-C=C-CH_3$

(h) no resonance forms—the charge must be on an atom next to a double or triple bond, or next to a non-bonded pair of electrons, in order for resonance to delocalize the charge

18

1-37

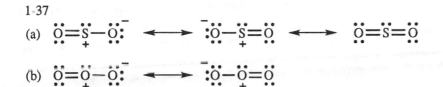

(a) $\ddot{\text{O}}=\underset{+}{\ddot{\text{S}}}-\ddot{\ddot{\text{O}}}\text{:}^-$ ⟷ $^-\text{:}\ddot{\ddot{\text{O}}}-\underset{+}{\ddot{\text{S}}}=\ddot{\text{O}}$ ⟷ $\ddot{\text{O}}=\ddot{\text{S}}=\ddot{\text{O}}$

(b) $\ddot{\text{O}}=\underset{+}{\ddot{\text{O}}}-\ddot{\ddot{\text{O}}}\text{:}^-$ ⟷ $^-\text{:}\ddot{\ddot{\text{O}}}-\underset{+}{\ddot{\text{O}}}=\ddot{\text{O}}$

(c) The last resonance form of SO_2 has no equivalent form in O_3. Sulfur, a row three element, can have more than eight electrons around it because of d orbitals, whereas oxygen, a row two element, must adhere strictly to the octet rule.

1-38

(a)

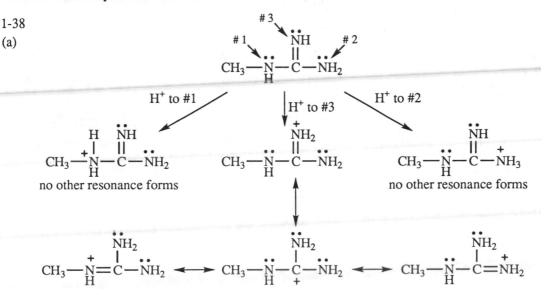

(b) Protonation at nitrogen #3 gives four resonance forms that delocalize the positive charge over all three nitrogens and a carbon—a very stable condition. Nitrogen #3 will be protonated preferentially, which we interpret as being more basic.

1-39

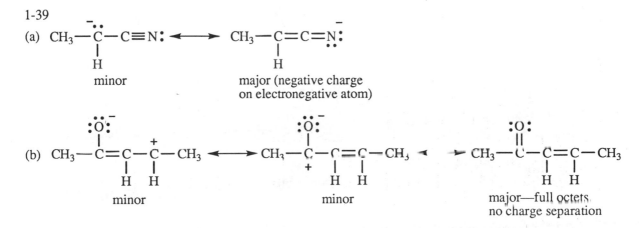

(a) $CH_3-\overset{\displaystyle H}{\underset{\displaystyle H}{\ddot{C}}}-C\equiv N\text{:}^-$ ⟷ $CH_3-\overset{\displaystyle H}{\underset{\displaystyle H}{C}}=C=\ddot{N}\text{:}^-$

   minor   major (negative charge on electronegative atom)

(b) $CH_3-\overset{:\ddot{\text{O}}:^-}{\underset{\displaystyle H}{C}}=\overset{+}{\underset{\displaystyle H}{C}}-C-CH_3$ ⟷ $CH_3-\overset{:\ddot{\text{O}}:^-}{\underset{+}{C}}-\overset{\displaystyle H}{C}=\overset{\displaystyle H}{C}-CH_3$ ⟷ $CH_3-\overset{:\text{O}:}{C}-\overset{\displaystyle H}{C}=\overset{\displaystyle H}{C}-CH_3$

   minor   minor   major—full octets no charge separation

19

1-39 continued

(c)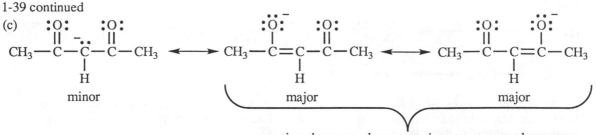

minor   major   major

negative charge on electronegative atoms—equal energy

(d)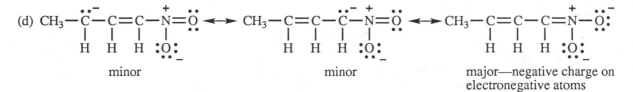

minor   minor   major—negative charge on electronegative atoms

NOTE: the two structures below are resonance forms also, with the double bonds in the NO_2 in different positions from the first two structures in part (d). Usually, chemists omit drawing these forms **with the understanding that their presence is implied!** *The importance of understanding resonance forms cannot be overemphasized.*

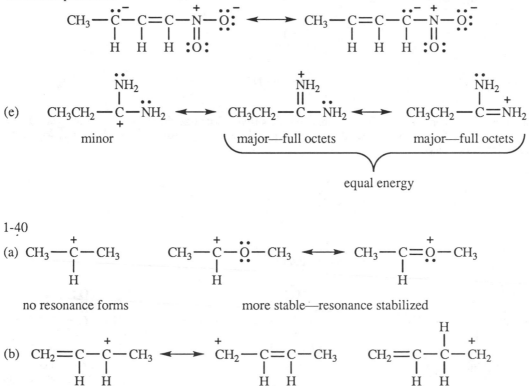

(e) CH_3CH_2—C—NH_2 ⟷ CH_3CH_2—C—NH_2 ⟷ CH_3CH_2—C=NH_2

minor   major—full octets   major—full octets

equal energy

1-40

(a) CH_3—C—CH_3 CH_3—C—O—CH_3 ⟷ CH_3—C=O—CH_3

no resonance forms   more stable—resonance stabilized

(b) CH_2=C—C—CH_3 ⟷ CH_2—C=C—CH_3 CH_2=C—C—CH_2

more stable—resonance stabilized   no resonance forms

20

1-40 continued

(c)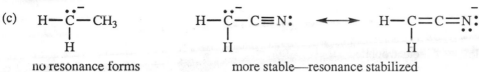

no resonance forms   more stable—resonance stabilized

(d)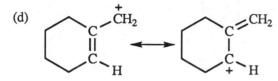

more stable—resonance stabilized   no resonance forms

(e)

$CH_3-\overset{\cdot\cdot}{N}-CH_3$ $CH_3-\overset{+}{N}-CH_3$ $CH_3-\overset{H}{\underset{}{C}}-CH_3$

$CH_3-\underset{+}{C}-CH_3$ $CH_3-\underset{}{C}-CH_3$ $CH_3-\underset{+}{C}-CH_3$

more stable—resonance stabilized   no resonance forms

1-41 These pK_a values from the text, Table 1-5, provide the answers.

least acidic			most acidic
NH_3 <	CH_3OH <	CH_3COOH <	H_2SO_4
33	15.5	4.74	< 0

1-42 Conjugate bases of the weakest acids will be the strongest bases. The pK_a values of the conjugate acids are listed here. (The relative order of the first two cannot be determined without the pK_a value of protonated acetic acid.)

least basic				most basic
HSO_4^-, CH_3COOH <	CH_3COO^- <	CH_3O^- <	$NaOH$ <	$^-NH_2$
< 0 ?	from 4.74	from 15.5	from 15.7	from 33

1-43

(a) $pK_a = -\log_{10} K_a = -\log_{10}(5.2 \times 10^{-5}) = $ **4.3** for phenylacetic acid

for propionic acid, pK_a 4.87: $K_a = 10^{-4.87} = $ **1.35×10^{-5}**

(b) phenylacetic acid is 3.8 times stronger than propionic acid

$$\frac{5.2 \times 10^{-5}}{1.35 \times 10^{-5}} = 3.8$$

(c) 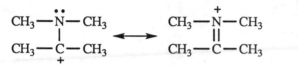

         weaker acid   stronger acid

Equilibrium favors the weaker acid and base. In this reaction, **reactants** are favored.

1-44 The newly formed bond is shown in bold.

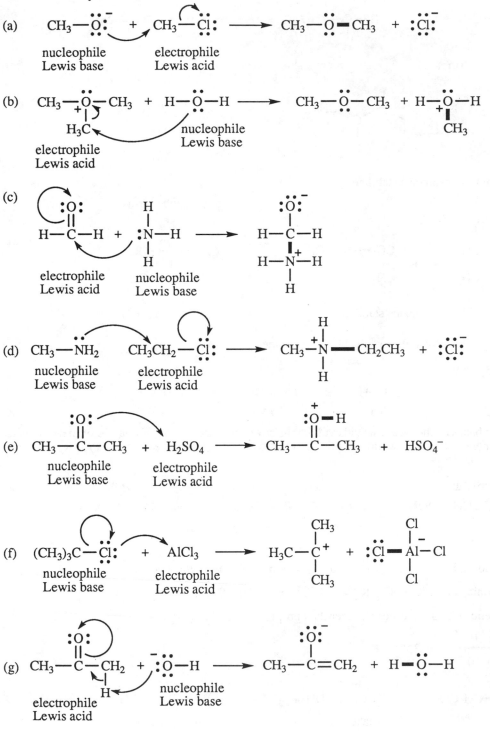

(a) nucleophile Lewis base electrophile Lewis acid

(b) electrophile Lewis acid nucleophile Lewis base

(c) electrophile Lewis acid nucleophile Lewis base

(d) nucleophile Lewis base electrophile Lewis acid

(e) nucleophile Lewis base electrophile Lewis acid

(f) nucleophile Lewis base electrophile Lewis acid

(g) electrophile Lewis acid nucleophile Lewis base

1-44 continued

(h) $CH_2{=}CH_2$ + BF_3 $\longrightarrow$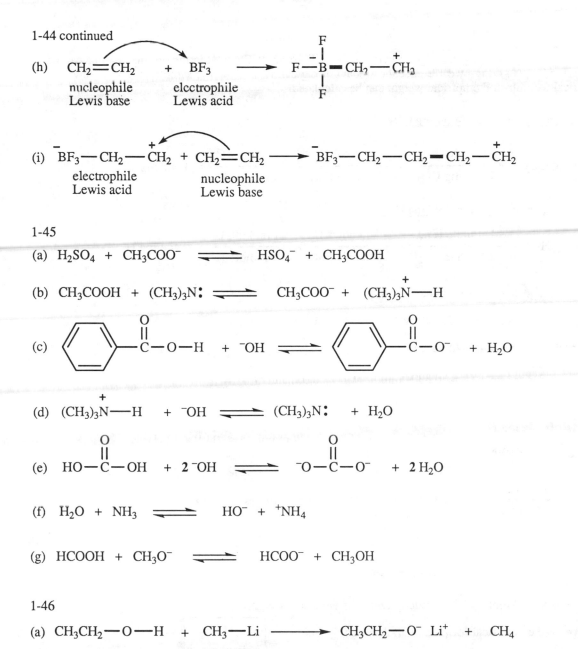

nucleophile   elcctrophile
Lewis base   Lewis acid

(i) $^-BF_3{-}CH_2{-}\overset{+}{C}H_2$ + $CH_2{=}CH_2$ $\longrightarrow$ $^-BF_3{-}CH_2{-}CH_2{-}CH_2{-}\overset{+}{C}H_2$

electrophile   nucleophile
Lewis acid   Lewis base

1-45

(a) H_2SO_4 + CH_3COO^- $\rightleftharpoons$ HSO_4^- + CH_3COOH

(b) CH_3COOH + $(CH_3)_3N{:}$ $\rightleftharpoons$ CH_3COO^- + $(CH_3)_3\overset{+}{N}{-}H$

(c) $\langle\!\!\langle\rangle\!\!\rangle{-}\overset{\overset{O}{\|}}{C}{-}O{-}H$ + ^-OH $\rightleftharpoons$ $\langle\!\!\langle\rangle\!\!\rangle{-}\overset{\overset{O}{\|}}{C}{-}O^-$ + H_2O

(d) $(CH_3)_3\overset{+}{N}{-}H$ + ^-OH $\rightleftharpoons$ $(CH_3)_3N{:}$ + H_2O

(e) $HO{-}\overset{\overset{O}{\|}}{C}{-}OH$ + $2\ ^-OH$ $\rightleftharpoons$ $^-O{-}\overset{\overset{O}{\|}}{C}{-}O^-$ + $2\,H_2O$

(f) H_2O + NH_3 $\rightleftharpoons$ HO^- + $^+NH_4$

(g) $HCOOH$ + CH_3O^- $\rightleftharpoons$ $HCOO^-$ + CH_3OH

1-46

(a) $CH_3CH_2{-}O{-}H$ + $CH_3{-}Li$ $\longrightarrow$ $CH_3CH_2{-}O^-\ Li^+$ + CH_4

(b) The conjugate acid of CH_3Li is CH_4. Table 1-5 gives the pK_a of CH_4 as > 40, one of the weakest acids known. The conjugate base of one of the weakest acids known must be one of the strongest bases known.

From the amounts of CO_2 and H_2O generated, the milligrams of C and H in the original sample can be determined, thus giving by difference the amount of oxygen in the 5.00 mg sample. From these values, the empirical formula and empirical weight can be calculated.

(a) how much carbon in 14.54 mg CO_2

$$14.54 \text{ mg CO}_2 \text{ x } \frac{1 \text{ mmole CO}_2}{44.01 \text{ mg CO}_2} \text{ x } \frac{1 \text{ mmole C}}{1 \text{ mmole CO}_2} \text{ x } \frac{12.01 \text{ mg C}}{1 \text{ mmole C}} = 3.968 \text{ mg C}$$

how much hydrogen in 3.97 mg H_2O

$$3.97 \text{ mg H}_2O \text{ x } \frac{1 \text{ mmole H}_2O}{18.016 \text{ mg H}_2O} \text{ x } \frac{2 \text{ mmoles H}}{1 \text{ mmole H}_2O} \text{ x } \frac{1.008 \text{ mg H}}{1 \text{ mmole H}} = 0.444 \text{ mg H}$$

how much oxygen in 5.00 mg estradiol

$5.00 \text{ mg estradiol } - 3.968 \text{ mg C } - 0.444 \text{ mg H } = 0.59 \text{ mg O}$

calculate empirical formula

$$\frac{3.968 \text{ mg C}}{12.01 \text{ mg/mole}} = 0.3304 \text{ mmoles C } \div 0.037 \text{ mmoles } = 8.93 \approx 9 \text{ C}$$

$$\frac{0.444 \text{ mg H}}{1.008 \text{ mg/mole}} = 0.440 \text{ mmoles H } \div 0.037 \text{ mmoles } = 11.9 \approx 12 \text{ H}$$

$$\frac{0.59 \text{ mg O}}{16.00 \text{ mg/mole}} = 0.037 \text{ mmoles O } \div 0.037 \text{ mmoles } = 1 \text{ O}$$

empirical formula = $\boxed{C_9H_{12}O}$ $\implies$ empirical weight = 136

(b) molecular weight = 272, exactly twice the empirical weight

twice the empirical formula = molecular formula = $\boxed{C_{18}H_{24}O_2}$

CHAPTER 2—STRUCTURE AND PROPERTIES OF ORGANIC MOLECULES

2-1 The fundamental principle of organic chemistry is that a molecule's chemical and physical properties depend on the molecule's structure: the structure-function or structure-reactivity correlation. It is essential that you understand the three-dimensional nature of organic molecules, and there is no better device to assist you than a molecular model set. You are strongly encouraged to use models regularly when reading the text and working the problems.

(a) requires use of models

(b)

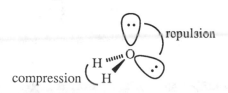

The wedge bonds represent bonds coming out of the plane of the paper toward you.
The dashed bonds represent bonds going behind the plane of the paper.

2-2 The hybridization of oxygen is sp^3 since it has two sigma bonds and two pairs of nonbonding electrons. The reason that the bond angle of 104.5° is less than the perfect tetrahedral angle of 109.5° is that the lone pairs in the two sp^3 orbitals are repelling each other more strongly than the electron pairs in the sigma bonds, thereby compressing the bond angle.

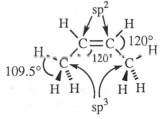

2-3 Each double-bonded carbon is sp^2 hybridized with bond angles about 120°; geometry around the sp^2 carbons is trigonal planar, that is, all four carbons, the two hydrogens on the sp^2 carbons, and one hydrogen on each of the sp^3 carbons are all in one plane. Each carbon on the end is sp^3 hybridized with tetrahedral geometry and bond angles about 109°.

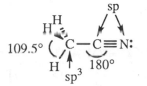

2-4 The hybridization of the nitrogen and the triple-bonded carbon are sp, giving linear geometry (C—C—N are linear) and a bond angle around the triple-bonded carbon of 180°. The CH$_3$ carbon is sp^3 hybridized, tetrahedral, with bond angles about 109°.

2-5

(a) linear, bond angle 180°

sp²   sp   sp²

(b) all atoms are sp³; tetrahedral geometry and bond angles of 109° around each atom

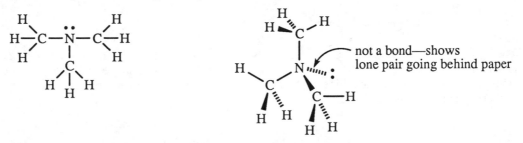

not a bond—shows lone pair coming out of paper

not a bond—shows lone pair going behind paper

(c) all atoms are sp³; tetrahedral geometry and bond angles of 109° around each atom

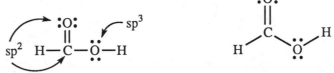

not a bond—shows lone pair going behind paper

(d) trigonal planar around the carbon, bond angles 120°; tetrahedral around the single-bonded oxygen, bond angle 109°

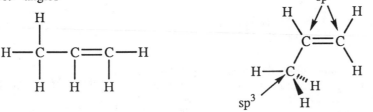

(e) carbon and nitrogen both sp, linear, bond angle 180°

H—C≡N:

(f) trigonal planar around the sp² carbons, bond angles 120°; around the sp³ carbon, tetrahedral geometry and 109° angles

(g) trigonal planar, bond angle about 120°

(the other resonance form of ozone shows that BOTH end oxygens must be sp²—see Solved Problem 2-8)

sp²   sp³ ?

26

2-6 Carbon-2 is sp hybridized. If the p orbitals making the pi bond between C-1 and C-2 are in the plane of the paper (putting the hydrogens in front of and behind the paper), then the other p orbital on C-2 must be perpendicular to the plane of the paper, making the pi bond between C-2 and C-3 perpendicular to the paper. This necessarily places the hydrogens on C-3 in the plane of the paper. (Models will surely help.)

model of *perpendicular* π bonds

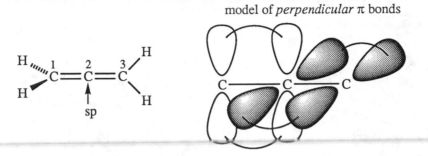

2-7 For clarity, electrons in sigma bonds are not shown.

(a) carbon and oxygen are both sp² hybridized

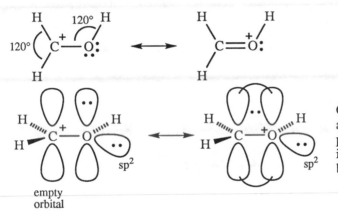

One pair of electrons on oxygen is always in an sp² orbital. The other pair of electrons is shown in a p orbital in the first resonance form, and in a pi bond in the second resonance form.

(b) oxygen and both carbons are sp² hybridized

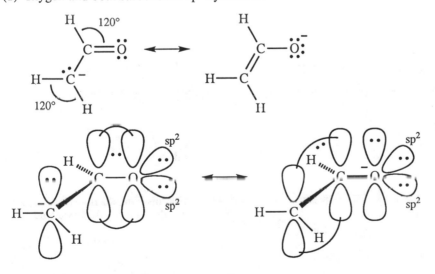

27

2-7 continued

(c) the nitrogen and the carbon bonded to it are sp hybridized; the left carbon is sp^2

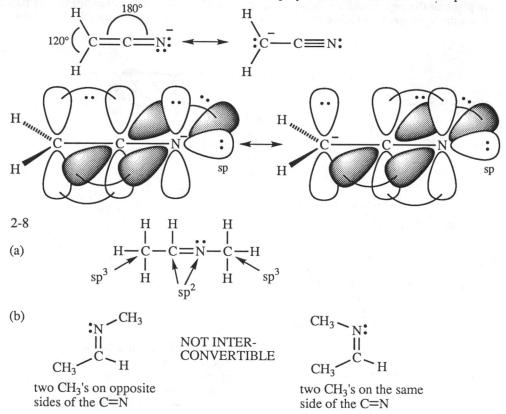

2-8

(a)

$$H-\overset{\overset{\displaystyle H}{|}}{\underset{\underset{\displaystyle H}{|}}{C}}-\overset{\overset{\displaystyle H}{|}}{C}=\ddot{N}-\overset{\overset{\displaystyle H}{|}}{\underset{\underset{\displaystyle H}{|}}{C}}-H$$

sp^3 sp^2 sp^3

(b)

NOT INTER-CONVERTIBLE

two CH₃'s on opposite sides of the C=N

two CH₃'s on the same side of the C=N

(c) the CH₃ on the N is on the same side as another CH₃ no matter how it is drawn—only one possible structure

2-9

(a)

cis   and   trans

(f)

and

"cis" and "trans" not defined for this example

(b) no cis-trans isomerism
(c) no cis-trans isomerism
(d) no cis-trans isomerism

(e)

cis   and   trans

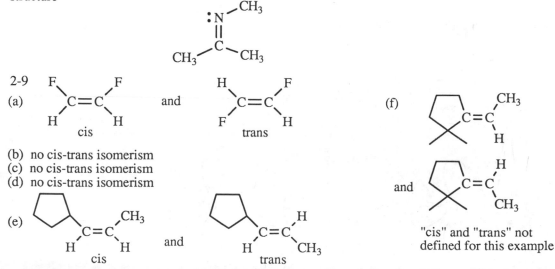

2-10 Models will be helpful here.
(a) constitutional isomers—the carbon skeleton is different
(b) cis-trans isomers—the first is trans, the second is cis
(c) constitutional isomers—the bromines are on different carbons in the first structure, on the same carbon in the second structure
(d) same compound—just flipped over
(e) same compound—just rotated
(f) same compound—just rotated
(g) not isomers—different molecular formulas
(h) constitutional isomers—the double bond has changed position
(i) same compound—just reversed
(j) constitutional isomers—the CH_3 groups are in different relative positions
(k) constitutional isomers—the double bond is in a different position relative to the CH_3

2-11
(a) $2.4 \, D \;=\; 4.8 \times \delta \times 1.21 \, Å$

   $\delta \;=\; 0.41$, or 41% of a positive charge on carbon and 41% of a negative charge on oxygen

(b)

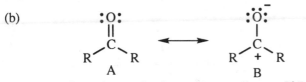

Resonance form A must be the major contributor. If B were the major contributor, the value of the charge separation would be between 0.5 and 1.0. Even though B is "minor", it is quite significant, explaining in part the high polarity of the C=O.

2-12

Both NH_3 and NF_3 have a pair of nonbonding electrons on the nitrogen. In NH_3, the *direction* of polarization of the N—H bonds is *toward* the nitrogen; thus, all three bond polarities and the lone pair polarity reinforce each other. In NF_3, on the other hand, the direction of polarization of the N—F bonds is *away* from the nitrogen; the three bond polarities cancel the lone pair polarity, so the net result is a very small *molecular* dipole moment.

polarities reinforce;
large dipole moment

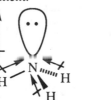

polarities oppose;
small dipole moment

2-13 Some magnitudes of dipole moments are difficult to predict; however, the direction of the dipole should be straightforward, in most cases. Actual values of molecular dipole moments are given in parentheses. (Each halogen atom has three nonbonded electron pairs, not shown below.)
   The C—H is usually considered non-polar.

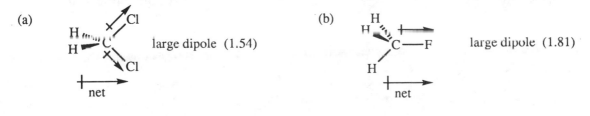

(a)   large dipole (1.54)

(b)   large dipole (1.81)

2-13 continued

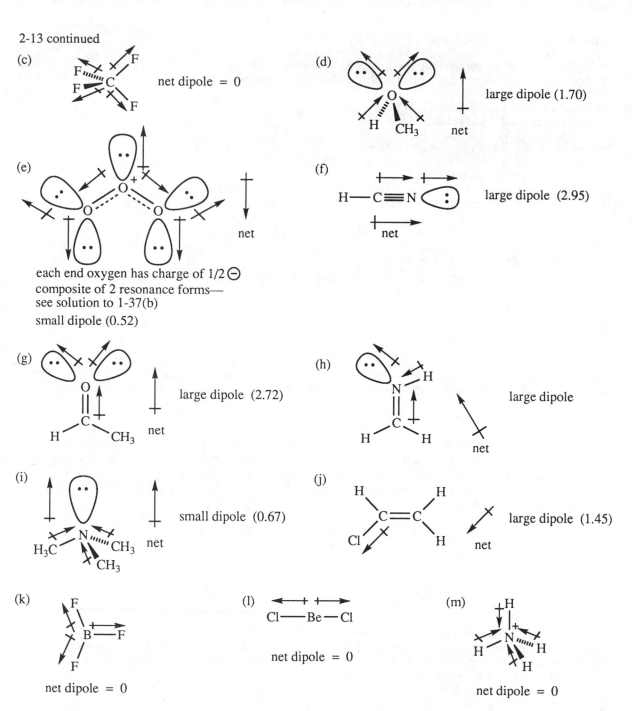

(c) net dipole = 0

(d) large dipole (1.70)

(e) each end oxygen has charge of 1/2 ⊖
composite of 2 resonance forms—
see solution to 1-37(b)
small dipole (0.52)

(f) H—C≡N large dipole (2.95)

(g) large dipole (2.72)

(h) large dipole

(i) small dipole (0.67)

(j) large dipole (1.45)

(k) net dipole = 0

(l) Cl——Be—Cl net dipole = 0

(m) net dipole = 0

In (k) through (m), the symmetry of the molecule allows the individual bond dipoles to cancel.

2-14 With chlorines on the same side of the double bond, the bond dipole moments reinforce each other, resulting in a large net dipole. With chlorines on opposite sides of the double bond, the bond dipole moments exactly cancel each other, resulting in a zero net dipole.

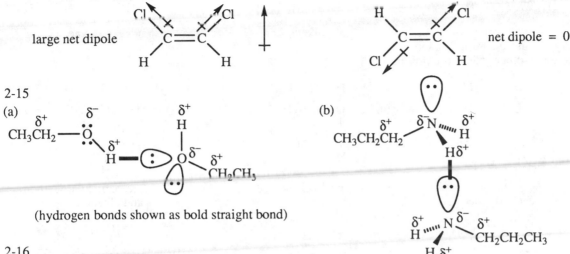

large net dipole   net dipole = 0

2-15
(a)

(hydrogen bonds shown as bold straight bond)

(b)

2-16

(a) $(CH_3)_2CHCH_2CH_2CH(CH_3)_2$ has less branching and boils at a higher temperature than $(CH_3)_3CC(CH_3)_3$.

(b) $CH_3(CH_2)_5CH_2OH$ can form hydrogen bonds and will boil at a much higher temperature than $CH_3(CH_2)_6CH_3$ which cannot form hydrogen bonds.

(c) $HOCH_2(CH_2)_4CH_2OH$ can form hydrogen bonds at both ends and has no branching; it will boil at a much higher temperature than $(CH_3)_3CCH(OH)CH_3$.

(d) $(CH_3CH_2CH_2)_2NH$ has an N—H bond and can form hydrogen bonds; it will boil at a higher temperature than $(CH_3CH_2)_3N$ which cannot form hydrogen bonds.

(e) Compound A has two N—H bonds and will form more hydrogen bonds than compound B will; A also has a higher molecular weight than B; A will boil at a higher temperature than B.

2-17

(a) $CH_3CH_2OCH_2CH_3$ can form hydrogen bonds with water and is more soluble than $CH_3CH_2CH_2CH_2CH_3$ which cannot form hydrogen bonds with water.

(b) $CH_3CH_2NHCH_3$ is more water soluble because it can form hydrogen bonds; $CH_3CH_2CH_2CH_3$ cannot form hydrogen bonds.

(c) CH_3CH_2OH is more soluble in water. The polar O—H group forms hydrogen bonds with water, overcoming the resistance of the non-polar CH_3CH_2 group toward entering the water. In $CH_3CH_2CH_2CH_2OH$, however, the hydrogen bonding from only one OH group cannot carry a four-carbon chain into the water; this substance is only slightly soluble in water.

(d) Both compounds form hydrogen bonds with water at the double-bonded oxygen, but only the smaller molecule (CH_3COCH_3) dissolves. The cyclic compound has too many non-polar CH_2 groups to dissolve.

2-18

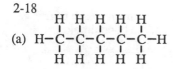

(a) alkane
(Usually, we use the term "alkane" only when no other groups are present.)

(b) 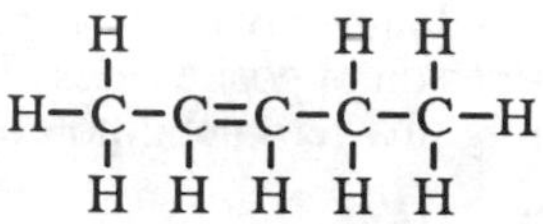 alkene

(c) 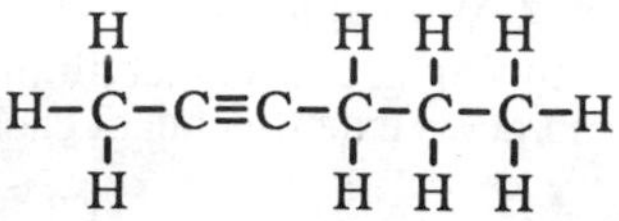 alkyne

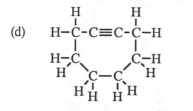

(d) cycloalkyne

(e) 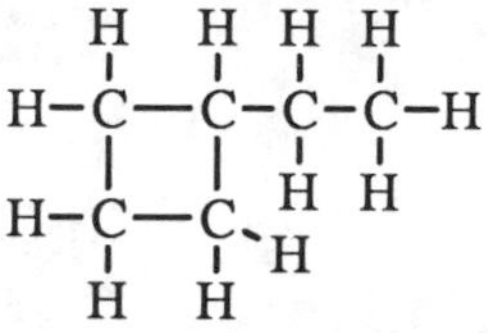 cycloalkane

(f) 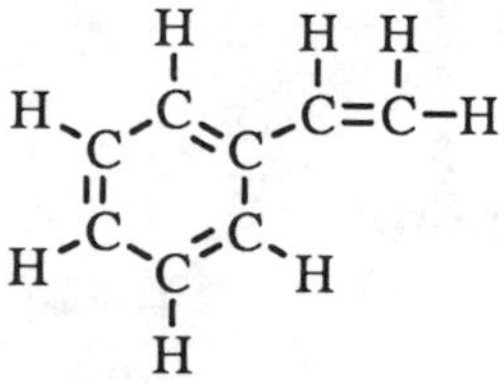 aromatic hydrocarbon and alkene

(g) 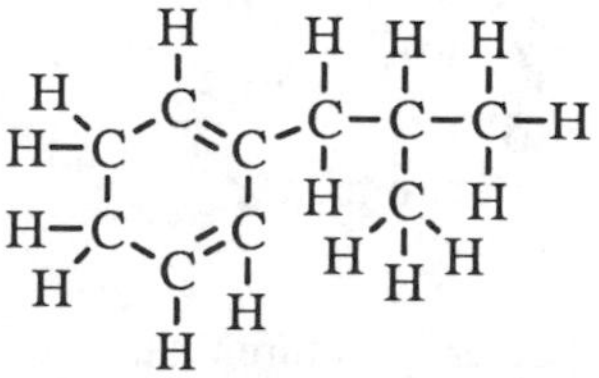 cycloalkene

(h) 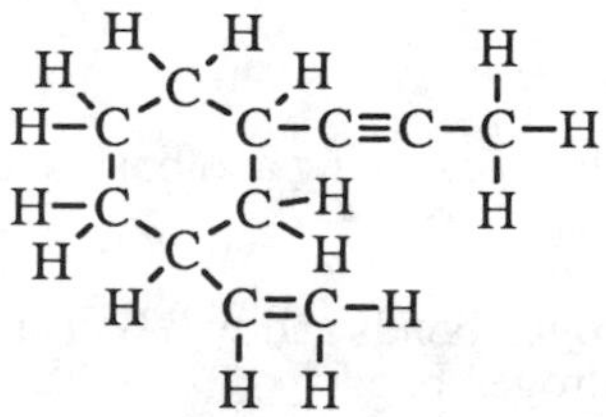 alkyne, alkene, cycloalkane

(i) 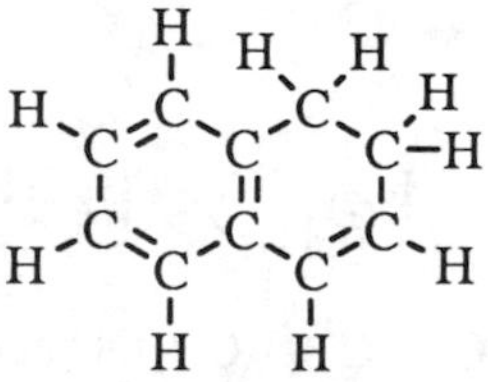 aromatic hydrocarbon and cycloalkene

2-19

(a) 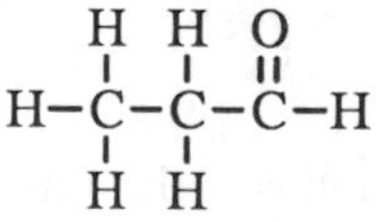 aldehyde

(b) 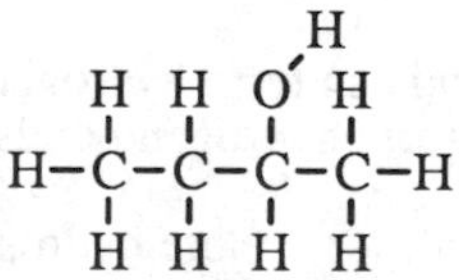 alcohol

(c) 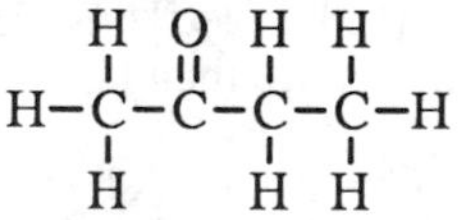 ketone

(d) 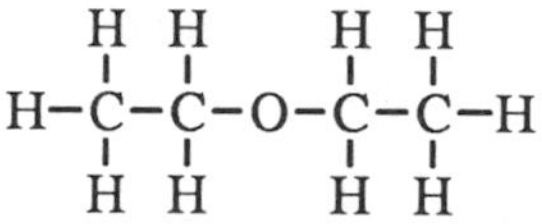 ether

(e) 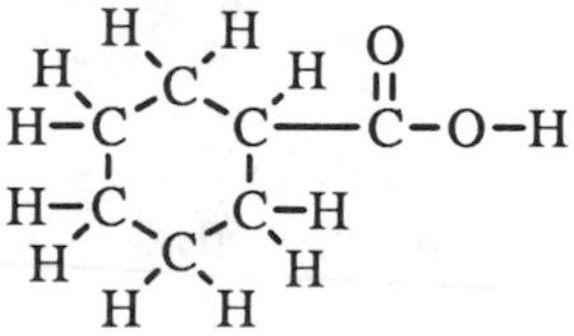 carboxylic acid

(f) 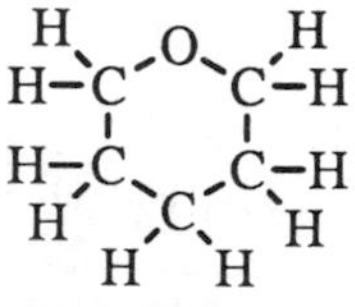 ether

(g) 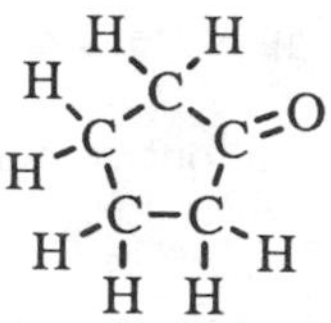 ketone

(h) 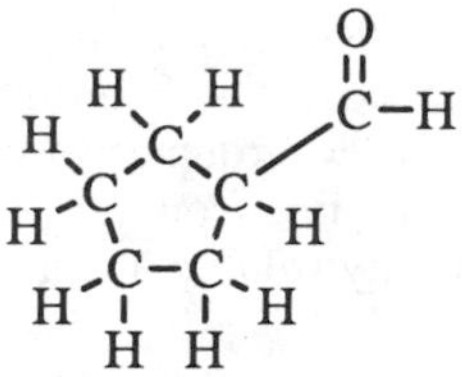 aldehyde

(i) 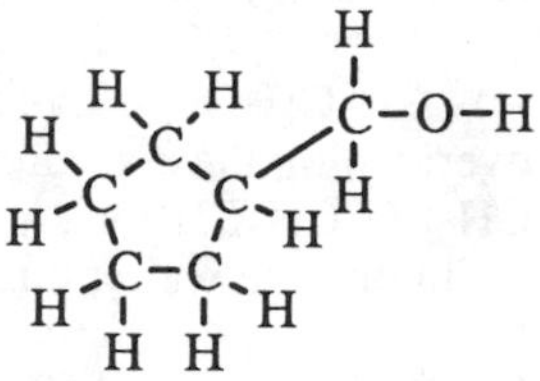 alcohol

2-20

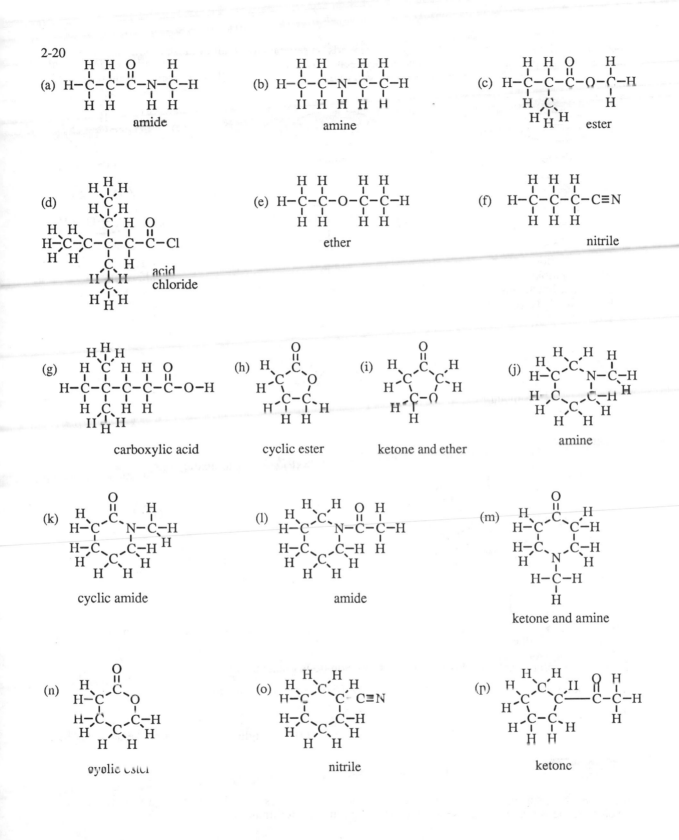

(a) amide

(b) amine

(c) ester

(d) acid chloride

(e) ether

(f) nitrile

(g) carboxylic acid

(h) cyclic ester

(i) ketone and ether

(j) amine

(k) cyclic amide

(l) amide

(m) ketone and amine

(n) cyclic ester

(o) nitrile

(p) ketone

2-21 When the identity of a functional group depends on several atoms, all of those atoms should be circled. For example, an ether is an oxygen between two carbons, so the oxygen and both carbons should be circled. A ketone is a carbonyl group between two other carbons, so all those atoms should be circled.

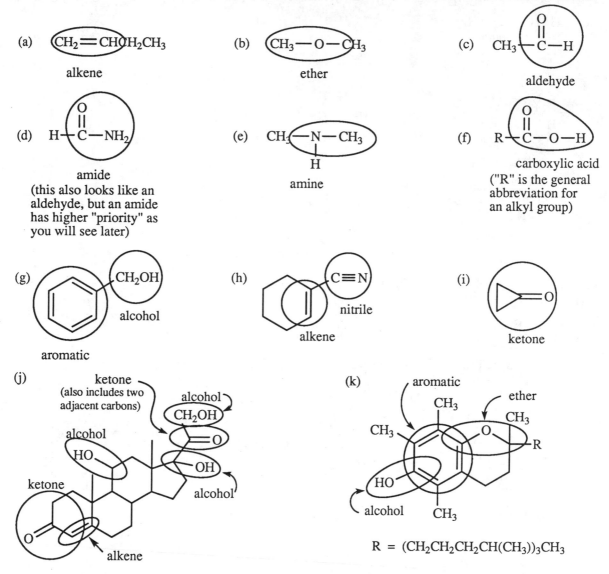

(a) $CH_2 = CHCH_2CH_3$

alkene

(b) $CH_3 - O - CH_3$

ether

(c) $CH_3 - C - H$ (with O double bonded to C)

aldehyde

(d) $H - C - NH_2$ (with O double bonded to C)

amide
(this also looks like an
aldehyde, but an amide
has higher "priority" as
you will see later)

(e) $CH_3 - N - CH_3$ with H below N

amine

(f) $R - C - O - H$ (with O double bonded to C)

carboxylic acid
("R" is the general
abbreviation for
an alkyl group)

(g) CH_2OH

alcohol

aromatic

(h) $C \equiv N$

nitrile

alkene

(i) $= O$

ketone

(j)
ketone
(also includes two
adjacent carbons)

alcohol
CH_2OH

alcohol
HO

$= O$

OH
alcohol

ketone
$O =$

alkene

(k)
aromatic
CH_3

ether
CH_3

CH_3

O
R

HO

alcohol CH_3

$R = (CH_2CH_2CH_2CH(CH_3))_3CH_3$

2-22 Please refer to solution 1-19, page 12 of this Solutions Manual.

2-23 Models show that the tetrahedral geometry of CH_2Cl_2 precludes stereoisomers.

2-24
(a)

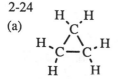

(b) Cyclopropane must have 60° bond angles compared with the usual sp^3 bond angle of 109.5° in an acyclic molecule.

(c) Like a bent spring, bonds that deviate from their normal angles or positions are highly strained. Cyclopropane is reactive because breaking the ring relieves the strain.

2-25

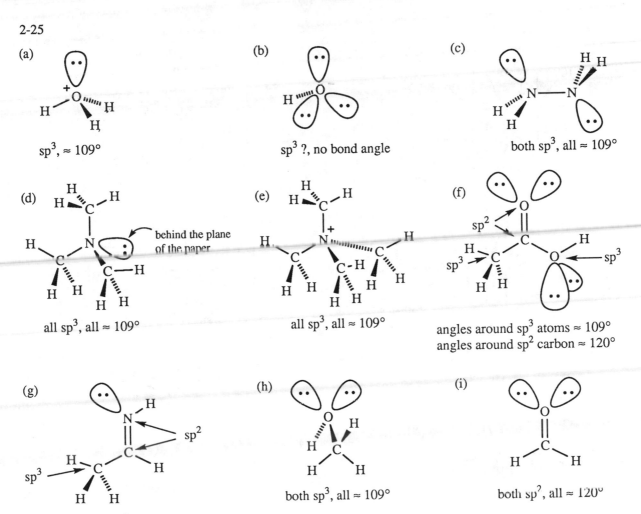

(a)

H—O⁺···H
  |
  H

sp³, ≈ 109°

(b)

sp³ ?, no bond angle

(c)

both sp³, all ≈ 109°

(d)

behind the plane
of the paper

all sp³, all ≈ 109°

(e)

all sp³, all ≈ 109°

(f)

sp²

sp³   sp³

angles around sp³ atoms ≈ 109°
angles around sp² carbon ≈ 120°

(g)

sp²

sp³

angles around sp³ atom ≈ 109°
angles around sp² atoms ≈ 120°

(h)

both sp³, all ≈ 109°

(i)

both sp², all ≈ 120°

2-26 For clarity in these pictures, hydrogens bonded to sp³ atoms are not labeled although their bonds are shown. These bonds are s-sp³ overlap.

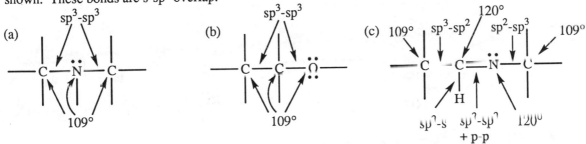

(a)

sp³-sp³

—C—N̈—C—

109°

(b)

sp³-sp³

—C—C—Ö—

109°

(c) 109°   sp³-sp²   120°   sp²-sp³   109°

—C—C=N̈—C—
     |
     H

sp²-s   sp²-sp²   120°
+ p-p

35

(d)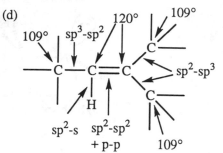

(e)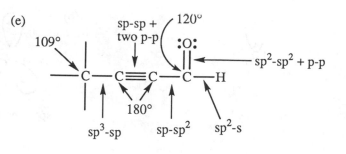

(f)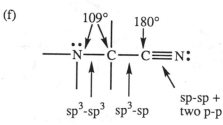

(g)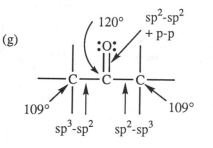

(h)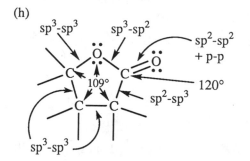

(i)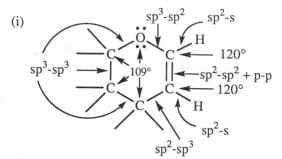

2-27 The second resonance form of formamide is a minor but significant resonance contributor. It shows that the nitrogen-carbon bond has some double bond character, requiring that the nitrogen be sp^2 hybridized with bond angles approaching 120°.

2-28

(a) The major resonance contributor shows a carbon-carbon double bond, suggesting that both carbons are sp^2 hybridized with trigonal planar geometry. The CH_3 carbon is sp^3 hybridized with tetrahedral geometry.

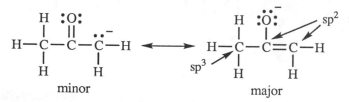

2-28 continued

(b) The major resonance contributor shows a carbon-nitrogen double bond, suggesting that all three carbons and the nitrogen are sp² hybridized with trigonal planar geometry.

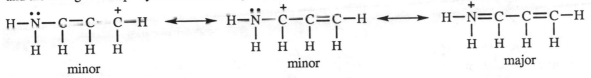

2-29   In (c) and (d), the unshadowed p orbitals are vertical and parallel. The shadowed p orbitals are perpendicular and horizontal.

2-30

(b) The coplanar atoms in the structures to the left and below are marked with asterisks.

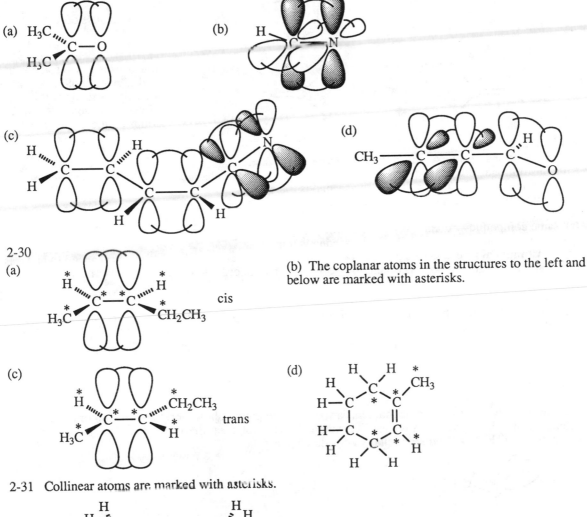

2-31   Collinear atoms are marked with asterisks.

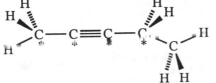

2-32

(a) no cis-trans isomerism

(b)
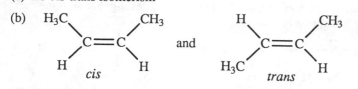
and
and *cis* *trans*

(c) no cis-trans isomerism

(d) Theoretically, cyclopentene could show cis-trans isomerism. In reality, the *trans* form is too unstable to exist because of the necessity of stretched bonds and deformed bond angles. *trans*-Cyclopentene has never been detected.

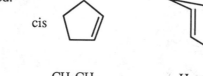

cis

"trans"—not possible because of ring strain

(e)

these are cis-trans isomers, but the designation of cis and trans to specific structures is not defined because of four different groups on the double bond

2-33

(a) constitutional isomers—the carbon skeletons are different
(b) constitutional isomers—the position of the chlorine atom has changed
(c) cis-trans isomers—the first is cis, the second is trans
(d) constitutional isomers—the carbon skeletons are different
(e) cis-trans isomers—the first is trans, the second is cis
(f) same compound—rotation of the first structure gives the second
(g) cis-trans isomers—the first is cis, the second is trans
(h) constitutional isomers—the position of the double bond relative to the ketone has changed (while it is true that the first double bond is cis and the second is trans, in order to have cis-trans isomers, the rest of the structure must be identical)

2-34 CO_2 is linear; its bond dipoles cancel, so it has no net dipole. SO_2 is bent, so its bond dipoles do not cancel.

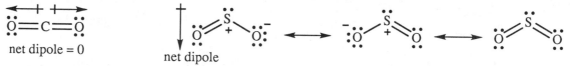

net dipole = 0

net dipole

2-35 Some magnitudes of dipole moments are difficult to predict; however, the direction of the dipole should be straightforward in most cases. Actual values of molecular dipole moments are given in parentheses. (The C—H bond is usually considered non-polar.)

(a)

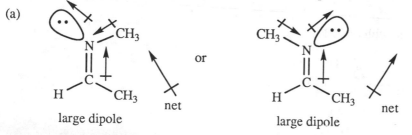

large dipole   net   or   large dipole   net

2-35 continued

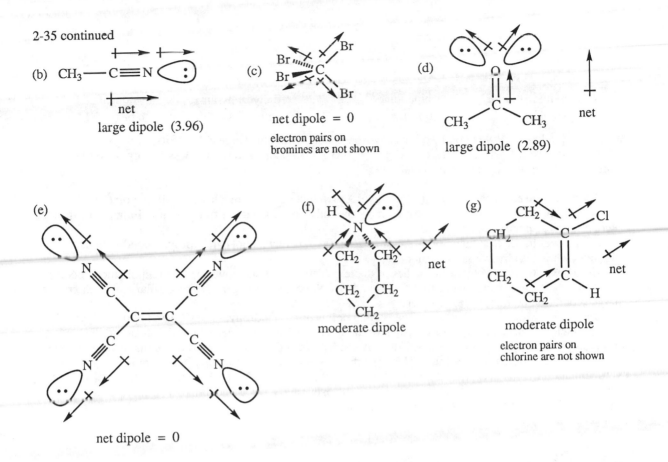

(b) CH₃—C≡N net

large dipole (3.96)

(c) net dipole = 0

electron pairs on bromines are not shown

(d) net

large dipole (2.89)

(e) net dipole = 0

(f) net

moderate dipole

(g) net

moderate dipole

electron pairs on chlorine are not shown

2-36 Diethyl ether and 1-butanol each have one oxygen, so each can form hydrogen bonds with water (water supplies the H for hydrogen bonding with diethyl ether); their water solubilities should be similar. The boiling point of 1-butanol is much higher because these molecules can hydrogen bond with each other, thus requiring more energy to separate one molecule from another. Diethyl ether molecules cannot hydrogen bond with each other, so it is relatively easy to separate them.

$CH_3CH_2—O—CH_2CH_3$

diethyl ether

can hydrogen bond with water
cannot hydrogen bond with itself

$CH_3CH_2CH_2CH_2—OH$

1-butanol

can hydrogen bond with water
can hydrogen bond with itself

2-37

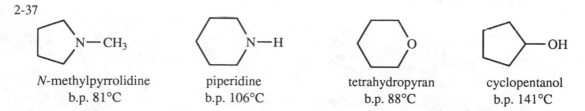

N-methylpyrrolidine
b.p. 81°C

piperidine
b.p. 106°C

tetrahydropyran
b.p. 88°C

cyclopentanol
b.p. 141°C

(a) Piperidine has an N—H bond, so it can hydrogen bond with other molecules of itself. N-Methylpyrrolidine has no N—H, so it cannot hydrogen bond and will require less energy (lower boiling point) to separate one molecule from another.

(b) Two effects need to be explained: 1) Why does cyclopentanol have a higher boiling point than tetrahydropyran? and 2) Why do the oxygen compounds have a greater difference in boiling points than the analogous nitrogen compounds?

The answer to the first question is the same as in (a): cyclopentanol can hydrogen bond with its neighbors while tetrahydropyran cannot.

The answer to the second question lies in the text, Table 2-1, that shows the bond dipole moments for C—O and H—O are much greater than C—N and H—N; bonds to oxygen are more polarized, with greater charge separation than bonds to nitrogen.

How is this reflected in the data? The boiling points of tetrahydropyran (88°C) and N-methylpyrrolidine (81°C) are close; tetrahydropyran molecules would have a slightly stronger dipole-dipole attraction, and tetrahydropyran is a little less "branched" than N-methylpyrrolidine, so it is reasonable that tetrahydropyran boils at a slightly higher temperature. The large difference comes when comparing the boiling points of cyclopentanol (141°C) and piperidine (106°C). The greater polarity of O—H versus N—H is reflected in a more negative oxygen (more electronegative than nitrogen) and a more positive hydrogen, resulting in a much stronger intermolecular attraction. The conclusion is that hydrogen bonding due to O—H is much stronger than that due to N—H.

2-38

(a) can hydrogen bond with itself and with water
(b) can hydrogen bond only with water
(c) can hydrogen bond with itself and with water
(d) can hydrogen bond only with water
(e) cannot hydrogen bond
(f) cannot hydrogen bond

(g) can hydrogen bond only with water
(h) can hydrogen bond with itself and with water
(i) can hydrogen bond only with water
(j) can hydrogen bond only with water
(k) can hydrogen bond only with water
(l) can hydrogen bond with itself and with water

2-39 Higher-boiling compounds are listed.

(a) $CH_3CH(OH)CH_3$ can form hydrogen bonds with other identical molecules

(b) $CH_3CH_2CH_2CH_2CH_3$ has a higher molecular weight than $CH_3CH_2CH_2CH_3$

(c) $CH_3CH_2CH_2CH_2CH_3$ has less branching than $(CH_3)_2CHCH_2CH_3$

(d) $CH_3CH_2CH_2CH_2CH_2Cl$ has a higher molecular weight AND dipole-dipole interaction compared with $CH_3CH_2CH_2CH_2CH_3$

2-40

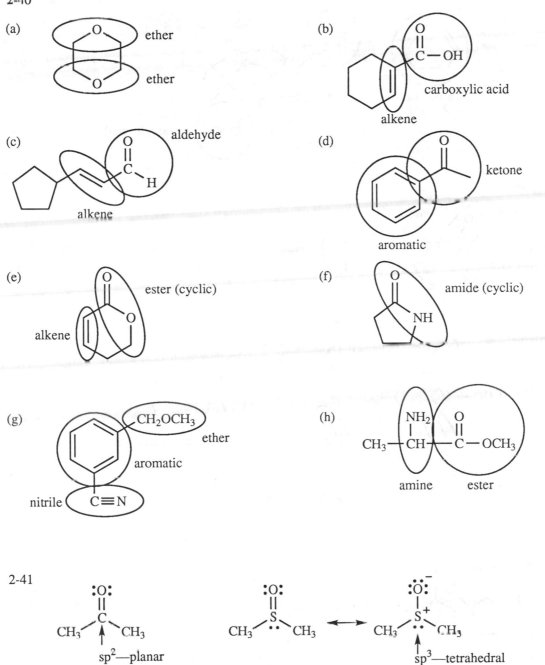

(a) ether / ether

(b) carboxylic acid / alkene

(c) aldehyde / alkene

(d) ketone / aromatic

(e) ester (cyclic) / alkene

(f) amide (cyclic)

(g) CH_2OCH_3 ether / aromatic / nitrile $C\equiv N$

(h) $CH_3-CH-C-OCH_3$ amine ester

2-41

sp^2—planar

sp^3—tetrahedral

The key to this problem is understanding that sulfur has a *lone pair of electrons*. The second resonance form shows four pairs of electrons around the sulfur, an electronic configuration requiring sp^3 hybridization. Sulfur in DMSO cannot be sp^2 like carbon in acetone, so we would expect sulfur's geometry to be pyramidal (tetrahedral).

2-42

(a) penicillin G

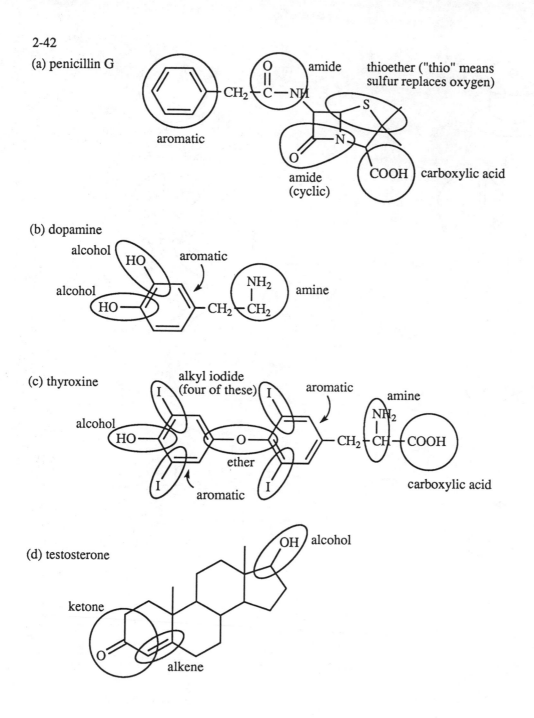

(b) dopamine

(c) thyroxine

(d) testosterone

3-1

(a) C_nH_{2n+2} where n = 30 gives $C_{30}H_{62}$

(b) C_nH_{2n+2} where n = 44 gives $C_{44}H_{90}$

> Note to the student: The IUPAC system of nomenclature has a well defined set of rules determining how structures are named. You will find a summary of these rules as an Appendix in this Solutions Manual.

3-2 Separate numbers from letters with hyphens.

(a) 3-methylpentane (always find the longest possible chain!)
(b) 5-ethyl-2-methyl-4-propylheptane ("When there are two longest chains of equal length, use the chain with the greater number of substituents.")
(c) 4-isopropyl-2-methyldecane

3-3 This Solutions Manual will present line formulas where a question asks for an answer including a structure. If you use condensed structural formulas instead, be sure that you are able to "translate" one structure type into the other.

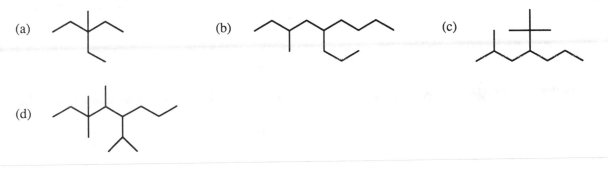

(a)

(b)

(c)

(d)

3-4 Separate numbers from numbers with commas.

(a) 2-methylbutane
(b) 2,2-dimethylpropane
(c) 3-ethyl-2-methylhexane
(d) 2,4-dimethylhexane
(e) 3-ethyl-2,2,4,5-tetramethylhexane
(f) 4-*t*-butyl-3-methylheptane

3-5 (Hints: *systematize* your approach to these problems. For the isomers of a six carbon formula, for example, start with the isomer containing all six carbons in a straight chain, then the isomers containing a five-carbon chain, then a four-carbon chain, *etc.* Carefully check your answers to AVOID DUPLICATE STRUCTURES.)

(a)

n-hexane

2-methylpentane

3-methylpentane

2,2-dimethylbutane

2,3-dimethylbutane

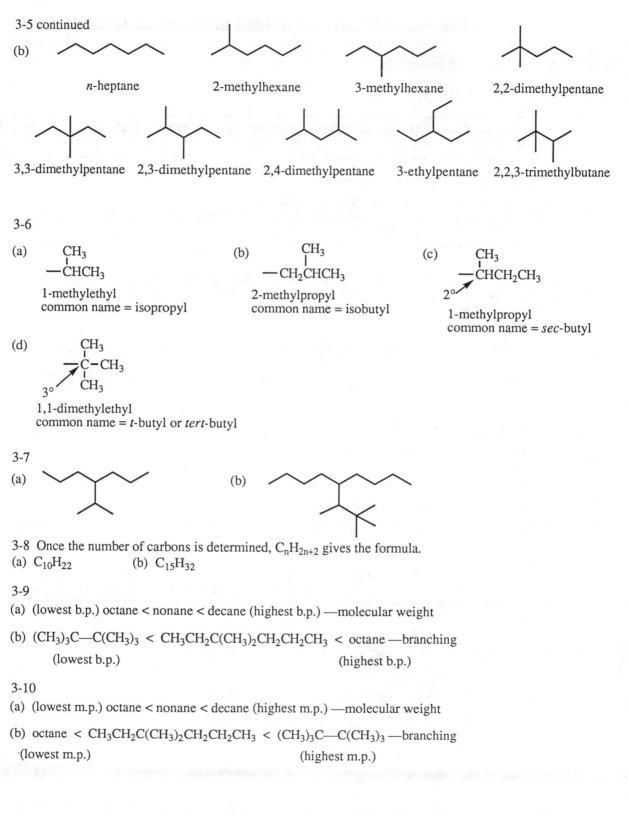

3-5 continued

(b)

n-heptane   2-methylhexane   3-methylhexane   2,2-dimethylpentane

3,3-dimethylpentane   2,3-dimethylpentane   2,4-dimethylpentane   3-ethylpentane   2,2,3-trimethylbutane

3-6

(a)
CH_3
$—CHCH_3$
1-methylethyl
common name = isopropyl

(b)
CH_3
$—CH_2CHCH_3$
2-methylpropyl
common name = isobutyl

(c)
CH_3
$—CHCH_2CH_3$
$2°$
1-methylpropyl
common name = *sec*-butyl

(d)
CH_3
$—C—CH_3$
$3°$ CH_3
1,1-dimethylethyl
common name = *t*-butyl or *tert*-butyl

3-7

(a)

(b)

3-8 Once the number of carbons is determined, C_nH_{2n+2} gives the formula.
(a) $C_{10}H_{22}$ (b) $C_{15}H_{32}$

3-9
(a) (lowest b.p.) octane < nonane < decane (highest b.p.) —molecular weight

(b) $(CH_3)_3C—C(CH_3)_3$ < $CH_3CH_2C(CH_3)_2CH_2CH_2CH_3$ < octane —branching
   (lowest b.p.)   (highest b.p.)

3-10
(a) (lowest m.p.) octane < nonane < decane (highest m.p.) —molecular weight

(b) octane < $CH_3CH_2C(CH_3)_2CH_2CH_2CH_3$ < $(CH_3)_3C—C(CH_3)_3$ —branching
   (lowest m.p.)   (highest m.p.)

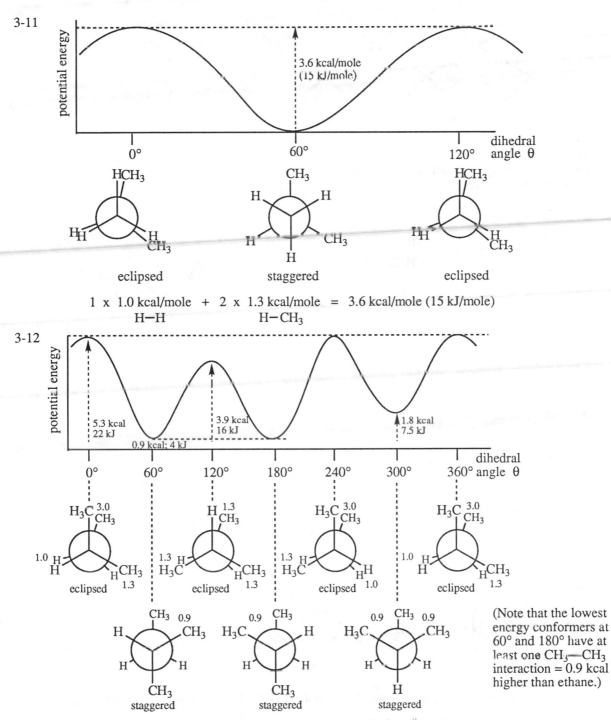

3-11

potential energy

3.6 kcal/mole
(15 kJ/mole)

0°   60°   120°   dihedral
                              angle θ

eclipsed   staggered   eclipsed

1 x 1.0 kcal/mole + 2 x 1.3 kcal/mole = 3.6 kcal/mole (15 kJ/mole)
      H—H   H—CH₃

3-12

potential energy

5.3 kcal
22 kJ
          0.9 kcal; 4 kJ
                    3.9 kcal
                    16 kJ
                              1.8 kcal
                              7.5 kJ

0°   60°   120°   180°   240°   300°   360° dihedral
                                              angle θ

eclipsed 1.3   eclipsed 1.3   eclipsed 1.0   eclipsed 1.3

staggered   staggered   staggered

(Note that the lowest energy conformers at 60° and 180° have at least one CH₃—CH₃ interaction = 0.9 kcal higher than ethane.)

Relative energies on the graph above were calculated using these values from the text:

0.9 kcal/mole for a CH_3–CH_3 gauche (staggered) interaction; 1.0 kcal/mole for a H–H eclipsed interaction; 1.3 kcal/mole for a H–CH_3 eclipsed interaction; 3.0 kcal/mole for a CH_3–CH_3 eclipsed interaction. These values are noted on each structure and are summed to give the energy value on the graph.

3-13

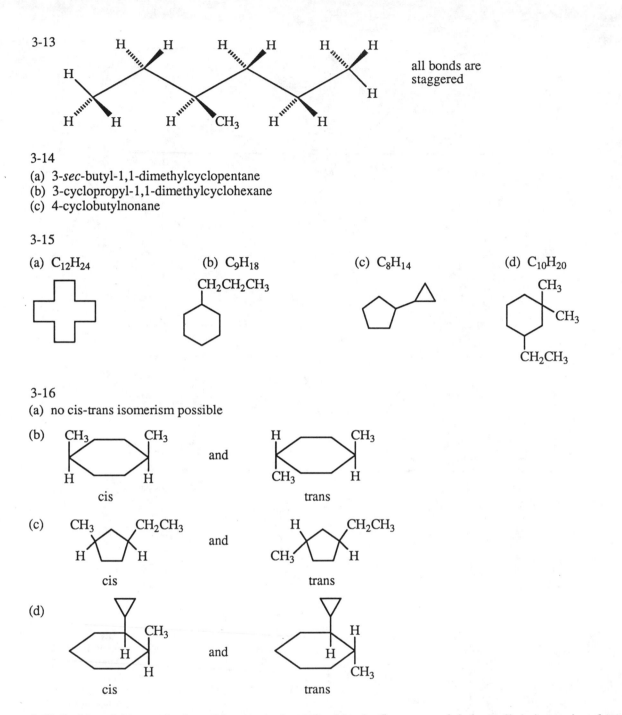

all bonds are
staggered

3-14

(a) 3-*sec*-butyl-1,1-dimethylcyclopentane
(b) 3-cyclopropyl-1,1-dimethylcyclohexane
(c) 4-cyclobutylnonane

3-15

(a) $C_{12}H_{24}$

(b) C_9H_{18}

$CH_2CH_2CH_3$

(c) C_8H_{14}

(d) $C_{10}H_{20}$

CH_3

CH_3

CH_2CH_3

3-16
(a) no cis-trans isomerism possible

(b)

CH_3 CH_3

H H

cis

and

H CH_3

CH_3 H

trans

(c)

CH_3 CH_2CH_3

H H

cis

and

H CH_2CH_3

CH_3 H

trans

(d)

CH_3

H

H

cis

and

H

H

CH_3

trans

3-17 In (a) and (b), numbering of the ring is determined by the first group *alphabetically* being assigned to ring carbon 1.

(a) *cis*-1-methyl-3-propylcyclobutane
(b) *trans*-1-*t*-butyl-3-ethylcyclohexane
(c) *trans*-1,2-dimethylcyclopropane

46

3-18 Combustion of the cis isomer gives off more energy, so *cis*-1,2-dimethylcyclo- propane must start at a higher energy than the trans isomer. The Newman projection of the cis isomer shows the two methyls are eclipsed with each other; in the trans isomer, the methyls are still eclipsed, but with hydrogens, not each other—a lower energy.

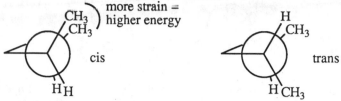

3-19 *trans*-1,2-Dimethylcyclobutane is more stable than *cis* because the two methyls can be farther apart when trans, as shown in the Newman projections.

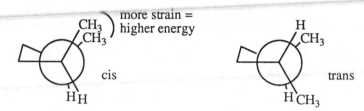

In the 1,3-dimethylcyclobutanes, however, the *cis* allows the methyls to be farther from other atoms and therefore more stable than the *trans*.

3-20

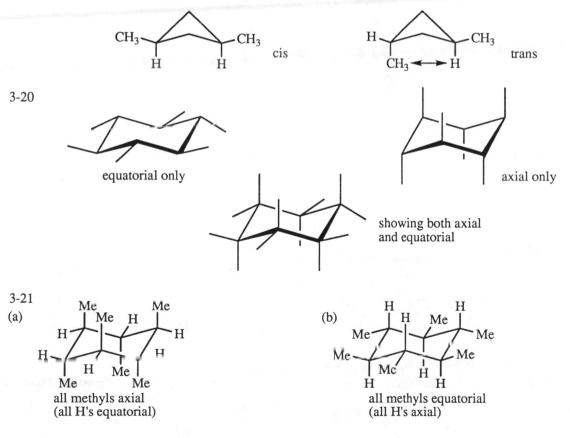

equatorial only

axial only

showing both axial and equatorial

3-21
(a)

Me Me
Me H Me
H   H
H   H
H Me
Me   Me
all methyls axial
(all H's equatorial)

(b)

H   H
H   Me
Me   Me
Me   Me
Me   H H
H   H
all methyls equatorial
(all H's axial)

47

3-22 Carbons 4 and 6 are behind the circles.

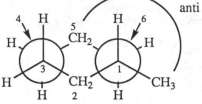

3-23 The isopropyl group can rotate so that its hydrogen is near the axial hydrogens on carbons 3 and 5, similar to a methyl group's hydrogen, and therefore similar to a methyl group in energy. The *t*-butyl group, however, must point a methyl group toward the hydrogens on carbons 3 and 5, giving severe diaxial interactions, causing the energy of this conformer to jump dramatically.

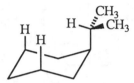

isopropylcyclohexane

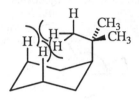

t-butylcyclohexane

3-24 The most stable conformers have substituents equatorial.

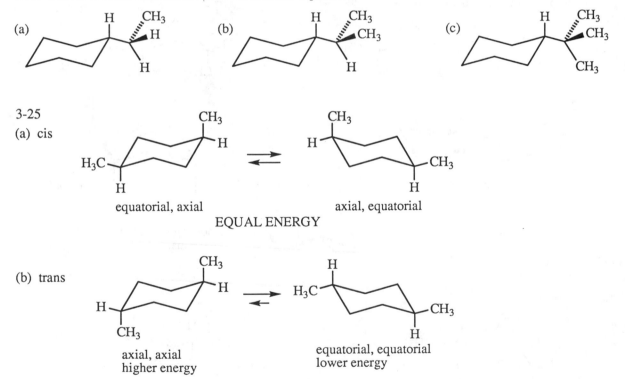

(a)

(b)

(c)

3-25

(a) cis

equatorial, axial   axial, equatorial

EQUAL ENERGY

(b) trans

axial, axial
higher energy

equatorial, equatorial
lower energy

(c) The trans isomer is more stable because BOTH substituents can be in the preferred equatorial positions.

3-26

Positions	cis	trans
1,2	(e,a) or (a,e)	(e,e) or (a,a)
1,3	(e,e) or (a,a)	(e,a) or (a,e)
1,4	(e,a) or (a,e)	(e,e) or (a,a)

3-27 The more stable conformer places the larger group equatorial.

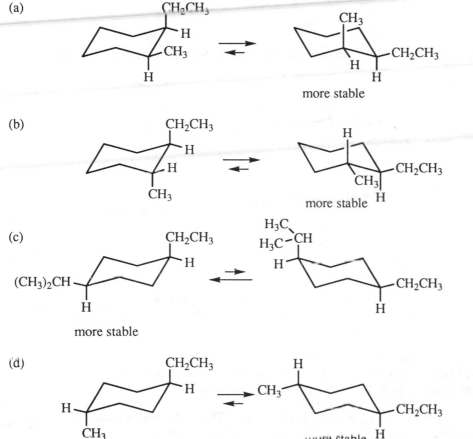

(a)

more stable

(b)

more stable

(c)

more stable

(d)

more stable

3-28 The key to determining cis and trans around a cyclohexane ring is to see whether a substituent group is "up" or "down" *relative to the H at the same carbon.* Two "up" groups or two "down" groups will be cis, one "up" and one "down" will be trans. This works independent of the conformation the molecule is in!

(a) *cis*-1,3-dimethylcyclohexane
(b) *cis*-1,4-dimethylcyclohexane
(c) *trans*-1,2-dimethylcyclohexane

(d) *cis*-1,3-dimethylcyclohexane
(e) *cis*-1,3-dimethylcyclohexane
(f) *trans*-1,4-dimethylcyclohexane

3-29

(a)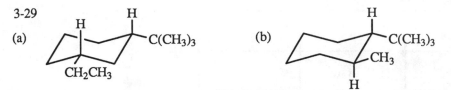

(b)

(c) Bulky substituents like *t*-butyl adopt equatorial rather than axial positions, even if that means altering the conformation of the ring. The twist boat conformation allows both bulky substituents to be "equatorial".

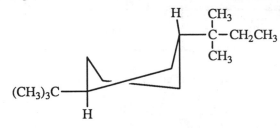

3-30
(a) bicyclo[3.1.0]hexane   (b) bicyclo[3.2.1]octane   (c) bicyclo[2.2.2]octane   (d) bicyclo[3.1.1]heptane

3-31 Using models is essential for this problem.

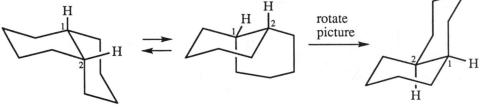

from text Figure 3-29

3-32   Please refer to solution 1-19, page 12 of this Solutions Manual.

3-33
(a) The third structure is 2-methylpropane (isobutane). The other four structures are all *n*-butane. Remember that a compound's identity is determined by how the atoms are connected, not by the position of the atoms when a structure is drawn on a page.

(b) The two structures at the top-left and bottom-left are both *cis*-2-butene. The two structures at the top-center and bottom-center are both 1-butene. The unique structure at the upper right is *trans*-2-butene. The unique structure at the lower right is 2-methylpropene.

(c) The first two structures are both *cis*-1,2-dimethylcyclopentane. The next two structures are both *trans*-1,2-dimethylcyclopentane. The last structure is different from all the others, *cis*-1,3-dimethylcyclopentane.

3-34 Line formulas are shown.

(a) 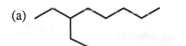   (b) 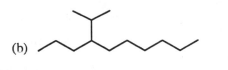   (c)

3-34 continued

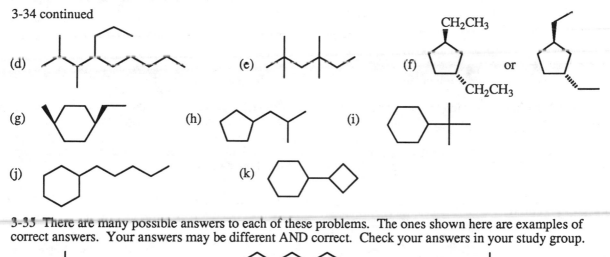

(d)

(e)

(f) CH$_2$CH$_3$ or   CH$_2$CH$_3$
      CH$_2$CH$_3$

(g)

(h)

(i)

(j)

(k)

3-35 There are many possible answers to each of these problems. The ones shown here are examples of correct answers. Your answers may be different AND correct. Check your answers in your study group.

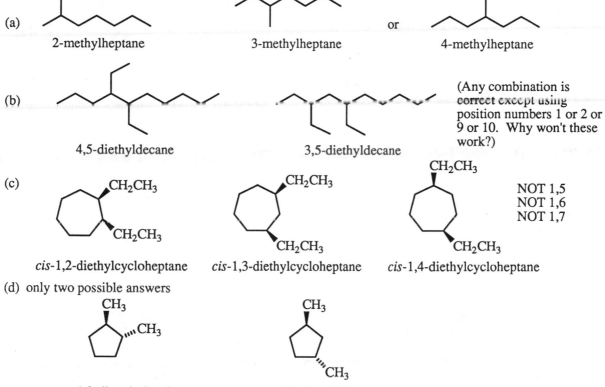

(a)

2-methylheptane

3-methylheptane   or   4-methylheptane

(b)

4,5-diethyldecane

3,5-diethyldecane

(Any combination is correct except using position numbers 1 or 2 or 9 or 10. Why won't these work?)

(c) CH$_2$CH$_3$

CH$_2$CH$_3$

cis-1,2-diethylcycloheptane

CH$_2$CH$_3$

CH$_2$CH$_3$

cis-1,3-diethylcycloheptane

CH$_2$CH$_3$

CH$_2$CH$_3$

cis-1,4-diethylcycloheptane

NOT 1,5
NOT 1,6
NOT 1,7

(d) only two possible answers

CH$_3$

...CH$_3$

trans-1,2-dimethylcyclopentane

CH$_3$

...CH$_3$

trans 1,3 dimethylcyclopentane

(e)

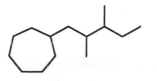

(2,3-dimethylpentyl)cycloheptane

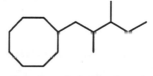

(2,3-dimethylpentyl)cyclooctane

other ring sizes are possible

3-36   HO−CH₃   HO−CH₂CH₂CH₃   HO−CH₂CH₂CH₂CH₂CH₃

       HO−CH₂CH₃   HO−CH₂CH₂CH₂CH₃   HO−CH₂CH₂CH₂CH₂CH₂CH₃

3-37
(a) 3-ethyl-2,2,6-trimethylheptane   (e) bicyclo[4.1.0]heptane
(b) 3-ethyl-2,6,7-trimethyloctane   (f) *cis*-1-ethyl-3-propylcyclopentane
(c) 3,7-diethyl-2,2,8-trimethyldecane   (g) (1,1-diethylpropyl)cyclohexane
(d) 2-ethyl-1,1-dimethylcyclobutane   (h) *cis*-1-ethyl-4-isopropylcyclodecane

3-38 There are eighteen isomers of C₈H₁₈. Here are eight of them. Yours may be different from the ones shown. An easy way to compare is to name yours and see if the names match.

n-octane

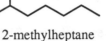

2-methylheptane

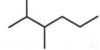

2,3-dimethylhexane

3-ethylhexane

2,2,4-trimethylpentane   2,3,4-trimethylpentane

3-ethyl-3-methylpentane

2,2,3,3-tetramethylbutane

3-39
(a)

correct name: 3-methylhexane
(longer chain)

(b)

correct name: 3-ethyl-2-methylhexane
(more branching with this numbering)

(c)

correct name: 3-methylhexane
(begin numbering at end closest to substituent)

(d)

correct name: 2,2-dimethylbutane (include
a position number for each substituent,
regardless of redundancies)

(e)

correct name: *sec*-butylcyclohexane or
(1-methylpropyl)cyclohexane
(the longer chain or ring is the base name)

(f)

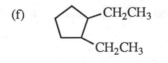

correct name: 1,2-diethylcyclopentane
(position numbers are the lowest possible)

3-40

(a) *n*-Octane has a higher boiling point than 2,2,3-trimethylpentane because linear molecules boil higher than branched molecules of the same molecular weight (increased van der Waals interaction)

(b) 2-Methylnonane has a higher boiling point than *n*-heptane because it has a significantly higher molecular weight than *n*-heptane.

(c) *n*-Nonane boils higher than 2,2,5-trimethylhexane for the same reason as in (a).

3-41   The point of attachment is shown by the bond at the left of each structure.

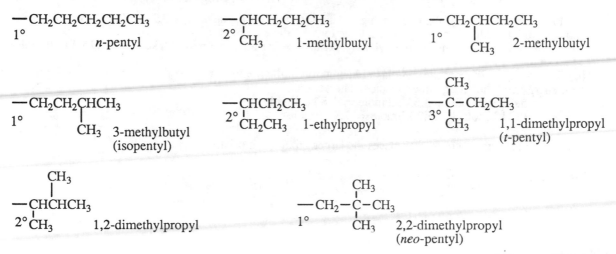

—$CH_2CH_2CH_2CH_2CH_3$
1°
   n-pentyl

—$CHCH_2CH_2CH_3$
2° CH_3 1-methylbutyl

—$CH_2CHCH_2CH_3$
1° CH_3 2-methylbutyl

—$CH_2CH_2CHCH_3$
1° CH_3 3-methylbutyl
         (isopentyl)

—$CHCH_2CH_3$
2° CH_2CH_3 1-ethylpropyl

CH_3
—C–CH_2CH_3
3° CH_3 1,1-dimethylpropyl
         (*t*-pentyl)

CH_3
—$CHCHCH_3$
2° CH_3 1,2-dimethylpropyl

CH_3
—CH_2–C–CH_3
1° CH_3 2,2-dimethylpropyl
         (*neo*-pentyl)

3-42 In each case, put the largest groups on adjacent carbons in anti positions to make the most stable conformations.

(a) 3-methylpentane

C-2 is the front carbon with H, H, and CH_3
C-3 is the back carbon with H, CH_3, and CH_2CH_3

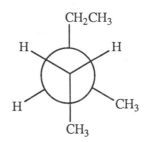

(b) 3,3-dimethylhexane

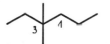

C-3 is the front carbon with CH_3, CH_3, and CH_2CH_3
C-4 is the back carbon with H, H, and CH_2CH_3

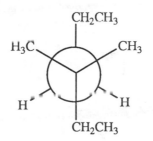

3-43
(a)
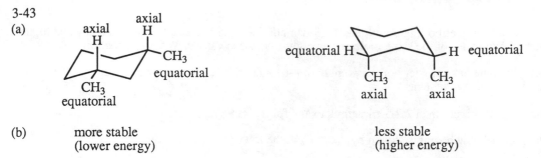

(b)　　　　more stable　　　　　　　　　　　　　less stable
　　　　　　(lower energy)　　　　　　　　　　　　(higher energy)

(c) From Section 3-14 of the text, each gauche interaction raises the energy 0.9 kcal/mole (3.8 kJ/mole), and each axial methyl has two gauche interactions, so the energy is:
2 methyls × 2 interactions per methyl × 0.9 kcal/mole per interaction = 3.6 kcal/mole (15.2 kcal/mole)

(d) The steric strain from the 1,3-diaxial interaction of the methyls must be the difference between the total energy and the energy due to gauche interactions:
　　　　5.4 kcal/mole − 3.6 kcal/mole = 1.8 kcal/mole
　　　　(23 kJ/mole − 15.2 kJ/mole = 7.8 kJ/mole)

3-44 The more stable conformer places the larger group equatorial.

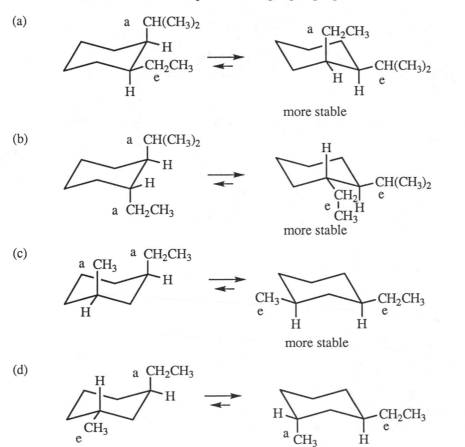

3-44 continued

(e)

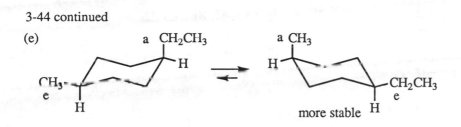

more stable

3-45 (Using models is essential to this problem.)

In both *cis*- and *trans*-decalin, the cyclohexane rings can be in chair conformations. The relative energies will depend on the number of axial substituents.

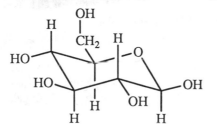

trans
no axial substituents
MORE STABLE

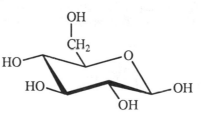

cis
one axial substituent

3-46 chair form of glucose—all substituents equatorial

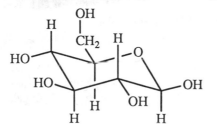

(without ring H's shown)

4-1

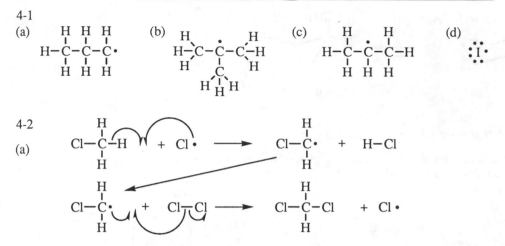

4-2

(a)

(b) Free-radical halogenation substitutes a halogen atom for a hydrogen. Even if a molecule has only one type of hydrogen, substitution of the first of these hydrogens forms a new compound. Any remaining hydrogens in this product can compete with the initial reactant for the available halogen. Thus, chlorination of methane, CH_4, produces all possible substitution products: CH_3Cl, CH_2Cl_2, $CHCl_3$, and CCl_4.

If a molecule has different types of hydrogens, the reaction can generate a mixture of the possible substitution products.

(c) Production of CCl_4 or CH_3Cl can be controlled by altering the ratio of CH_4 to Cl_2. To produce CCl_4, use an excess of Cl_2 and let the reaction proceed until all C—H bonds have been replaced with C—Cl bonds. Producing CH_3Cl is more challenging because the reaction tends to proceed past the first substitution. By using a very large excess of CH_4 to Cl_2, perhaps 100 to 1 or even more, a chlorine atom is more likely to find a CH_4 molecule than it is to find a CH_3Cl, so only a small amount of CH_4 is transformed to CH_3Cl by the time the Cl_2 runs out, with almost no CH_2Cl_2 being produced.

4-3

(a) This mechanism requires that one photon of light be added for each CH_3Cl generated, a quantum yield of 1. The actual quantum yield is several hundred or thousand. The high quantum yield suggests a chain reaction, but this mechanism is not a chain; it has no propagation steps.

(b) This mechanism conflicts with at least two experimental observations. First, the energy of light required to break a H–CH_3 bond is 104 kcal/mole (435 kJ/mole, from Table 4-2); the energy of light determined by experiment to initiate the reaction is only 60 kcal/mole of photons (251 kJ/mole of photons), much less than the energy needed to break this H–C bond. Second, as in (a), each CH_3Cl produced would require one photon of light, a quantum yield of 1, instead of the actual number of several hundred or thousand. As in (a), there is no provision for a chain process, since all the radicals generated are also consumed in the mechanism.

4-4

(a) The twelve hydrogens of cyclohexane are all on equivalent 2° carbons. Replacement of any one of the twelve will lead to the same product, chlorocyclohexane. *n*-Hexane, however, has hydrogens in three different positions: on carbon-1 (equivalent to carbon-6), carbon-2 (equivalent to carbon 5), and carbon-3 (equivalent to carbon-4). Monochlorination of *n* hexane will produce a mixture of all three possible isomers: 1-, 2 , and 3-chlorohexane.

(b) The best conversion of cyclohexane to chlorocyclohexane would require the ratio of cyclohexane/chlorine to be a large number. If the ratio were small, as the concentration of chlorocyclohexane increased during the reaction, chlorine would begin to substitute for a second hydrogen of chlorocyclohexane, generating unwanted products. The goal is to have chlorine attack a molecule of cyclohexane before it ever encounters a molecule of chlorocyclohexane, so the concentration of cyclohexane should be kept high.

4-5

(a) $K_{eq} = e^{-\Delta G°/RT}$ (converted from kcal)

$= e^{-(-500 \text{ cal/mole})/((1,987 \text{ cal/kelvin-mole}) \cdot (298 \text{ kelvin}))}$

$= e^{500/592} = e^{0.845} = \boxed{2.3}$

(b) $K_{eq} = 2.3 = \dfrac{[CH_3SH][HBr]}{[CH_3Br][H_2S]}$

	$[CH_3Br]$	$[H_2S]$	$[CH_3SH]$	$[HBr]$
initial concentrations:	1	1	0	0
final concentrations	$1-x$	$1-x$	x	x

$K_{eq} = 2.3 = \dfrac{x \cdot x}{(1-x)(1-x)} = \dfrac{x^2}{1-2x+x^2} \implies x^2 = 2.3x^2 - 4.6x + 2.3$

$0 = 1.3x^2 - 4.6x + 2.3 \implies x = 0.60, \ 1-x = 0.40$

(using quadratic equation)

$\boxed{[CH_3SH] = [HBr] = 0.60 \text{ M}; \ [CH_3Br] = [H_2S] = 0.40 \text{ M}}$

4-6   2 acetone $\rightleftharpoons$ diacetone

Assume that the initial concentration of acetone is 1 molar, and 5% of the acetone is converted to diacetone. NOTE THE MOLE RATIO.

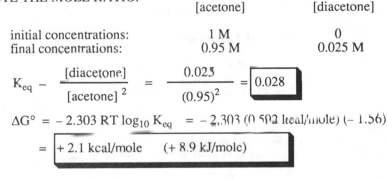

	[acetone]	[diacetone]
initial concentrations:	1 M	0
final concentrations:	0.95 M	0.025 M

$K_{eq} = \dfrac{[diacetone]}{[acetone]^2} = \dfrac{0.025}{(0.95)^2} = \boxed{0.028}$

$\Delta G° = -2.303 \, RT \log_{10} K_{eq} = -2.303 \, (0.592 \text{ kcal/mole})(-1.56)$

$= \boxed{+2.1 \text{ kcal/mole} \quad (+8.9 \text{ kJ/mole})}$

4-7 ΔS° will be negative since two molecules are combined into one, a loss of freedom of motion. Since ΔS° is negative, – TΔS° is positive; but ΔG° is a large negative number since the reaction goes to completion. Therefore, ΔH° must also be a large negative number, necessarily larger in value than ΔG°. We can explain this by formation of two strong C–H bonds (98 kcal/mole each) after breaking a strong H–H bond (104 kcal/ mole) and a WEAKER C=C pi bond.

4-8

(a) ΔS° is positive—one molecule became two smaller molecules with greater freedom of motion
(b) ΔS° is negative—two smaller molecules combined into one larger molecule with less freedom of motion
(c) ΔS° cannot be predicted since the number of molecules in reactants and products is the same

4-9 (a)   initiation   (1) $Cl-Cl \xrightarrow{\text{hv}}$ 2 Cl •

                              (2)   Cl • + H–CH$_2$CH$_3$ ⟶ H–Cl + • CH$_2$CH$_3$
         propagation
                              (3)   Cl–Cl + • CH$_2$CH$_3$ ⟶ Cl–CH$_2$CH$_3$ + Cl •

                              (4)   Cl • + • Cl ⟶ Cl–Cl
        termination (5)   Cl • + • CH$_2$CH$_3$ ⟶ Cl–CH$_2$CH$_3$
                              (6)   CH$_3$CH$_2$ • + • CH$_2$CH$_3$ ⟶ CH$_3$CH$_2$CH$_2$CH$_3$

(b)   step (1)   break Cl–Cl   ΔH° = + 58 kcal/mole (+ 242 kJ/mole)

       step (2)   break H–CH$_2$CH$_3$ ΔH° = + 98 kcal/mole   (+ 410 kJ/mole)
                    make H–Cl   ΔH° = – 103 kcal/mole (– 431 kJ/mole)
                          step (2)   **ΔH° = –5 kcal/mole (–21 kJ/mole)**

       step (3)   break Cl–Cl   ΔH° = + 58 kcal/mole (+242 kJ/mole)
                    make Cl–CH$_2$CH$_3$ ΔH° = – 81 kcal/mole (– 339 kJ/mole)
                          step (3)   **ΔH° = –23 kcal/mole (–97 kJ/mole)**

  (c) ΔH° for the reaction is the sum of the ΔH° values of the individual propagation steps:

                   – 5 kcal/mole + – 23 kcal/mole = **– 28 kcal/mole**
                   (– 21 kJ/mole + – 97 kJ/mole = **– 118 kJ/mole**)

4-10 (a)   initiation   (1) $Br-Br \xrightarrow{\text{hv}}$ 2 Br •

                              (2)   Br • + H–CH$_3$ ⟶ H–Br + • CH$_3$
         propagation
                              (3)   Br–Br + • CH$_3$ ⟶ Br–CH$_3$ + Br •

       step (1)   break Br–Br   ΔH° = + 46 kcal/mole (+ 192 kJ/mole)

       step (2)   break H–CH$_3$ ΔH° = + 104 kcal/mole   (+ 435 kJ/mole)
                    make H–Br   ΔH° = – 88 kcal/mole (– 368 kJ/mole)
                          step (2)   **ΔH° = +16 kcal/mole (+67 kJ/mole)**

       step (3)   break Br–Br   ΔH° = + 46 kcal/mole (+ 192 kJ/mole)
                    make Br–CH$_3$ ΔH° = – 70 kcal/mole (– 293 kJ/mole)
                          step (3)   **ΔH° = –24 kcal/mole (–101 kJ/mole)**

4-10 continued

(b) ΔH° for the reaction is the sum of the ΔH° values of the individual propagation steps:

$$+ 16 \text{ kcal/mole } + -24 \text{ kcal/mole } = -8 \text{ kcal/mole}$$
$$(+ 67 \text{ kJ/mole } + \quad 101 \text{ kJ/mole } = -34 \text{ kJ/mole})$$

4-11

(a) first order: the exponent of [$(CH_3)_3CCl$] in the rate law = 1
(b) zeroth order: [CH_3OH] does not appear in the rate law (its exponent is zero)
(c) first order: the sum of the exponents in the rate law = 1

4-12

(a) there is a linear relationship between the rate and [CH_3Cl], so the exponent = 1 and the rate equation is first order with respect to chloromethane
(b) there is a linear relationship between the rate and [^-CN], so the exponent = 1 and the rate equation is first order with respect to the cyanide ion
(c) rate = k_r [CH_3Cl] [^-CN]
(d) overall, second order: the sum of the exponents in the rate law = 2

4-13

(a) the reaction rate depends on neither [ethylene] nor [hydrogen], so it is zeroth order in both species. The overall reaction must be zeroth order.
(b) rate = k_r
(c) The rate law does not depend on the concentration of the reactants. It must depend, therefore, on the only other chemical present, the catalyst. Apparently, whatever is happening on the surface of the catalyst determines the rate, regardless of the concentrations of the two gases. Increasing the surface area of the catalyst, or simply adding more catalyst, would speed the reaction.

4-14

(a)

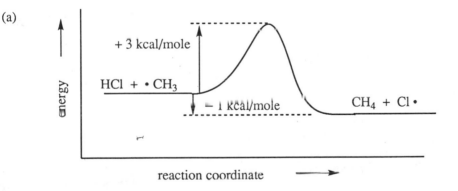

(b) E_a = + 3 kcal/mole   (+ 13 kJ/mole)
(c) ΔH° = − 1 kcal/mole (− 4 kJ/mole)

4-15

(a)

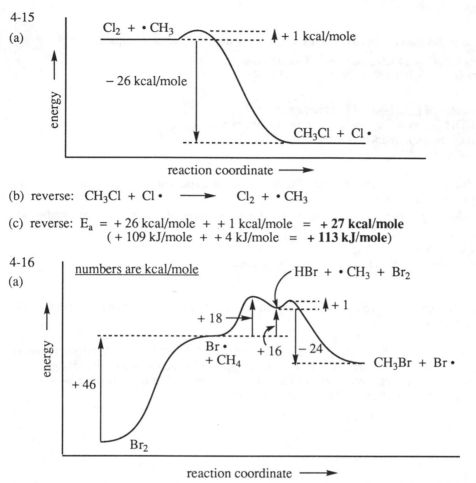

(b) reverse: $CH_3Cl + Cl \cdot \longrightarrow Cl_2 + \cdot CH_3$

(c) reverse: $E_a = +26 \text{ kcal/mole} + +1 \text{ kcal/mole} = \textbf{+27 kcal/mole}$
$(+109 \text{ kJ/mole} + +4 \text{ kJ/mole} = \textbf{+113 kJ/mole})$

4-16

(a)

(b) The step leading to the highest energy transition state is rate-determining. In this mechanism, the first propagation step is rate-determining:

$$Br \cdot + CH_4 \longrightarrow HBr + \cdot CH_3$$

(c)   (1) $\begin{bmatrix} \delta \cdot & \delta \cdot \\ Br\text{-------}Br \end{bmatrix}^{\ddagger}$ ("$\delta \cdot$" means partial radical character on the atom)

(2) $\begin{bmatrix} \overset{\displaystyle H}{\underset{\displaystyle H}{\overset{|}{\underset{|}{H-C}}}}\text{-------}H\text{------}Br \end{bmatrix}^{\ddagger}$ with $\delta \cdot$ and $\delta \cdot$

(3) $\begin{bmatrix} \overset{\displaystyle H}{\underset{\displaystyle H}{\overset{|}{\underset{|}{H-C}}}}\text{-------}Br\text{------}Br \end{bmatrix}^{\ddagger}$ with $\delta \cdot$ and $\delta \cdot$

(d) $\Delta H°$ for the reaction is the sum of the $\Delta H°$ values of the individual propagation steps (refer to the solution to 4-10 (a) and (b)):

$$+16 \text{ kcal/mole} + -24 \text{ kcal/mole} = \textbf{-8 kcal/mole}$$
$$(+67 \text{ kJ/mole} + -101 \text{ kJ/mole} = \textbf{-34 kJ/mole})$$

4-17

(a) initiation (1) I—I $\xrightarrow{hv}$ 2 I·

propagation $\begin{cases} (2) & \text{I· + H—CH}_3 \longrightarrow \text{H—I + ·CH}_3 \\ (3) & \text{I—I + ·CH}_3 \longrightarrow \text{I—CH}_3 + \text{I·} \end{cases}$

step (1) break I–I $\Delta H° = +36$ kcal/mole $(+151$ kJ/mole)

step (2) break H–CH$_3$ $\Delta H° = +104$ kcal/mole $(+435$ kJ/mole)
          make H–I $\underline{\Delta H° = -\ 71 \text{ kcal/mole}\ \ (-297 \text{ kJ/mole})}$
          step (2) $\Delta H° = +33$ **kcal/mole (+138 kJ/mole)**

step (3) break I–I $\Delta H° = +36$ kcal/mole $(+151$ kJ/mole)
          make I–CH$_3$ $\underline{\Delta H° = -\ 56 \text{ kcal/mole}\ \ (-234 \text{ kJ/mole})}$
          step (3) $\Delta H° = -20$ **kcal/mole (–83 kJ/mole)**

(b) $\Delta H°$ for the reaction is the sum of the $\Delta H°$ values of the individual propagation steps:

+ 33 kcal/mole + – 20 kcal/mole = **+ 13 kcal/mole**
(+ 138 kJ/mole + – 83 kJ/mole = **+ 55 kJ/mole**)

(c) Iodination of methane is unfavorable for both kinetic and thermodynamic reasons. Kinetically, the rate of the first propagation step must be very slow because it is very endothermic; the activation energy must be at least + 33 kcal/mole. Thermodynamically, the overall reaction is endothermic, so an equilibrium would favor reactants, not products; there is no energy decrease to drive the reaction to products.

4-18 Propane has six primary hydrogens and two secondary hydrogens, a ratio of 3 : 1. If primary and secondary hydrogens were replaced by chlorine at equal rates, the chloropropane isomers would reflect the same 3 : 1 ratio, that is, 75% 1-chloropropane and 25% 2-chloropropane.

4-19

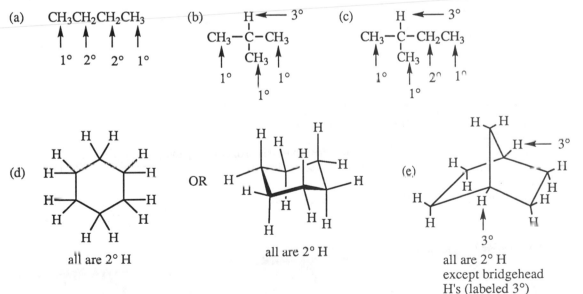

61

4-20

3° H abstraction

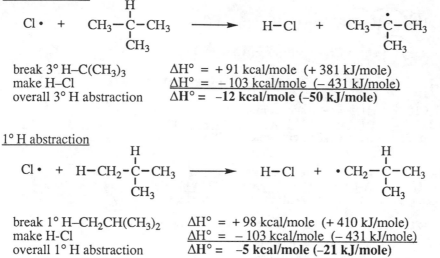

break 3° H–C(CH$_3$)$_3$ $\Delta H° = +91$ kcal/mole (+ 381 kJ/mole)
make H–Cl $\Delta H° = -103$ kcal/mole (– 431 kJ/mole)
overall 3° H abstraction $\Delta H° = $ **–12 kcal/mole (–50 kJ/mole)**

1° H abstraction

Cl•   +   H–CH$_2$–C(H)(CH$_3$)–CH$_3$ ⟶   H–Cl   +   •CH$_2$–C(H)(CH$_3$)–CH$_3$

break 1° H–CH$_2$CH(CH$_3$)$_2$ $\Delta H° = +98$ kcal/mole (+ 410 kJ/mole)
make H-Cl $\Delta H° = -103$ kcal/mole (– 431 kJ/mole)
overall 1° H abstraction $\Delta H° = $ **–5 kcal/mole (–21 kJ/mole)**

(Note: $\Delta H°$ for abstraction of a 1° H from both ethane and propane are + 98 kcal/mole (+ 410 kJ/mole). It is reasonable to use this same value for abstraction of the 1° H in isobutane.)

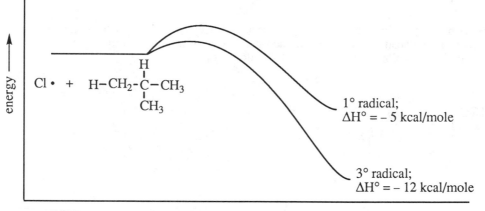

Since $\Delta H°$ for forming the 3° radical is more negative than $\Delta H°$ for forming the 1° radical, it is reasonable to infer that the activation energy leading to the 3° radical is lower than the activation energy leading to the 1° radical.

4-21

2-Methylbutane can produce four mono-chloro isomers. To calculate the relative amount of each in the product mixture, multiply the numbers of hydrogens which could lead to that product times the reactivity for that type of hydrogen. Each relative amount divided by the sum of all the amounts will provide the percent of each in the product mixture.

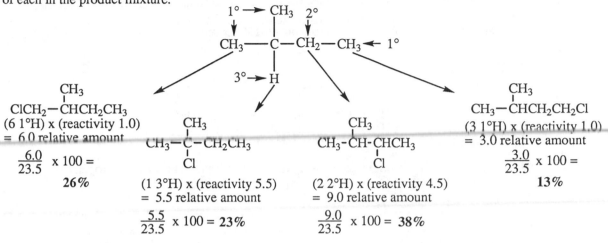

CH_3
$ClCH_2-CHCH_2CH_3$
(6 1°H) x (reactivity 1.0)
= 6.0 relative amount

$\frac{6.0}{23.5}$ x 100 = **26%**

CH_3
$CH_3-C-CH_2CH_3$
         Cl
(1 3°H) x (reactivity 5.5)
= 5.5 relative amount

$\frac{5.5}{23.5}$ x 100 = **23%**

CH_3
$CH_3-CH-CHCH_3$
            Cl
(2 2°H) x (reactivity 4.5)
= 9.0 relative amount

$\frac{9.0}{23.5}$ x 100 = **38%**

CH_3
$CH_3-CHCH_2CH_2Cl$
(3 1°H) x (reactivity 1.0)
= 3.0 relative amount

$\frac{3.0}{23.5}$ x 100 = **13%**

4-22

(a) When *n*-heptane is burned, only 1° and 2° radicals can be formed (from either C—H or C—C bond cleavage). These are high energy, unstable radicals which rapidly form other products. When isooctane (2,2,4-trimethylpentane, below) is burned, 3° radicals can be formed from either C—H or C—C bond cleavage. The 3° radicals are lower in energy than 1° or 2°, relatively stable, with lowered reactivity. Slower combustion translates to less "knocking."

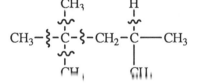

isooctane (2,2,4-trimethylpentane)

Any indicated bond cleavage will produce a 3° radical.

(b)

CH_3
$H_3C-C-O-H$ + •R → $H_3C-C-O•$ + R—H
CH_3 some
                alkyl
                radical
CH_3
CH_3

When the alcohol hydrogen is abstracted from *t*-butyl alcohol, a relatively stable *t*-butoxy radical ($(CH_3)_3C-O•$) is produced. This low energy radical is slower to react than alkyl radicals, moderating the reaction and producing less "knocking."

4-23

(a) <u>1° H abstraction</u>

$$F\cdot \ + \ CH_3CH_2CH_3 \longrightarrow H{-}F \ + \ \cdot CH_2CH_2CH_3$$

break 1° H–CH$_2$CH$_2$CH$_3$	$\Delta H° = +\ 98$ kcal/mole $(+\ 410$ kJ/mole$)$
make H–F	$\Delta H° = -\ 136$ kcal/mole $(-\ 569$ kJ/mole$)$
overall 1° H abstraction	$\Delta H° = -\ 38$ **kcal/mole** $(-159$ **kJ/mole**$)$

<u>2° H abstraction</u>

$$F\cdot \ + \ CH_3CH_2CH_3 \longrightarrow H{-}F \ + \ CH_3\overset{\bullet}{C}HCH_3$$

break 2° H–CH(CH$_3$)$_2$	$\Delta H° = +\ 95$ kcal/mole $(+\ 397$ kJ/mole$)$
make H–F	$\Delta H° = -\ 136$ kcal/mole $(-\ 569$ kJ/mole$)$
overall 2° H abstraction	$\Delta H° = -\ 41$ **kcal/mole** $(-172$ **kJ/mole**$)$

(b) Fluorination is extremely exothermic and is likely to be indiscriminate in which hydrogens are abstracted. (In fact, C—C bonds are also broken during fluorination.)

(c) Free radical fluorination is extremely exothermic. In exothermic reactions, the transition states resemble the starting materials more than the products, so while the 1° and 2° radicals differ by about 3 kcal/mole (13 kJ/mole), the transition states will differ by only a tiny amount. For fluorination, then, the rate of abstraction for 1° and 2° hydrogens will be virtually identical. Product ratios will depend on statistical factors only.

Fluorination is difficult to control, but if propane were monofluorinated, the product mixture would reflect the ratio of the types of hydrogens: six 1° H to two 2° H, or 3 : 1 ratio, giving 75% 1-fluoropropane and 25% 2-fluoropropane.

4-24

(a) $\underbrace{C_2H_5{-}D \ + \ Cl_2 \longrightarrow}\ \underbrace{C_2H_5{-}Cl \ + \ DCl}_{7\%}, \ + \ \underbrace{C_2H_4DCl \ + \ HCl}_{93\%}$

D replacement: 7% ÷ 1 D = 7 (reactivity factor)
H replacement: 93% ÷ 5 H = 18.6 (reactivity factor)
relative reactivity of H : D abstraction = 18.6 ÷ 7 = 2.7

Each hydrogen is abstracted 2.7 times faster than deuterium.

(b) In both reactions of chlorine with either methane or ethane, the first propagation step is rate-determining. The reaction of chlorine atom with methane is *endothermic* by 1 kcal/mole (4 kJ/mole), while for ethane this step is *exothermic* by 5 kcal/mole (21 kJ/mole). By the Hammond Postulate, differences in activation energy are most pronounced in *endothermic* reactions where the transition states most resemble the products. Therefore, a change in the methane molecule causes a greater change in its transition state energy than the same change in the ethane molecule causes in its transition state energy. Deuterium will be abstracted more slowly in both methane and ethane, but the rate effect will be more pronounced in methane than in ethane.

4-25

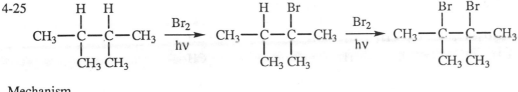

Mechanism

initiation   Br—Br $\xrightarrow{\text{hv}}$ 2 Br•

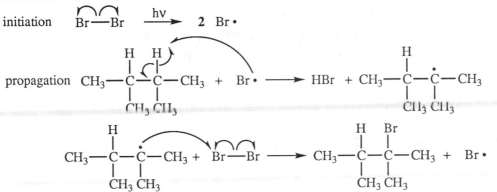

4-26 The triphenylmethyl cation is so stable because of the delocalization of the charge. The more resonance forms a species has, the more stable it will be.

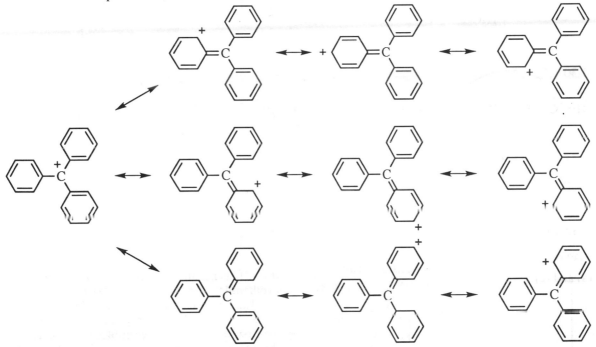

(Note: these resonance forms do not include the simple benzene resonance forms as shown below; they are significant, but repetitive. Refer to the note in the solution to 1-39(d), p. 20 of this Solutions Manual.)

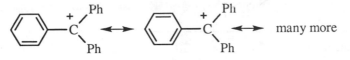

4-27 most stable (c) > (b) > (a) least stable

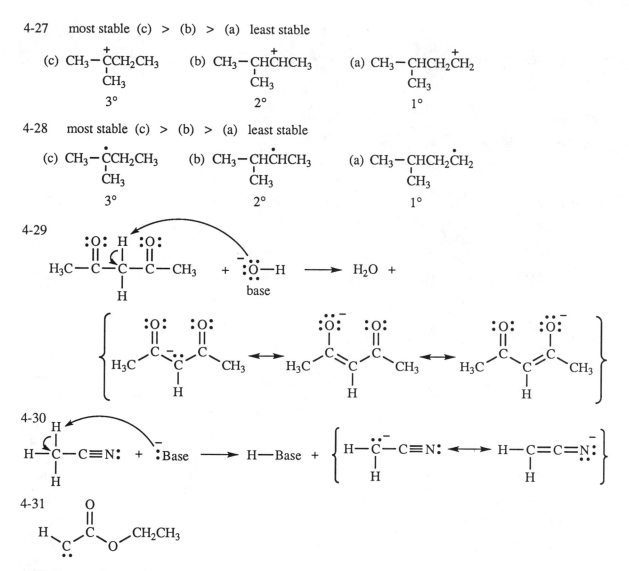

(c) $CH_3-\overset{+}{\underset{CH_3}{C}}CH_2CH_3$ (b) $CH_3-\overset{+}{\underset{CH_3}{CH}}CHCH_3$ (a) $CH_3-\underset{CH_3}{CH}CH_2\overset{+}{C}H_2$

    3°   2°   1°

4-28 most stable (c) > (b) > (a) least stable

(c) $CH_3-\overset{\bullet}{\underset{CH_3}{C}}CH_2CH_3$ (b) $CH_3-\overset{\bullet}{\underset{CH_3}{CH}}CHCH_3$ (a) $CH_3-\underset{CH_3}{CH}CH_2\overset{\bullet}{C}H_2$

    3°   2°   1°

4-29

4-30

4-31

4-32 Please refer to solution 1-19, page 12 of this Solutions Manual.

4-33

(a) and (c)

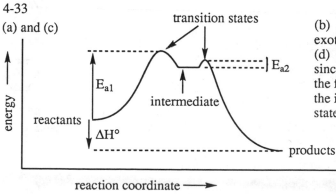

(b) ΔH° is negative (decreases), so the reaction is exothermic.

(d) The first transition state determines the rate since it is the highest energy point. The *structure* of the first transition state resembles the *structure* of the intermediate since the *energy* of the transition state is closest to the *energy* of the intermediate.

4-34

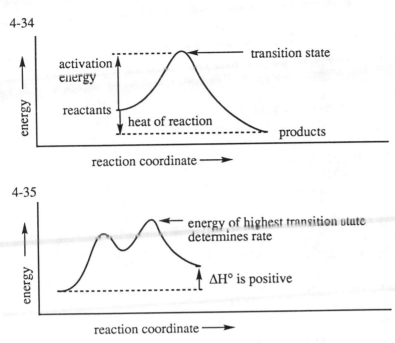

4-35

4-36
The rate law is first order with respect to the concentrations of hydrogen ion and of *t*-butyl alcohol, zeroth order with respect to the concentration of chloride ion, second order overall.
rate $= k_r \ [(CH_3)_3COH] \ [H^+]$

4-37

(a) $CH_3CH_2CH(CH_3)_2$

    $1° \quad 2° \quad 3° \quad 1°$

(b) $(CH_3)_3CCH_2C(CH_3)_3$

    $1° \quad 2° \quad 1°$

(c)

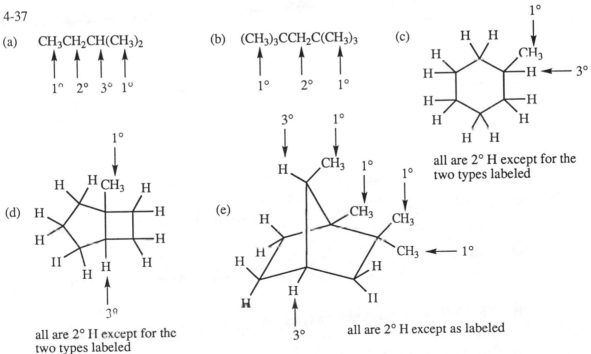

all are 2° H except for the two types labeled

(d)

all are 2° H except for the two types labeled

(e)

all are 2° H except as labeled

67

4-38

(a) break H–CH$_2$CH$_3$ and I–I, make I–CH$_2$CH$_3$ and H–I
kcal/mole: $(+98 + +36) + (-53 + -71) = +10$ kcal/mole
kJ/mole: $(+410 + +151) + (-222 + -297) = +42$ kJ/mole

(b) break CH$_3$CH$_2$–Cl and H–I, make CH$_3$CH$_2$–I and H–Cl
kcal/mole: $(+81 + +71) + (-53 + -103) = -4$ kcal/mole
kJ/mole: $(+339 + +297) + (-222 + -431) = -17$ kJ/mole

(c) break (CH$_3$)$_3$C–OH and H–Cl, make (CH$_3$)$_3$C–Cl and H–OH
kcal/mole: $(+91 + +103) + (-79 + -119) = -4$ kcal/mole
kJ/mole: $(+381 + +431) + (-331 + -498) = -17$ kJ/mole

(d) break CH$_3$CH$_2$–CH$_3$ and H–H, make CH$_3$CH$_2$–H and H–CH$_3$
kcal/mole: $(+85 + +104) + (-98 + -104) = -13$ kcal/mole
kJ/mole: $(+356 + +435) + (-410 + -435) = -54$ kJ/mole

(e) break CH$_3$CH$_2$–OH and H–Br, make CH$_3$CH$_2$–Br and H–OH
kcal/mole: $(+91 + +88) + (-68 + -119) = -8$ kcal/mole
kJ/mole: $(+381 + +368) + (-285 + -498) = -34$ kJ/mole

4-39 Numbers are bond dissociation energies in kcal/mole (kJ/mole).

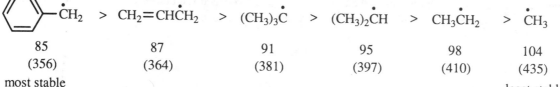

| 85 | 87 | 91 | 95 | 98 | 104 |
| (356) | (364) | (381) | (397) | (410) | (435) |

most stable   least stable

4-40

(a) 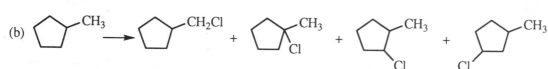 Only one product; chlorination would work.

(b)

Chlorination would produce four constitutional isomers and would not be a good method to make only one of these.

(c)

$$CH_3-\overset{\overset{H}{|}}{C}-\overset{\overset{H}{|}}{\underset{\underset{CH_3}{|}}{C}}-CH_3 \longrightarrow CH_3-\overset{\overset{H}{|}}{C}-\overset{\overset{Cl}{|}}{\underset{\underset{CH_3}{|}}{C}}-CH_3 + CH_3-\overset{\overset{H}{|}}{C}-\overset{\overset{H}{|}}{\underset{\underset{CH_3}{|}}{C}}-CH_2Cl$$

Chlorination would produce two constitutional isomers and would not be a good method to make only one of these.

(d)

$$CH_3-\overset{\overset{CH_3}{|}}{C}-\overset{\overset{CH_3}{|}}{\underset{\underset{CH_3}{|}}{C}}-CH_3 \longrightarrow CH_3-\overset{\overset{CH_3}{|}}{C}-\overset{\overset{CH_3}{|}}{\underset{\underset{CH_3}{|}}{C}}-CH_2Cl$$ Only one product; chlorination would work.

4-41

initiation   (1)   Cl—Cl $\xrightarrow{h\nu}$ 2 Cl·

propagation

(2) Cl· + H—C ⟶ H—Cl + ·C

(3) Cl—Cl + ·C ⟶ Cl—C + Cl·

Termination steps are any two radicals combining.

4-42

(a) $CH_2=CH-\overset{\bullet}{C}H_2 \longleftrightarrow \overset{\bullet}{C}H_2-CH=CH_2$

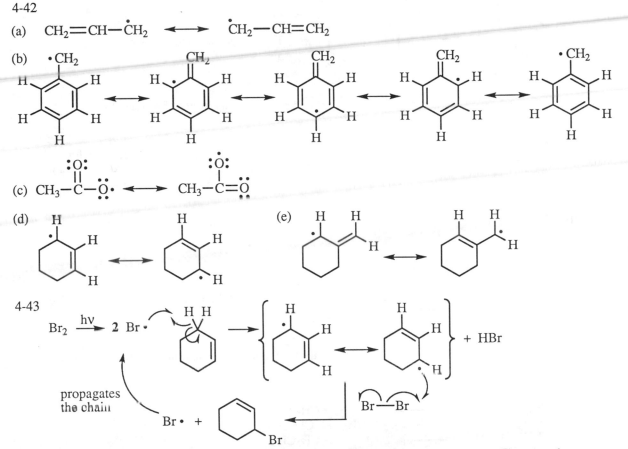

(b)

(c) $CH_3-\overset{:\overset{..}{O}:}{\underset{||}{C}}-\overset{..}{\underset{..}{O}}\overset{..}{\cdot} \longleftrightarrow CH_3-\overset{:\overset{..}{O}:}{C}=\overset{..}{\underset{..}{O}}$

(d)

(e)

4-43

$Br_2 \xrightarrow{h\nu}$ 2 Br·

+ HBr

propagates the chain

Br· +

Cyclohexene has three types of hydrogens which could be abstracted by bromine atom. They are shown below, labeled with their bond dissociation energy in kcal/mole (kJ/mole). Bromine atom will abstract the hydrogen with the lowest bond dissociation energy at the fastest rate, in this case, the allylic hydrogen. The allylic hydrogen is most easily abstracted because the radical produced is stabilized by resonance.

aliphatic: 95 (397) ⟶

⟵ 87 (364): allylic

⟵ 108 (452): vinylic

69

4-44 Where mixtures are possible, only the major product is shown.

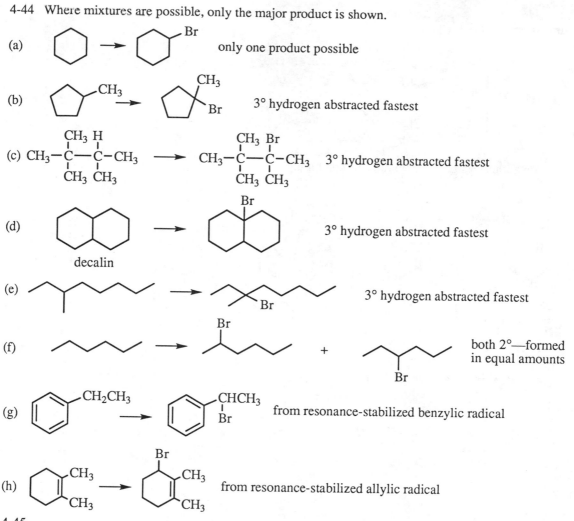

(a) only one product possible

(b) 3° hydrogen abstracted fastest

(c) $CH_3-\overset{\overset{\displaystyle CH_3}{|}}{\underset{\underset{\displaystyle CH_3}{|}}{C}}-\overset{\overset{\displaystyle H}{|}}{\underset{\underset{\displaystyle CH_3}{|}}{C}}-CH_3 \longrightarrow CH_3-\overset{\overset{\displaystyle CH_3}{|}}{\underset{\underset{\displaystyle CH_3}{|}}{C}}-\overset{\overset{\displaystyle Br}{|}}{\underset{\underset{\displaystyle CH_3}{|}}{C}}-CH_3$ 3° hydrogen abstracted fastest

(d) decalin 3° hydrogen abstracted fastest

(e) 3° hydrogen abstracted fastest

(f) both 2°—formed in equal amounts

(g) from resonance-stabilized benzylic radical

(h) from resonance-stabilized allylic radical

4-45

(a) As CH_3Cl is produced, it can compete with CH_4 for available $Cl\cdot$, generating CH_2Cl_2. This can generate $CHCl_3$, *etc.*

propagation steps

$$CH_4 + Cl\cdot \longrightarrow HCl + \cdot CH_3$$
$$\cdot CH_3 + Cl_2 \longrightarrow ClCH_3 + Cl\cdot$$
$$ClCH_3 + Cl\cdot \longrightarrow HCl + \cdot CH_2Cl$$
$$\cdot CH_2Cl + Cl_2 \longrightarrow CH_2Cl_2 + Cl\cdot$$
$$CH_2Cl_2 + Cl\cdot \longrightarrow HCl + \cdot CHCl_2$$
$$\cdot CHCl_2 + Cl_2 \longrightarrow CHCl_3 + Cl\cdot$$
$$CHCl_3 + Cl\cdot \longrightarrow HCl + \cdot CCl_3$$
$$\cdot CCl_3 + Cl_2 \longrightarrow CCl_4 + Cl\cdot$$

(b) To maximize CH_3Cl and minimize formation of polychloromethanes, the ratio of methane to chlorine must be kept high (see problem 4-2).

To guarantee that all hydrogens are replaced with chlorine to produce CCl_4, the ratio of chlorine to methane must be kept high.

4-46

n-Pentane can produce three mono-chloro isomers. To calculate the relative amount of each in the product mixture, multiply the numbers of hydrogens which could lead to that product times the reactivity for that type of hydrogen. Each relative amount divided by the sum of all the amounts will provide the percent of each in the product mixture.

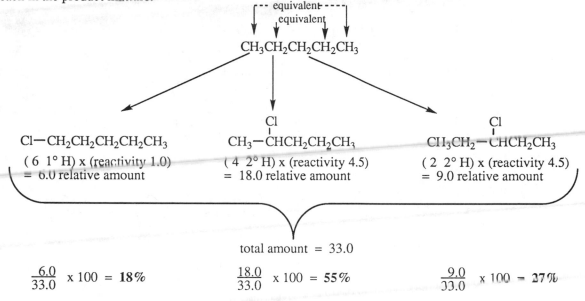

total amount $= 33.0$

$\dfrac{6.0}{33.0}$ x 100 = **18%**

$\dfrac{18.0}{33.0}$ x 100 = **55%**

$\dfrac{9.0}{33.0}$ x 100 = **27%**

4-47

(a) The second propagation step in the chlorination of methane is highly exothermic ($\Delta H° = -26$ kcal/mole (-109 kJ/mole)). The transition state resembles the reactants, that is, the Cl–Cl bond will be slightly stretched and the Cl–CH$_3$ bond will just be starting to form.

$$\overset{\delta\bullet}{Cl} \text{------} Cl \text{--------------} \overset{\delta\bullet}{CH_3}$$
stronger   weaker

(b) The second propagation step in the bromination of methane is highly exothermic ($\Delta H° = -24$ kcal/mole (-101 kJ/mole)). The transition state resembles the reactants, that is, the Br–Br bond will be slightly stretched and the Br–CH$_3$ bond will just be starting to form.

$$\overset{\delta\bullet}{Br} \text{------} Br \text{--------------} \overset{\delta\bullet}{CH_3}$$
stronger   weaker

4-48

Two mechanisms are possible depending on whether HO • reacts with chlorine or with cyclopentane.

Mechanism 1

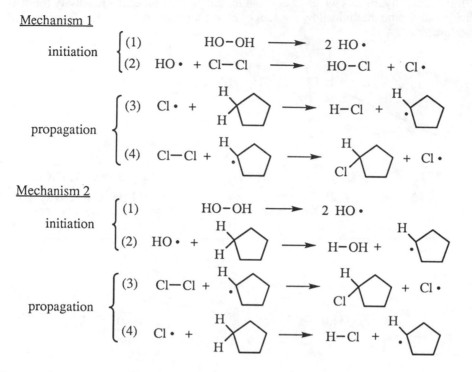

initiation
- (1) HO−OH ⟶ 2 HO •
- (2) HO • + Cl−Cl ⟶ HO−Cl + Cl •

propagation
- (3) Cl • + (cyclopentane) ⟶ H−Cl + (cyclopentyl radical)
- (4) Cl−Cl + (cyclopentyl radical) ⟶ (chlorocyclopentane) + Cl •

Mechanism 2

initiation
- (1) HO−OH ⟶ 2 HO •
- (2) HO • + (cyclopentane) ⟶ H−OH + (cyclopentyl radical)

propagation
- (3) Cl−Cl + (cyclopentyl radical) ⟶ (chlorocyclopentane) + Cl •
- (4) Cl • + (cyclopentane) ⟶ H−Cl + (cyclopentyl radical)

The energies of the propagation steps determine which mechanism is followed. The bond dissociation energy of HO—Cl is about 50 kcal/mole, making initiation step (2) in mechanism 1 *endothermic*. In mechanism 2, initiation step (2) is *exothermic* by about 24 kcal/mole; mechanism 2 is preferred.

4-49

$$Cl-\overset{\overset{\displaystyle Cl}{|}}{\underset{\underset{\displaystyle H}{|}}{C}}-H \ + \ \ddot{\underset{..}{O}}-H \ \longrightarrow \ H_2O \ + \ Cl-\overset{..}{\underset{\underset{\displaystyle H}{|}}{C}}{:}^- \ \longrightarrow \ \overset{\displaystyle Cl}{\underset{\underset{\displaystyle H}{|}}{C}}{:} \ + \ {:}\ddot{\underset{..}{Cl}}{:}^-$$

a carbene

4-50 This critical equation is the key to this problem: $\Delta G = \Delta H - T \Delta S$

At 1400 K, the equilibrium constant is 1; therefore:

$$K_{eq} = 1 \ \Longrightarrow \ \Delta G = 0 \ \Longrightarrow \ \Delta H = T \Delta S$$

Assuming ΔH is about the same at 1400 K as it is at calorimeter temperature:

$$\Delta S = \frac{\Delta H}{T} = \frac{-32.7 \text{ kcal/mole}}{1400 \text{ K}} = \frac{-32,700}{1400} \text{ cal/kelvin-mole}$$

$$= -23 \ \text{cal/kelvin-mole}$$
$$= -23 \ \text{e.u. ("entropy units")}$$

This is a large *decrease* in entropy, consistent with two molecules combining into one.

4-51

Assume that chlorine atoms (radicals) are still generated in the initiation reaction. Focus on the propagation steps. Bond dissociation energies are given below the bonds, in kcal/mole (kJ/mole).

$$Cl\cdot \ + \ H-CH_3 \ \longrightarrow \ H-Cl \ + \ \cdot CH_3 \qquad \Delta H = +1 \text{ kcal/mole } (+4 \text{ kJ/mole})$$

104 (435)   103 (431)

$$Cl-Cl \ + \ \cdot CH_3 \ \longrightarrow \ Cl-CH_3 \ + \ Cl\cdot \qquad \Delta H = -26 \text{ kcal/mole } (-109 \text{ kJ/mole})$$

58 (242)   84 (351)

What happens when the radical species react with iodine?

$$Cl\cdot \ + \ I-I \ \longrightarrow \ I-Cl \ + \ \cdot I \qquad \Delta H = -14 \text{ kcal/mole } (-60 \text{ kJ/mole})$$

36 (151)   50 (211)

$$I-I \ + \ \cdot CH_3 \ \longrightarrow \ I-CH_3 \ + \ I\cdot \qquad \Delta H = -20 \text{ kcal/mole } (-83 \text{ kJ/mole})$$

36 (151)   56 (234)

Compare the second reaction in each pair: methyl radical reacting with chlorine is more exothermic than methyl radical reacting with iodine; this does not explain how iodine prevents the chlorination reaction. Compare the first reaction in each pair: chlorine atom reacting with iodine is very exothermic whereas chlorine atom reacting with methane is slightly endothermic. Here is the key: chlorine atoms will be scavenged by iodine before they have a chance to react with methane. Without chlorine atoms, the reaction dies.

4-52

Mechanism 1:

(a) $Cl\cdot \ + \ O_3 \ \longrightarrow \ ClO\cdot \ + \ O_2$

(b) $2 \ ClO\cdot \ \longrightarrow \ Cl-O-O-Cl$

(c) $Cl-O-O-Cl \ \xrightarrow{h\nu} \ O_2 \ + \ 2 \ Cl\cdot$

The biggest problem in Mechanism 1 lies in step (b). The concentration of Cl atoms is very small, so at any given time, the concentration of ClO will be very small. The probability of two ClO radicals finding each other to form ClOOCl is virtually zero. Even though this mechanism shows a catalytic cycle with Cl• starting the mechanism and being regenerated at the end, the middle step makes it highly unlikely.

Mechanism 2:

(d) $O_3 \ \xrightarrow{h\nu} \ O_2 \ + \ O$

Step (d) is the "light" reaction that occurs naturally in daylight. At night, the reaction reverses and regenerates ozone.

(e) $Cl\cdot \ + \ O_3 \ \longrightarrow \ ClO\cdot \ + \ O_2$

(f) $ClO\cdot \ + \ O \ \longrightarrow \ O_2 \ + \ Cl\cdot$

Step (f) is the crucial step. A low concentration of ClO *will* find a relatively high concentration of O atoms because the "light" reaction is producing O atoms in relative abundance. Cl• is regenerated and begins propagation step (e), continuing the catalytic cycle.

Mechanism 2 is believed to be the dominant mechanism in ozone depletion. Mechanism 1 can be discounted because of the low probability of step (b) occurring, where two species in very low, catalytic concentration are required to find each other in order for the step to occur.

Note to the student: Stereochemistry is the study of molecular structure and reactions in three dimensions. Molecular models will be especially helpful in this chapter.

5-1 The chiral objects are the spring, the writing desk, the screw-cap bottle, the rifle and the knotted rope; the spring, the bottle top, and the rope each have a twist in one direction, and the rifle and desk are clearly made for right-handed users. All the other objects are achiral and would feel equivalent to right- or left-handed users.

5-2

(a) cis

(b) trans

(c) cis first, then trans

(d)

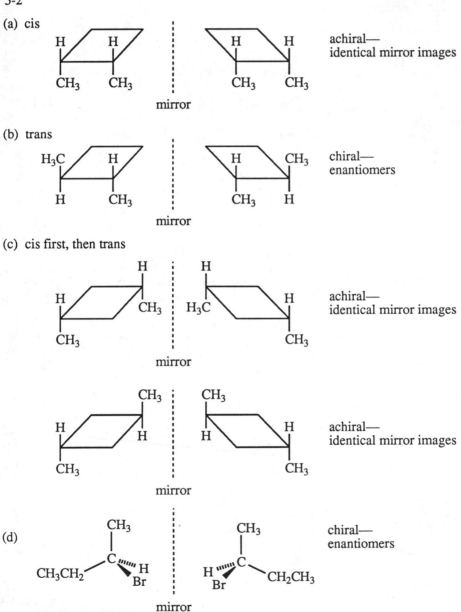

5-2 continued

(e)

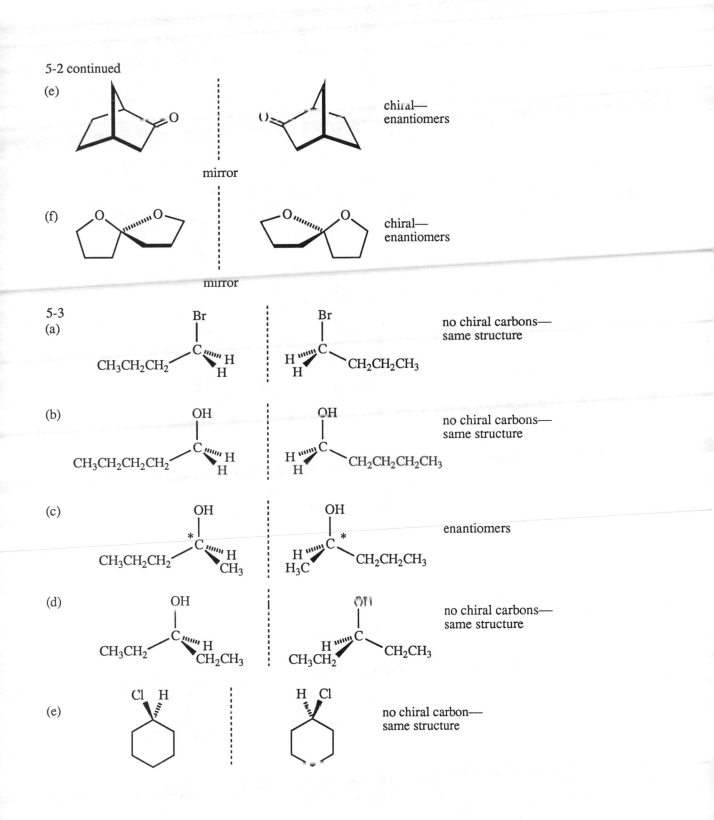

chiral—
enantiomers

mirror

(f)

chiral—
enantiomers

mirror

5-3

(a)

Br | C | H | H ; CH$_3$CH$_2$CH$_2$

Br | C | H | H ; CH$_3$CH$_2$CH$_2$

no chiral carbons—
same structure

(b)

OH | C | H | H ; CH$_3$CH$_2$CH$_2$CH$_2$

no chiral carbons—
same structure

(c)

OH | *C | H | CH$_3$; CH$_3$CH$_2$CH$_2$

OH | C* | H$_3$C | CH$_2$CH$_2$CH$_3$

enantiomers

(d)

OH | C | H | CH$_2$CH$_3$; CH$_3$CH$_2$

OH | C | H | CH$_2$CH$_3$; CH$_3$CH$_2$

no chiral carbons—
same structure

(e)

Cl H ; H Cl

no chiral carbon—
same structure

75

5-3 continued

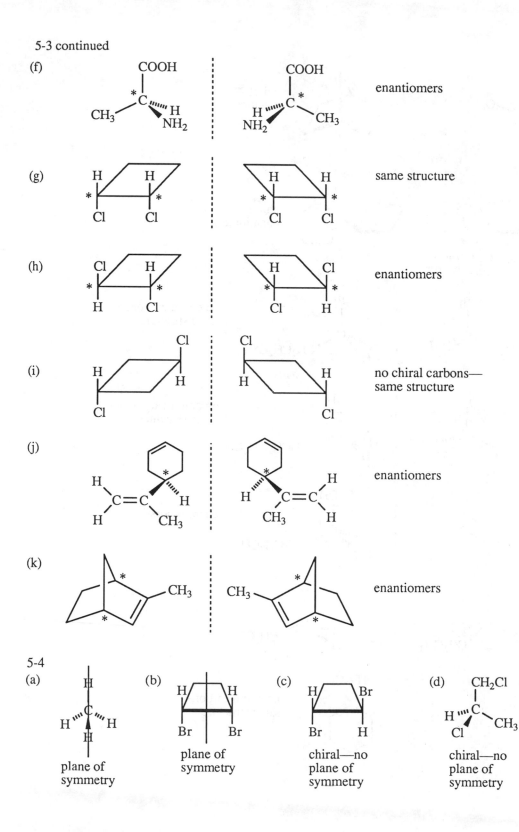

(f)   enantiomers

(g)   same structure

(h)   enantiomers

(i)   no chiral carbons—
same structure

(j)   enantiomers

(k)   enantiomers

5-4

(a) plane of symmetry   (b) plane of symmetry   (c) chiral—no plane of symmetry   (d) chiral—no plane of symmetry

5-4 continued

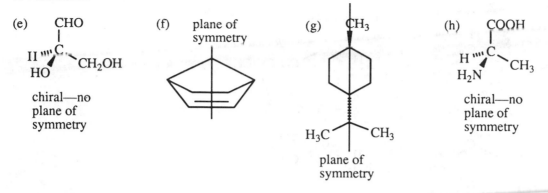

(e) CHO
H ""C" CH₂OH
HO

chiral—no
plane of
symmetry

(f) plane of
symmetry

(g) CH₃

H₃C   CH₃

plane of
symmetry

(h) COOH
H ""C" CH₃
H₂N

chiral—no
plane of
symmetry

5-5   ALWAYS place the 4th priority group away from you. Then determine if the sequence 1→2→3 is
clockwise (R) or counter-clockwise (S).

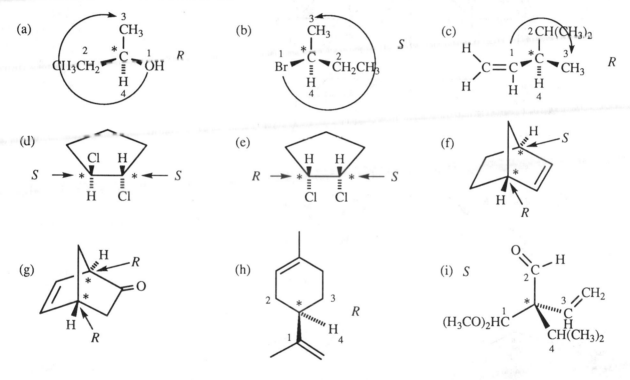

(a) 3 CH₃
2 *C 1 R
CH₃CH₂ OH
H
4

(b) 3 CH₃
1 *C 2   S
Br CH₂CH₃
H
4

(c) 2 CH(CH₃)₂
H 1 *C 3   R
C=C CH₃
H   H   H
4

(d) Cl   H
S → *   * ← S
H   Cl

(e) H   H
R → *   * ← S
Cl   Cl

(f) H — S
*
*
H
R

(g) H — R
*
*   O
H
R

(h)
2   *   3   R
1   H
4

(i) S
O — H
C
2
*   CH₂
3
1   C
(H₃CO)₂HC   H
CH(CH₃)₂
4

Part (i) deserves some explanation. The difference between groups 1 and 2 hinge on what is on the "extra"
oxygen.

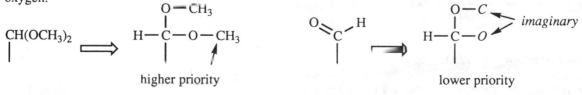

CH(OCH₃)₂   O — CH₃
| ⟹   H — C — O — CH₃
|

higher priority

O=C—H
| ⟹
|

O — C
H — C — O   imaginary
|

lower priority

77

5-6 There are no chiral carbons in 5-3 (a), (b), (d), (e), or (i).

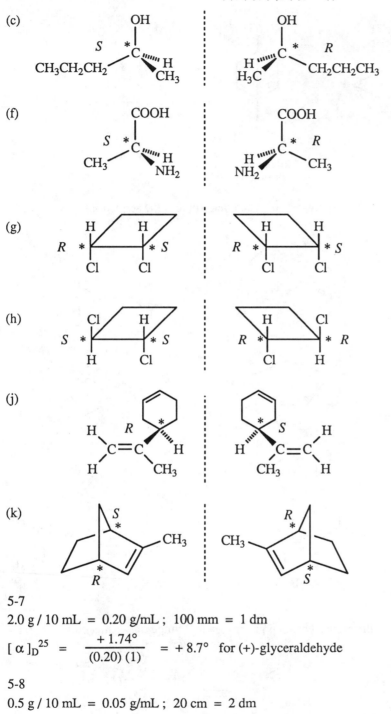

(c)

(f)

(g)

(h)

(j)

(k)

5-7

2.0 g / 10 mL = 0.20 g/mL ; 100 mm = 1 dm

$$[\,\alpha\,]_D^{25} \;=\; \frac{+1.74°}{(0.20)\,(1)} \;=\; +8.7°\;\text{ for (+)-glyceraldehyde}$$

5-8

0.5 g / 10 mL = 0.05 g/mL ; 20 cm = 2 dm

$$[\,\alpha\,]_D^{25} \;=\; \frac{-5.0°}{(0.05)\,(2)} \;=\; -50°\;\text{ for (–)-epinephrine}$$

5-9

Measure using a solution of about half the concentration of the first. The value will be either + 90° or – 90°, which gives the sign of the rotation.

5-10

Whether a sample is dextrorotatory (abbreviated "(+)") or levorotatory (abbreviated "(–)") is determined experimentally by a polarimeter. Except for the molecule glyceraldehyde, there is no direct, universal correlation between direction of optical rotation ((+) and (–)) and designation of configuration (R and S). In other words, one dextrorotatory compound might have R configuration while a different dextrorotatory compound might have S configuration.

(a) Yes, both of these are determined experimentally: the (+) or (–) by the polarimeter and the smell by the nose.

(b) No, R or S cannot be determined by either the polarimeter or the nose.

(c) The drawings show that (+)-carvone from caraway has the S configuration and (–)-carvone from spearmint has the R configuration.

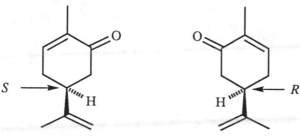

(+)-carvone (caraway seed)   (–)-carvone (spearmint)

(For fun, ask your instructor if you can smell the two enantiomers of carvone. Some people are unable, presumably for genetic reasons, to distinguish the fragrance of the two enantiomers.)

5-11

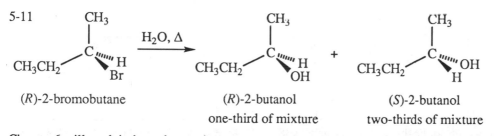

(R)-2-bromobutane   (R)-2-butanol   (S)-2-butanol
                              one-third of mixture   two-thirds of mixture

Chapter 6 will explain how these mixtures come about. For this problem, the S enantiomer accounts for 66.7% of the 2-butanol in the mixture and the rest, 33.3%, is the R enantiomer. Therefore, the excess of one enantiomer over the racemic mixture must be 33.3% of the S, the enantiomeric excess. (All of the R is "canceled" by an equal amount of the S, algebraically as well as in optical rotation.)

The optical rotation of pure (S)-2-butanol is + 13.5°. The optical rotation of this mixture is:
$$33.3\% \times +13.5° = +4.5°$$

5-12 The rotation of pure (+)-2-butanol is $+13.5°$.

$$\frac{\text{observed rotation}}{\text{rotation of pure enantiomer}} = \frac{+0.45°}{+13.5°} \times 100\% = 3.3\% \text{ optical purity}$$
$$= 3.3\% \text{ e.e.} = \text{excess of (+) over (−)}$$

To calculate percentages of (+) and (−):   (two equations in two unknowns)

$(+) + (−) = 100\%$ ⟹ $(−) = 100\% − (+)$

$(+) − (−) = 3.3\%$ ⟹ $(+) − (100\% − (+)) = 3.3\%$

$\mathbf{2}\ (+) = 103.3\%$

$(+) = 51.6\%$ (rounded)
$(−) = 48.4\%$

5-13

(a)

chiral—
optically active

$H_3C - \overset{\underset{\displaystyle |}{H}}{\underset{\displaystyle |}{\overset{*}{C}}} - Cl$
$|$
Br

(b) plane includes Br and Cl

plane of symmetry containing
Br—C—C—Cl; not optically active

$Br - \overset{\underset{\displaystyle |}{H}}{\underset{\displaystyle |}{C}} - \overset{\underset{\displaystyle |}{H}}{\underset{\displaystyle |}{C}} - Cl$

no chiral centers

(c)

chiral—
optically active

$ClH_2C - \overset{\underset{\displaystyle |}{H}}{\underset{\displaystyle |}{\overset{*}{C}}} - Cl$
$|$
Br

(d)

plane of symmetry—not optically
active despite the presence of two
chiral carbons

(e)

no plane of symmetry—
optically active
(other chair form is
equivalent—no plane of
symmetry)

(f)

plane of symmetry through
C-1 and C-4—not
optically active

no chiral centers

Part (2) Predictions of optical activity based on chiral centers give the same answers as predictions based on the most symmetric conformation.

5-14

(a)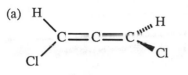

no chiral carbons, but the molecule is chiral (an allene); the drawing below is a three-dimensional picture of the allene in (a) showing there is no plane of symmetry

(b)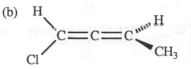

no chiral carbons, but the molecule is chiral (an allene)

(c)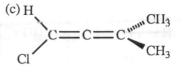

no chiral carbons; this allene has a plane of symmetry between the two methyls (the plane of the paper), including all the other atoms; not a chiral molecule

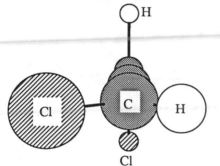

(d)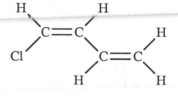

planar molecule—no chiral carbons; not a chiral molecule

(e)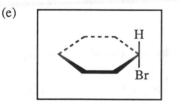

plane of symmetry bisecting the molecule; no chiral carbons; not a chiral molecule

(f)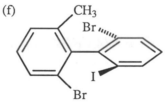

no chiral carbons, but the molecule is chiral due to restricted rotation; the drawing below is a three-dimensional picture showing that the rings are perpendicular (hydrogens are not shown)

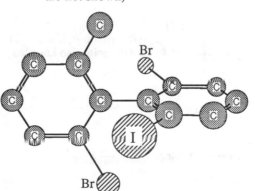

(g)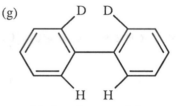

no chiral carbons, and the groups are not large enough to restrict rotation; not a chiral compound

(h)

Cl

two chiral carbons; a chiral compound

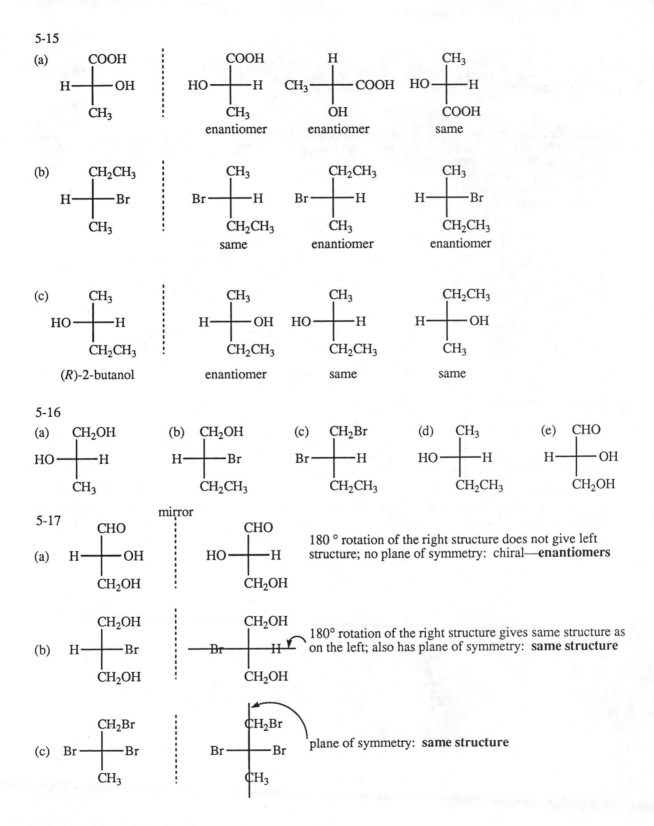

5-15

(a) enantiomer · enantiomer · same

(b) same · enantiomer · enantiomer

(c) (R)-2-butanol · enantiomer · same · same

5-16

(a) (b) (c) (d) (e)

5-17

mirror

(a) 180° rotation of the right structure does not give left structure; no plane of symmetry: chiral—**enantiomers**

(b) 180° rotation of the right structure gives same structure as on the left; also has plane of symmetry: **same structure**

(c) plane of symmetry: **same structure**

5-17 continued   mirror

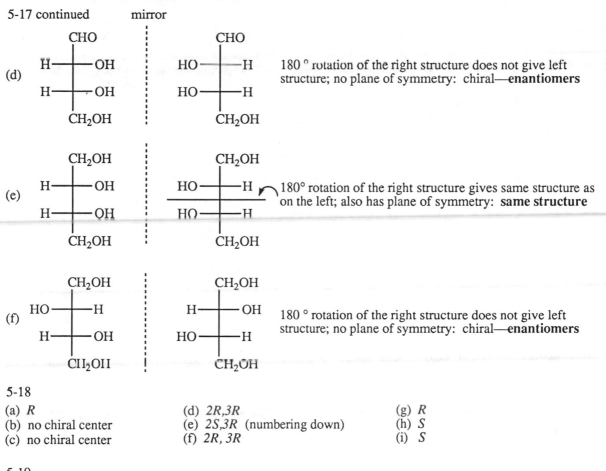

(d) 180 ° rotation of the right structure does not give left structure; no plane of symmetry: chiral—**enantiomers**

(e) 180° rotation of the right structure gives same structure as on the left; also has plane of symmetry: **same structure**

(f) 180 ° rotation of the right structure does not give left structure; no plane of symmetry: chiral—**enantiomers**

5-18

(a) *R*	(d) *2R,3R*	(g) *R*
(b) no chiral center	(e) *2S,3R* (numbering down)	(h) *S*
(c) no chiral center	(f) *2R, 3R*	(i) *S*

5-19

(a) enantiomers—both chiral centers inverted
(b) diastereomers—only one chiral center inverted
(c) diastereomers—only one chiral center inverted
(d) constitutional isomers—C=C shifted position

(e) enantiomers—chiral, mirror images
(f) diastereomers—only one chiral center inverted
(g) enantiomers—all chiral centers inverted
(h) same compound—superimposable mirror images

5-20

(a)

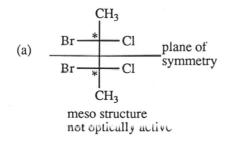

meso structure
not optically active

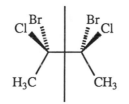

plane of symmetry

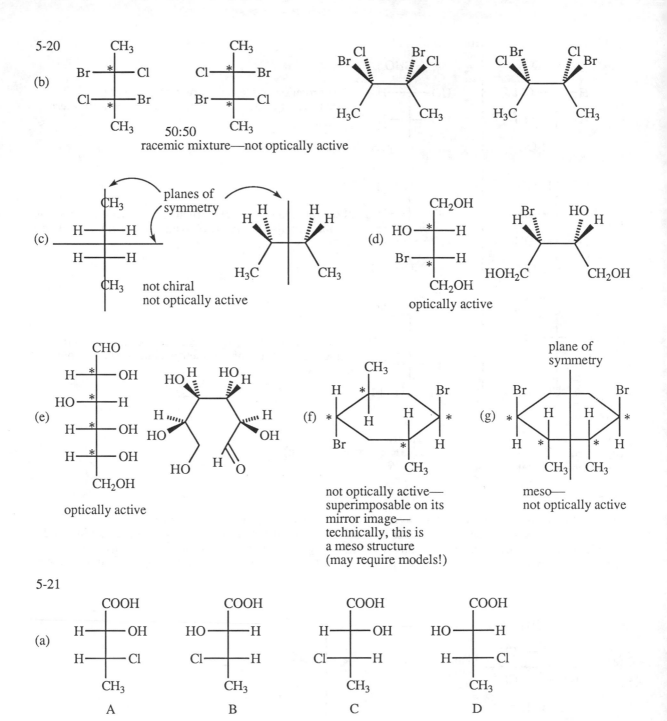

5-20

(b)

CH_3 — Br—*—Cl / Cl—*—Br — CH_3

CH_3 — Cl—*—Br / Br—*—Cl — CH_3

50:50
racemic mixture—not optically active

(c)

planes of symmetry

CH_3 — H—H / H—H — CH_3

not chiral
not optically active

(d)

CH_2OH — HO—*—H / Br—*—H — CH_2OH

optically active

(e)

CHO — H—*—OH / HO—*—H / H—*—OH / H—*—OH — CH_2OH

optically active

(f)

CH_3 ... Br ... CH_3

not optically active—
superimposable on its
mirror image—
technically, this is
a meso structure
(may require models!)

(g)

plane of symmetry

Br ... CH_3 CH_3 ... Br

meso—
not optically active

5-21

(a)

COOH	COOH	COOH	COOH
H—OH	HO—H	H—OH	HO—H
H—Cl	Cl—H	Cl—H	H—Cl
CH_3	CH_3	CH_3	CH_3
A	B	C	D

enantiomers: A and B; C and D
diastereomers: A and C, A and D, B and C, B and D

5-21 continued

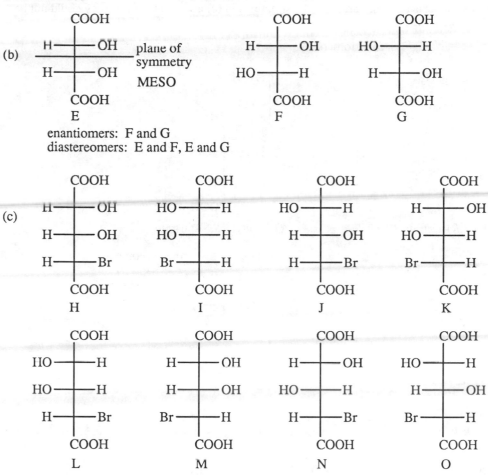

(b)

COOH
H——OH   plane of
H——OH   symmetry
COOH   MESO
E

COOH
H——OH
HO——H
COOH
F

COOH
HO——H
H——OH
COOH
G

enantiomers: F and G
diastereomers: E and F, E and G

(c)

COOH
H——OH
H——OH
H——Br
COOH
H

COOH
HO——H
HO——H
Br——H
COOH
I

COOH
HO——H
H——OH
H——Br
COOH
J

COOH
H——OH
HO——H
Br——H
COOH
K

COOH
HO——H
HO——H
H——Br
COOH
L

COOH
H——OH
H——OH
Br——H
COOH
M

COOH
H——OH
HO——H
H——Br
COOH
N

COOH
HO——H
H——OH
Br——H
COOH
O

enantiomers: H and I, J and K, L and M, N and O
diastereomers: any pair which is not enantiomeric

(d)

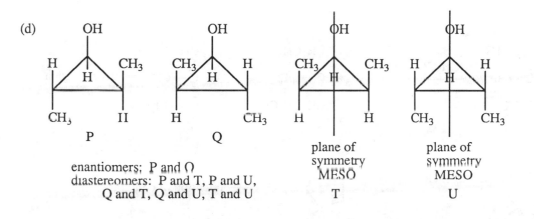

OH
H   CH₃
    H
CH₃   H
P

OH
CH₃   H
    H
H   CH₃
Q

OH
CH₃   CH₃
    H
H   H
plane of
symmetry
MESO
T

OH
H   H
    H
CH₃   CH₃
plane of
symmetry
MESO
U

enantiomers: P and Q
diastereomers: P and T, P and U,
    Q and T, Q and U, T and U

5-22 Any diasteromeric pair could be separated by a physical process like distillation or crystallization. Diastereomers are found in parts (a), (b), and (d). The structures in (c) are enantiomers; they could not be separated by normal physical means.

5-23

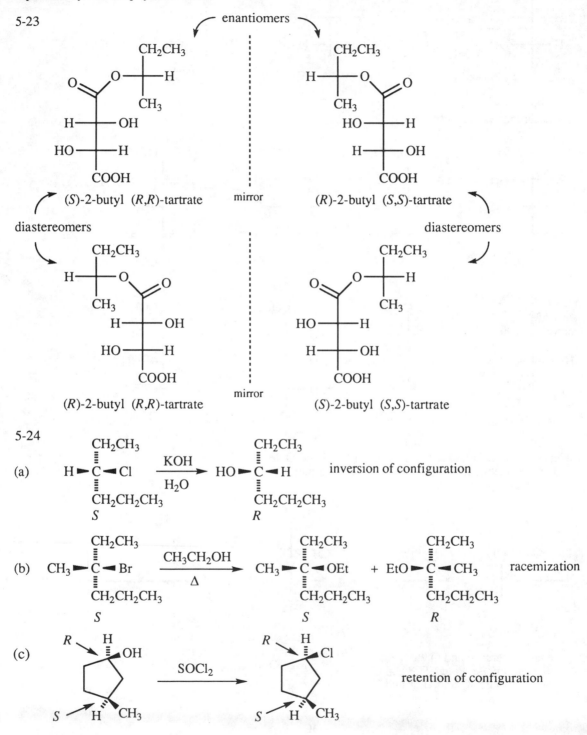

5-24

(a) inversion of configuration

(b) racemization

(c) retention of configuration

5-25 The Cahn-Ingold-Prelog priorities of the groups are the circled numbers in (a).

(a)

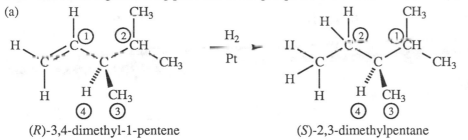

(R)-3,4-dimethyl-1-pentene   (S)-2,3-dimethylpentane

(b) The reaction did not occur at the chiral center, so the configuration of the chiral center has not changed— the reaction went with retention. What did change was the priority of the groups in the R,S system of nomenclature, which resulted in a change of *designation* of configuration. The designations R and S must be used *with thought* to determine whether configuration has changed in a reaction.

(c) There is no general correlation between R and S designation and the physical property of optical rotation. Professor Wade's poetic couplet makes an important point: do not confuse an object and its properties with the name for that object. (Scholars of Shakespeare have come to believe that this quote from Juliet is a veiled reference to designation of R,S configuration versus optical rotation of a chiral molecule.)

5-26 Please refer to solution 1-19, page 12 of this Solutions Manual.

5-27

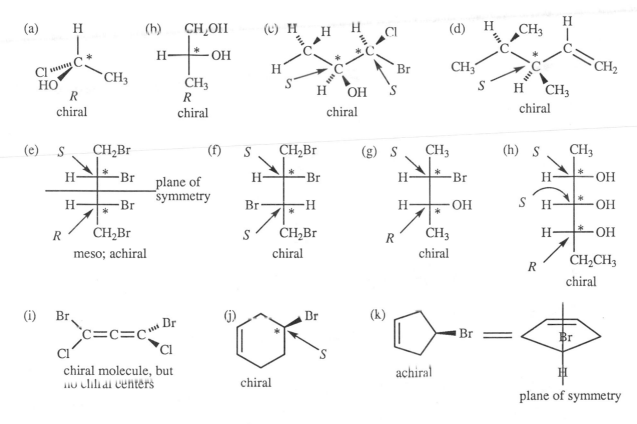

5-27 continued

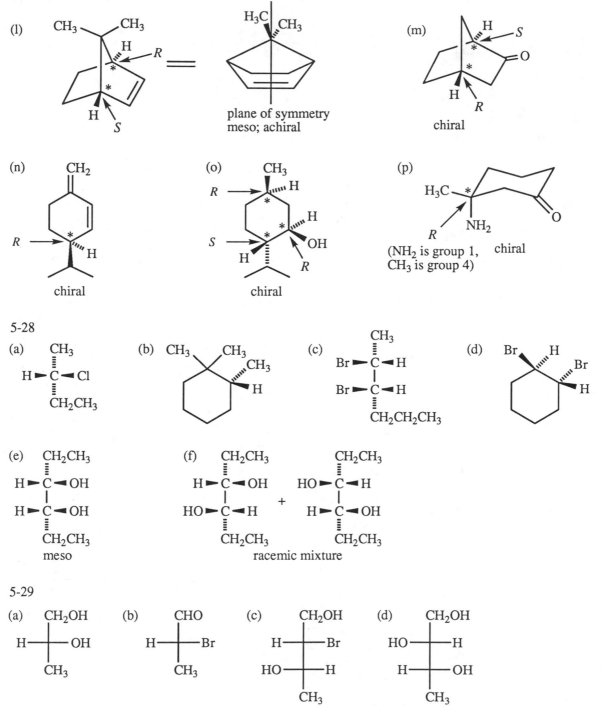

(l)

CH₃   CH₃
      H   R
      *
      *
   H
      S

==

H₃C   CH₃

plane of symmetry
meso; achiral

(m)

      H
      * ← S
      *
      ‖O
   H
      R
chiral

(n)

CH₂

R →   *
      H
chiral

(o)

CH₃
R →   H
      *
         H
      *
S →
   H   OH
      R
chiral

(p)

H₃C   *
   R   NH₂
         ‖O
(NH₂ is group 1,   chiral
CH₃ is group 4)

5-28

(a)   CH₃
      H►C◄Cl
      CH₂CH₃

(b)   CH₃   CH₃
            CH₃
            H

(c)   CH₃
      Br►C◄H
      Br►C◄H
      CH₂CH₂CH₃

(d)   Br   H
            Br
            H

(e)   CH₂CH₃
   H►C◄OH
   H►C◄OH
   CH₂CH₃
   meso

(f)   CH₂CH₃   CH₂CH₃
   H►C◄OH   HO►C◄H
   HO►C◄H   +   H►C◄OH
   CH₂CH₃   CH₂CH₃
            racemic mixture

5-29

(a)   CH₂OH
   H——OH
   CH₃

(b)   CHO
   H——Br
   CH₃

(c)   CH₂OH
   H——Br
   HO——H
   CH₃

(d)   CH₂OH
   HO——H
   H——OH
   CH₃

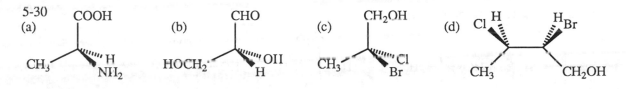

5-30
(a) [structure: COOH, CH₃, H, NH₂]
(b) [structure: CHO, HOCH₂, OH, H]
(c) [structure: CH₂OH, CH₃, Cl, Br]
(d) [structure: Cl, H, H, Br, CH₃, CH₂OH]

5-31
(a) same (meso)—plane of symmetry, superimposable
(b) enantiomers—both chiral centers inverted
(c) enantiomers—both chiral centers inverted
(d) enantiomers

(e) enantiomers—both chiral centers inverted
(f) diastereomers—only one chiral center inverted
(g) enantiomers—both chiral centers inverted
(h) same compound—rotate the right structure 180° around a horizontal axis and it becomes the left structure

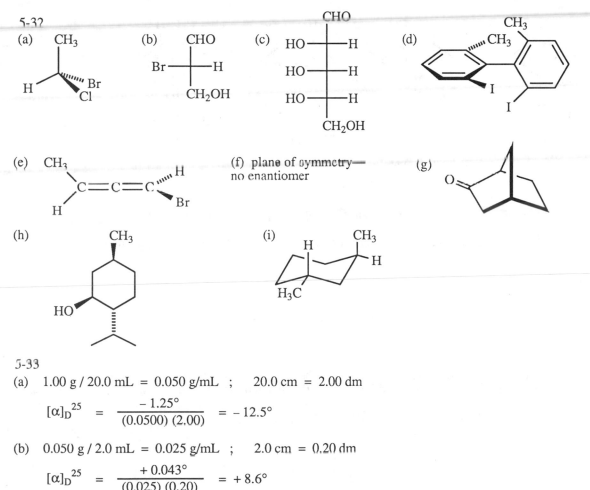

5-32
(a) [structure: CH₃, H, Br, Cl]
(b) [structure: CHO, Br, H, CH₂OH]
(c) [structure: CHO, HO—H, HO—H, HO—H, CH₂OH]
(d) [structure: biphenyl with CH₃, CH₃, I, I]

(e) [structure: CH₃, H, C=C=C, H, Br]
(f) plane of symmetry—no enantiomer
(g) [structure: bicyclic ketone, O]

(h) [structure: cyclohexane with CH₃, HO, isopropyl]
(i) [structure: cyclohexane, H, CH₃, H, H₃C]

5-33
(a) 1.00 g / 20.0 mL = 0.050 g/mL ;   20.0 cm = 2.00 dm

$$[\alpha]_D^{25} = \frac{-1.25°}{(0.0500)\,(2.00)} = -12.5°$$

(b) 0.050 g / 2.0 mL = 0.025 g/mL ;   2.0 cm = 0.20 dm

$$[\alpha]_D^{25} = \frac{+0.043°}{(0.025)\,(0.20)} = +8.6°$$

5-34 The 40% of the mixture that is (–)-tartaric acid will cancel the optical rotation of the 40% of the mixture that is (+)-tartaric acid, leaving only 20% of the mixture as excess (+)-tartaric acid to give measurable optical rotation. The specific rotation will therefore be only 20% of the rotation of pure (+)-tartaric acid: +12.0° x 20% = +2.4°

5-35 all structures in this problem are chiral

(a)

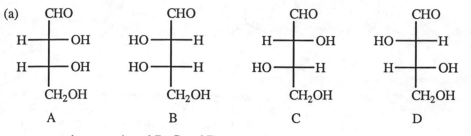

A   B   C   D

enantiomers: A and B; C and D
diastereomers: A and C; A and D; B and C; B and D

(b)

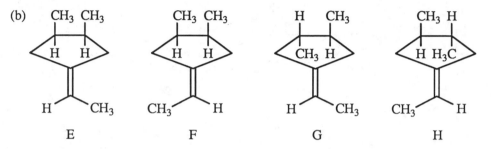

E   F   G   H

enantiomers: E and F; G and H
diastereomers: E and G; E and H; F and G; F and H

5-36

(a) 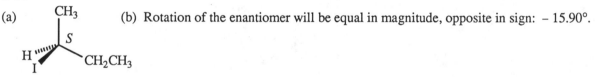   (b) Rotation of the enantiomer will be equal in magnitude, opposite in sign: $-15.90°$.

(c) The rotation $-7.95°$ is what percent of $-15.90°$?

$$\frac{-7.95°}{-15.90°} \times 100\% = 50\% \text{ e.e.}$$

There is 50% excess of (R)-2-iodobutane over the racemic mixture; that is, another 25% must be R and 25% must be S. The total composition is 75% (R)-$(-)$-2-iodobutane and 25% (S)-$(+)$-2-iodobutane.

5-37

(a)

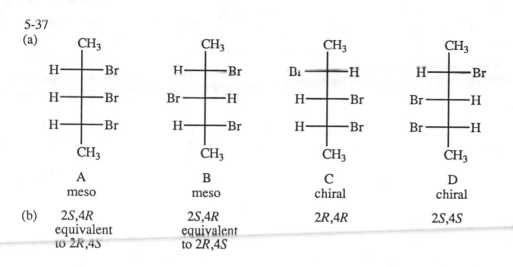

A	B	C	D
meso	meso	chiral	chiral

(b)   2S,4R   2S,4R   2R,4R   2S,4S
     equivalent   equivalent
     to 2R,4S   to 2R,4S

(c) This is one of the challenging types of problems in stereochemistry. Each C-3 in structures A and B is chiral, because the substituent groups are different (*R* versus *S*), yet structures A and B have planes of symmetry passing through C-3. In structures C and D, C-3 is not chiral because both substituent groups are either *R* or *S*, yet the molecules are chiral. Chemists classify these carbons as "pseudochiral" because they contradict our notions of chiral carbons and chiral molecules.

5-38

(a) The product has no chiral centers.

(b) The product is an example of a chiral compound with no chiral centers. Like the allenes, it is classified as an "extended tetrahedron"; that is, it has four groups that extend from the rigid molecule in four different directions. (A model will help.) In this structure, the plane containing the COOH and carbons of the double bond is perpendicular to the plane bisecting the OH and H and carbon that they are on.

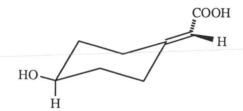

(c) By using a chiral enzyme to reduce the ketone to the alcohol, an excess of one enantiomer was produced, so the product was optically active. If achiral conditions like H_2 and a metal catalyst were used, a racemic mixture would be produced and it would not be optically active.

5-39 See Problem 5-37 for a similar situation.

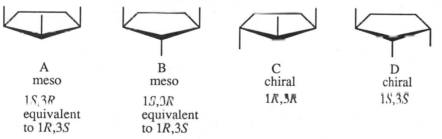

A	B	C	D
meso	meso	chiral	chiral
1S,3R	1S,3R	1R,3R	1S,3S
equivalent	equivalent		
to 1R,3S	to 1R,3S		

C and D are enantiomers. All other pairs are diastereomers.

6-1 In problems like part (a), draw out the whole structure to detect double bonds.

(a) vinyl halide
(b) alkyl halide
(c) aryl halide
(d) alkyl halide
(e) vinyl halide
(f) aryl halide

6-2

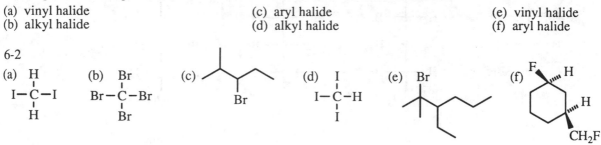

6-3 IUPAC name; common name; degree of halogen-bearing carbon

(a) 1-chloro-2-methylpropane; isobutyl chloride; 1°
(b) diiodomethane; methylene iodide; methyl
(c) 1,1-dichloroethane; no common name; 1°
(d) 2-bromo-1,1,1-trichloroethane; no common name; all 1°
(e) trichloromethane; chloroform; methyl
(f) 2-bromo-2-methylpropane; *t*-butyl bromide; 3°
(g) 2-bromobutane; *sec*-butyl bromide; 2°
(h) 1-chloro-2-methylbutane; no common name; 1°
(i) *cis*-1-bromo-2-chlorocyclobutane; no common name; both 2°
(j) 3-bromo-4-methylhexane; no common name; 2°
(k) 4-fluoro-1,1-dimethylcyclohexane; no common name; 2°
(l) *trans*-1,3-dichlorocyclopentane; no common name; all 2°

6-4

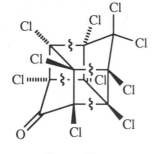
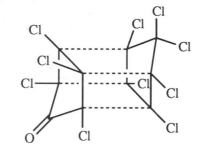

Kepone®

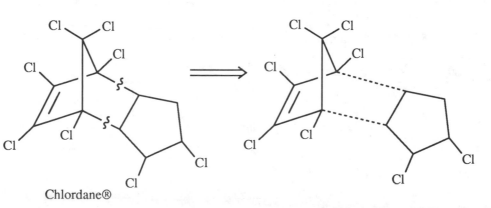

Chlordane®

6-5 From the text, Section 2-11, the bond dipole moment depends not only on bond length but also on charge separation, which in turn depends on the difference in electronegativities of the two atoms connected by the bond. Because chlorine's electronegativity (3.2) is significantly higher than iodine's (2.7), the C—Cl bond dipole is greater than that of C—I, despite C—I being a longer bond.

6-6

(a) *n*-Butyl bromide has a higher molecular weight and less branching, and boils at a higher temperature than isopropyl bromide.

(b) *t*-Butyl bromide has a higher molecular weight and a larger halogen, and despite its greater branching, boils at a higher temperature than isopropyl chloride.

(c) *n*-Butyl bromide has a higher molecular weight and a larger halogen, and boils at a higher temperature than *n*-butyl chloride.

6-7 From Table 3-1, the density of hexane is 0.66; it will float on the water layer (d 1.00). From Table 6-2, the density of chloroform is 1.50; water will float on the chloroform.

6-8

(a) $CH_3-\overset{\underset{\textstyle CH_3}{|}}{\underset{\textstyle CH_3}{\overset{\textstyle CH_3}{C}}}-CH_3$ $\xrightarrow[hv]{Cl_2}$ $CH_3-\overset{\underset{\textstyle CH_3}{|}}{\underset{\textstyle CH_3}{\overset{\textstyle CH_3}{C}}}-CH_2Cl$

has only one type of hydrogen—only one monochlorine isomer can be produced

(b) $CH_3-\overset{\underset{\textstyle CH_2CH_3}{|}}{\underset{\textstyle CH_2CH_3}{\overset{\textstyle CH_3}{C}}}-H$ $\xrightarrow[hv]{Br_2}$ $CH_3-\overset{\underset{\textstyle CH_2CH_3}{|}}{\underset{\textstyle CH_2CH_3}{\overset{\textstyle CH_3}{C}}}-Br$

Bromination has a strong preference for abstracting hydrogens (like 3°) that give stable radical intermediates.

(c) ⟨benzene⟩—CH(H)CH₂CH₂CH₃ $\xrightarrow[hv]{Br_2}$ ⟨benzene⟩—CH(Br)CH₂CH₂CH₃

(or NBS)

Bromine atom will abstract the hydrogen giving the most stable radical; in this case, the radical intermediate will be stabilized by resonance with the benzene ring.

6-9   propagation steps

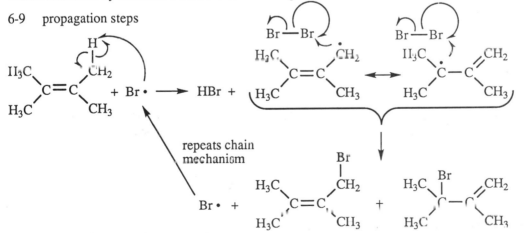

repeats chain mechanism

The resonance-stabilized allylic radical intermediate has radical character on both the 1° and 3° carbons, so bromine can bond to either of these carbons producing two isomeric products.

6-10
(a) substitution—Br is replaced
(b) elimination—H and OH are lost
(c) elimination—both Br atoms are lost

6-11

(a) $CH_3(CH_2)_4CH_2$—OCH_2CH_3 (b) $CH_3(CH_2)_4CH_2$—CN (c) $CH_3(CH_2)_4CH_2$—OH

6-12 The rate law is first order in both CH_3I and hydroxide. If the concentration of CH_3I is lowered to one-fifth the original value, the rate must decrease to one-fifth; if the concentration of hydroxide is doubled, the rate must also double. Thus, the rate must decrease to 0.02 mole/L per second:

$$\text{rate} = \underbrace{(0.05 \text{ mole/L per second})}_{\text{original rate}} \times \underbrace{\frac{(0.1 \text{ M})}{(0.5 \text{ M})}}_{\substack{\text{change in} \\ CH_3I}} \times \underbrace{\frac{(2.0 \text{ M})}{(1.0 \text{ M})}}_{\substack{\text{change in} \\ NaOH}} = \underbrace{0.02 \text{ mole/L per second}}_{\text{new rate}}$$

A completely different way to answer this problem is to solve for the rate constant k, then put in new values for the concentrations.

rate $= k$ [CH_3I] [$NaOH$] $\Longrightarrow$ 0.05 mole L^{-1} sec^{-1} $= k$ (0.5 mol L^{-1}) (1.0 mol L^{-1}) $\Longrightarrow$
rate constant $k = $ 0.1 L mol^{-1} sec^{-1}
rate $= k$ [CH_3I] [$NaOH$] $=$ (0.1 L mol^{-1} sec^{-1}) (0.1 mol L^{-1}) (2.0 mol L^{-1}) $=$ 0.02 mole L^{-1} sec^{-1}

6-13 Organic and inorganic products are shown here for completeness.

(a) $(CH_3)_3C$—O—CH_2CH_3 + KBr

(b) $HC{\equiv}C$—$CH_2CH_2CH_2CH_3$ + NaCl

(c) $(CH_3)_2CHCH_2$—$\overset{+}{N}H_3$ $\overset{-}{Br}$ $\xrightarrow{NH_3}$ $(CH_3)_2CHCH_2$—NH_2 + $\overset{+}{N}H_4$ $\overset{-}{Br}$

(d) CH_3CH_2—CN + NaI

(e) 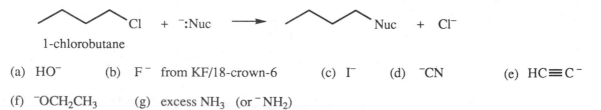 $\diagup I$ + NaCl

(f) $\diagup F$ + KCl (18-crown-6 is the catalyst and does not change; CH_3CN is the solvent)

6-14 All reactions in this problem follow the same pattern; the only difference is the nucleophile ($^-$:Nuc). Only the nucleophile is listed below. (Cations like Na^+ or K^+ accompany the nucleophile but are simply spectator ions and do not take part in the reaction; they are not shown here.)

$\diagup\!\!\diagdown\!\!\diagup Cl$ + $^-$:Nuc $\longrightarrow$ $\diagup\!\!\diagdown\!\!\diagup Nuc$ + Cl^-
1-chlorobutane

(a) HO^- (b) F^- from KF/18-crown-6 (c) I^- (d) ^-CN (e) $HC{\equiv}C^-$

(f) $^-OCH_2CH_3$ (g) excess NH_3 (or $^-NH_2$)

94

6-15

(a) $(CH_3CH_2)_2NH$ is a better nucleophile—less hindered

(b) $(CH_3)_2S$ is a better nucleophile—S is larger, more polarizable than O

(c) PH_3 is a better nucleophile—P is larger, more polarizable than N

(d) CH_3S^- is a better nucleophile—anions are better than neutral atoms of the same element

(e) $(CH_3)_3N$ is a better nucleophile—less electronegative than oxygen, better able to donate an electron pair

(f) $CH_3CH_2CH_2O^-$ is a better nucleophile—less branching, less steric hindrance

(g) I^- is a better nucleophile—larger, more polarizable

6-16 A mechanism must show *electron movement.*

$$CH_3 - \overset{..}{\underset{..}{O}} - CH_3 \ + \ H^+ \ \rightleftharpoons \ CH_3 - \overset{\overset{H}{|}}{\underset{..}{O}} - CH_3 \ + \ :\overset{..}{\underset{..}{Br}}: \ \longrightarrow \ CH_3 - \overset{\overset{H}{|}}{\underset{..}{O}}: \ + \ CH_3 - \overset{..}{\underset{..}{Br}}:$$

Protonation converts OCH_3 to a good leaving group.

6-17 The type of carbon with the halide, and relative leaving group ability of the halide, determine the reactivity.

methyl iodide > methyl chloride > ethyl chloride > isopropyl bromide >> neopentyl bromide, *t*-butyl iodide

Predicting the relative order of neopentyl bromide and *t*-butyl iodide would be difficult because both would be extremely slow.

6-18 In all cases, the less hindered structure is the better S_N2 substrate.

(a) 2-methyl-1-iodopropane (1° versus 3°)

(b) cyclohexyl bromide (2° versus 3°)

(c) isopropyl bromide (no substituent on neighboring carbon)

(d) 2-chlorobutane (even though this is a 2° halide, it is easier to attack than the 1° neopentyl type in 2,2-dimethyl-1-chlorobutane—see below)

(e) isopropyl iodide (same reason as in (d))

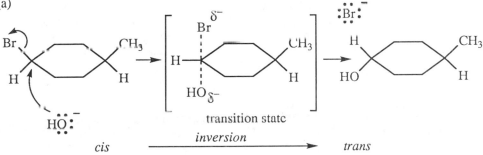

a neopentyl halide— hindered to backside attack by neighboring methyl groups

6-19 All S_N2 reactions occur with inversion of configuration at carbon.

(a)

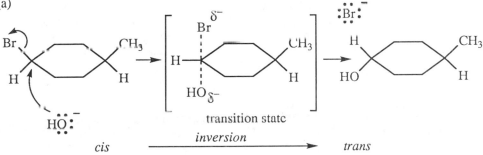

cis   *inversion*   trans

6-19 continued

(b)

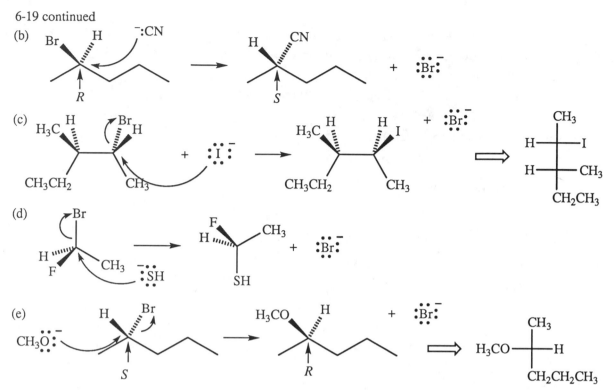

(c)

(d)

(e)

6-20

(a) The best leaving groups are the weakest bases. Bromide ion is so weak it is not considered at all basic; it is an excellent leaving group. Fluoride is moderately basic, by far the most basic of the halides. It is a terrible leaving group. Bromide is many orders of magnitude better than fluoride in leaving group ability.

(b)

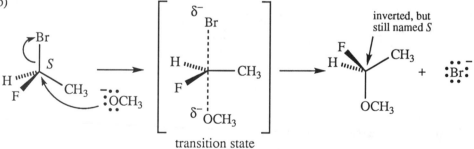

transition state

(c) As noted on the structure above, the configuration is inverted, but because the oxygen has a lower priority than the bromide it replaced, the *designation* of configuration is misleading. Refer to the solution to problem 5-25, p. 87, for the caution about confusing absolute configuration with the *designation* of configuration.

(d) The result is perfectly consistent with the S_N2 mechanism. Even though both the reactant and the product have the S designation, the configuration has been inverted: the nomenclature priority of fluorine changes from second (after bromine) in the reactant to first (before oxygen) in the product. While the designation may be misleading, the structure shows with certainty that an inversion has occurred.

96

6-21 The structure that can form the more stable carbocation will undergo S_N1 faster.

(a) 2-bromopropane: will form a 2° carbocation
(b) 2-bromo-2-methylbutane: will form a 3° carbocation
(c) 3-bromocyclohexene: will form an allylic (resonance-stabilized) carbocation
(d) 2-bromopropane: will form a 2° carbocation

6-22 Ionization is the rate-determining step in S_N1. Anything that stabilizes the intermediate will speed the reaction. Both of these compounds form resonance-stabilized intermediates.

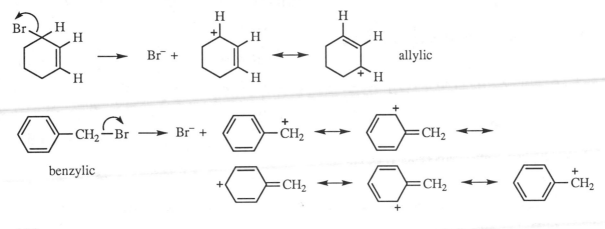

6-23

(a) 2-iodo-2-methylbutane is faster than *t*-butyl chloride (iodide is a better leaving group than chloride)
(b) 2-bromo-2-methylbutane (3°) is faster than ethyl iodide (1°); although iodide is a somewhat better leaving group, the difference between 3° and 1° carbocation stability dominates
(c) 3-bromocyclohexene (2°) is faster than *n*-propyl bromide (1°) (3-bromocyclohexene can form a resonance-stabilized intermediate as shown in Problem 6-22)
(d) cyclohexyl bromide (2°) is faster than methyl iodide

6-24

6-25 It is important to analyze the structure of carbocations to consider if migration of any groups from adjacent carbons will lead to a more stable carbocation. As a general rule, if rearrangement would lead to a more stable carbocation, a carbocation will rearrange. (Beginning with this problem, only those unshared electons pairs involved in a particular step will be shown.)

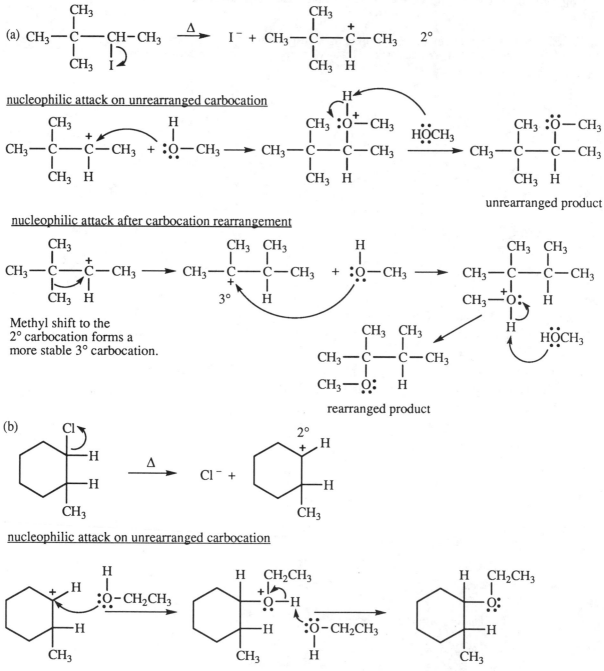

(a)

nucleophilic attack on unrearranged carbocation

unrearranged product

nucleophilic attack after carbocation rearrangement

Methyl shift to the 2° carbocation forms a more stable 3° carbocation.

rearranged product

(b)

nucleophilic attack on unrearranged carbocation

unrearranged product

6-25 (b) continued

nucleophilic attack after carbocation rearrangement

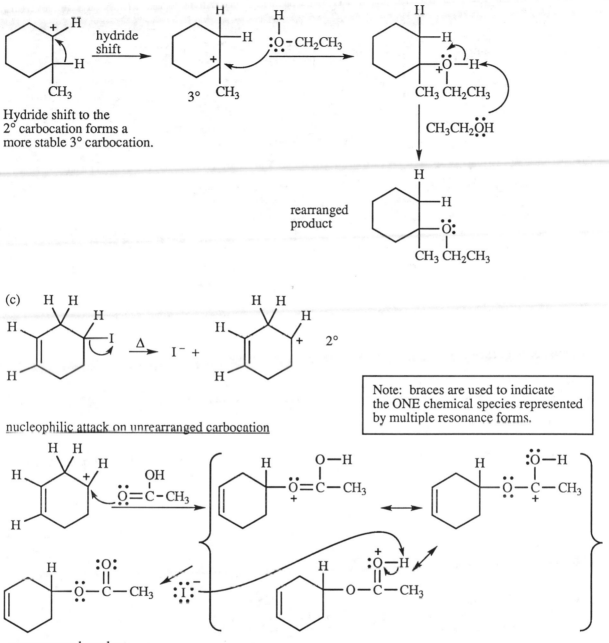

Hydride shift to the
2° carbocation forms a
more stable 3° carbocation.

rearranged
product

(c)

Note: braces are used to indicate
the ONE chemical species represented
by multiple resonance forms.

nucleophilic attack on unrearranged carbocation

unrearranged product

99

6-25 (c) continued

nucleophilic attack after carbocation rearrangement

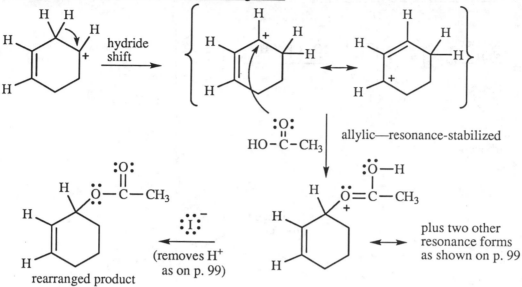

allylic—resonance-stabilized

plus two other
resonance forms
as shown on p. 99

rearranged product

(removes H⁺
as on p. 99)

Comments on 6-25(c)

(1) The hydride shift to a 2° carbocation generates an allylic, resonance-stabilized 2° carbocation.

(2) The double-bonded oxygen of acetic acid is more nucleophilic because of the resonance forms it can have after attack. (See Solved Problem 1-5 and Problem 1-15 in the text.)

(3) Attack on only one carbon of the allylic carbocation is shown. In reality, both positive carbons would be attacked in equal amounts, but they would give the identical product *in this case*. In other compounds, however, attack on the different carbons might give different products. ALWAYS CONSIDER ALL POSSIBILITIES.

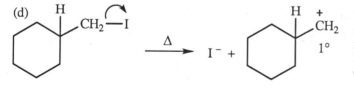

The 1° carbocation initially formed is very unstable; some chemists believe that rearrangement occurs at the same time as the leaving group leaves. At most, the 1° carbocation has a very short lifetime.

hydride shift followed by nucleophilic attack

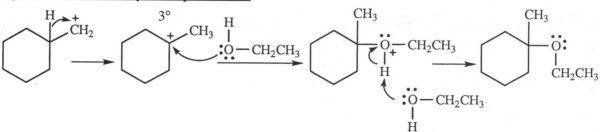

100

6-25 (d) continued

alkyl migration (ring expansion) followed by nucleophilic attack

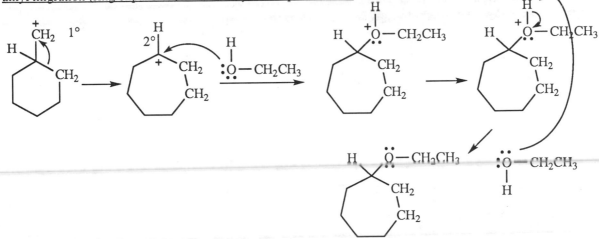

6-26
(a) The mechanism for the substitution of OH by Cl using $SOCl_2$ more closely resembles an S_N1 mechanism. The leaving group leaves before the nucleophile (chloride ion) attacks the positively-charged carbon. There is a striking similarity with S_N2, however: there is no racemization at carbon; that is, this substitution is also *stereospecific*, although it goes with retention rather than the typical inversion of configuration characteristic of S_N2.
(b) The nucleophile, chloride ion, comes from the leaving group. (HCl is also produced in the reaction and it is possible that some chloride ion is present in the solution, but HCl is a gas and quickly leaves the heated solution, so the concentration of chloride in the solution is very low.) Ionization is the slow step, but as soon as the leaving group leaves, the nucleophile quickly attacks in a fast step. In other words, before the leaving group has gone very far, the chloride ion forms a bond to carbon *from the same face where the leaving group just left*.

6-27

(a) $CH_3-\overset{\overset{\displaystyle CH_3}{|}}{\underset{\underset{\displaystyle CH_2CH_3}{|}}{C}}-O-\overset{\overset{\displaystyle O}{||}}{C}-CH_3$

$S_N1, 3°$

(b) $CH_3-\underset{\underset{\displaystyle CH_3}{|}}{CH}CH_2OCH_3$

$S_N2, 1°$

(c)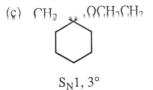

$S_N1, 3°$

(d)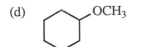

S_N1, weak nucleophile

(e)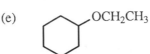

S_N2, strong nucleophile

101

6-28

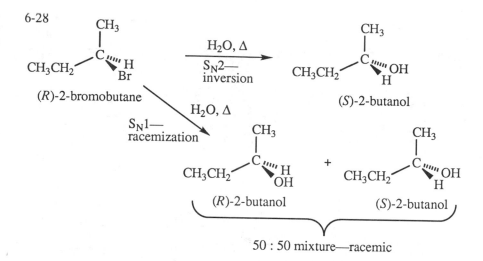

(R)-2-bromobutane

(S)-2-butanol

(R)-2-butanol   +   (S)-2-butanol

50 : 50 mixture—racemic

If S_N1, which gives racemization, occurs exactly twice as fast as S_N2, which gives inversion, then the racemic mixture (50 : 50 $R + S$) is 66.7% of the mixture and the rest, 33.3%, is the S enantiomer from S_N2. Therefore, the excess of one enantiomer over the racemic mixture must be 33.3%, the enantiomeric excess. (In the racemic mixture, the R and S "cancel" each other algebraically as well as in optical rotation.)

The optical rotation of pure (S)-2-butanol is + 13.5°. The optical rotation of this mixture is:
$$33.3\% \quad \times \quad + 13.5° \quad = \quad + 4.5°$$

6-29

(a) Methyl shift may occur simultaneously with ionization. The lifetime of 1° carbocations is exceedingly short; some chemists believe that they are only a transition state to the rearranged product.

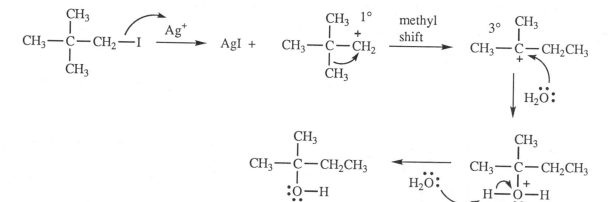

6-29 continued

(b) Alkyl shift may occur simultaneously with ionization.

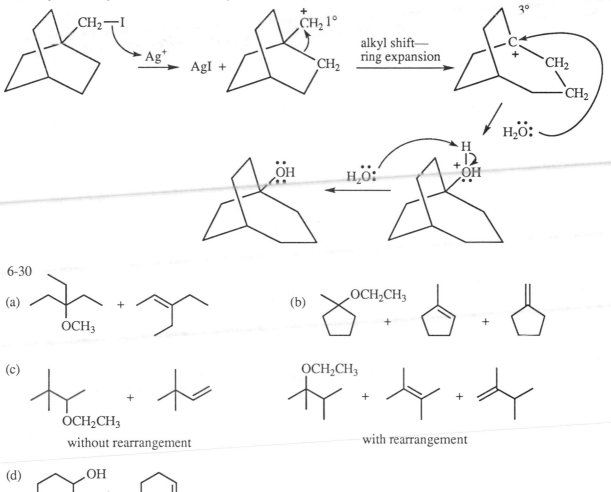

6-30

(a) [structure] + [structure]

(b) [structure] + [structure] + [structure]

(c) [structure] + [structure]

without rearrangement

[structure] + [structure] + [structure]

with rearrangement

(d) [structure] + [structure]

6-31

2-Bromopentane is a 2° halide and can undergo S_N2 by a strong nucleophile.

1-Bromo-1-methylcyclohexane is a 3° halide and cannot undergo S_N2; a strong nucleophile that is also basic will generate elimination products only.

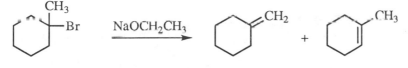

6-32

(a)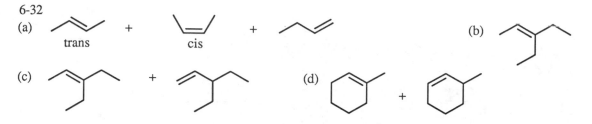

trans   cis

(b)

(c)   +

(d)

6-33 Determine the substitution pattern by counting the number of carbons bonded to the two sp^2 carbons of the alkene. (Alternatively, count the hydrogens on these two carbons and substract from 4.)

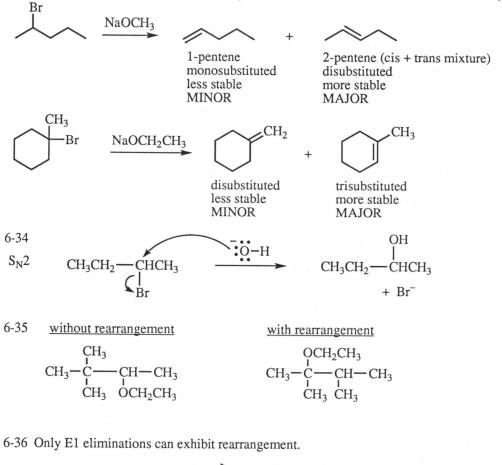

Br

NaOCH$_3$

1-pentene
monosubstituted
less stable
MINOR

+

2-pentene (cis + trans mixture)
disubstituted
more stable
MAJOR

CH$_3$

—Br   NaOCH$_2$CH$_3$

CH$_2$ CH$_3$

disubstituted
less stable
MINOR

+

trisubstituted
more stable
MAJOR

6-34

S$_N$2   CH$_3$CH$_2$—CHCH$_3$

:Ö—H

OH

CH$_3$CH$_2$—CHCH$_3$

Br

+ Br$^-$

6-35   without rearrangement

CH$_3$

CH$_3$—C——CH—CH$_3$

CH$_3$ OCH$_2$CH$_3$

with rearrangement

OCH$_2$CH$_3$

CH$_3$—C——CH—CH$_3$

CH$_3$ CH$_3$

6-36 Only E1 eliminations can exhibit rearrangement.

(a) E2 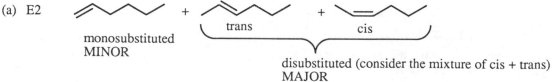 +   +

trans   cis

monosubstituted
MINOR

disubstituted (consider the mixture of cis + trans)
MAJOR

6-36 continued

(b) E1

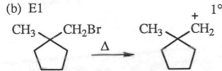

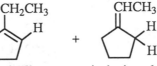 This short-lived 1° carbocation will undergo either a methyl shift or an alkyl shift (ring expansion), perhaps simultaneously with ionization.

methyl shift

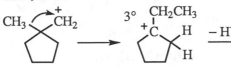

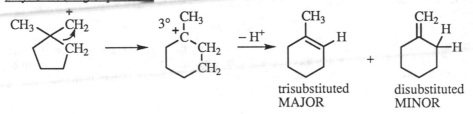

both alkenes are trisubstituted—
formed in about equal amounts

alkyl shift (ring expansion)

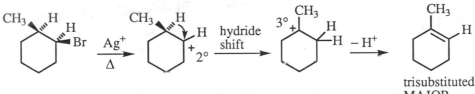

trisubstituted   disubstituted
MAJOR   MINOR

We cannot predict which rearrangement will predominate. All four alkenes will be produced, but of the 6-membered ring compounds, the trisubstituted alkene will be produced in greater amount than the disubstituted isomer.

(c) E1

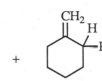

trisubstituted   disubstituted
MAJOR   MINOR

(d) E2

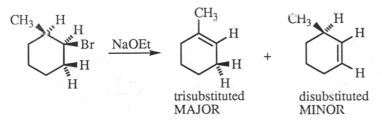

trisubstituted   disubstituted
MAJOR   MINOR

(Note: if the problem had specified *trans*-1-bromo-2-methylcyclohexane, only the second alkene of this pair would be correct. See text section 6-20.)

6-36 continued

(e) Methyl shift may occur simultaneously with ionization.

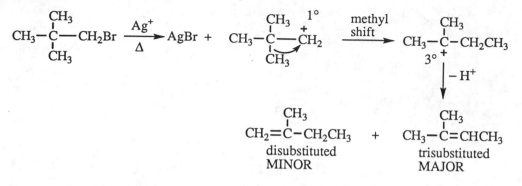

(f) no reaction if under E2 conditions; heating in a polar solvent with $NaOCH_3$ will produce the same product mixture as in 6-36(e)

6-37 In systems where free rotation is possible, the H to be abstracted by the base and the leaving group (Br here) must be anti-coplanar. The E2 mechanism is a concerted, one-step mechanism, so the arrangement of the other groups around the carbons in the starting material is retained in the product; there is no intermediate to allow time for rotation of groups. (Models will help.)

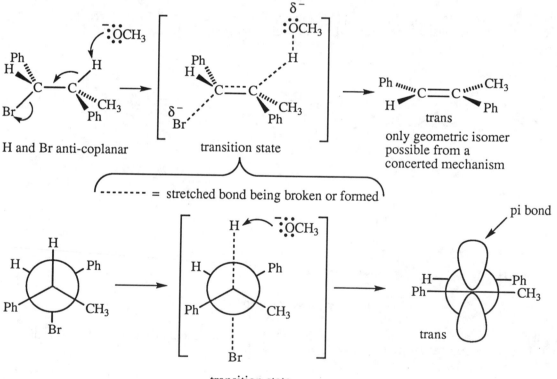

6-38

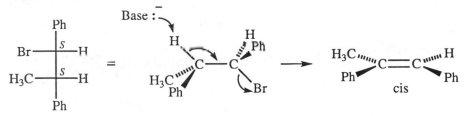

6-39

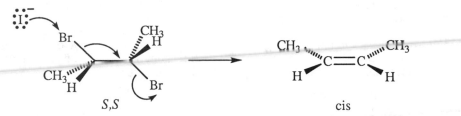

6-40

(a)

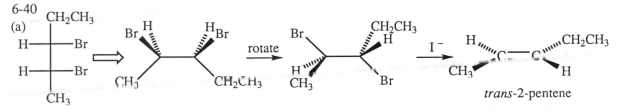

(b)

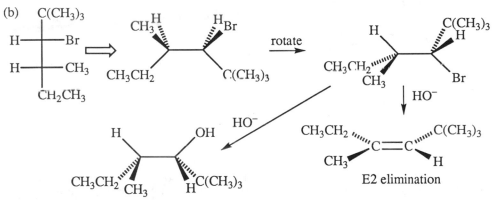

substitution product—since the substrate is a
neopentyl halide and highly hindered,
the S_N2 substitution is slow, and elimination is favored

(c)

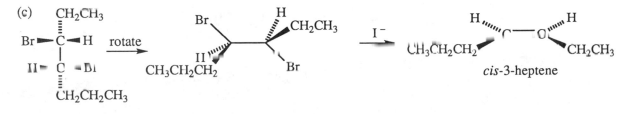

6-41 In the conformation where Br is axial, the only H anti-coplanar is at carbon-6. Only one alkene can be produced by E2: the LESS highly substituted alkene. Note that because of the stereochemical requirement of the E2 mechanism, the product is not a mixture and does not follow the Saytzeff Rule.

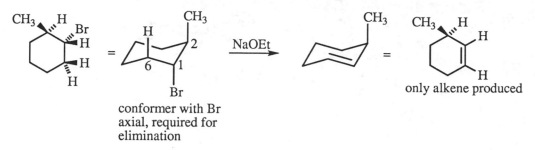

conformer with Br
axial, required for
elimination

only alkene produced

6-42 E2 elimination requires that the H and the leaving group be anti-coplanar; in a chair cyclohexane, this requires that the two groups be trans diaxial. However, when the bromine atom is in an axial position, there are no hydrogens in axial positions on adjacent carbons, so no elimination can occur.

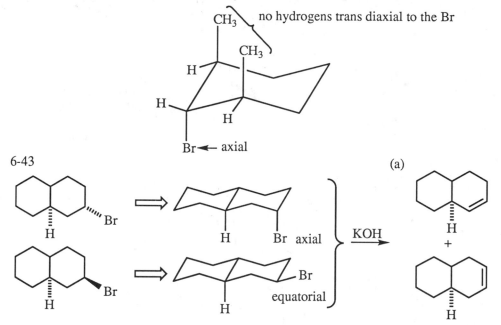

no hydrogens trans diaxial to the Br

Br ← axial

6-43

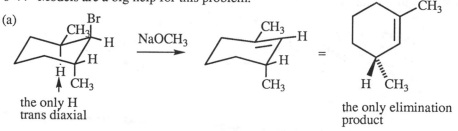

(a)

Br axial
Br equatorial
KOH

+

(b) Showing the chair form of the decalins makes the answer clear. The top isomer has the Br locked into an axial position—optimum for E2 elimination. The bottom isomer has Br equatorial where it is exceedingly slow to eliminate.

6-44 Models are a big help for this problem.

(a)

NaOCH₃

the only H
trans diaxial

the only elimination
product

6-44 continued

(b)

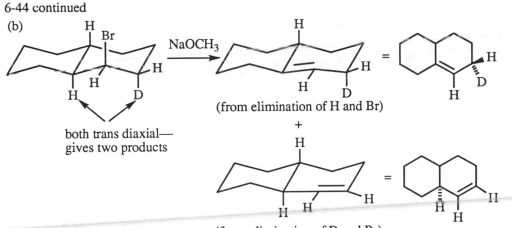

both trans diaxial—
gives two products

(from elimination of H and Br)

+

(from elimination of D and Br)

6-45

(a) Ethoxide is a strong base/nucleophile—second-order conditions. The 1° bromide favors substitution over elimination, so S_N2 will predominate over E2.

CH$_3$CH$_2$OCH$_2$CH$_3$
substitution—major

CH$_2$=CH$_2$
elimination—minor

(b) Ethoxide is a strong base/nucleophile—second order conditions. The 3° bromide is hindered and cannot undergo S_N2 by backside attack. E2 is the only route possible.

$$CH_3—\underset{\underset{CH_3}{|}}{C}=CH_2$$

(c) Ethoxide is a strong base/nucleophile—second-order conditions. The 2° bromide can undergo S_N2 or E2 reaction, the relative amounts depending on solvent and temperature.

$$CH_3—\underset{\underset{OCH_2CH_3}{|}}{CH}CH_3$$
substitution

CH$_3$—CH=CH$_2$
elimination

(d) Hydroxide is a strong base/nucleophile—second-order conditions. The 1° bromide is more likely to undergo S_N2 than E2, but both products will be observed.

$$CH_3—\underset{\underset{CH_3}{|}}{CH}CH_2OH$$
substitution—major

$$CH_3—\underset{\underset{CH_3}{|}}{C}=CH_2$$
elimination—minor

6-45 continued

(e) Silver nitrate in aqueous ethanol ionizes alkyl halides and there is no strong base present—first-order conditions. The isobutyl cation will rearrange to the *t*-butyl cation which in turn can be attacked by either water or ethanol as the nucleophile. S_N1 will predominate over E1 since there is no indication of high temperature.

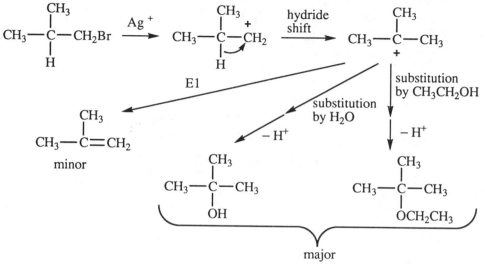

(f) Heating a 3° halide in methanol is quintessential first-order conditions, either E1 or S_N1 (solvolysis).

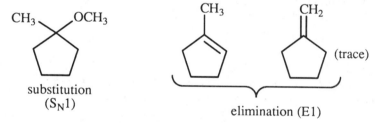

(g) Silver nitrate in methanol ionizes alkyl halides—first-order conditions. Simultaneous rearrangement by either hydride shift or by alkyl shift (ring expansion) produces carbocations that are likely to undergo S_N1 with methanol at room temperature; higher temperature would favor E1.

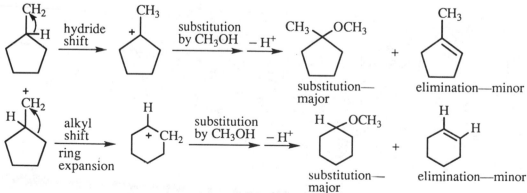

6-46 The stereochemical requirement of E2 elimination is anti-coplanar; in cyclohexanes, this translates to trans diaxial. Both dibromides are trans, but because the *t*-butyl group must be in an equatorial position, only the left molecule can have the bromines diaxial. The one on the right has both bromines locked into equatorial positions, from which they cannot undergo E2 elimination.

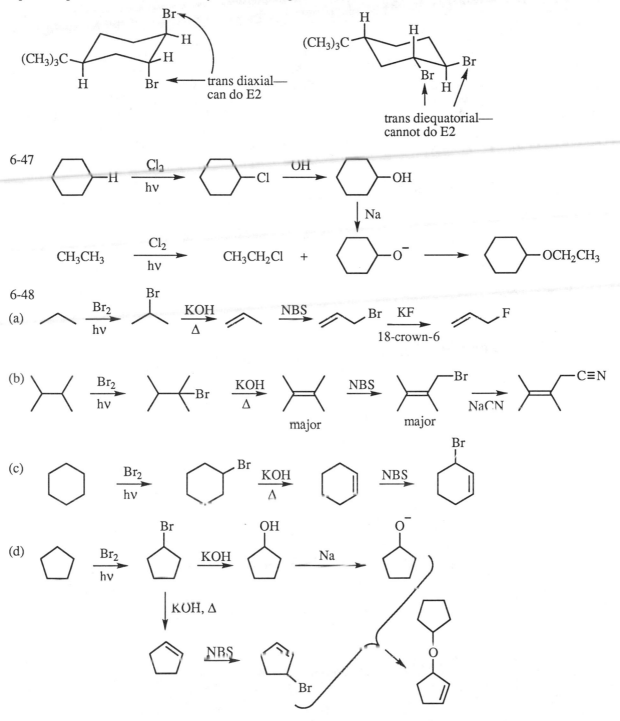

6-47

6-48

(a)

(b)

(c)

(d)

6-49 Please refer to solution 1-19, page 12, of this Solutions Manual.

6-50

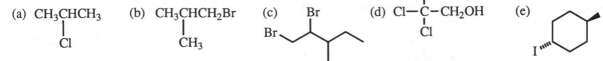

(a) CH₃CHCH₃ with Cl below   (b) CH₃CHCH₂Br with CH₃ below   (c) structure with Br   (d) Cl–C–CH₂OH with Cl above and Cl below   (e) cyclohexane with I

6-51

(a) 2-bromo-2-methylpentane
(b) 1-chloro-1-methylcyclohexane
(c) 1,1-dichloro-3-fluorocycloheptane

(d) 4-(2-bromoethyl)-3-(fluoromethyl)-2-methylheptane
(e) 4,4-dichloro-5-cyclopropyl-1-iodoheptane
(f) *cis*-1,2-dichloro-1-methylcyclohexane

6-52 Ease of backside attack (less steric hindrance) decides which undergoes S_N2 faster in all these examples except (b).

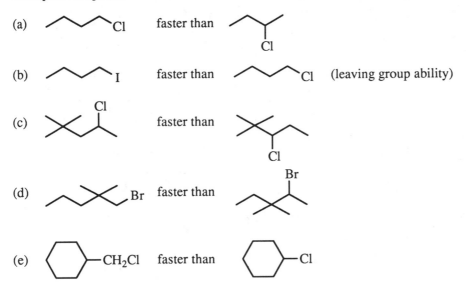

(a) [structure] CH₂Cl   faster than   [structure] Cl

(b) [structure] I   faster than   [structure] Cl   (leaving group ability)

(c) [structure with Cl]   faster than   [structure with Cl]

(d) [structure] Br   faster than   [structure with Br]

(e) [cyclohexane]–CH₂Cl   faster than   [cyclohexane]–Cl

6-53 Formation of the more stable carbocation decides which undergoes S_N1 faster in all these examples except (d).

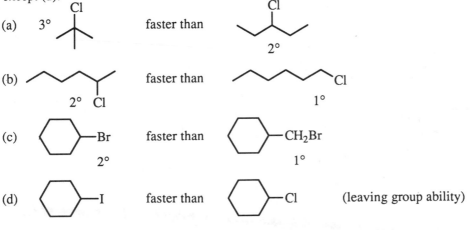

(a) 3° [structure with Cl]   faster than   2° [structure with Cl]

(b) [structure] 2° Cl   faster than   [structure] Cl 1°

(c) [cyclohexane]–Br 2°   faster than   [cyclohexane]–CH₂Br 1°

(d) [cyclohexane]–I   faster than   [cyclohexane]–Cl   (leaving group ability)

6-53 continued

(e) 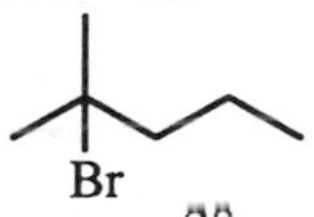
Br
3° faster than 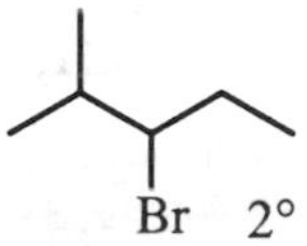
Br
2°

(f) 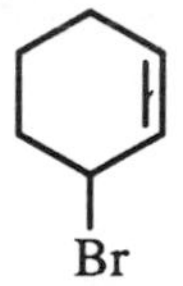
Br
2° allylic faster than 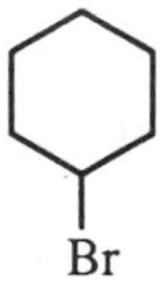
Br
2°

6-54 For S_N2, reactions should be designed such that the nucleophile attacks the least highly substituted alkyl halide. ("X" stands for a halide: Cl, Br, or I.)

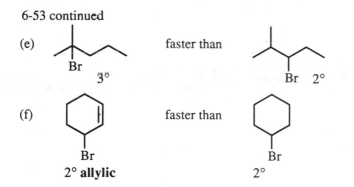

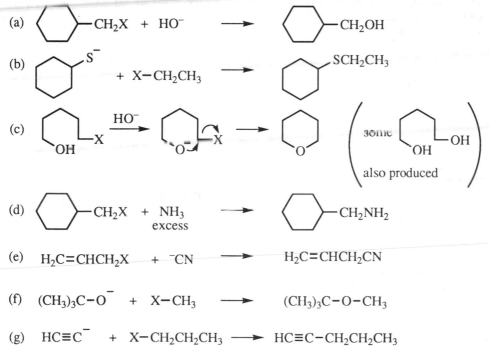

(a) 

CH_2X + HO^-
CH_2OH

(b) 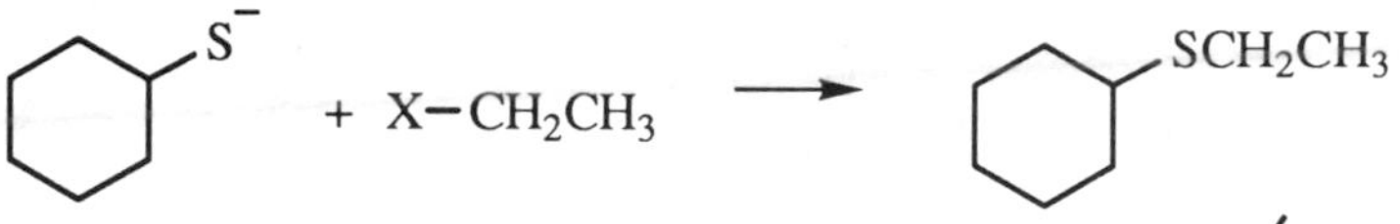

S^-
+ $X-CH_2CH_3$
SCH_2CH_3

(c) 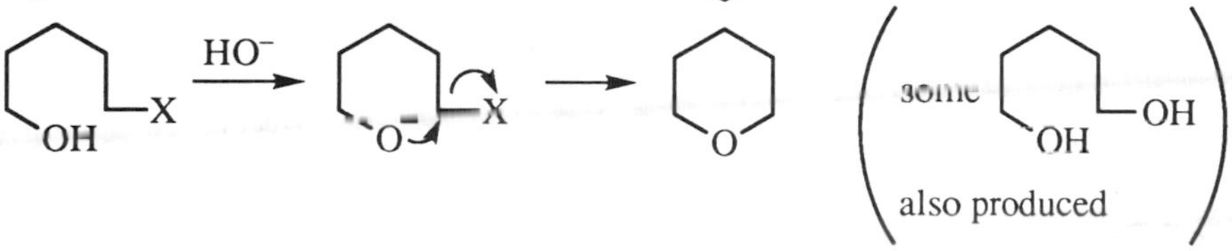

OH
X
HO^-
O^-
X
O
some
OH
OH
also produced

(d) 

CH_2X + NH_3
excess
CH_2NH_2

(e) $H_2C{=}CHCH_2X$ + ^-CN ⟶ $H_2C{=}CHCH_2CN$

(f) $(CH_3)_3C{-}O^-$ + $X{-}CH_3$ ⟶ $(CH_3)_3C{-}O{-}CH_3$

(g) $HC{\equiv}C^-$ + $X{-}CH_2CH_2CH_3$ ⟶ $HC{\equiv}C{-}CH_2CH_2CH_3$

6-55

(a) (1)

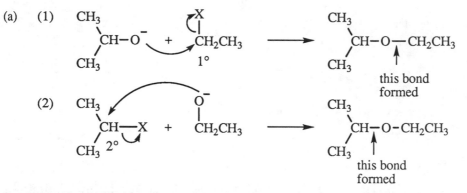

(2)

Synthesis (1) would give a better yield of the desired ether product. (1) uses S_N2 attack of a nucleophile on a 1° carbon, while (2) requires attack on a more hindered 2° carbon. Reaction (2) would give a lower yield of substitution, with more elimination.

(b) CANNOT DO S_N2 ON A 3° CARBON!

Better to do S_N2 on a methyl carbon:

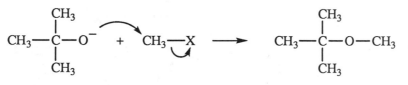

6-56

(a) S_N2—second order: reaction rate doubles
(b) S_N2—second order: reaction rate increases six times
(c) Virtually all reaction rates, including this one, increase with a temperature increase.

6-57 This is an S_N1 reaction; the rate law depends only on the substrate concentration, not on the nucleophile concentration.

(a) no change in rate
(b) the rate triples, dependent only on [t-butyl bromide]
(c) Virtually all reaction rates, including this one, increase with a temperature increase.

6-58 The key to this problem is that iodide ion is both an excellent nucleophile AND leaving group. Substitution on chlorocyclohexane is faster with iodide than with cyanide (see Table 6-3 for relative nucleophilicities). Once iodocyclohexane is formed, substitution by cyanide is much faster on iodocyclohexane than on chlorocyclohexane because iodide is a better leaving group than chloride. So two fast reactions involving iodide replace a slower single reaction, resulting in an overall rate increase.

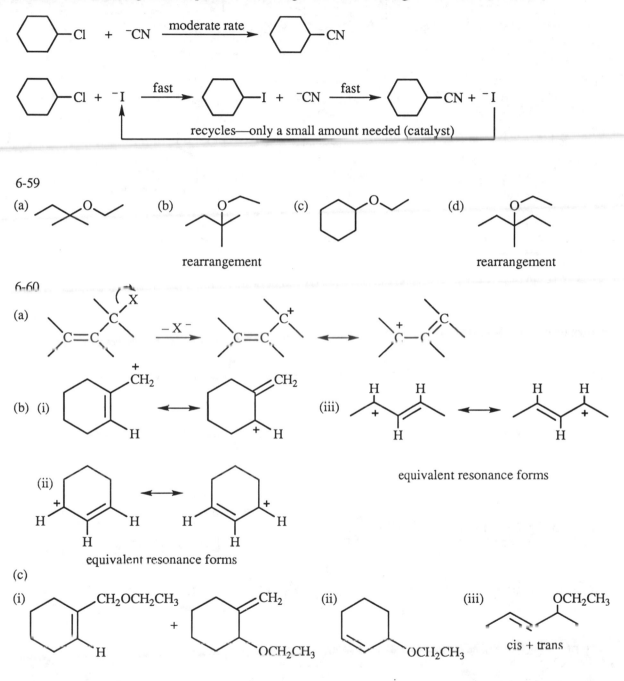

6-59

(a)

(b)

rearrangement

(c)

(d)

rearrangement

6-60

(a)

(b) (i)

(ii)

equivalent resonance forms

(iii)

equivalent resonance forms

(c)

(i)

+

(ii)

(iii)

cis + trans

115

6-61

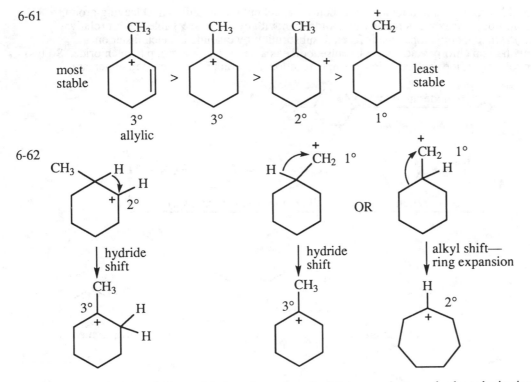

most stable > > > least stable

3° allylic   3°   2°   1°

6-62

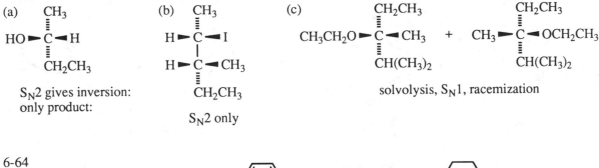

hydride shift   OR   hydride shift   alkyl shift— ring expansion

6-63 Reactions in (a) and (c) would also give some elimination products; only the substitution products are shown here.

(a)

CH_3

$HO - C - H$

CH_2CH_3

S_N2 gives inversion: only product:

(b)

CH_3

$H - C - I$

$H - C - CH_3$

CH_2CH_3

S_N2 only

(c)

CH_2CH_3

$CH_3CH_2O - C - CH_3$ + $CH_3 - C - OCH_2CH_3$

$CH(CH_3)_2$ $CH(CH_3)_2$

solvolysis, S_N1, racemization

6-64

(a) $CH_3CH_2OCH_2CH_3$

(b) 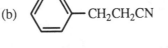 $-CH_2CH_2CN$

(c) 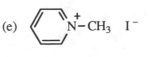 $-SCH_3$

(d) $CH_3(CH_2)_8CH_2I$

(e) 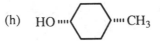 $N^+ - CH_3$ I^-

(f) $(CH_3)_3C - CH_2CH_2NH_2$

(g)

(h) $HO \cdots$ $\cdots CH_3$

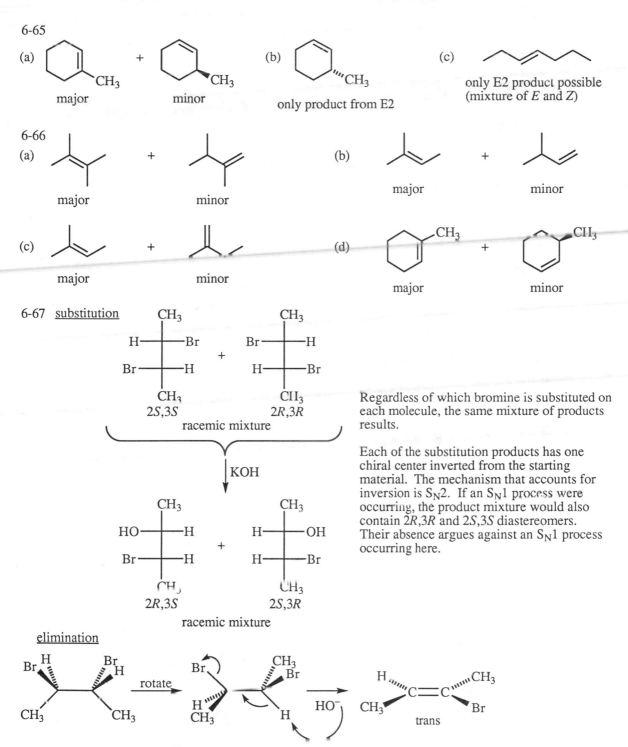

6-65

(a) major + minor

(b) only product from E2

(c) only E2 product possible (mixture of *E* and *Z*)

6-66

(a) major + minor

(b) major + minor

(c) major + minor

(d) major + minor

6-67 substitution

CH₃
H——Br
Br——H
CH₃
2S,3S

+

CH₃
Br——H
H——Br
CH₃
2R,3R

racemic mixture

Regardless of which bromine is substituted on each molecule, the same mixture of products results.

Each of the substitution products has one chiral center inverted from the starting material. The mechanism that accounts for inversion is S_N2. If an S_N1 process were occurring, the product mixture would also contain 2R,3R and 2S,3S diastereomers. Their absence argues against an S_N1 process occurring here.

↓ KOH

CH₃
HO——H
Br——H
CH₃
2R,3S

+

CH₃
H——OH
H——Br
CH₃
2S,3R

racemic mixture

elimination

The other enantiomer gives the same product (you should prove this to yourself).

The absence of cis product is evidence that only the E2 elimination is occurring, not E1.

6-68

(a) $\dfrac{+15.58°}{+15.90°}$ x 100% = 98% of original optical activity = 98% e.e.

Thus, 98% of the *S* enantiomer and 2% racemic mixture gives an overall composition of 99% *S* and 1% *R*.

(b) The 1% of radioactive iodide has produced exactly 1% of the *R* enantiomer. Each substitution must occur with inversion, a classic S_N2 mechanism.

6-69

(a) An S_N2 mechanism with inversion will convert *R* to its enantiomer, *S*. An accumulation of excess *S* does not occur because it can also react with bromide, regenerating *R*. The system approaches a racemic mixture at equilibrium.

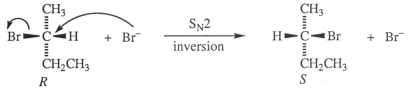

(b) In order to undergo substitution and therefore inversion, HO^- would have to be the leaving group, but HO^- is never a leaving group in S_N2. No reaction can occur.

(c) Once the OH is protonated, it can leave as H_2O. Racemization occurs in the S_N1 mechanism because of the planar, achiral carbocation intermediate which "erases" all stereochemistry of the starting material. Racemization occurs in the S_N2 mechanism by establishing an equilibrium of *R* and *S* enantiomers, as explained in 6-69(a).

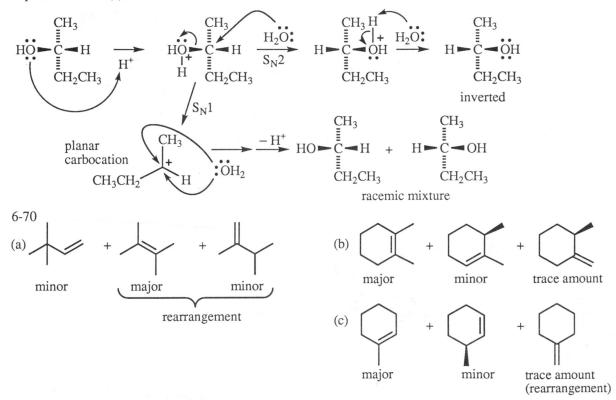

6-70

118

6-71 The allylic carbocation has two resonance forms showing that two carbons share the positive charge. The ethanol nucleophile can attack either of these carbons, giving the S_N1 products; or loss of an adjacent H will give the E1 product.

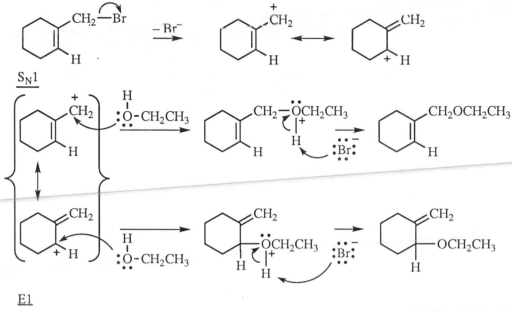

S_N1

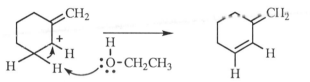

E1

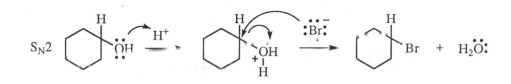

6-72

(a)

(b) S_N1

S_N2

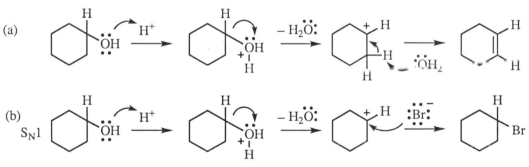

119

6-73

(a) $CH_3CH_2CH=CH_2$

(b) No reaction—the two bromines are not trans and therefore cannot be trans-diaxial.

(c) 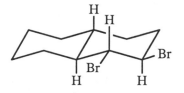 (d)

(e) No reaction—the bromines are trans, but they are diequatorial because of the locked conformation of the trans-decalin system. E2 can occur only when the bromines are trans diaxial.

bromines are diequatorial—
cannot undergo E2

6-74 NBS generates bromine which produces bromine radical. Bromine radical abstracts an allylic hydrogen, resulting in a resonance-stabilized allylic radical. The allylic radical can bond to bromine at either of the two carbons with radical character.

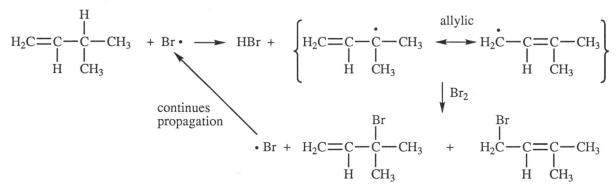

6-75 The bromine radical from NBS will abstract whichever hydrogen produces the most stable intermediate; in this structure, that is a benzylic hydrogen, giving the resonance-stabilized benzylic radical.

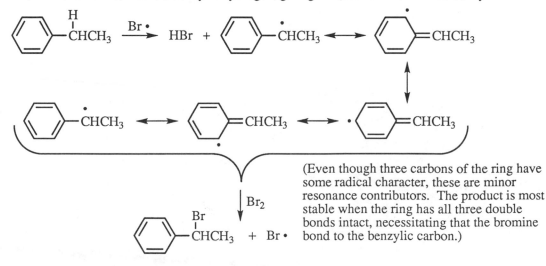

(Even though three carbons of the ring have some radical character, these are minor resonance contributors. The product is most stable when the ring has all three double bonds intact, necessitating that the bromine bond to the benzylic carbon.)

6-76 Two related factors could explain this observation. First, as carbocation stability increases, the leaving group will be less tightly held by the carbocation for stabilization; the more stable carbocations are more "free" in solution, meaning more exposed. Second, more stable carbocations will have longer lifetimes, allowing the leaving group to drift off in the solvent, leading to more possibility for the incoming nucleophile to attack from the side that the leaving group just left.

The less stable carbocations hold tightly to their leaving groups, preventing nucleophiles from attacking this side. Backside attack with inversion is the preferred stereochemical route in this case.

6-77

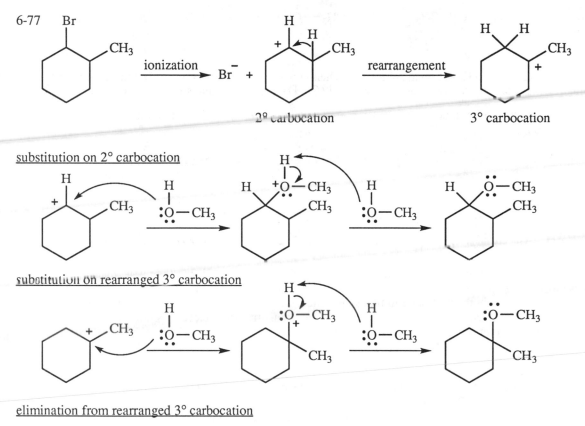

substitution on 2° carbocation

substitution on rearranged 3° carbocation

elimination from rearranged 3° carbocation

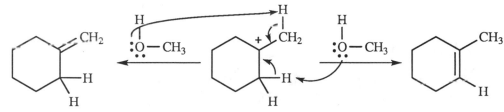

6-78

(a) E2—
one step

$$H_2C\text{—}CH\text{-}CH_3 \longrightarrow H_2C=CH\text{-}CH_3 + H_2O + Br^-$$

with Br leaving group, H abstracted by $^-:\overset{..}{O}H$

S_N2—
one step

$$CH_3\text{—}CH\text{-}CH_3 \longrightarrow CH_3\text{—}CH\text{-}CH_3 + Br^-$$

with Br leaving group, $^-:\overset{..}{O}H$ attacking; product OH

(b) In the E2 reaction, a C—H bond is broken. When D is substituted for H, a C—D bond is broken, slowing the reaction. In the S_N2 reaction, no C—H (C—D) bond is broken, so the rate is unchanged.

(c) These are first-order reactions. The slow, rate-determining step is the first step in each mechanism.

E1 $H_2C\text{—}CH\text{-}CH_3 \xrightarrow{\text{slow}} :\overset{..}{\underset{..}{Br}}:^- \quad H_2C\text{—}\overset{+}{CH}\text{-}CH_3 \xrightarrow{\text{fast}} H_2C=CH\text{-}CH_3$

S_N1 $CH_3\text{—}CH\text{-}CH_3 \xrightarrow{\text{slow}} Br^- + CH_3\text{—}\overset{+}{CH}\text{-}CH_3 \xrightarrow[\text{fast}]{H_2\overset{..}{O}:} CH_3\text{—}\overset{+}{\underset{|}{CH}}\text{-}CH_3 \xrightarrow[\text{fast}]{H_2\overset{..}{O}:} CH_3\text{—}CH\text{-}CH_3$

The only mechanism of these two involving C—H bond cleavage is the E1, but the C—H cleavage does NOT occur in the slow, rate-determining step. Kinetic isotope effects are observed only when C—H (C—D) bond cleavage occurs in the rate-determining step. Thus, we would expect to observe *no change in rate* for the deuterium-substituted molecules in the E1 or S_N1 mechanisms. (In fact, this technique of measuring isotope effects is one of the most useful tools chemists have for determining what mechanism a reaction follows.)

6-79 Both products are formed through E2 reactions. The difference is whether a D or an H is removed by the base. As explained in Problem 6-78, C—D cleavage can be up to 7 times slower than C—H cleavage, so the product from C—H cleavage should be formed about 7 times as fast. This rate preference is reflected in the 7 : 1 product mixture. ("Ph" is the abbreviation for a benzene ring.)

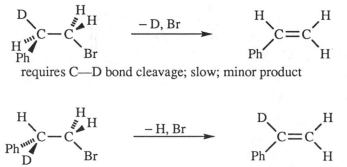

requires C—D bond cleavage; slow; minor product

requires C—H bond cleavage; 7 times faster; major product

6-80 The energy, and therefore the structure, of the transition state determines the rate of a reaction. Any factor which lowers the energy of the transition state will speed the reaction.

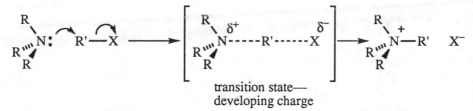

transition state—
developing charge

This example of S_N2 is unusual in that the nucleophile is a neutral molecule—it is not negatively charged. The transition state is beginning to show the positive and negative charges of the products (ions), so the transition state is more charged than the reactants. The polar transition state will be stabilized in a more polar solvent through dipole-dipole interactions, so the rate of reaction will be enhanced in a polar solvent.

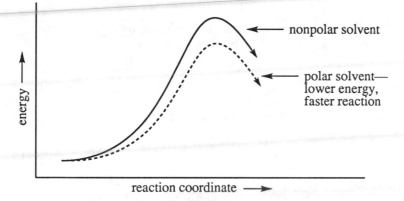

← nonpolar solvent

← polar solvent—
lower energy,
faster reaction

energy →

reaction coordinate →

6-81 The problem is how to explain this reaction:

<u>facts</u>

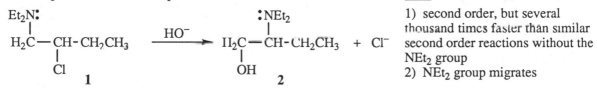

1) second order, but several thousand times faster than similar second order reactions without the NEt_2 group
2) NEt_2 group migrates

Solution

Clearly, the NEt_2 group is involved. The nitrogen is a nucleophile and can do an internal nucleophilic substitution (S_Ni), a very fast reaction for entropy reasons because two different molecules do not have to come together.

Et_2N:
H_2C—CH-CH_2CH_3 $\xrightarrow[\text{very fast}]{S_Ni}$
Cl
1

$$\begin{array}{c} Et \backslash \overset{+}{N} / Et \\ / \backslash \end{array}$$
H_2C—CH-CH_2CH_3 + Cl^-
3

(turn quickly to the next page)

6-81 continued

The slower step is attack of HO⁻ on intermediate **3**; the N is a good leaving group because it is positively charged. Where will HO⁻ attack **3**? On the less substituted carbon, in typical S_N2 fashion.

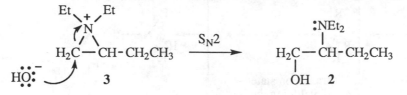

This overall reaction is fast because of the *neighboring group assistance* in forming **3**. It is second order because the HO⁻ group and **3** collide in the slow step (not the *only* step, however). And the NEt₂ group "migrates", although in two steps.

6-82 The symmetry of this molecule is crucial.

(a)

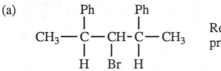

Regardless of which adjacent H is removed by *t*-butoxide, the product will be 2,4-diphenyl-2-pentene.

(b) Here are a Newman projection, a three-dimensional representation and a Fischer projection of the required diastereomer. On both carbons 2 and 4, the H has to be anti-coplanar with the bromine while leaving the other groups to give the same product. Not coincidentally, the correct diastereomer is a meso structure.

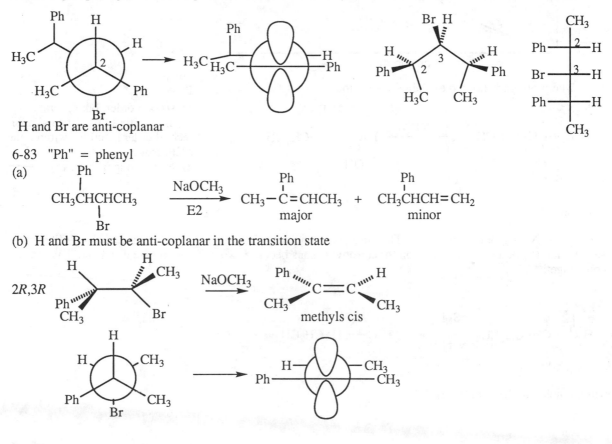

H and Br are anti-coplanar

6-83 "Ph" = phenyl

(a)

$$CH_3CHCHCH_3 \xrightarrow[E2]{NaOCH_3} CH_3-C=CHCH_3 \; + \; CH_3CHCH=CH_2$$

with Ph substituents, major + minor

(b) H and Br must be anti-coplanar in the transition state

2R,3R

$\xrightarrow{NaOCH_3}$ methyls cis

124

6-83 continued

(c)

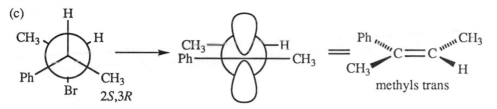

(d) The *2S,3S* is the mirror image of *2R,3R;* it would give the mirror image of the alkene that *2R,3R* produced (with two methyl groups cis). The alkene product is planar, not chiral, so its mirror image is the same: the *2S,3S* and the *2R,3R* give the same alkene.

6-84 All five products (boxed) come from rearranged carbocations. Rearrangement, which may occur simultaneously with ionization, can occur by hydride shift to the 3° methylcyclopentyl cation, or by ring expansion to the cyclohexyl cation.

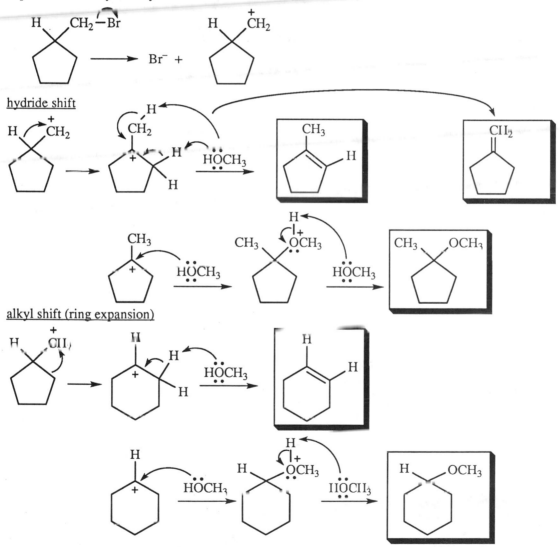

hydride shift

alkyl shift (ring expansion)

7-1 The number of elements of unsaturation in a hydrocarbon formula is given by:

$$\frac{2(\#C) + 2 - (\#H)}{2}$$

(a) $C_6H_{12} \Rightarrow \dfrac{2(6) + 2 - (12)}{2}$ = 1 element of unsaturation

(b) Many examples are possible. Yours may not match these, but all must have either a double bond or a ring, that is, one element of unsaturation.

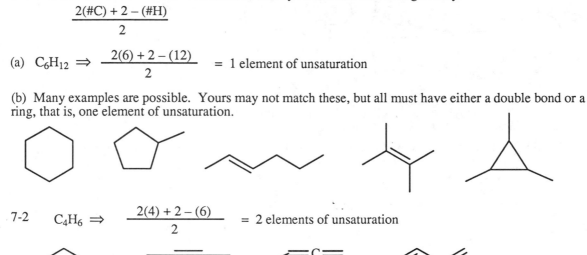

7-2 $C_4H_6 \Rightarrow \dfrac{2(4) + 2 - (6)}{2}$ = 2 elements of unsaturation

7-3   Hundreds of examples of C_4H_6NOCl are possible. Yours may not match these, but all must contain two elements of unsaturation.

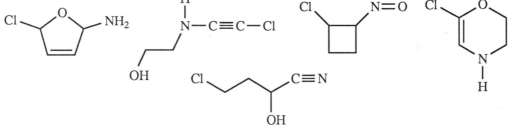

7-4   Many examples of these formulas are possible. Yours may not match these, but correct answers must have the same number of elements of unsaturation.

(a) $C_3H_4Cl_2 \Rightarrow C_3H_6$ = **1**

(b) $C_4H_8O \Rightarrow C_4H_8$ = **1**

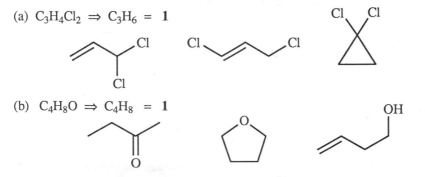

7-4 continued

(c) $C_4H_4O_2 \Rightarrow C_4H_4 = 3$

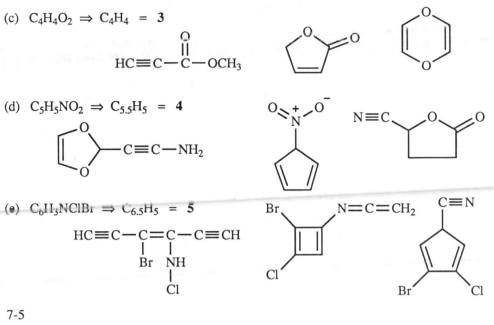

(d) $C_5H_5NO_2 \Rightarrow C_{5.5}H_5 = 4$

(e) $C_6H_3NClBr \Rightarrow C_{6.5}H_5 = 5$

7-5

(a) 4-methyl-1-pentene
(b) 2-ethyl-1-hexene
(c) 1,4-pentadiene
(d) 1,2,4-pentatriene
(e) 2,5-dimethyl-1,3-cyclopentadiene

(f) 4-vinyl-1-cyclohexene ("1" is optional)
(g) 3-phenyl-1-propene ("1" is optional)
(h) *trans*-3,4-dimethyl-1-cyclopentene ("1" is optional)
(i) 7-methylene-1,3,5-cycloheptatriene

7-6 (b), (e), and (f) do not show *cis,trans* isomerism

(a)

(Z)-3-hexene
(*cis*-3-hexene)

(E)-3-hexene
(*trans*-3-hexene)

(c)

(2Z,4Z)-2,4-hexadiene
cis, cis 2,4-hexadiene

(2Z,4E)-2,4-hexadiene
cis, trans-2,4-hexadiene

(2E,4E)-2,4-hexadiene
trans, trans-2,4-hexadiene

(d)

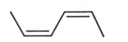

(Z)-3-methyl-2-pentene

(E)-3-methyl-2-pentene

"*Cis*" and "*trans*" are not clear for this example;
"*E*" and *Z* are unambiguous.

127

7-7

(a)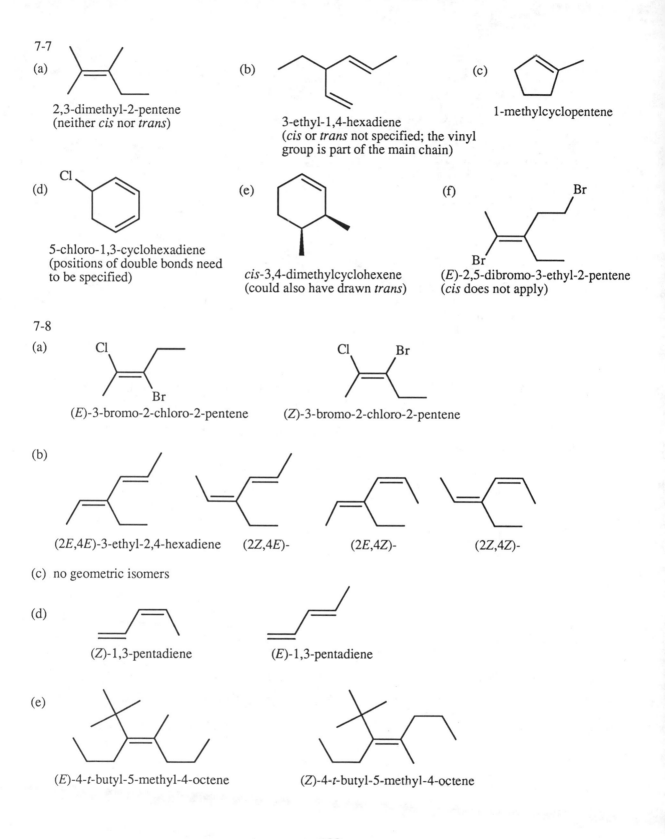

2,3-dimethyl-2-pentene
(neither *cis* nor *trans*)

(b)

3-ethyl-1,4-hexadiene
(*cis* or *trans* not specified; the vinyl
group is part of the main chain)

(c)

1-methylcyclopentene

(d)

5-chloro-1,3-cyclohexadiene
(positions of double bonds need
to be specified)

(e)

cis-3,4-dimethylcyclohexene
(could also have drawn *trans*)

(f)

(*E*)-2,5-dibromo-3-ethyl-2-pentene
(*cis* does not apply)

7-8

(a)

(*E*)-3-bromo-2-chloro-2-pentene   (*Z*)-3-bromo-2-chloro-2-pentene

(b)

(2*E*,4*E*)-3-ethyl-2,4-hexadiene   (2*Z*,4*E*)-   (2*E*,4*Z*)-   (2*Z*,4*Z*)-

(c) no geometric isomers

(d)

(*Z*)-1,3-pentadiene   (*E*)-1,3-pentadiene

(e)

(*E*)-4-*t*-butyl-5-methyl-4-octene   (*Z*)-4-*t*-butyl-5-methyl-4-octene

128

7-8 continued

(f)

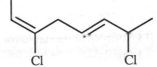

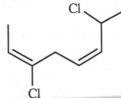

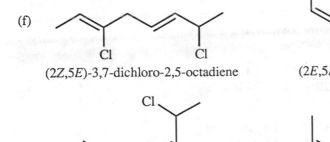

(2Z,5E)-3,7-dichloro-2,5-octadiene   (2E,5E)-3,7-dichloro-2,5-octadiene

(2Z,5Z)-3,7-dichloro-2,5-octadiene   (2E,5Z)-3,7-dichloro-2,5-octadiene

(g) no geometric isomers (an *E* double bond would be too highly strained)

(h)

(Z)-cyclodecene   (E)-cyclodecene

(i)

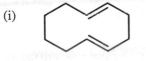   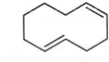   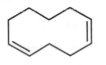

(1E,5E)-1,5-cyclodecadiene   (1Z,5E)-1,5-cyclodecadiene   (1Z,5Z)-1,5-cyclodecadiene

7-9 From Table 7-1, approximate heats of hydrogenation can be determined for similarly substituted alkenes. The energy difference is approximately 1.4 kcal/mole (6 kJ/mole), the more highly substituted alkene being more stable.

gem-disubstituted
28.0 kcal/mole
(117 kJ/mole)

tetrasubstituted
26.6 kcal/mole
(111 kJ/mole)

7-10 Use the relative values in Figure 7-7.

(a)   2 x (*trans*-disubstituted – *cis*-disubstituted) =
       2 x (5.2 – 4.2) = 2 kcal/mole more stable for *trans,trans*
       [2 x (22 – 18) = 8 kJ/mole]

(b)   *gem*-disubstituted – monosubstituted = 4.8 – 2.7 = 2.1 kcal/mole
                                              (20 – 11 = 9 kJ/mole)
       2-methyl-1-butene is more stable

(c)   trisubstituted – *gem*-disubstituted = 5.9 – 4.8 = 1.1 kcal/mole
                                             (25 – 20 = 5 kJ/mole)
       2-methyl-2-butene is more stable

(d)   tetrasubstituted – *gem*-disubstituted = 6.2 – 4.8 = 1.4 kcal/mole
                                               (26 – 20 = 6 kJ/mole)
       2,3-dimethyl-2-butene is more stable

7-11
(a)

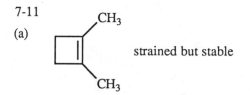

strained but stable

(b) could not exist—ring size must be 8 atoms or greater to include *trans* double bond

(c)

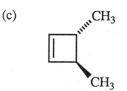

Despite the ambiguity of this name, we must assume that the *trans* refers to the two methyls since this compound can exist, rather than the *trans* referring to the alkene, a molecule which could not exist.

(d)

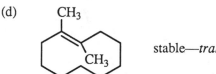

stable—*trans* in 10-membered ring

(e) unstable at room temperature—cannot have *trans* alkene in 7-membered ring (possibly isolable at very low temperature—this type of experiment is one of the challenges chemists attack with gusto)

(f) stable—alkene not at bridgehead

(g) unstable—violation of Bredt's Rule (alkene at bridgehead in 6-membered ring)

(h) stable—alkene at bridgehead in 8-membered ring

7-12
(a) The dibromo compound should boil at a higher temperature because of its much larger molecular weight.
(b) The *cis* should boil at a higher temperature than the *trans* as the *trans* has a zero dipole moment and therefore no dipole-dipole interactions.
(c) 1,2-Dichlorocyclohexene should boil at a higher temperature because of its much larger molecular weight.

7-13

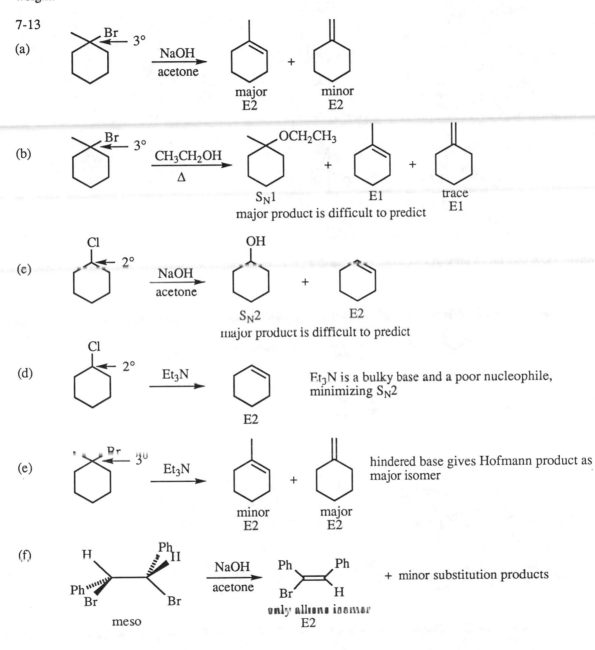

7-13 continued

(g)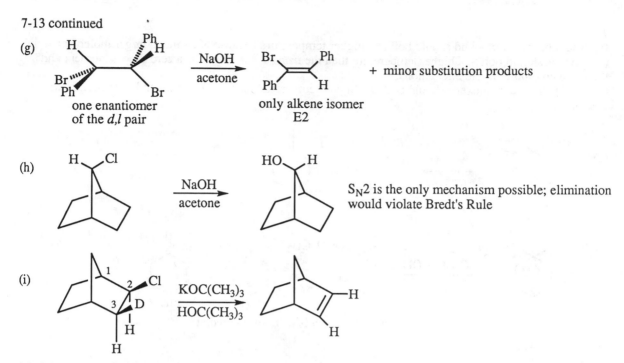

NaOH
acetone

+ minor substitution products

only alkene isomer
E2

one enantiomer
of the *d,l* pair

(h)

NaOH
acetone

S_N2 is the only mechanism possible; elimination
would violate Bredt's Rule

(i)

KOC(CH₃)₃
HOC(CH₃)₃

Models show that the H on C-3 cannot be anti-coplanar with the Cl on C-2. Thus, this E2 elimination must occur with a *syn*-coplanar orientation.

7-14

(a)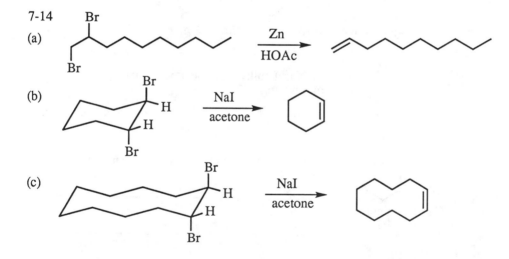

Zn
HOAc

(b)

NaI
acetone

(c)

NaI
acetone

(d)

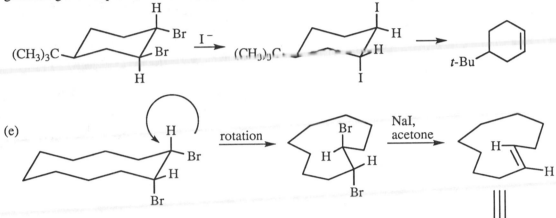

The correct answer is "no reaction" because the two bromines are not coplanar (neither the preferred anti nor the possible syn orientations) and are locked in this conformation by the *t*-butyl group. Given time, however, the excellent nucleophile iodide can displace bromide by backside attack (S_N2), eventually generating anti- coplanar *trans*-diiodo, which could then eliminate, as shown below.

(e)

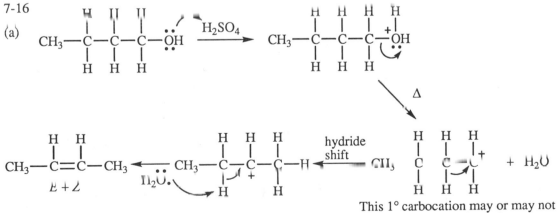

In the first conformation shown, the bromine atoms are not coplanar and cannot eliminate. Rotation in this large ring can place the two bromines anti-coplanar, generating a *trans* alkene in the ring. (Use models!)

7-15
(a) basic and nucleophilic mechanism: $Ba(OH)_2$ is a strong base
(b) acidic and electrophilic mechanism: the catalyst is H^+
(c) free radical chain reaction: the catalyst is a peroxide that initiates free radical reactions
(d) acidic and electrophilic mechanism: the catalyst BF_3 is a strong Lewis acid

7-16
(a)

CH₃—C—C—C—OH →(H₂SO₄)→ CH₃—C—C—C—OH⁺ →(Δ)

CH₃—C=C—CH₃ ← CH₃—C—C—C—H ←(hydride shift) CH₃ C C C⁺ + H₂O

E + Z

This 1° carbocation may or may not exist. It is shown for clarity.

7-16 continued

(b)

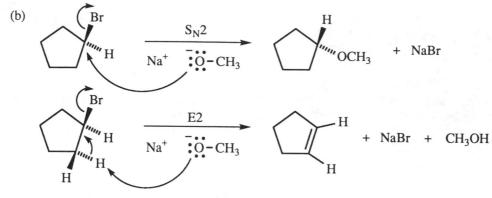

7-17

(a)

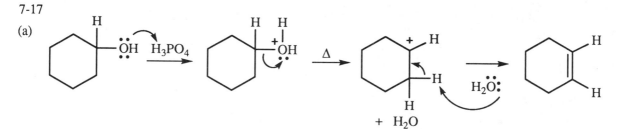

(b)

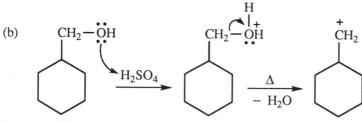

This 1° carbocation may or may not exist. It is shown for clarity.

Two possible rearrangements

1. Hydride shift

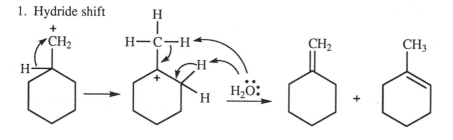

2. Alkyl shift—ring expansion

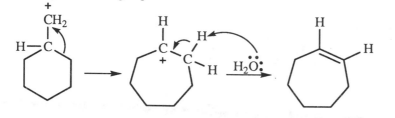

7-17 continued

(c)

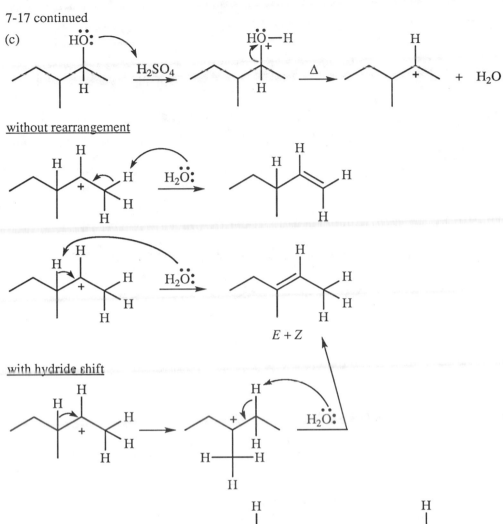

without rearrangement

with hydride shift

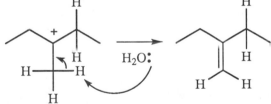

7-18   (a) $\Delta G°$ $= \Delta H° - T\Delta S°$

$= +27{,}600 \text{ cal/mol} - 298 \text{ K } (28.0 \text{ cal/K} \cdot \text{mol})$

$= +19{,}300 \text{ cal/mol} = +19.3 \text{ kcal/mol}$

$\Delta G°$ is positive, the reaction is **disfavored**

(b) ΔG_{1000} $= +27{,}600 \text{ cal/mol} - 1273 \text{ K } (28.0 \text{ cal/K} \cdot \text{mol})$

$= -8000 \text{ cal/mol} = -8.0 \text{ kcal/mol}$

ΔG is negative, the reaction is **favored**

7-19 Please refer to solution 1-19, page 12 of this Solutions Manual.

7-20

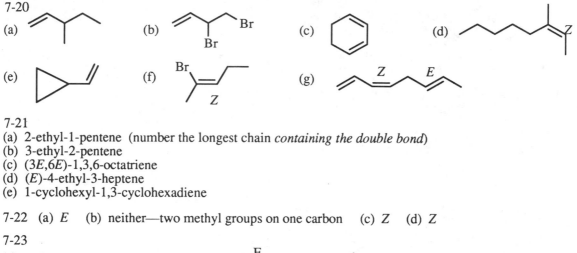

7-21
(a) 2-ethyl-1-pentene (number the longest chain *containing the double bond*)
(b) 3-ethyl-2-pentene
(c) (3E,6E)-1,3,6-octatriene
(d) (E)-4-ethyl-3-heptene
(e) 1-cyclohexyl-1,3-cyclohexadiene

7-22 (a) *E* (b) neither—two methyl groups on one carbon (c) *Z* (d) *Z*

7-23

(a)

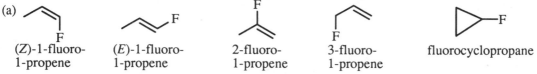

| (Z)-1-fluoro-
1-propene | (E)-1-fluoro-
1-propene | 2-fluoro-
1-propene | 3-fluoro-
1-propene | fluorocyclopropane |

(b) Cholesterol, $C_{27}H_{46}O$, has five elements of unsaturation. If only one of those is a pi bond, the other four must be rings.

7-24

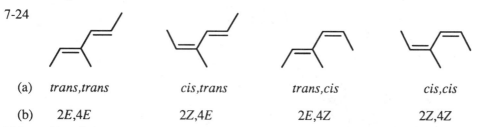

(a)	*trans,trans*	*cis,trans*	*trans,cis*	*cis,cis*
(b)	2E,4E	2Z,4E	2E,4Z	2Z,4Z

This problem is intended to show the difficulty of using *cis-trans* nomenclature with any but the simplest alkenes. *Cis* and *trans* are ambiguous: the first alkene in part (a) is *cis* if the two similar substituents are considered, but *trans* if the chain is considered. The *E-Z* nomenclature is unambiguous and is preferred for all four of these isomers.

7-25 (a), (d), and (e) have no geometric isomers

(b)

| *trans*-2-pentene
(E)-2-pentene | *cis*-2-pentene
(Z)-2-pentene | | (c) | *trans*-3-hexene
(E)-3-hexene | *cis*-3-hexene
(Z)-3-hexene |

(f)

| *trans,trans*-2,4-hexadiene
(2E,4E)-2,4-hexadiene | *cis,trans*-2,4-hexadiene
(2Z,4E)-2,4-hexadiene | *cis,cis*-2,4-hexadiene
(2Z,4Z)-2,4-hexadiene |

7 26

(a)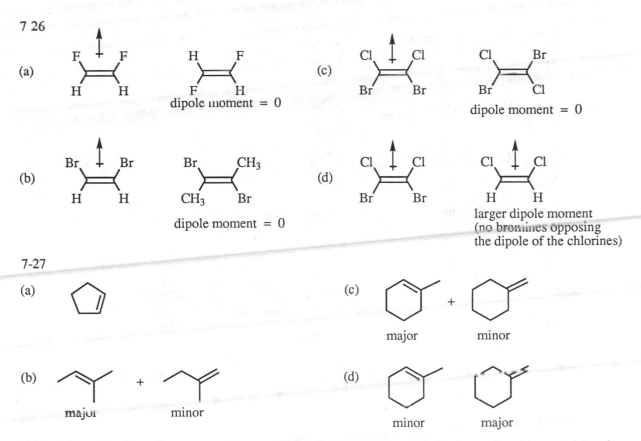

F—F (arrow up), H H

H—F, F—H   dipole moment = 0

(c) Cl—Cl (arrow up), Br Br

Cl—Br, Br—Cl   dipole moment = 0

(b) Br—Br (arrow up), H H

Br—CH₃, CH₃—Br   dipole moment = 0

(d) Cl—Cl (arrow up), Br Br

Cl—Cl (arrow up), H H   larger dipole moment (no bromines opposing the dipole of the chlorines)

7-27

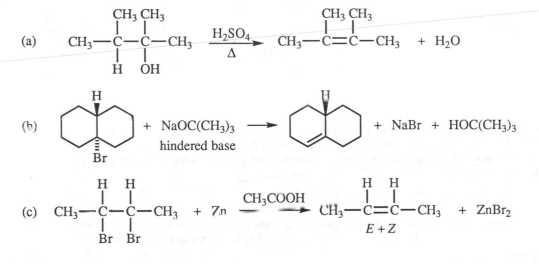

(a)

(c)   major + minor

(b)   major + minor

(d)   minor   major

7-28 Only major alkene isomers are shown. Minor alkene isomers would also be produced in parts (a) and (b).

(a) CH_3—$\overset{\overset{CH_3}{|}}{\underset{\underset{H}{|}}{C}}$—$\overset{\overset{CH_3}{|}}{\underset{\underset{OH}{|}}{C}}$—$CH_3$ $\xrightarrow[\Delta]{H_2SO_4}$ CH_3—$\overset{\overset{CH_3}{|}}{C}$=$\overset{\overset{CH_3}{|}}{C}$—$CH_3$ + H_2O

(b) [decalin with Br] + NaOC(CH₃)₃ hindered base ⟶ [decalin alkene] + NaBr + HOC(CH₃)₃

(c) CH_3—$\overset{\overset{H}{|}}{\underset{\underset{Br}{|}}{C}}$—$\overset{\overset{H}{|}}{\underset{\underset{Br}{|}}{C}}$—$CH_3$ + Zn $\xrightarrow{CH_3COOH}$ CH_3—$\overset{\overset{H}{|}}{C}$=$\overset{\overset{H}{|}}{C}$—$CH_3$ + ZnBr₂
$E + Z$

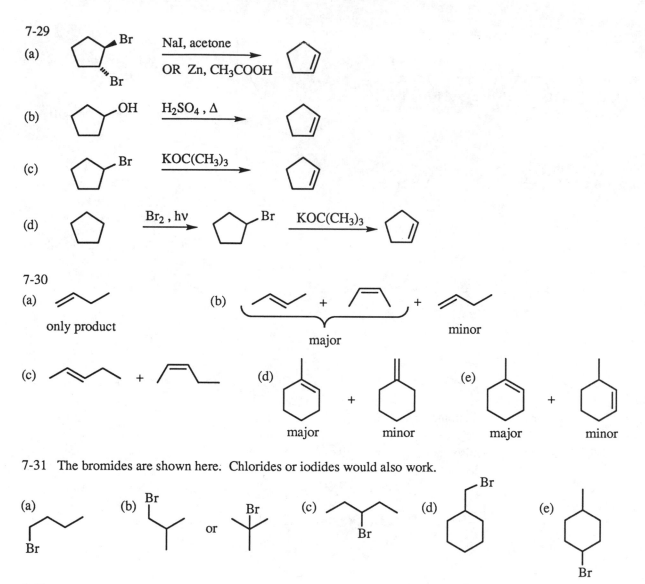

7-31 The bromides are shown here. Chlorides or iodides would also work.

7-32

(a) There are two reasons why alcohols do not dehydrate with strong base. The potential leaving group, hydroxide, is itself a strong base and therefore a terrible leaving group. Second, the strong base deprotonates the —OH faster than any other reaction can occur, consuming the base and making the leaving group anionic and therefore even worse.

$$ROH + \ ^-OC(CH_3)_3 \rightleftharpoons RO^- + HOC(CH_3)_3$$

(b) A halide is already a decent leaving group. Since halides are extremely weak bases, the halogen atom is not easily protonated, and even if it were, the leaving group ability is not significantly enhanced. The hard step is to remove the adjacent H, something only a strong base can do—and strong bases will not be present under strong acid conditions.

7-33

(a) 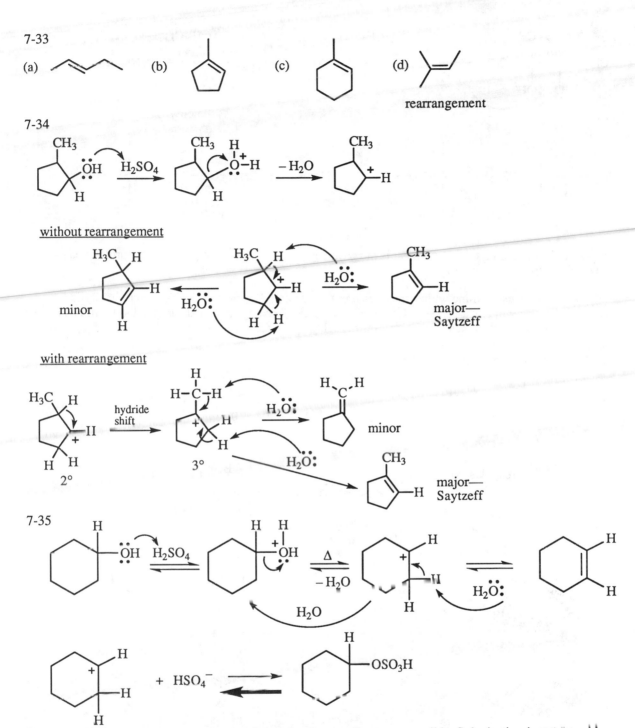 (b) (c) (d)

rearrangement

7-34

without rearrangement

with rearrangement

7-35

E1 works well because only one carbocation and only one alkene are possible. Substitution is not a problem here. The only nucleophiles are water, which would simply form starting material by a reverse of the dehydration, and bisulfate anion. Bisulfate anion is an extremely weak base and poor nucleophile; if it did attack the carbocation, the unstable product would quickly re-ionize, with no net change, back to the carbocation.

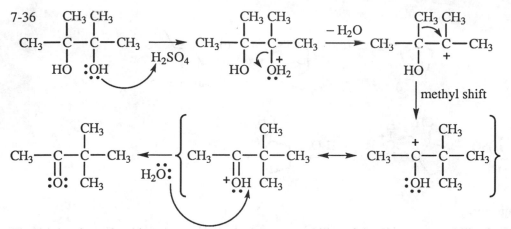

7-36

The driving force for this rearrangement is the great stability of the resonance-stabilized, protonated carbonyl group.

7-37 NBS generates bromine which produces bromine radical. Bromine radical abstracts an allylic hydrogen, resulting in a resonance-stabilized allylic radical. The allylic radical can bond to bromine at either of the two carbons with radical character. See the solution to problem 6-74.

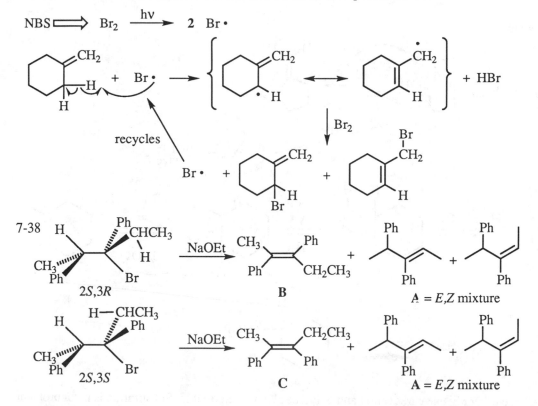

E2 dehydrohalogenation requires anti-coplanar arrangement of H and Br, so specific cis-trans isomers (**B** or **C**) are generated depending on the stereochemistry of the starting material. Removing a hydrogen from C-4 (achiral) will give about the same mixture of *cis* and *trans* (**A**) from either diastereomer.

7-39

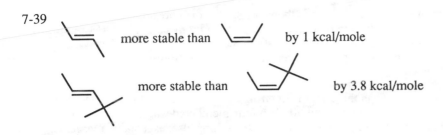

more stable than   by 1 kcal/mole

more stable than   by 3.8 kcal/mole

Steric crowding by the *t*-butyl group is responsible for the energy difference. In *cis*-2-butene, the two methyl groups have only slight interaction. However, in the 4,4-dimethyl- 2-pentenes, the larger size of the *t*-butyl group crowds the methyl group in the *cis* isomer, increasing its strain and therefore its energy.

7-40

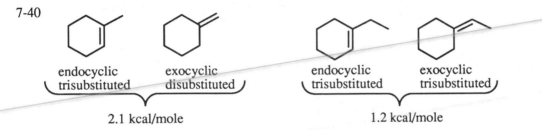

| endocyclic trisubstituted | exocyclic disubstituted | | endocyclic trisubstituted | exocyclic trisubstituted |

2.1 kcal/mole   1.2 kcal/mole

A standard principle of science is to compare experiments which differ by only one variable. Changing more than one variable clouds the interpretation, possibly to the point of invalidating the experiment.

The first set of structures compares endo and exocyclic double bonds, but the degree of substitution on the alkene is also different, so this comparison is not valid—we are not isolating simply the exo or endocyclic effect.

The second pair is a much better measure of endo versus exocyclic stability because both alkenes are trisubstituted, so the degree of substitution plays no part in the energy values. Thus, 1.2 kcal/mole is a better value.

7-41 In E2, the two groups to be eliminated must be coplanar. In conformationally mobile systems like acyclic molecules, or in cyclohexanes, anti-coplanar is the preferred orientation where the H and leaving group are 180° apart. In rigid systems like norbornanes, however, SYN-coplanar (angle 0°) is the only possible orientation and E2 will occur, although at a slower rate than anti-coplanar.

The structure having the H and the Cl syn-coplanar is the *trans*, which undergoes the E2 elimination. (It is possible that the *other* H and Cl eliminate from the *trans* isomer; the results from this reaction cannot distinguish between these two possibilities.)

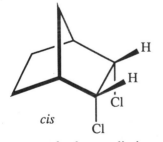

cis

extremely slow to eliminate—
H and Cl not coplanar

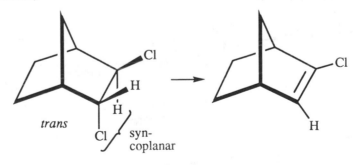

trans   syn-coplanar

7-42 Breaking a C—D bond requires more energy and is therefore slower than breaking a C—H bond. Since breaking a C—H bond is required in the rate-determining step of the E2 mechanism but not in the S$_N$2 mechanism, substituting D for H will slow the E2 but will not affect the S$_N$2. The S$_N$2 rate has not changed, but more S$_N$2 product is generated since the competing E2 reaction is so much slower (roughly 7 times slower with D than with H).

7-43 It is interesting to note that even though three-membered rings are more strained than four-membered rings, three-membered rings are far more common in nature than four-membered rings. It seems that whatever rearrangement will occur from a four-membered ring to something else, especially a larger ring, will happen quickly.

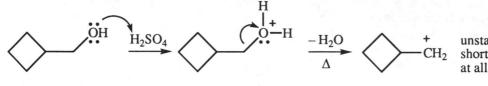

unstable 1° carbocation—
short lifetime if it exists
at all

without rearrangement

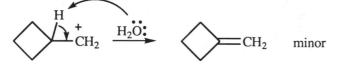

minor

with rearrangement—hydride shift

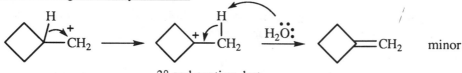

minor

3° carbocation, but
still in a strained
4-membered ring

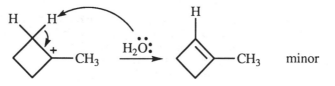

minor

with rearrangement—alkyl shift—ring expansion

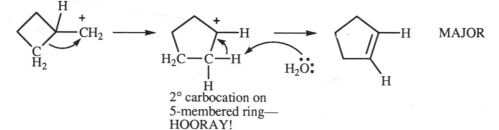

MAJOR

2° carbocation on
5-membered ring—
HOORAY!

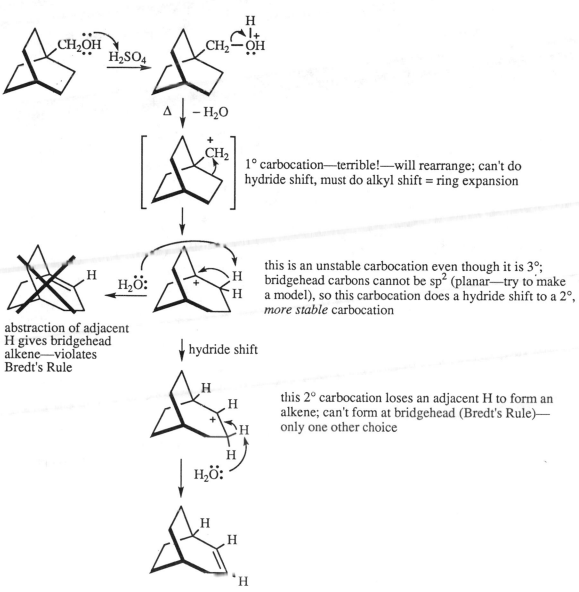

1° carbocation—terrible!—will rearrange; can't do hydride shift, must do alkyl shift = ring expansion

this is an unstable carbocation even though it is 3°; bridgehead carbons cannot be sp² (planar—try to make a model), so this carbocation does a hydride shift to a 2°, *more stable* carbocation

abstraction of adjacent H gives bridgehead alkene—violates Bredt's Rule

hydride shift

this 2° carbocation loses an adjacent H to form an alkene; can't form at bridgehead (Bredt's Rule)—only one other choice

143

8-1 *Major* products are produced in greatest amount; they are not necessarily the *only* products produced.

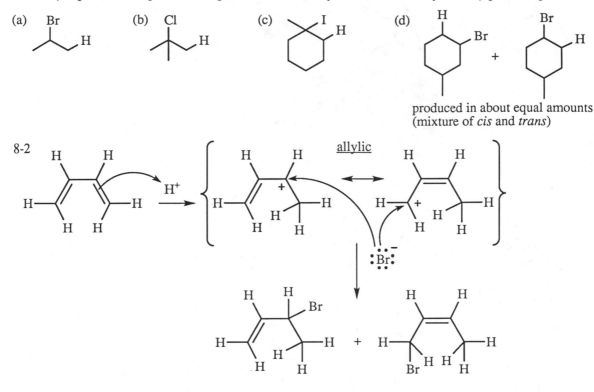

produced in about equal amounts
(mixture of *cis* and *trans*)

3-bromo-1-butene   1-bromo-2-butene

Because the allylic carbocation has partial positive charge at two carbons, the bromide nucleophile can bond at either electrophilic carbon, giving two products.

8-3 Recall that the abbreviation for a phenyl substituent is "Ph".

8-4

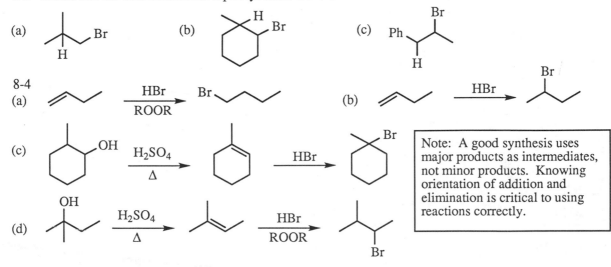

Note: A good synthesis uses major products as intermediates, not minor products. Knowing orientation of addition and elimination is critical to using reactions correctly.

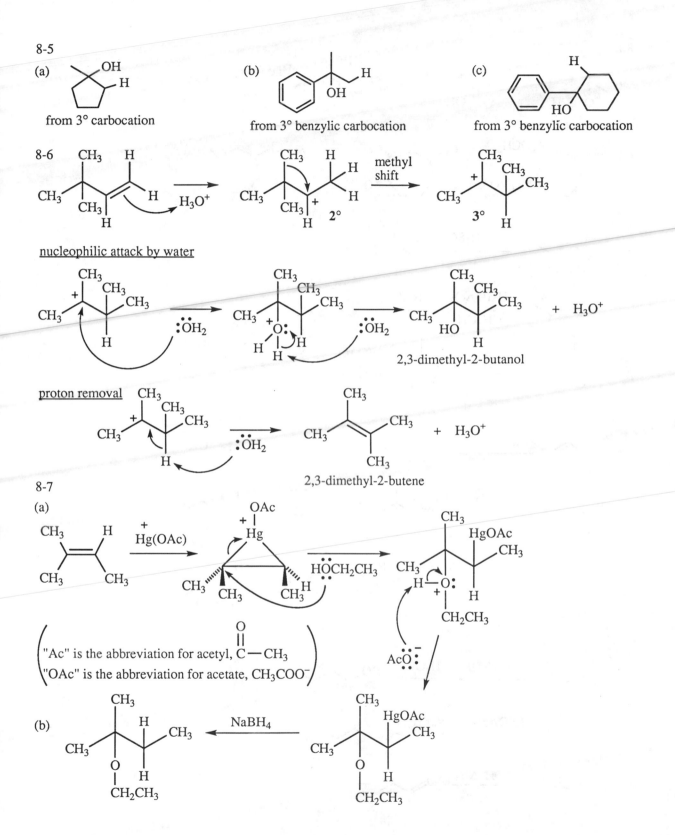

8-5

(a) from 3° carbocation

(b) from 3° benzylic carbocation

(c) from 3° benzylic carbocation

8-6

methyl shift

nucleophilic attack by water

2,3-dimethyl-2-butanol

proton removal

2,3-dimethyl-2-butene

8-7

(a)

$$\left(\begin{array}{l}\text{"Ac" is the abbreviation for acetyl, } \overset{O}{\underset{\|}{C}}-CH_3 \\ \text{"OAc" is the abbreviation for acetate, } CH_3COO^-\end{array}\right)$$

(b)

145

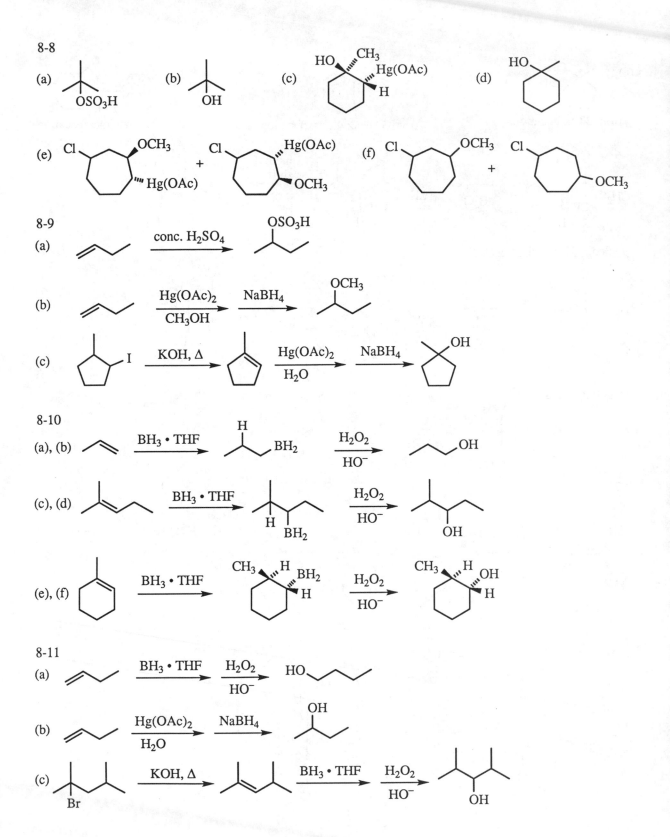

8-8

(a), (b), (c), (d)

8-9

(a), (b), (c)

8-10

(a), (b), (c), (d), (e), (f)

8-11

(a), (b), (c)

8-12 Instead of borane attacking the bottom face of 1-methylcyclopentene, it is equally likely to attack the top face, leading to the enantiomer.

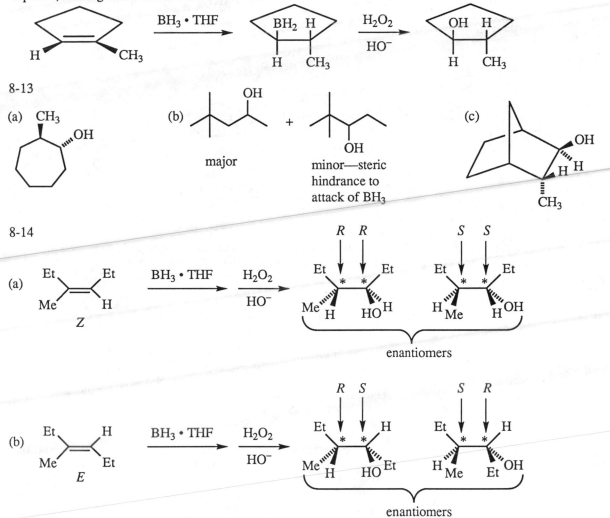

8-13

(a) CH₃

(b)

major

+

minor—steric
hindrance to
attack of BH₃

(c)

8-14

(a)

The enantiomeric pair produced from the Z-alkene is diastereomeric with the other enantiomeric pair produced from the E-alkene. These answers underscore that hydroboration-oxidation is stereospecific, that is, each alkene gives a specific set of stereoisomers, not a random mixture.

8-15

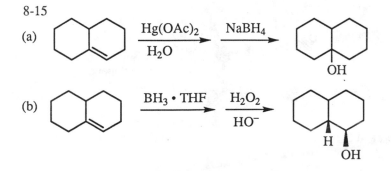

147

8-15 continued

(c)

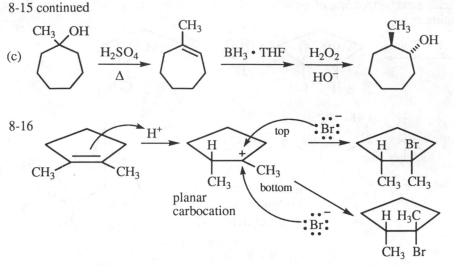

8-16

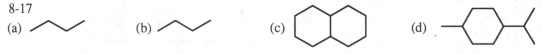

The *planar* carbocation is responsible for non-stereoselectivity. The bromide nucleophile can attack from the top or bottom, leading to a mixture of stereoisomers. The addition is therefore a mixture of syn and anti addition.

8-17

(a) (b) (c) (d)

8-18 Limonene, $C_{10}H_{16}$, has three elements of unsaturation. Upon catalytic hydrogenation, the product, $C_{10}H_{20}$, has one element of unsaturation. Two elements of unsaturation have been removed by hydrogenation—these must have been pi bonds, either two double bonds or one triple bond. The one remaining unsaturation must be a ring. Thus, limonene must have one ring and either two double bonds or one triple bond. (The structure of limonene is shown in the text in Problem 8-17(d), and the hydrogenation product is shown above in the solution to 8-17(d).)

8-19 The BINAP ligand is an example of a conformationally hindered biphenyl as described in text section 5-9A and Figure 5-16. The groups are too large to permit rotation around the single bond connecting the rings, so the molecules are locked into one chiral twist or its mirror image.

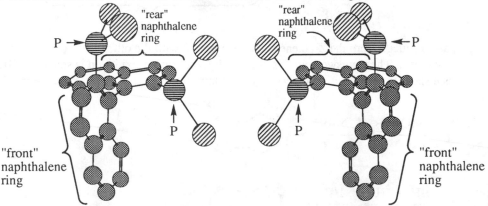

These simplified three-dimensional drawings of the enantiomers show that the two naphthalene rings are twisted almost perpendicular to each other, and the large —P(Ph)$_2$ substituents prevent interconversion of these mirror images.

8-20 Methylene inserts into the circled bonds.

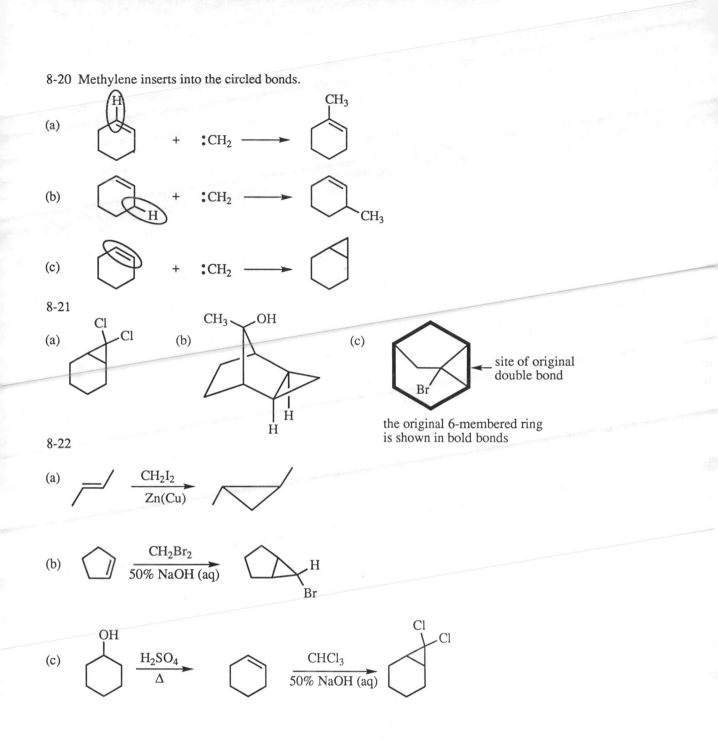

8-21

(a)

(b)

(c)

site of original
double bond

the original 6-membered ring
is shown in bold bonds

8-22

(a)

(b)

(c)

8-23 During bromine addition to either the *cis*- or *trans*-alkene, two new chiral centers are being formed. Neither alkene (nor bromine) is optically active, so the product cannot be optically active.

The *cis*-2-butene gives two chiral products, a racemic mixture. However, *trans*-2-butene, because of it symmetry, gives only one *meso* product which can never be chiral. The "optical *in*activity" is built into this symmetric molecule.

This can be seen by following what happens to the configuration of the chiral centers from the intermediates to products, below. (The key lies in the symmetry of the intermediate and *inversion* of configuration when bromide attacks.)

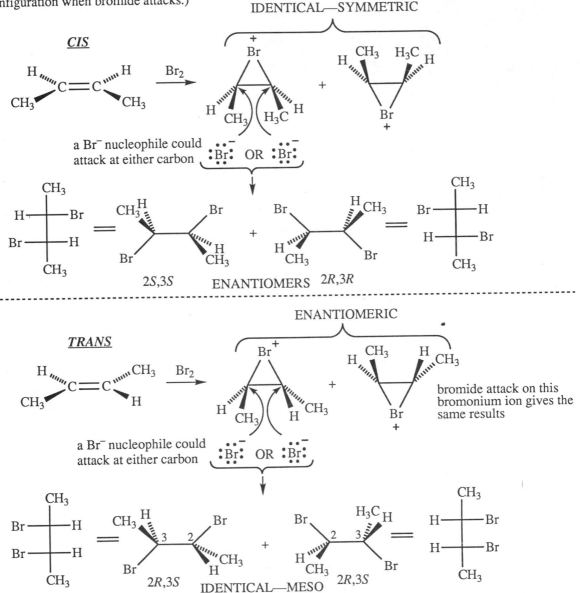

CONCLUSION: anti addition of a symmetric reagent to a symmetric *cis*-alkene gives racemic product, while anti addition to a *trans*-alkene gives meso product. (We will see shortly that syn addition to a *cis*-alkene gives meso product, and syn addition to a *trans*-alkene gives racemic product. Stay tuned.)

8-24 Enantiomers of chiral products are also produced but not shown.

(a)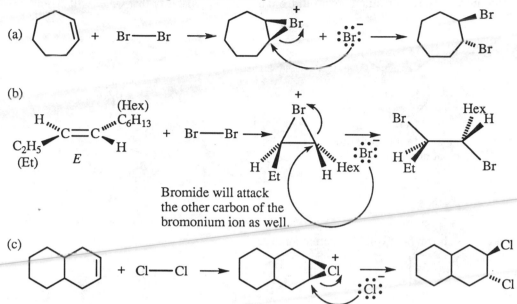

(b)

H$_{\text{""}}$C=C$_{\text{""}}$C$_6$H$_{13}$ (Hex)

C$_2$H$_5$ (Et) *E* H

+ Br—Br →

Bromide will attack
the other carbon of the
bromonium ion as well.

(c)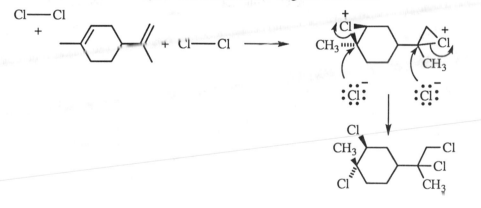

(d) Three new chiral centers are produced in this reaction. All stereoisomers will be produced with the restriction that the two adjacent chlorines on the ring must be *trans*.

Cl—Cl
+

+ Cl—Cl →

8-25 The *trans* product results from water attacking the bromonium ion from the face opposite the bromine. Equal amounts of the two enantiomers result from the equal probability that water will attack either C-1 or C-2.

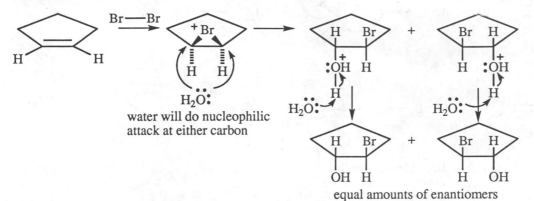

equal amounts of enantiomers

8-26 The chiral products shown here will be racemic mixtures.

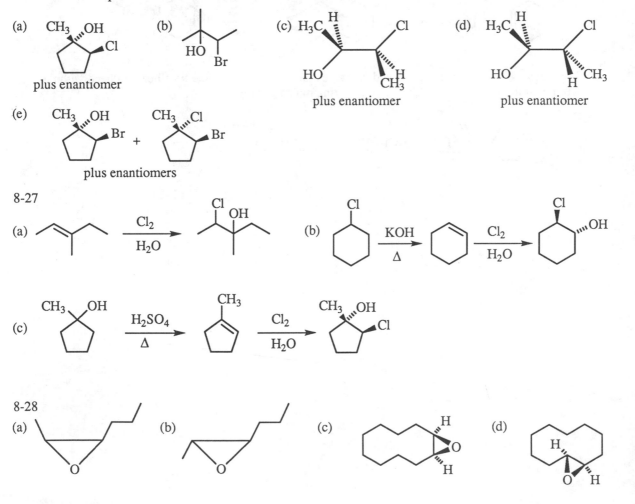

8-29

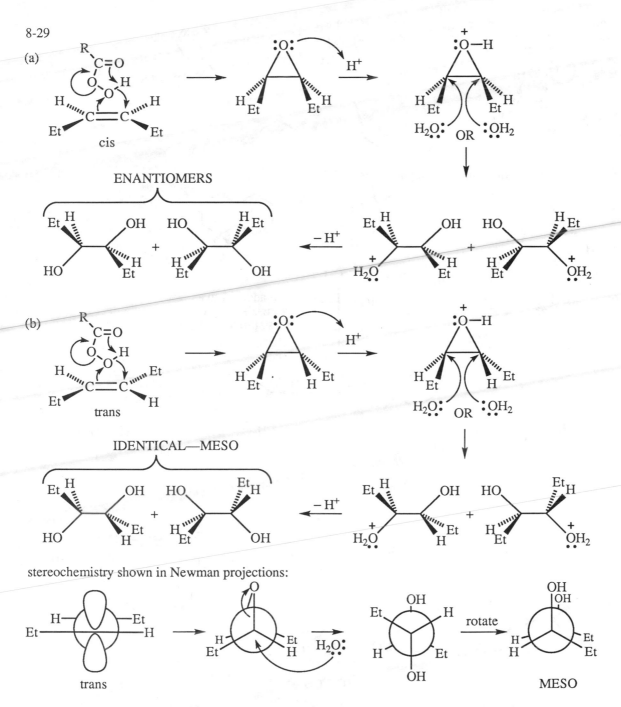

(a)

ENANTIOMERS

(b)

IDENTICAL—MESO

stereochemistry shown in Newman projections:

trans

MESO

Remember the lesson from Problem 8-23: anti addition of a symmetric reagent to a symmetric *cis*-alkene gives racemic product, while anti addition to a *trans*-alkene gives meso product.

8-30

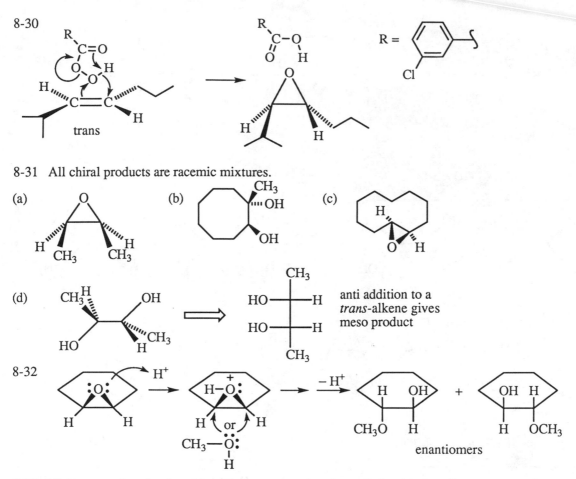

8-31 All chiral products are racemic mixtures.

(a)

(b)

(c)

(d) anti addition to a *trans*-alkene gives meso product

8-32

enantiomers

8-33 All these reactions begin with achiral reagents; therefore, all the chiral products are racemic.

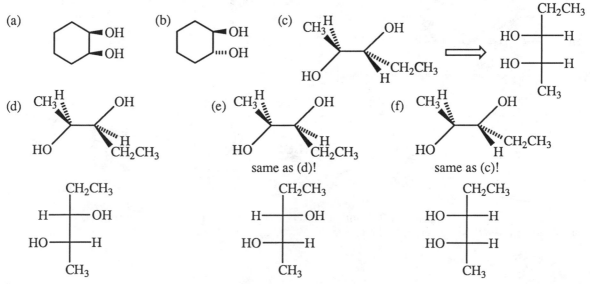

(a)

(b)

(c)

(d)

(e) same as (d)!

(f) same as (c)!

Refer to the observation in the solution to Problem 8-34 on the next page.

8-34

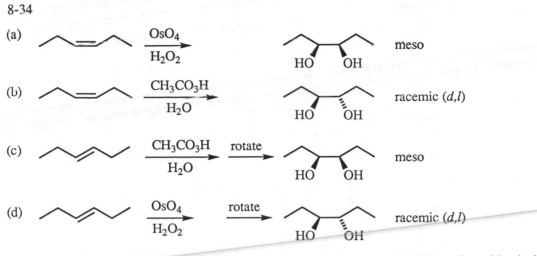

(a) [cis-butene-like] $\xrightarrow[H_2O_2]{OsO_4}$ [diol] meso

(b) $\xrightarrow[H_2O]{CH_3CO_3H}$ [diol] racemic (d,l)

(c) $\xrightarrow[H_2O]{CH_3CO_3H}$ rotate → [diol] meso

(d) $\xrightarrow[H_2O_2]{OsO_4}$ rotate → [diol] racemic (d,l)

Have you noticed yet? For symmetric alkenes and symmetric reagents (addition of two identical X groups):

cis-alkene + **syn** addition → meso

cis-alkene + **anti** addition → racemic

trans-alkene + **syn** addition → racemic

trans-alkene + **anti** addition → meso

> Assume that cis/syn/meso are "same", and trans/anti/racemic are "opposite". Then any combination can be predicted, just like math!
>
> +1 x +1 = +1
> +1 x −1 = −1
> −1 x +1 = −1
> −1 x −1 = +1

8-35 Solve these ozonolysis problems by working backwards, that is, by "reattaching" the two carbons of the new carbonyl groups into alkenes. Here's a hint. When you cut a circular piece of string, you still have only one piece. When you cut a linear piece of string, you have two pieces. Same with molecules. If ozonolysis forms only one product with two carbonyls, the alkene had to have been in a ring. If ozonolysis gives two molecules, the alkene had to have been in a chain.

(a) two carbonyls from ozonolysis are in a chain, so alkene had to have been in a ring

(b) two carbonyls from ozonolysis are in two different products, so alkene had to have been in a chain, not a ring

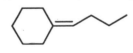

(c) two carbonyls from ozonolysis are in two different products, so alkene had to have been in a chain, not a ring

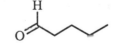

E or Z of alkene cannot be determined from products

8-36

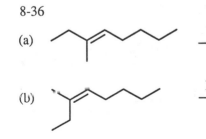

(a) [alkene] $\xrightarrow{O_3}$ $\xrightarrow{Me_2S}$ [ketone] + [aldehyde]

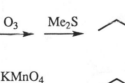

(b) [alkene] $\xrightarrow[\Delta]{KMnO_4}$ [ketone] + [carboxylic acid] (c)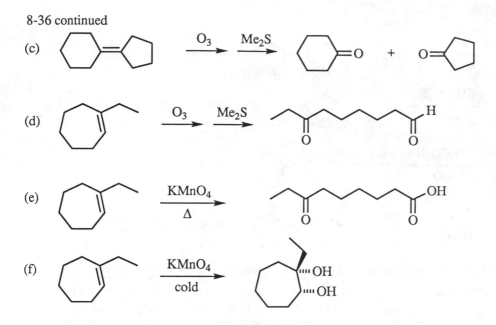

(d)

(e)

(f)

8-37 The representation for a generic acid will be H—B, where B is the conjugate base.

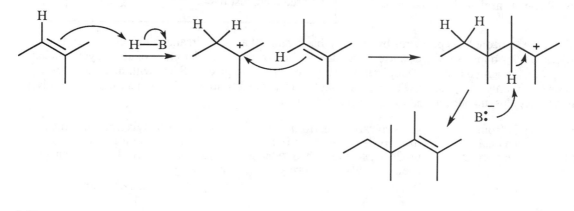

8-38

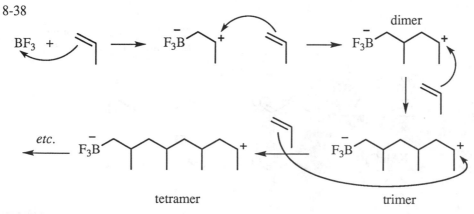

dimer

etc.

tetramer   trimer

8-39

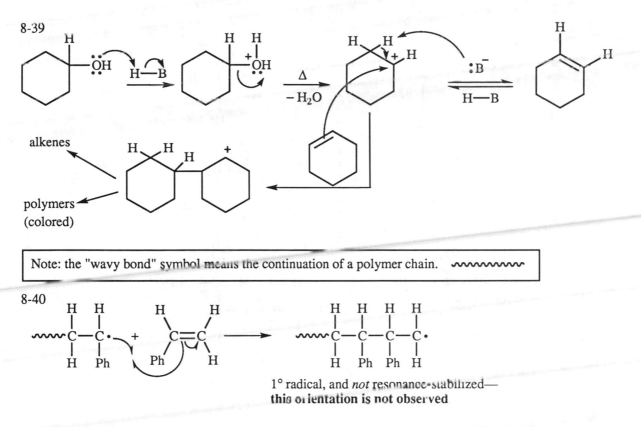

alkenes

polymers
(colored)

Note: the "wavy bond" symbol means the continuation of a polymer chain. ∿∿∿∿∿

8-40

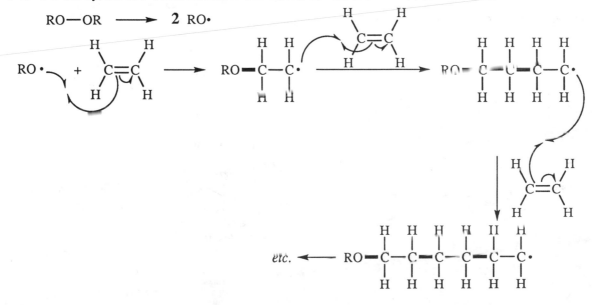

1° radical, and *not* resonance-stabilized—
this orientation is not observed

Orientation of addition always generates the more stable intermediate; the energy difference between a 1° radical (shown above) and a benzylic radical is huge. The phenyl substituents must necessarily be on alternating carbons because the orientation of attack is always the same—not a random process.

8-41 For clarity, the new bonds formed in this mechanism are shown in bold.

8-42

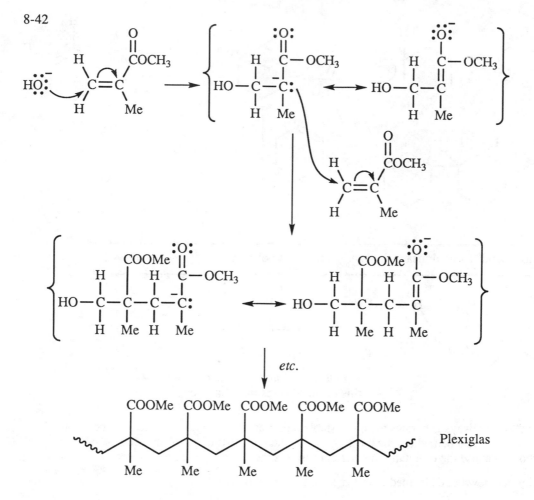

8-43 Please refer to solution 1-19, page 12 of this Solutions Manual.

8-44

(a)

(b)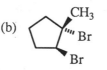

(c) 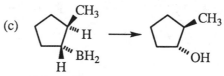 intermediate

(d)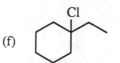

(e)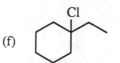

(f)
peroxides do not
affect HCl addition

(g)

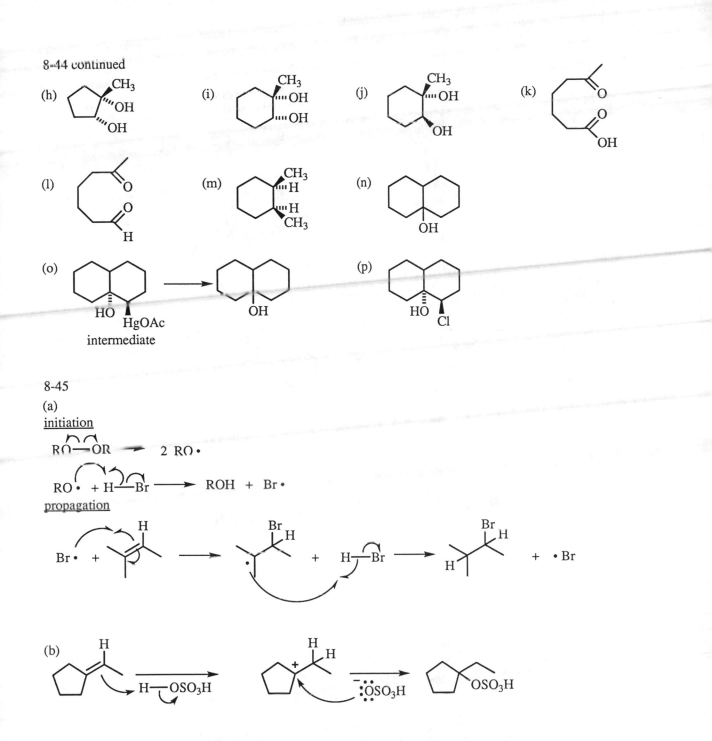

(h)

(i)

(j)

(k)

(l)

(m)

(n)

(o)

intermediate

(p)

8-45

(a)

initiation

propagation

(b)

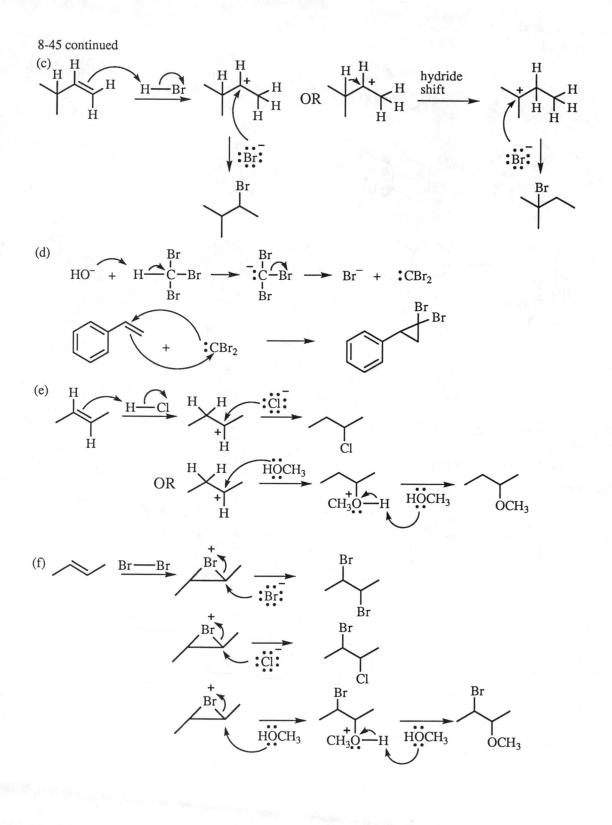

(g)

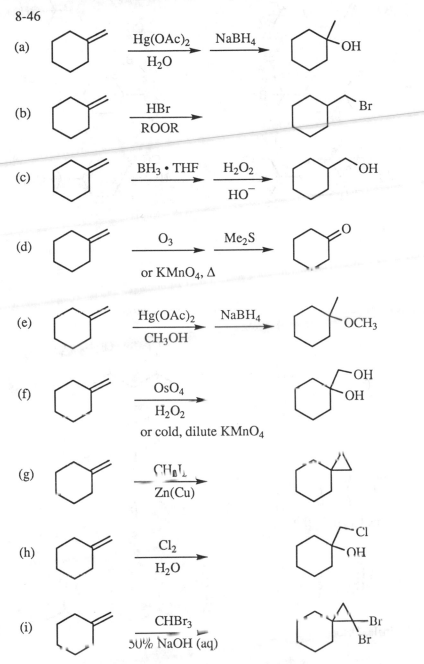

Recall that "Ph" is the abbreviation for phenyl.

8-46

(a) Hg(OAc)$_2$ / H$_2$O   NaBH$_4$ → (product with OH)

(b) HBr / ROOR → (product with Br)

(c) BH$_3$ • THF   H$_2$O$_2$ / HO$^-$ → (product with OH)

(d) O$_3$ Me$_2$S   or KMnO$_4$, Δ → (product with O)

(e) Hg(OAc)$_2$ / CH$_3$OH   NaBH$_4$ → (product with OCH$_3$)

(f) OsO$_4$ / H$_2$O$_2$ or cold, dilute KMnO$_4$ → (product with OH, OH)

(g) CH$_2$I$_2$ / Zn(Cu) → (product)

(h) Cl$_2$ / H$_2$O → (product with Cl, OH)

(i) CHBr$_3$ / 50% NaOH (aq) → (product with Br, Br)

161

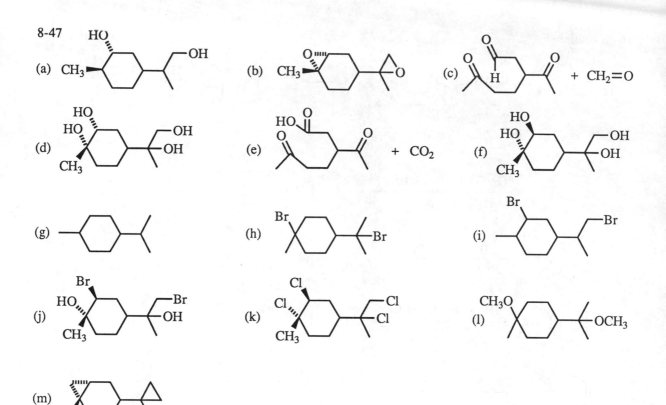

8-47

(a), (b), (c), (d), (e), (f), (g), (h), (i), (j), (k), (l), (m)

8-48 Each monomer has two carbons in the backbone, so the substituents on the monomer will repeat every two carbons in the polymer.

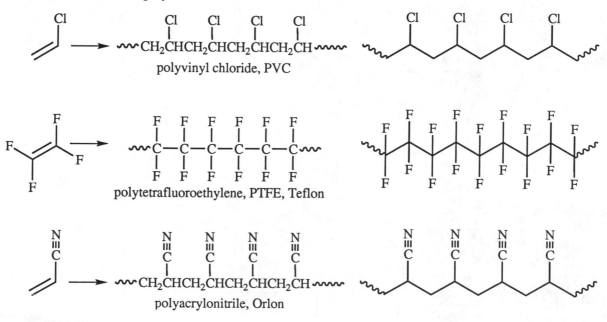

polyvinyl chloride, PVC

polytetrafluoroethylene, PTFE, Teflon

polyacrylonitrile, Orlon

8 49 Without divinylbenzene, individual chains of polystyrene are able to slide past one another. Addition of divinylbenzene during polymerization forms bridges, or "crosslinks", between chains, adding strength and rigidity to the polymer. Divinylbenzene and similar molecules are called crosslinking agents.

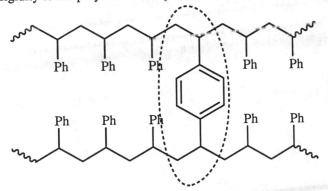

two polystyrene chains crosslinked by a divinylbenzene monomer shown in the dashed oval

8-50 A peroxy radical is shown as the initiator. Newly formed bonds are shown in bold.

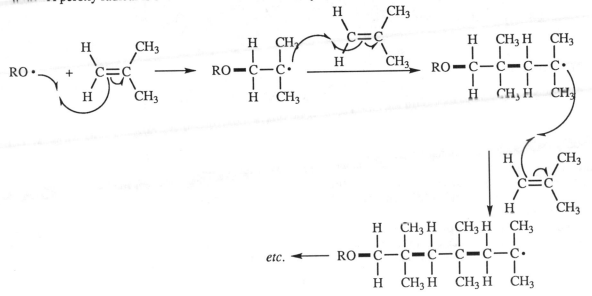

etc.

8-51

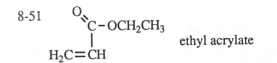

ethyl acrylate

8-52 In each case, the compound (boxed) that produces the more stable carbocation is more reactive.

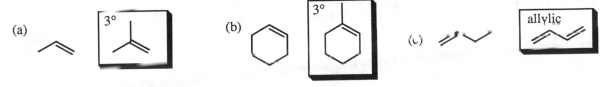

(a) 3°

(b) 3°

(c) allylic

8-53 Once the bromonium ion is formed, it can be attacked by either nucleophile, bromide or chloride, leading to the mixture of products.

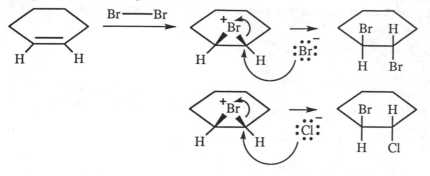

8-54 Two possible orientations of attack of bromine radical are possible:

(A) anti-Markovnikov

(B) Markovnikov

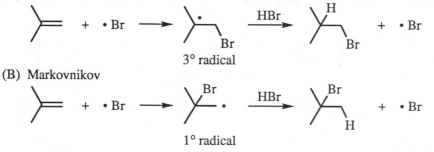

The first step in the mechanism is endothermic and rate determining. The 3° radical produced in anti-Markovnikov attack (A) of bromine radical is several kcal/mole more stable than the 1° radical generated by Markovnikov attack (B). The Hammond Postulate tells us that it is reasonable to assume that the activation energy for anti-Markovnikov addition is lower than for Markovnikov addition. This defines the first half of the energy diagram.

The relative stabilities of the final products are somewhat difficult to predict. (Remember that stability of final products does not necessarily reflect relative stabilities of intermediates; this is why a thermodynamic product can be different from a kinetic product.) From bond dissociation energies (kcal/mole) in Table 4-2:

anti-Markovnikov		Markovnikov	
H to 3° C	91	H to 1° C	98
Br to 1° C	68	Br to 3° C	65
	159 kcal/mole		163 kcal/mole

If it takes more energy to break bonds in the Markovnikov product, it must be lower in energy, therefore, more stable—OPPOSITE OF STABILITY OF THE INTERMEDIATES! Thus, it is the anti-Markovnikov product that is the kinetic product, not the thermodynamic product; the anti-Markovnikov product is obtained since its rate-determining step has the lower activation energy.

Now we are ready to construct the energy diagram; see the next page.

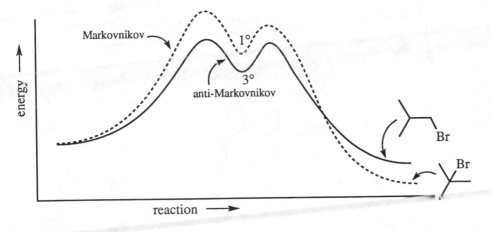

8-55 Recall these facts about ozonolysis: each alkene cleaved by ozone produces two carbonyl groups; an alkene in a chain produces two separate products; an alkene in a ring produces one product in which the two carbonyls are connected.

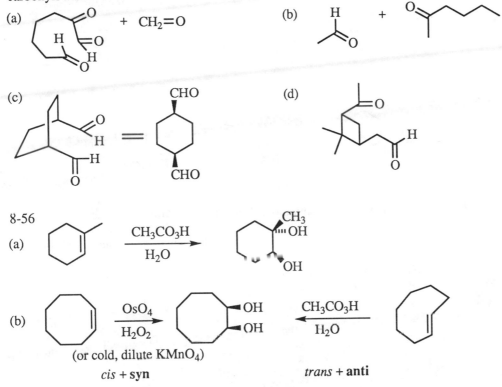

(a)   + CH₂=O

(b)   H   +   O

(c)   CHO
                  =
                    CHO

(d)

8-56

(a)

$$\xrightarrow{\begin{array}{c}CH_3CO_3H\\H_2O\end{array}}$$

(b)

$$\xrightarrow{\begin{array}{c}OsO_4\\H_2O_2\end{array}}$$

(or cold, dilute KMnO₄)

cis + **syn**

OH
OH

$$\xleftarrow{\begin{array}{c}CH_3CO_3H\\H_2O\end{array}}$$

trans + **anti**

165

8-56 continued

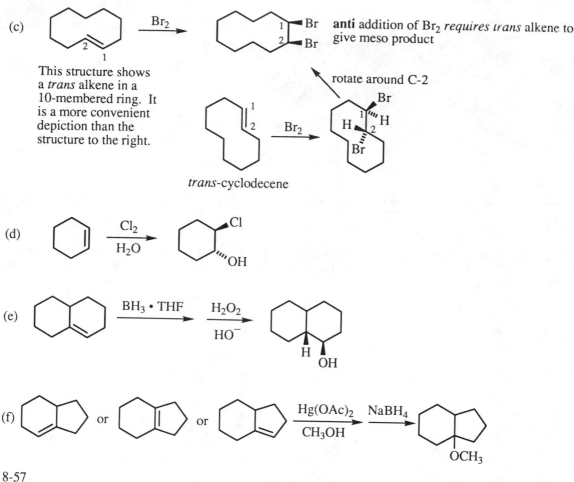

(c) This structure shows a *trans* alkene in a 10-membered ring. It is a more convenient depiction than the structure to the right.

trans-cyclodecene

anti addition of Br_2 *requires trans* alkene to give meso product

rotate around C-2

(d)

(e)

(f)

8-57

A) Unknown X, C_5H_9Br, has one element of unsaturation. X reacts with neither bromine nor $KMnO_4$, so the unsaturation in X cannot be an alkene; it must be a ring.

B) Upon treatment with strong base, X loses H and Br to give Y, C_5H_8, which does react with bromine and $KMnO_4$; it must have an alkene and a ring. Only one isomer is formed.

C) Ozonolysis of Y gives Z, $C_5H_8O_2$, which contains all the original carbons, so the alkene cleaved in the ozonolysis had to be in the ring.

There are several possible answers:

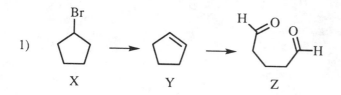

1)

more possibilities on the next page

8-57 continued

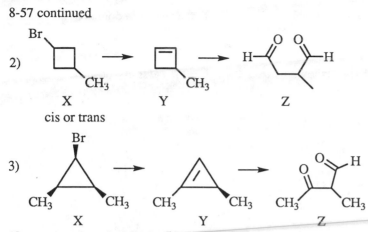

2)

X
cis or trans
Y
Z

3)

X
Y
Z

(On paper, example 3 looks fine. In reality, cyclopropane rings are so strained that they react with bromine or $KMnO_4$.)

Examples of structures that fit the molecular formula for X but not the chemistry

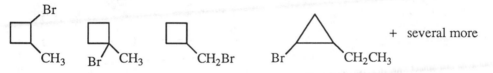

+ several more

8-58 The clue to the structure of α-pinene is the ozonolysis. Working backwards shows the alkene position.

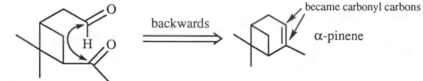

became carbonyl carbons

backwards

α-pinene

after ozonolysis, the two carbonyls are still connected; the alkene must have been in a ring, so reconnect the two carbonyl carbons with a double bond

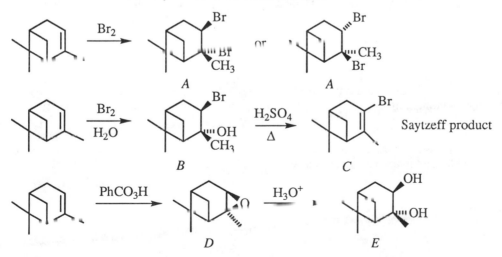

167

8-59 The two products from permanganate oxidation must have been connected by a double bond at the carbonyl carbons. Whether the alkene was *E* or *Z* cannot be determined by this experiment.

$CH_3(CH_2)_{12}CH=CH(CH_2)_7CH_3$

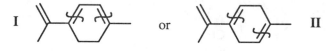

shown here as *Z* which is the naturally-occurring isomer

8-60

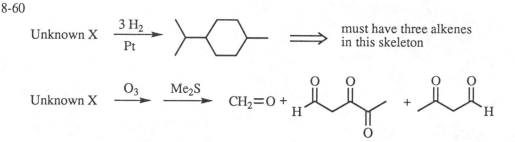

Unknown X $\xrightarrow[\text{Pt}]{\text{3 H}_2}$ [structure] ⟹ must have three alkenes in this skeleton

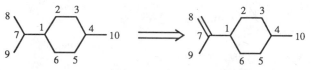

Unknown X $\xrightarrow{O_3} \xrightarrow{Me_2S}$ $CH_2=O$ + [structures]

There are several ways to attack a problem like this. One is the trial-and-error method, that is, put double bonds in all possible positions until the ozonolysis products match. There are times when the trial-and-error method is useful (as in simple problems where the number of possibilities is few), but this is not one of them.

Let's try logic. Analyze the ozonolysis products carefully—what do you see? There are only two methyl groups, so one of the three terminal carbons in the skeleton (C-8, C-9, or C-10) has to be a $=CH_2$. Do we know which terminal carbon has the double bond? Yes, we can deduce that. If C-10 were double-bonded to C-4, then after ozonolysis, C-8 and C-9 must still be attached to C-7. However, in the ozonolysis products, there is no branched chain, that is, no combination of C-8 + C-9 + C-7 + C-1. What if C-7 had a double bond to C-1? Then we would have acetone, CH_3COCH_3, as an ozonolysis product—we don't. Thus, we can't have a double bond from C-4 to C-10. One of the other terminal carbons (C-8) must have a double bond to C-7.

[structures with numbered carbons 8, 9, 7, 1, 2, 3, 4, 5, 6, 10 ⟹ structure]

The other two double bonds have to be in the ring, but where? The products do not have branched chains, so double bonds must appear at both C-1 and C-4. There are only two possibilities for this requirement.

I [structure] or [structure] **II**

Ozonolysis of **I** would give fragments containing one carbon, two carbons, and seven carbons. Ozonolysis of **II** would give fragments containing one carbon, four carbons, and five carbons. Aha! Our mystery structure must be **II**.

(Editorial comment: Science is more than a collection of facts. The application of observation and logic to solve problems by *deduction* and *inference* are critical scientific skills, ones that distinguish humans from algae.)

8-61 In this type of problem, begin by determining which bonds are broken and which are formed. These will always give clues as to what is happening.

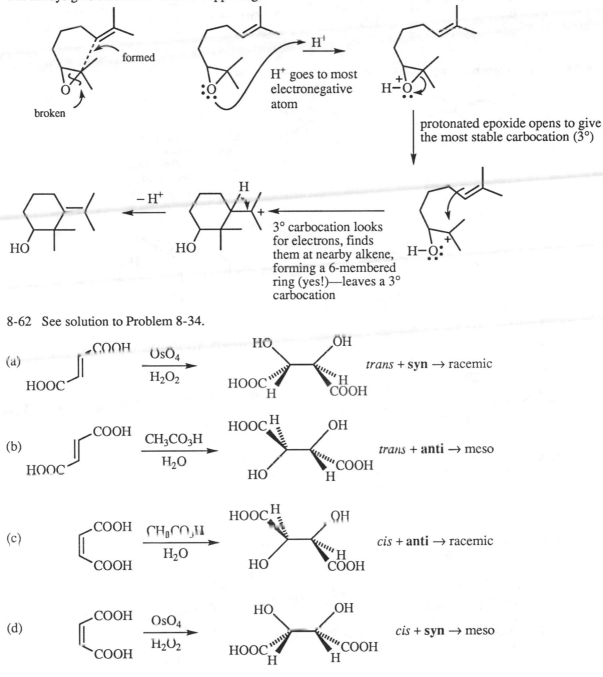

8-62 See solution to Problem 8-34.

(a) $HOOC$ —CH=CH— $COOH$ $\xrightarrow{OsO_4}{H_2O_2}$ product *trans* + **syn** → racemic

(b) $\xrightarrow{CH_3CO_3H}{H_2O}$ product *trans* + **anti** → meso

(c) $\xrightarrow{CH_3CO_3H}{H_2O}$ product *cis* + **anti** → racemic

(d) $\xrightarrow{OsO_4}{H_2O_2}$ product *cis* + **syn** → meso

8-63

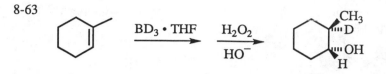

8-64 By now, these rearrangements should not be so "unexpected".

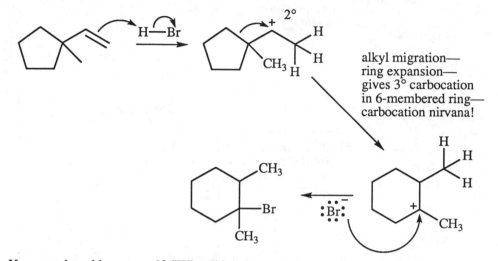

alkyl migration—
ring expansion—
gives 3° carbocation
in 6-membered ring—
carbocation nirvana!

You must be asking yourself, "Why didn't the methyl group migrate?" To which you answered by drawing the carbocation that would have been formed:

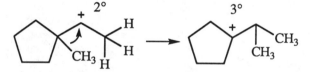

The new carbocation is indeed 3°, but it is only in a 5-membered ring, not quite as stable as in a 6-membered ring. In all probability, some of the product from methyl migration would be formed, but the 6-membered ring would be the major product.

8-65 Each alkene will produce two carbonyls upon ozonolysis or permanganate oxidation. Oxidation of the unknown generated four carbonyls, so the unknown must have had two alkenes. There is only one possibility for their positions.

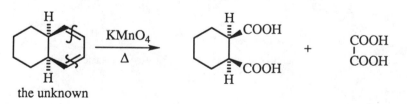

170

8-66

(a) Fumarase catalyzes the addition of H and OH, a hydration reaction.

(b) Fumaric acid is planar and cannot be chiral. Malic acid does have a chiral center and is chiral. The enzyme-catalyzed reaction produces only the S enantiomer, so the product must be optically active.

(c) One of the fundamental rules of stereochemistry is that optically inactive starting materials produce optically inactive products. Sulfuric-acid-catalyzed hydration would produce a racemic mixture of malic acid, that is, equal amounts of R and S.

(d) If the product is optically active, then either the starting materials or the catalyst were chiral. We know that water and fumaric acid are not chiral, so we must infer that fumarase is chiral.

(e) The D and the OD are on the "same side" of the Fischer projection (sometimes called the "erythro" stereoisomer). These are produced from either: (1) syn addition to *cis* alkenes, or (2) anti addition to *trans* alkenes. We know that fumaric acid is *trans*, so the addition of D and OD must necessarily be anti.

(f) Hydroboration is a syn addition.

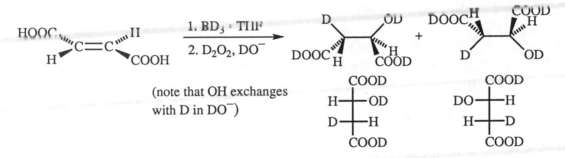

(note that OH exchanges with D in DO⁻)

As expected, *trans* alkene plus syn addition puts the two groups on the "opposite" side of the Fischer projection (sometimes called "threo").

8-67

(a)

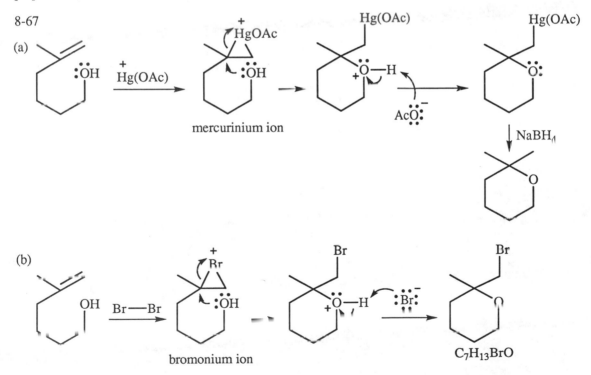

mercurinium ion

(b)

bromonium ion

$C_7H_{13}BrO$

8-68 The addition of BH_3 to an alkene is reversible. Given heat and time, the borane will eventually "walk" its way to the end of the chain through a series of addition-elimination cycles. The most stable alkylborane has the boron on the end carbon; eventually, the series of equilibria lead to that product which is oxidized to the primary alcohol.

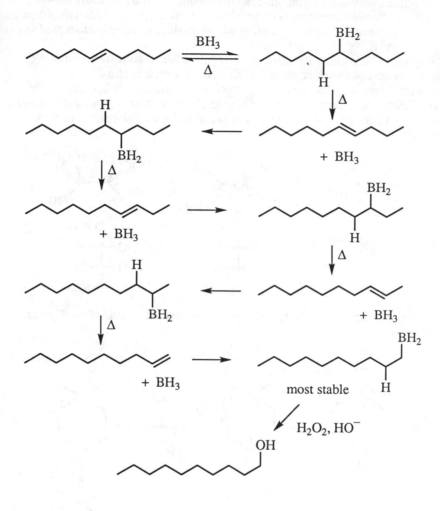

8-69 First, we explain *how* the mixture of stereoisomers results, then *why*.

We have seen many times that the bridged halonium ion permits attack of the nucleophile only from the opposite side.

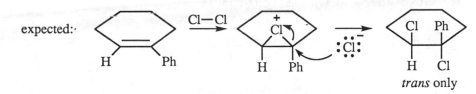

A mixture of *cis* and *trans* could result only if attack of chloride were possible from both top and bottom, something possible only if a *carbocation* existed at this carbon.

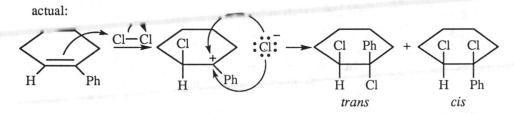

Why does a carbocation exist here? Not only is it 3°, *it is also next to a benzene ring (benzylic) and therefore resonance-stabilized*. This resonance stabilization would be forfeited in a halonium ion intermediate.

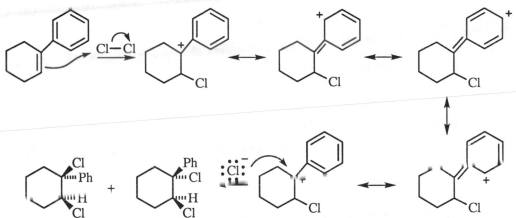

173

9-1 Other structures are possible in each case.

(a) C_6H_{10} $CH_3CH_2C\equiv CCH_2CH_3$ $CH_3CH_2CH_2CH_2C\equiv CH$

(b) C_8H_{12} (cyclohexyl)$-C\equiv CH$ $HC\equiv C-CH_2CH=CHCH_2CH_2CH_3$

(c) C_7H_{10} (cyclopentyl)$-C\equiv CH$ $CH_2=CH-C\equiv C-CH_2CH_2CH_3$

9-2 The asterisk (*) denotes acetylenic hydrogens of terminal alkynes.

(a) $CH_3CH_2-C\equiv C-\overset{*}{H}$ $CH_3-C\equiv C-CH_3$
         1-butyne   2-butyne

(b) $CH_3CH_2CH_2-C\equiv C-\overset{*}{H}$ $CH_3CH_2-C\equiv C-CH_3$ $\overset{CH_3}{\overset{|}{CH_3CH}}-C\equiv C-\overset{*}{H}$
         1-pentyne   2-pentyne   3-methyl-1-butyne

9-3 This reaction is exothermic ($\Delta H° = -56$ kcal/mole) as well as having an increase in entropy. Thermodynamically, at 1500°C, an increase in entropy will have a large effect on ΔG (remember $\Delta G = \Delta H - T\Delta S$?). Kinetically, almost any activation energy barrier will be overcome at 1500°C. Acetylene would likely decompose into its elements:

$$C_2H_2 \xrightarrow{1500° C} 2\ C\ +\ H_2$$

9-4 Adding sodium amide to the mixture will produce the sodium salt of 1-hexyne, leaving 1-hexene untouched. Distillation will remove the 1-hexene, leaving the non-volatile salt behind.

$\left. \begin{array}{l} CH_2=CHCH_2CH_2CH_2CH_3 \\ H-C\equiv C-CH_2CH_2CH_2CH_3 \end{array} \right\} \xrightarrow{NaNH_2} \left\{ \begin{array}{ll} CH_2=CHCH_2CH_2CH_2CH_3 & \\ Na^+\ {}^-{:}C\equiv C-CH_2CH_2CH_2CH_3 & \text{non-volatile salt} \end{array} \right.$

9-5 The key to this problem is to understand that *a proton donor will react only with the conjugate base of a weaker acid*. See Appendix 2 at the end of this Solutions Manual.

(a) $H-C\equiv C-H\ +\ NaNH_2\ \longrightarrow\ H-C\equiv C{:}^-\ Na^+\ +\ NH_3$

(b) $H-C\equiv C-H\ +\ CH_3Li\ \longrightarrow\ H-C\equiv C{:}^-\ Li^+\ +\ CH_4$

(c) no reaction: $NaOCH_3$ is not a strong enough base

(d) no reaction: NaOH is not a strong enough base

(e) $H-C\equiv C{:}^-\ Na^+\ +\ CH_3OH\ \longrightarrow\ H-C\equiv C-H\ +\ NaOCH_3$ (opposite of (c))

(f) $H-C\equiv C{:}^-\ Na^+\ +\ H_2O\ \longrightarrow\ H-C\equiv C-H\ +\ NaOH$ (opposite of (d))

(g) no reaction : $H-C\equiv C{:}^-\ Na^+$ is not a strong enough base

(h) no reaction: $NaNH_2$ is not a strong enough base

(i) $CH_3OH\ +\ NaNH_2\ \longrightarrow\ NaOCH_3\ +\ NH_3$

9-6

$$H-C{\equiv}C-H \xrightarrow{\text{NaNH}_2} H-C{\equiv}C{:}^- \ Na^+ \xrightarrow{\text{CH}_3\text{CH}_2\text{Br}} H-C{\equiv}C-CH_2CH_3$$

$$\downarrow \text{NaNH}_2$$

$$CH_3(CH_2)_5-C{\equiv}C-CH_2CH_3 \xleftarrow{\text{CH}_3(\text{CH}_2)_5\text{Br}} Na^+ \ {:}^-C{\equiv}C-CH_2CH_3$$

9-7

(a) $H-C{\equiv}C-H \xrightarrow[\text{2) CH}_3\text{CH}_2\text{CH}_2\text{CH}_2\text{Br}]{\text{1) NaNH}_2} H-C{\equiv}C-CH_2CH_2CH_2CH_3$

(b) $H-C{\equiv}C-H \xrightarrow[\text{2) CH}_3\text{CH}_2\text{CH}_2\text{Br}]{\text{1) NaNH}_2} H-C{\equiv}C-CH_2CH_2CH_3 \xrightarrow[\text{2) CH}_3\text{I}]{\text{1) NaNH}_2} CH_3-C{\equiv}C-CH_2CH_2CH_3$

(c) $H-C{\equiv}C-H \xrightarrow[\text{2) CH}_3\text{CH}_2\text{Br}]{\text{1) NaNH}_2} H-C{\equiv}C-CH_2CH_3 \xrightarrow[\text{2) CH}_3\text{CH}_2\text{Br}]{\text{1) NaNH}_2} CH_3CH_2-C{\equiv}C-CH_2CH_3$

(d) cannot be synthesized by an S_N2 reaction—would require attack on a 2° alkyl halide

$$CH_3-C{\equiv}C{:}^- \ Na^+ \ + \ \overset{\overset{\displaystyle Br \ \ 2°}{|}}{\underset{\underset{\displaystyle CH_3}{|}}{CHCH_2CH_3}} \ \times\!\!\!\to \ CH_3-C{\equiv}C-\underset{\underset{\displaystyle CH_3}{|}}{CHCH_2CH_3}$$

$$\text{not produced}$$

(e) $H-C{\equiv}C-H \xrightarrow[\text{2) CH}_3\text{I}]{\text{1) NaNH}_2} CH_3-C{\equiv}C-H \xrightarrow[\underset{\underset{\displaystyle CH_3}{|}}{\text{2) BrCH}_2\text{CHCH}_3}]{\text{1) NaNH}_2} CH_3-C{\equiv}C-CH_2\underset{\underset{\displaystyle CH_3}{|}}{CHCH_3}$

(f) $H-C{\equiv}C-H \xrightarrow[\text{2) Br(CH}_2)_8\text{Br}]{\text{1) NaNH}_2}$

$$\underset{Br \diagup\!\!\diagdown\!\!\diagup\!\!\diagdown\!\!\diagup}{H-C{\equiv}C-CH_2}$$

$$\downarrow \text{NaNH}_2$$

$$\underset{Br\diagup\!\!\diagdown\!\!\diagup\!\!\diagdown\!\!\diagup}{Na^+ \ {:}^-C{\equiv}C-CH_2}$$

$$H_2C-C{\equiv}C-CH_2 \longleftarrow$$

Intramolecular cyclization of large rings must be carried out in dilute solution so the last S_N2 displacement will be *intra*molecular and not *inter*molecular.

9-8

(a) $H-C\equiv C-H$ $\xrightarrow[\text{2) CH}_3\text{I}]{\text{1) NaNH}_2}$ $CH_3-C\equiv C-H$ $\xrightarrow[\substack{\text{2) HCCH}_2\text{CH}_2\text{CH}_3\\ \text{3) H}_2\text{O}}]{\text{1) NaNH}_2}$ $CH_3-C\equiv C-\underset{\text{OH}}{\underset{|}{C}}HCH_2CH_2CH_3$

(b) $H-C\equiv C-H$ $\xrightarrow[\text{2) H}_2\text{C=O}]{\text{1) NaNH}_2}$ $\xrightarrow{\text{H}_2\text{O}}$ $H-C\equiv C-CH_2OH$

(c) $H-C\equiv C-H$ $\xrightarrow[\text{2) Ph}\overset{\text{O}}{\underset{}{\diagdown}}\text{CH}_3]{\text{1) NaNH}_2}$ $\xrightarrow{\text{H}_2\text{O}}$ $H_3C-\underset{\text{Ph}}{\underset{|}{\overset{\text{OH}}{\overset{|}{C}}}}-C\equiv C-H$

(d) $H-C\equiv C-H$ $\xrightarrow[\text{2) CH}_3\text{I}]{\text{1) NaNH}_2}$ $CH_3-C\equiv C-H$ $\xrightarrow[\substack{\text{2) CH}_3\text{CCH}_2\text{CH}_3\\ \text{3) H}_2\text{O}}]{\text{1) NaNH}_2}$ $CH_3-C\equiv C-\underset{\text{CH}_3}{\underset{|}{\overset{\text{OH}}{\overset{|}{C}}}}CH_2CH_3$

9-9

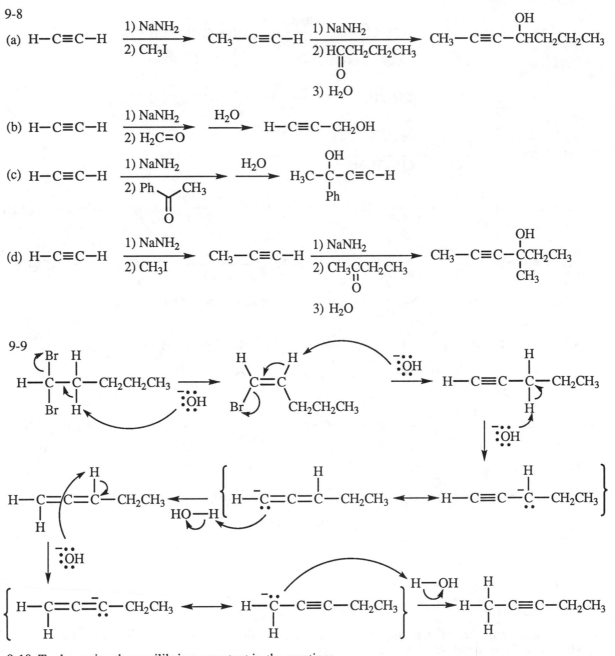

9-10 To determine the equilibrium constant in the reaction:

terminal alkyne $\rightleftharpoons$ internal alkyne $\qquad \Delta G = -RT \ln K_{eq}$

$\Delta G = -4.0$ kcal/mole
$ (-17.0$ kJ/mole$)$

$$K_{eq} = e^{\left(-\frac{\Delta G}{RT}\right)} = e^{\left(\frac{-(-4.0)}{(0.00198)(473)}\right)} = e^{4.27} = 72$$

$$\frac{[\text{internal}]}{[\text{terminal}]} = \frac{72}{1} = \frac{98.6\% \text{ internal}}{1.4\% \text{ terminal}}$$

176

9-11 (a) This isomerization is the reverse of the mechanism in the solution to 9-9.

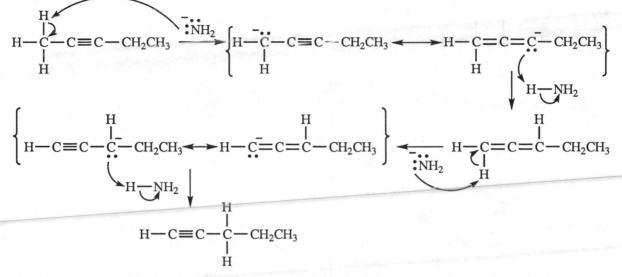

(b) All steps in part (a) are reversible. With a weaker base like KOH, an equilibrium mixture of 1-pentyne and 2-pentyne would result. With the strong base $NaNH_2$, however, the final terminal alkyne is deprotonated to give the acetylide ion:

$$H-C\equiv C-CH_2CH_2CH_3 \; + \; NaNH_2 \longrightarrow Na^+ \; \overset{-}{:}C\equiv C-CH_2CH_2CH_3 \; + \; NH_3$$

Because 1-pentyne is about 10 pK units more acidic than ammonia, this deprotonation is *not reversible*. The acetylide ion is produced and can't go back. Le Châtelier's Principle tells us that the reaction will try to replace the 1-pentyne that is being removed from the reaction mixture, so eventually all of the 2-pentyne will be drawn into the 1-pentyne anion "sink".

(c) Using the weaker base KOH at 200°C will restore the equilibrium between the two alkyne isomers with 2-pentyne predominating.

9-12

(a)

(1)
$$H_3C-\underset{\underset{H}{|}}{\overset{\overset{Br}{|}}{C}}-\underset{\underset{H}{|}}{\overset{\overset{Br}{|}}{C}}-CH_3 \quad \xrightarrow[\Delta]{KOH} \quad H_3C-C\equiv C-CH_3$$

(2)
$$CH_3CH=CHCH_3 \quad \xrightarrow{Br_2, \; CCl_4} \quad CH_3\underset{\underset{Br}{|}}{CH}-\underset{\underset{Br}{|}}{CH}CH_3$$

(b)

(1)
$$H-\underset{\underset{H}{|}}{\overset{\overset{Br}{|}}{C}}-\underset{\underset{H}{|}}{\overset{\overset{Br}{|}}{C}}-(CH_2)_5CH_3 \quad \xrightarrow[150°]{NaNH_2} \quad H-C\equiv C-(CH_2)_5CH_3$$

(2)
$$CH_2=CH(CH_2)_5CH_3 \quad \xrightarrow{Br_2, \; CCl_4} \quad H_2\underset{\underset{Br}{|}}{C}-\underset{\underset{Br}{|}}{C}H(CH_2)_5CH_3$$

9-12 continued

(c)

(1)

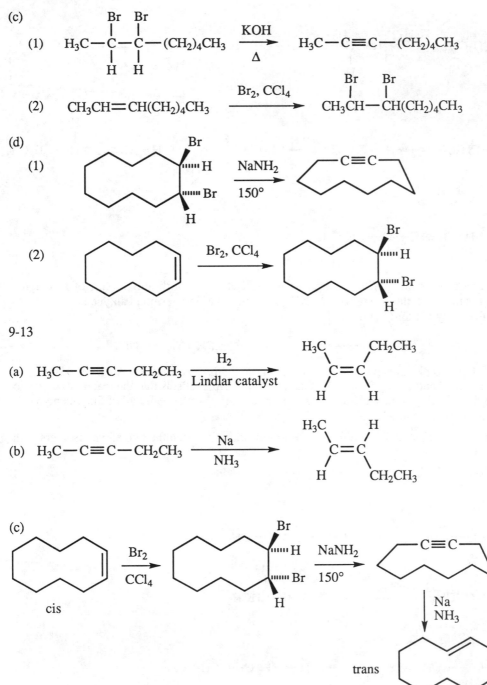

$$H_3C-\underset{\underset{H}{|}}{\overset{\overset{Br}{|}}{C}}-\underset{\underset{H}{|}}{\overset{\overset{Br}{|}}{C}}-(CH_2)_4CH_3 \xrightarrow[\Delta]{KOH} H_3C-C{\equiv}C-(CH_2)_4CH_3$$

(2) $CH_3CH{=}CH(CH_2)_4CH_3 \xrightarrow{Br_2, CCl_4} CH_3\overset{\overset{Br}{|}}{CH}-\overset{\overset{Br}{|}}{CH}(CH_2)_4CH_3$

(d)

(1) $\xrightarrow[150°]{NaNH_2}$

(2) $\xrightarrow{Br_2, CCl_4}$

9-13

(a) $H_3C-C{\equiv}C-CH_2CH_3 \xrightarrow[\text{Lindlar catalyst}]{H_2}$

(b) $H_3C-C{\equiv}C-CH_2CH_3 \xrightarrow[NH_3]{Na}$

(c) cis $\xrightarrow[CCl_4]{Br_2}$ $\xrightarrow[150°]{NaNH_2}$ $\xrightarrow[NH_3]{Na}$ trans

178

9-13 continued

(d)

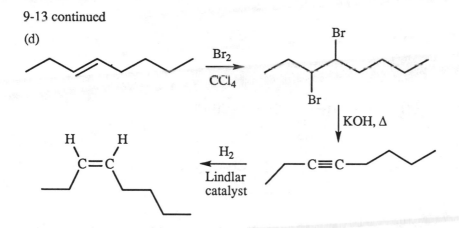

9-14 The goal is to add only one equivalent of bromine, always avoiding an excess of bromine. If the alkyne is added to the bromine, the first drops of alkyne will encounter a large excess of bromine. Instead, adding bromine to the alkyne will always ensure an excess of alkyne and should give a good yield of dibromo product.

9-15

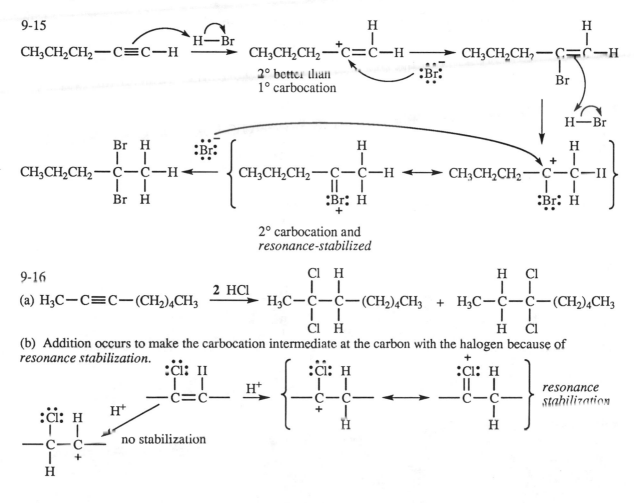

9-16

(a) $H_3C-C\equiv C-(CH_2)_4CH_3$ $\xrightarrow{\text{2 HCl}}$

(b) Addition occurs to make the carbocation intermediate at the carbon with the halogen because of *resonance stabilization.*

9-17

initiation:

$$RO-OR \xrightarrow{h\nu} 2\ RO\cdot$$

$$RO\cdot + H-Br \longrightarrow RO-H + Br\cdot$$

propagation:

$Br\cdot + H-C\equiv C-CH_2CH_2CH_3 \longrightarrow H-C=C-CH_2CH_2CH_3$
with Br substituent

$H-C=C-CH_2CH_2CH_3 + H-Br \longrightarrow H-C=C-CH_2CH_2CH_3 + Br\cdot$
with Br and H

The 2° radical is more stable than 1°. The anti-Markovnikov orientation occurs because the bromine radical attacks first, rather than the H^+ attacking first as in electrophilic (ionic) addition.

9-18

(a) $H-C\equiv C-(CH_2)_3CH_3 \xrightarrow{1\ eq.\ Cl_2} H-C=C-(CH_2)_3CH_3$ $E + Z$
with Cl, Cl

(b) $H-C\equiv C-(CH_2)_3CH_3 \xrightarrow[ROOR]{HBr} H-C=C-(CH_2)_3CH_3$ $E + Z$
with Br, H

(c) $H-C\equiv C-(CH_2)_3CH_3 \xrightarrow{HBr} H-C=C-(CH_2)_3CH_3$
with H, Br

(d) $H-C\equiv C-(CH_2)_3CH_3 \xrightarrow[CCl_4]{2\ Br_2} H-C-C-(CH_2)_3CH_3$
with Br, Br / Br, Br

(e) $H-C\equiv C-(CH_2)_3CH_3 \xrightarrow[\substack{Lindlar\\catalyst\\(or\ Na,\ NH_3)}]{H_2} H-C=C-(CH_2)_3CH_3 \xrightarrow{HBr} CH_3-CH-(CH_2)_3CH_3$
with H, H ; Br

(f) $H-C\equiv C-(CH_2)_3CH_3 \xrightarrow{2\ HBr} H-C-C-(CH_2)_3CH_3$
with H, Br / H, Br

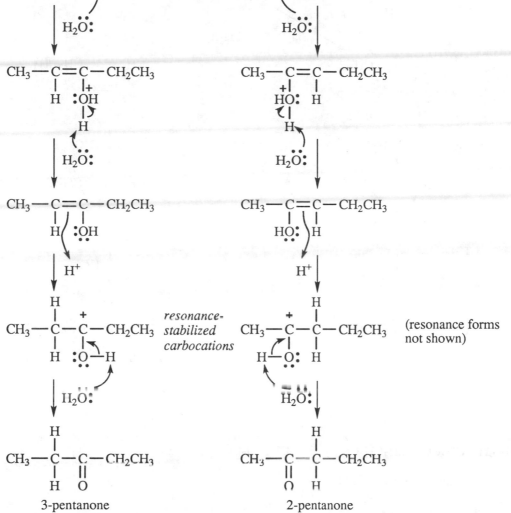

The role of the mercury catalyst is not shown in this mechanism. As a Lewis acid, it may act like the proton in the first step, helping to form vinyl cations.

9-20

(a) 2-Butyne is symmetric. Either orientation produces the same product.

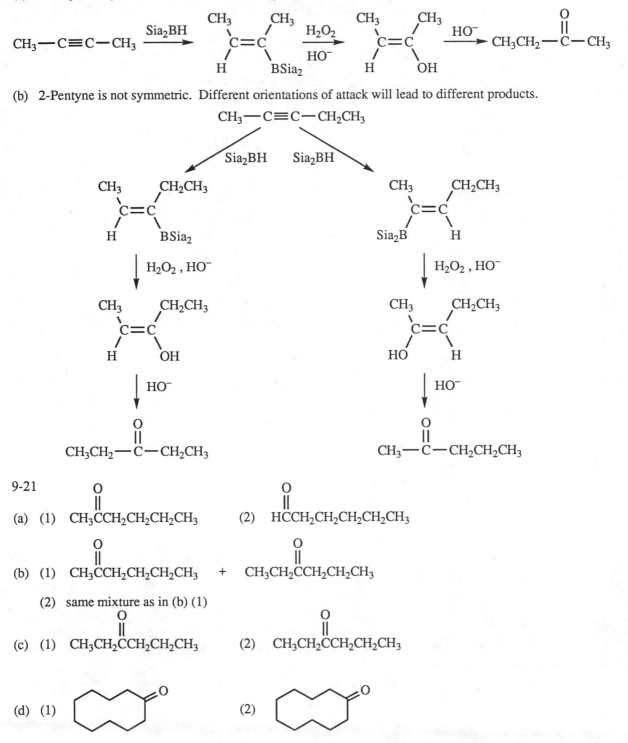

(b) 2-Pentyne is not symmetric. Different orientations of attack will lead to different products.

9-21

(a) (1) $CH_3\overset{O}{\overset{||}{C}}CH_2CH_2CH_2CH_3$ (2) $H\overset{O}{\overset{||}{C}}CH_2CH_2CH_2CH_2CH_3$

(b) (1) $CH_3\overset{O}{\overset{||}{C}}CH_2CH_2CH_2CH_3$ + $CH_3CH_2\overset{O}{\overset{||}{C}}CH_2CH_2CH_3$

    (2) same mixture as in (b) (1)

(c) (1) $CH_3CH_2\overset{O}{\overset{||}{C}}CH_2CH_2CH_3$ (2) $CH_3CH_2\overset{O}{\overset{||}{C}}CH_2CH_2CH_3$

(d) (1)   (2)

9-22

(a)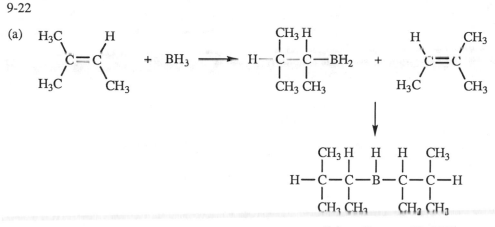

disiamylborane, Sia$_2$BH

(b) There is too much steric hindrance in Sia$_2$BH for the third B—H to add across another alkene. The reagent can add to alkynes because alkynes are linear and attack is not hindered by bulky substituents.

9-23

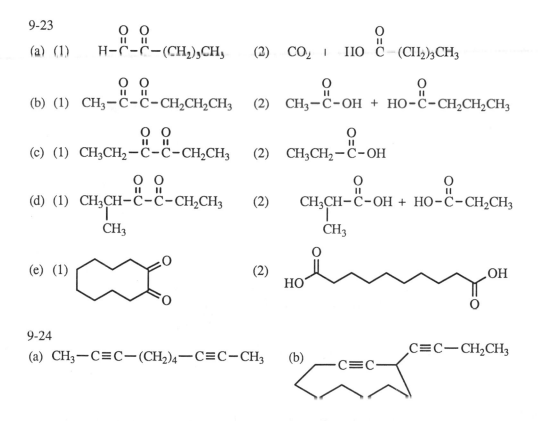

9-24

(a) $CH_3-C\equiv C-(CH_2)_4-C\equiv C-CH_3$ (b)

9-25 When proposing syntheses, begin by analyzing the target molecule, looking for smaller pieces that can be combined to make the desired compound. This is especially true for targets that have more carbons than the starting materials; immediately, you will know that a carbon-carbon bond forming reaction will be necessary.

People who succeed at synthesis *know the reactions*—there is no shortcut. Practice the reactions for each functional group until they become automatic.

(a) analysis of target

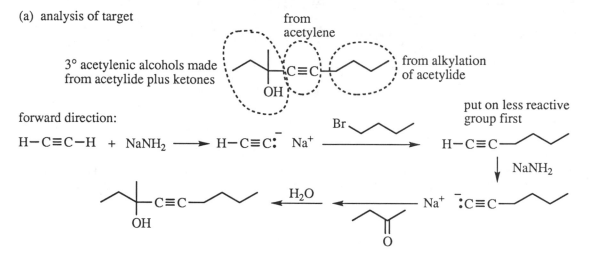

3° acetylenic alcohols made from acetylide plus ketones

from acetylene

from alkylation of acetylide

forward direction:

put on less reactive group first

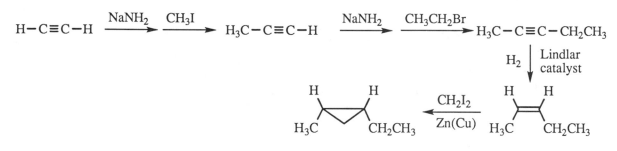

(b) analysis of target: cyclopropanes are made by carbene insertion into alkenes; stereochemistry of alkene must be cis to get cis substitution around cyclopropane; cis alkene comes from catalytic hydrogenation of an alkyne

$$H-C\equiv C-H \xrightarrow{\text{NaNH}_2 \quad \text{CH}_3\text{I}} H_3C-C\equiv C-H \xrightarrow{\text{NaNH}_2 \quad \text{CH}_3\text{CH}_2\text{Br}} H_3C-C\equiv C-CH_2CH_3$$

$$\xrightarrow[\text{catalyst}]{\text{H}_2 \mid \text{Lindlar}}$$

$$\xleftarrow[\text{Zn(Cu)}]{\text{CH}_2\text{I}_2}$$

(c) analysis of target: epoxides are made by direct epoxidation of alkenes; stereochemistry of alkene must be trans to get trans substitution around epoxide; trans alkene comes from sodium/ammonia reduction of an alkyne

$$H-C\equiv C-H \xrightarrow{\text{NaNH}_2 \quad \text{CH}_3\text{CH}_2\text{Br}} CH_3CH_2-C\equiv C-H \xrightarrow{\text{NaNH}_2 \quad \text{CH}_3\text{CH}_2\text{CH}_2\text{Br}}$$

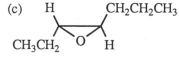

PhCOOH

$$\xleftarrow[\text{NH}_3]{\text{Na}} CH_3CH_2-C\equiv C-CH_2CH_2CH_3$$

9-26 Please refer to solution 1-19, page 12 of this Solutions Manual.

9-27

(a) $CH_3CH_2-C\equiv C-(CH_2)_4CH_3$ (b) $H_3C-C\equiv C-(CH_2)_4CH_3$ (c) $-C\equiv C-H$

(d) $-C\equiv C-H$ (e) $CH_3CH_2-C\equiv C-\underset{\underset{CH_3}{|}}{CH}CH_2CH_2CH_3$

(f)

(g) $CH_3\underset{\underset{OH}{|}}{CH}-C\equiv C-(CH_2)_3CH_3$

(h) $\underset{\underset{H}{|}}{H_3C}C=C\underset{\underset{H}{|}}{\overset{C\equiv C-CHCH_2CH_3}{\underset{CH_2CH_3}{|}}}$

(i) $H-C\equiv C-CH_2-C\equiv C-CH_2CH_3$ (j) $H-C\equiv C-CH=CH_2$ (k) $H-C\equiv C-\overset{CH_3}{\underset{\underset{H}{|}}{\overset{|}{C}}}\overset{H}{\underset{\,}{}}C=CH_2$

9-28

(a) ethylmethylacetylene   (c) *sec*-butyl-*n*-propylacetylene
(b) phenylacetylene   (d) *sec*-butyl-*t*-butylacetylene

9-29

(a) 4-phenyl-2-pentyne   (c) 2,6,6-trimethyl-3-heptyne   (e) 3-methyl-4-hexyn-3-ol
(b) 4,4-dibromo-2-pentyne (d) (*E*)-3-methyl-2-hepten-4-yne (f) 1-cycloheptylpropyne

9-30

(a)   terminal alkynes   internal alkynes

$H-C\equiv C-(CH_2)_3CH_3$ $CH_3-C\equiv C-CH_2CH_2CH_3$
     1-hexyne   2-hexyne

$H-C\equiv C-\underset{\underset{CH_3}{|}}{CH}CH_2CH_3$ $CH_3CH_2-C\equiv C-CH_2CH_3$
  3-methyl-1-pentyne   3-hexyne

$H-C\equiv C-CH_2\underset{\underset{CH_3}{|}}{CH}CH_3$ $CH_3-C\equiv C-\underset{\underset{CH_3}{|}}{CH}CH_3$
  4-methyl-1-pentyne   4-methyl-2-pentyne

$H-C\equiv C-\overset{\overset{CH_3}{|}}{\underset{\underset{CH_3}{|}}{C}}-CH_3$
  3,3-dimethyl-1-butyne

(b) All four terminal alkynes will form precipitates with cuprous ion.

185

9-31

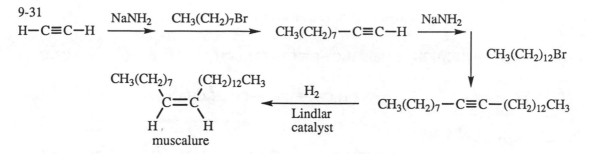

9-32 Terminal alkynes form precipitates with silver ion or cuprous ion. Filtration will remove the 1-decyne precipitate from 2-decyne.

$$H-C\equiv C-(CH_2)_7CH_3 \ + \ Cu^+ \ \xrightarrow{NH_3} \ Cu-C\equiv C-(CH_2)_7CH_3$$
$$\text{precipitate}$$

9-33

(a) $CH_2=\overset{\overset{\displaystyle Cl}{|}}{C}CH_2CH_2CH_3$

(b) $H_3C-\overset{\overset{\displaystyle Cl}{|}}{\underset{\underset{\displaystyle Cl}{|}}{C}}-CH_2CH_2CH_3$

(c) $CH_3CH_2CH_2CH_2CH_3$

(d) $CH_2=CHCH_2CH_2CH_3$

(e) $\underset{Br}{\overset{H}{\diagdown}}C=C\underset{CH_2CH_2CH_3}{\overset{Br}{\diagup}}$

(f) $H-\overset{\overset{\displaystyle Br}{|}}{\underset{\underset{\displaystyle Br}{|}}{C}}-\overset{\overset{\displaystyle Br}{|}}{\underset{\underset{\displaystyle Br}{|}}{C}}-CH_2CH_2CH_3$

(g) $H-\overset{\overset{\displaystyle O}{||}}{C}-\overset{\overset{\displaystyle O}{||}}{C}-CH_2CH_2CH_3$

(h) $CO_2 \ + \ HO-\overset{\overset{\displaystyle O}{||}}{C}-CH_2CH_2CH_3$

(i) $H_2C=CHCH_2CH_2CH_3$

(j) $Na^+ \ \ ^-\!:C\equiv C-CH_2CH_2CH_3$

(k) $Ag-C\equiv C-CH_2CH_2CH_3$

(l) $H_3C-\overset{\overset{\displaystyle O}{||}}{C}-CH_2CH_2CH_3$

(m) $H-\overset{\overset{\displaystyle O}{||}}{C}-CH_2CH_2CH_2CH_3$

9-34

(a) $H_3C-\overset{\overset{\displaystyle Br}{|}}{\underset{\underset{\displaystyle Br}{|}}{C}}-CH_2CH_3 \ \xrightarrow[150°]{NaNH_2} \ \xrightarrow{H_2O} \ H-C\equiv C-CH_2CH_3$

(b) $H_3C-\overset{\overset{\displaystyle Br}{|}}{\underset{\underset{\displaystyle Br}{|}}{C}}-CH_2CH_3 \ \xrightarrow[200°]{KOH} \ H_3C-C\equiv C-CH_3$

(c) $CH_3CH_2-C\equiv C-H \ \xrightarrow{NaNH_2} \ CH_3CH_2-C\equiv C:^- \ Na^+$

$\Big\downarrow CH_3CH_2CH_2CH_2Br$

$CH_3CH_2-C\equiv C-CH_2CH_2CH_2CH_3$

9-34 continued

(d)
$$H_3C \atop H \!\!\!> C=C <\!\!\! {H \atop CH_2CH_2CH_3} \xrightarrow[CCl_4]{Br_2} CH_3\overset{Br}{\underset{H}{C}}-\overset{Br}{\underset{H}{C}}CH_2CH_2CH_3 \xrightarrow[200°]{KOH} H_3C-C\equiv C-CH_2CH_2CH_3$$

(e)
$$H_3C \atop H \!\!\!> C=C <\!\!\! {H \atop CH_2CH_2CH_3} \xrightarrow[CCl_4]{Br_2} CH_3\overset{Br}{\underset{H}{C}}-\overset{Br}{\underset{H}{C}}CH_2CH_2CH_3 \xrightarrow[150°]{NaNH_2} H-C\equiv C-CH_2CH_2CH_2CH_3$$

(f)
$$\xrightarrow[\substack{Lindlar \\ catalyst}]{H_2}$$
cis

(g)
$$\xrightarrow[NH_3]{Na}$$
trans

(h)
$$H-C\equiv C-CH_2CH_2CH_2CH_3 \xrightarrow[\substack{H_2SO_4 \\ HgSO_4}]{H_2O} \left[CH_2=\overset{OH}{C}CH_2CH_2CH_2CH_3 \atop \text{unstable enol} \right] \longrightarrow H_3C-\overset{O}{C}CH_2CH_2CH_2CH_3$$

(i)
$$H-C\equiv C-CH_2CH_2CH_2CH_3 \xrightarrow[\substack{2)\ H_2O_2, \\ HO^-}]{1)\ Sia_2BH} \left[\overset{OH}{CH}=CHCH_2CH_2CH_2CH_3 \atop \text{unstable enol} \right] \longrightarrow H-\overset{O}{C}CH_2CH_2CH_2CH_2CH_3$$

(j)
$$H_3C \atop H \!\!\!> C=C <\!\!\! {H \atop CH_2CH_2CH_3} \xrightarrow[CCl_4]{Br_2} CH_3\overset{Br}{\underset{H}{C}}-\overset{Br}{\underset{H}{C}}CH_2CH_2CH_3 \xrightarrow[200°]{KOH} H_3C-C\equiv C-CH_2CH_2CH_3$$

$$\xrightarrow[\substack{Lindlar \\ catalyst}]{H_L} $$

$$H_3C \atop H \!\!\!> C=C <\!\!\! {CH_2CH_2CH_3 \atop H}$$

9-35

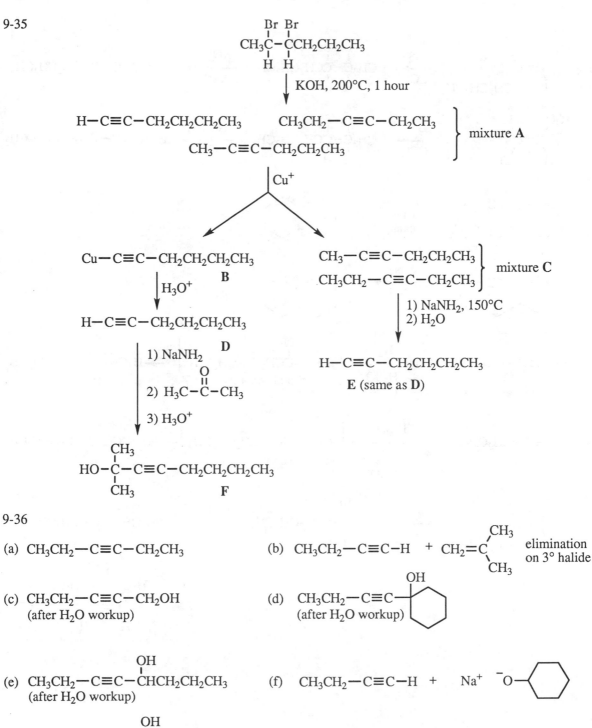

9-36

(a) $CH_3CH_2-C{\equiv}C-CH_2CH_3$

(b) $CH_3CH_2-C{\equiv}C-H$ + $CH_2{=}C\begin{smallmatrix}CH_3\\\\CH_3\end{smallmatrix}$ elimination on 3° halide

(c) $CH_3CH_2-C{\equiv}C-CH_2OH$
(after H_2O workup)

(d) $CH_3CH_2-C{\equiv}C-$ [cyclohexane with OH]
(after H_2O workup)

(e) $CH_3CH_2-C{\equiv}C-\overset{OH}{\underset{}{C}}HCH_2CH_2CH_3$
(after H_2O workup)

(f) $CH_3CH_2-C{\equiv}C-H$ + $Na^+\ {}^-O-$[cyclohexane]

(g) $CH_3CH_2-C{\equiv}C-\overset{OH}{\underset{CH_3}{C}}-CH_2CH_3$
(after H_2O workup)

9-37

(a) $HC\equiv C-H$ $\xrightarrow{NaNH_2}$ $HC\equiv C:^-$ Na^+ $\xrightarrow{CH_3CH_2CH_2CH_2Br}$ $HC\equiv C-CH_2CH_2CH_2CH_3$

(b) $HC\equiv C-H$ $\xrightarrow{NaNH_2}$ $HC\equiv C:^-$ Na^+ $\xrightarrow{CH_3CH_2CH_2Br}$ $HC\equiv C-CH_2CH_2CH_3$

$\xrightarrow[CH_3I]{NaNH_2}$

$H_3C-C\equiv C-CH_2CH_2CH_3$

(c) $H_3C-C\equiv C-CH_2CH_2CH_3$ from (b) $\xrightarrow[\text{Lindlar catalyst}]{H_2}$

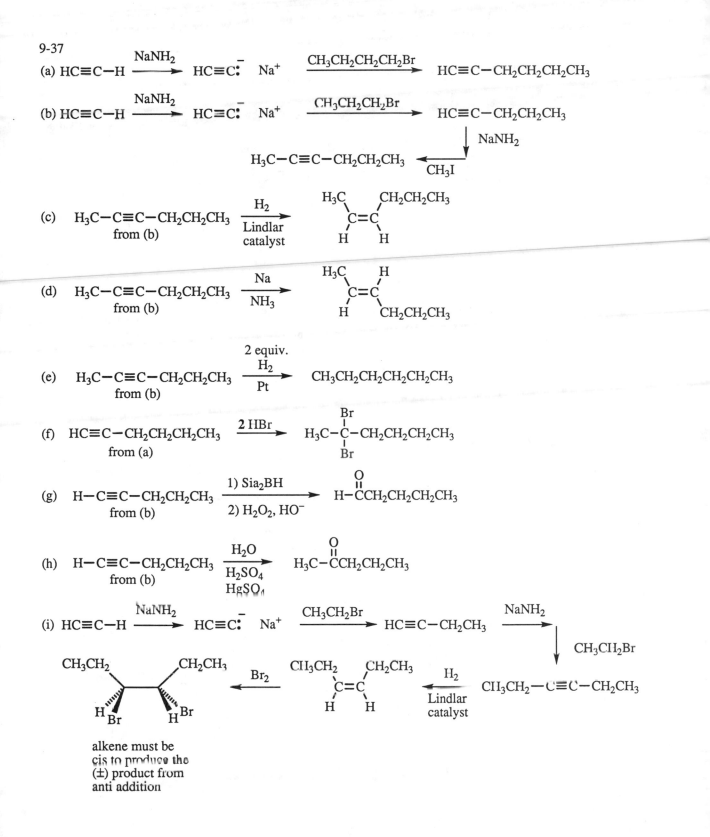

(d) $H_3C-C\equiv C-CH_2CH_2CH_3$ from (b) $\xrightarrow[NH_3]{Na}$

(e) $H_3C-C\equiv C-CH_2CH_2CH_3$ from (b) $\xrightarrow[Pt]{\text{2 equiv.}\ H_2}$ $CH_3CH_2CH_2CH_2CH_2CH_3$

(f) $HC\equiv C-CH_2CH_2CH_2CH_3$ from (a) $\xrightarrow{\text{2 HBr}}$ $H_3C-\overset{Br}{\underset{Br}{\overset{|}{\underset{|}{C}}}}-CH_2CH_2CH_2CH_3$

(g) $H-C\equiv C-CH_2CH_2CH_3$ from (b) $\xrightarrow[\text{2) }H_2O_2,\ HO^-]{\text{1) }Sia_2BH}$ $H-\overset{O}{\overset{||}{C}}CH_2CH_2CH_2CH_3$

(h) $H-C\equiv C-CH_2CH_2CH_3$ from (b) $\xrightarrow[\substack{H_2SO_4 \\ HgSO_4}]{H_2O}$ $H_3C-\overset{O}{\overset{||}{C}}CH_2CH_2CH_3$

(i) $HC\equiv C-H$ $\xrightarrow{NaNH_2}$ $HC\equiv C:^-$ Na^+ $\xrightarrow{CH_3CH_2Br}$ $HC\equiv C-CH_2CH_3$ $\xrightarrow{NaNH_2}$

$\xrightarrow{CH_3CH_2Br}$

$CH_3CH_2-C\equiv C-CH_2CH_3$ $\xrightarrow[\text{Lindlar catalyst}]{H_2}$

$\xleftarrow{Br_2}$

alkene must be
cis to produce the
(±) product from
anti addition

189

9-37 continued

(j) $HC\equiv C-H \xrightarrow{NaNH_2} HC\equiv C:^- Na^+ \xrightarrow{CH_3I} HC\equiv C-CH_3 \xrightarrow[\text{2) }CH_3I]{\text{1) }NaNH_2} H_3C-C\equiv C-CH_3$

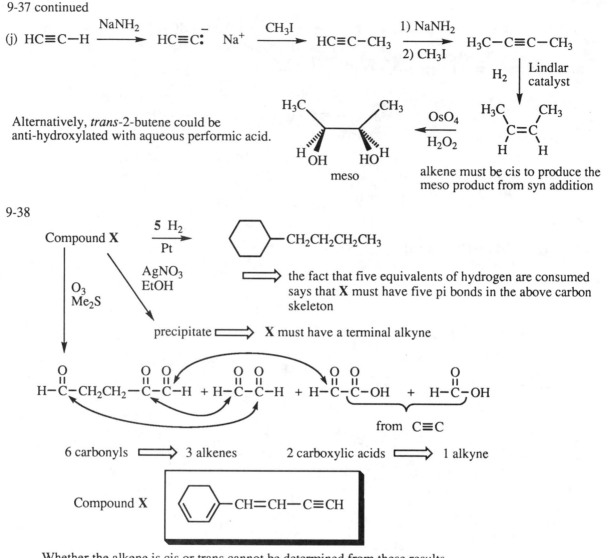

Alternatively, *trans*-2-butene could be anti-hydroxylated with aqueous performic acid.

meso

H_2 Lindlar catalyst

$\xrightarrow[\text{H}_2\text{O}_2]{\text{OsO}_4}$

alkene must be cis to produce the meso product from syn addition

9-38

Compound **X** $\xrightarrow[\text{Pt}]{5 \text{ H}_2}$ [cyclohexane]$-CH_2CH_2CH_2CH_3$

$\Rightarrow$ the fact that five equivalents of hydrogen are consumed says that **X** must have five pi bonds in the above carbon skeleton

O_3 / Me_2S

$\xrightarrow{\text{AgNO}_3 \text{ EtOH}}$ precipitate $\Rightarrow$ **X** must have a terminal alkyne

$H-\overset{O}{\underset{\|}{C}}-CH_2CH_2-\overset{O}{\underset{\|}{C}}-\overset{O}{\underset{\|}{C}}-H + H-\overset{O}{\underset{\|}{C}}-\overset{O}{\underset{\|}{C}}-H + H-\overset{O}{\underset{\|}{C}}-\overset{O}{\underset{\|}{C}}-OH + H-\overset{O}{\underset{\|}{C}}-OH$

from $C\equiv C$

6 carbonyls $\Rightarrow$ 3 alkenes   2 carboxylic acids $\Rightarrow$ 1 alkyne

Compound **X** | [phenyl]$-CH=CH-C\equiv CH$ |

Whether the alkene is cis or trans cannot be determined from these results.

9-39 Compound **Z**: precipitate with $Ag^+ \Rightarrow$ terminal alkyne

ozonolysis $\Rightarrow$ $CH_3(CH_2)_4-\overset{O}{\underset{\|}{C}}-H$ $CH_3-\overset{O}{\underset{\|}{C}}-CH_2-\overset{O}{\underset{\|}{C}}-OH$ $HO-\overset{O}{\underset{\|}{C}}-H$

from alkene   from alkyne

Compound **Z**: | $CH_3(CH_2)_4CH=\underset{\underset{CH_3}{|}}{C}-CH_2-C\equiv C-H$ | Whether the alkene is *E* or *Z* cannot be determined from this information.

190

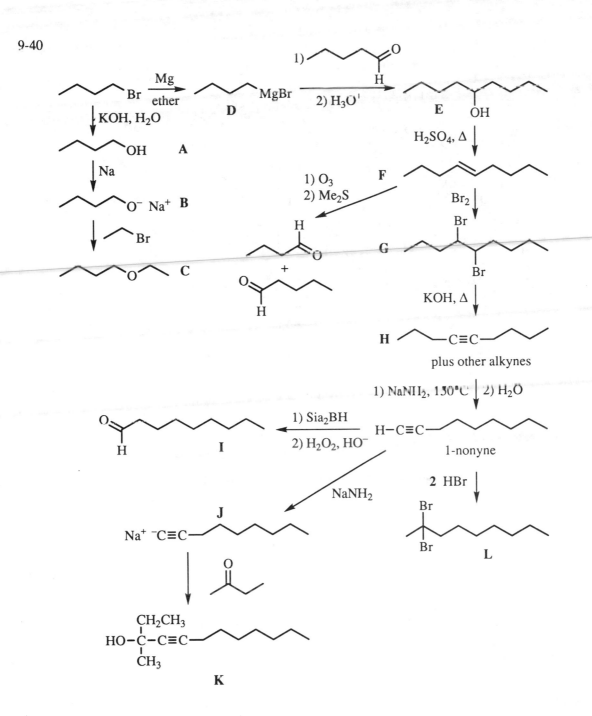

9-41

(a) $CH_3CH_2-C\equiv C-H$ $\xrightarrow[\text{2) } H_2O_2, HO^-]{\text{1) } Sia_2BH}$ $CH_3CH_2CH_2-\overset{\displaystyle O}{\overset{\|}{C}}-H$

(b)

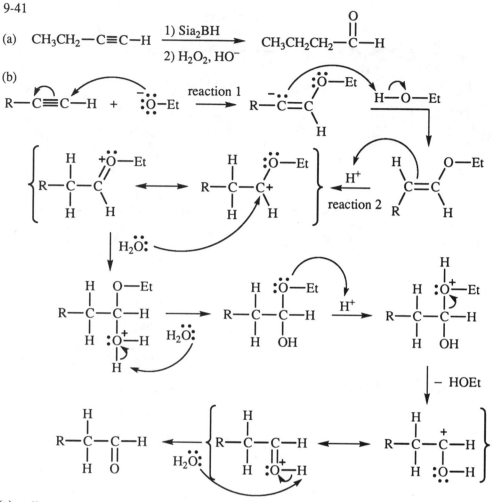

(c) <u>alkyne</u>

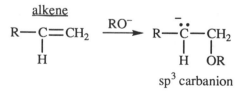

$R-C\equiv CH$ $\xrightarrow{RO^-}$ $R-\overset{..}{\overset{-}{C}}=CH$
$\qquad\qquad\qquad\qquad\qquad\quad |$
$\qquad\qquad\qquad\qquad\qquad\quad OR$
$\qquad\qquad\qquad\quad sp^2$ carbanion

An sp^2 carbanion is more stable than an sp^3 carbanion: the sp^2 carbanion has more s character and is closer to the positive nucleus. The sp^2 carbanion is easier to form because of its relative stability.

<u>alkene</u>

$R-\overset{\displaystyle}{\underset{\displaystyle |}{\underset{H}{C}}}=CH_2$ $\xrightarrow{RO^-}$ $R-\overset{\displaystyle ..}{\overset{-}{\underset{|}{\underset{H}{C}}}}-\underset{|}{\underset{OR}{CH_2}}$
$\qquad\qquad\qquad\qquad\quad sp^3$ carbanion

CHAPTER 10—STRUCTURE AND SYNTHESIS OF ALCOHOLS

10-1

(a) 2-phenyl-2-propanol
(b) 5-bromo-2-heptanol
(c) 4-methyl 3-cyclohexen-1-ol ("1" is optional)
(d) *trans*-2-methyl-1-cyclohexanol ("1" is optional)
(e) (*E*) 2-chloro-3-methyl-2-penten-1-ol
(f) (2*R*,3*S*)-2-bromo-3-hexanol

10-2 IUPAC name first, then common name.

(a) cyclopropanol; cyclopropyl alcohol
(b) 2-methyl-2-propanol; *t*-butyl alcohol
(c) 1-cyclobutyl-2-propanol; no common name
(d) 3-methyl-1-butanol; isopentyl alcohol
 (also isoamyl alcohol)

10-3

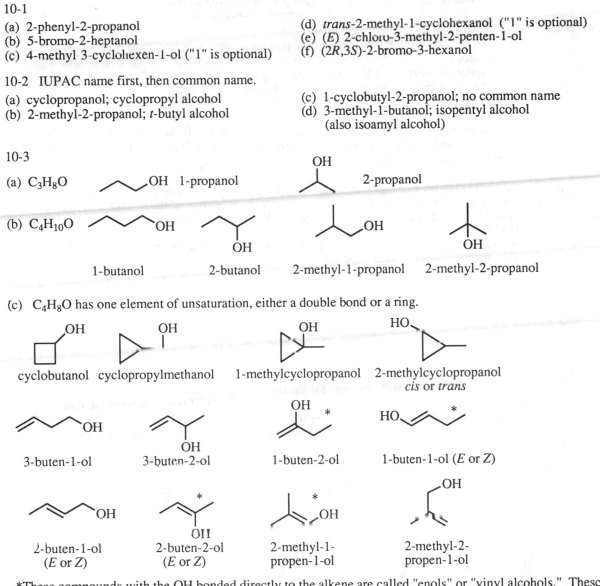

(a) C_3H_8O ⌃⌃OH 1-propanol

OH 2-propanol

(b) $C_4H_{10}O$ ⌃⌃OH

OH

OH

OH

1-butanol   2-butanol   2-methyl-1-propanol   2-methyl-2-propanol

(c) C_4H_8O has one element of unsaturation, either a double bond or a ring.

OH   OH   OH   HO

cyclobutanol   cyclopropylmethanol   1-methylcyclopropanol   2-methylcyclopropanol
 cis or *trans*

OH   OH *   HO *

3-buten-1-ol   3-buten-2-ol   1-buten-2-ol   1-buten-1-ol (*E* or *Z*)

OH   OH   *   OH

2-buten-1-ol   2-buten-2-ol   2-methyl-1-   2-methyl-2-
 (*E* or *Z*)   (*E* or *Z*)   propen-1-ol   propen-1-ol

*These compounds with the OH bonded directly to the alkene are called "enols" or "vinyl alcohols." These are unstable, although the structures are legitimate.

10-4

(a) 8,8-dimethyl-2,7-nonanediol
(b) 1,8-octanediol
(c) *cis*-2-cyclohexene-1,4-diol
(d) 3-cyclopentyl-2,4-heptanediol

193

10-5

(a) Cyclohexanol is more soluble because its alkyl group is more compact than in 1-hexanol.
(b) 4-Methylphenol is more soluble because its hydrocarbon portion is more compact than in 1-heptanol, and phenols form particularly strong hydrogen bonds with water.
(c) 3-Ethyl-3-hexanol is more soluble because its alkyl portion is more spherical than in 2-octanol.
(d) Cyclooctane-1,4-diol is more soluble because it has two OH groups which can hydrogen bond with water, whereas 2-hexanol has only one OH group. (The ratio of carbons to OH is 4 to 1 in the former compound and 6 to 1 in the latter; the smaller this ratio, the more soluble.)
(e) These are enantiomers and will have identical solubility.

10-6 Dimethylamine molecules can hydrogen bond among themselves so it takes more energy (higher temperature) to separate them from each other. Trimethylamine has no N-H and cannot hydrogen bond, so it takes less energy to separate these molecules from each other.

10-7 See Appendix 2 at the back of this Solutions Manual for a review of acidity and basicity.
(a) Methanol is more acidic than *t*-butyl alcohol. The greater the substitution, the lower the acidity.
(b) 1-Chloroethanol is more acidic because the electron-withdrawing chlorine atom is closer to the OH group than in 2-chloroethanol.
(c) 2,2-Dichloroethanol is more acidic because two electron-withdrawing chlorine atoms increase acidity more than just the one chlorine in 2-chloroethanol.

10-8

most acidic													least acidic
sulfuric acid	>>	2-chloroethanol	>	water	>	ethanol	>	t-butyl alcohol	>	ammonia	>	hexane
H_2SO_4		$ClCH_2CH_2OH$		H_2O	CH_3CH_2OH		$(CH_3)_3COH$			NH_3		C_6H_{14}

Sulfuric acid is one of the strongest acids known. On the other extreme, alkanes like hexane are the least acidic compounds. The N–H bond in ammonia is less acidic than any O–H bond. Among the four compounds with O–H bonds, the tertiary alcohols are the least acidic. Water is more acidic than most alcohols including ethanol. However, if a strong electron-withdrawing substituent like chlorine is near the alcohol group, the acidity increases enough so that it is more acidic than water. (Determining exactly where water appears in this list is the most difficult part.)

10-9 Resonance forms of phenoxide anion show the negative charge delocalized onto the ring only at carbons 2, 4, and 6:

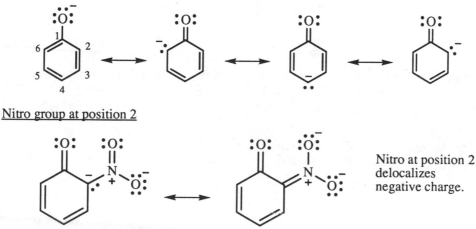

Nitro group at position 2

Nitro at position 2 delocalizes negative charge.

10-9 continued

Nitro group at position 3

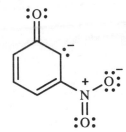

Nitro at position 3 cannot delocalize negative charge at position 2 or 4—no resonance stabilization.

Nitro group at position 4

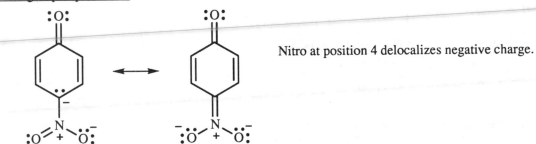

Nitro at position 4 delocalizes negative charge.

Only when the nitro group is at one of the negative carbons will the nitro have a stabilizing effect. Thus, 2-nitrophenol and 4-nitrophenol are substantially more acidic than phenol itself, but 3-nitrophenol is only slightly more acidic than phenol.

10-10

(a) Structure **A** is a phenol because the OH is bonded to a benzene ring. As a phenol, it will be acidic enough to react with sodium hydroxide to generate a phenoxide ion that will be fairly soluble in water. Structure **B** is a 2° benzylic alcohol.

(b) Both of these organic compounds will be soluble in an organic solvent like dichloromethane. Shaking this organic solution with aqueous sodium hydroxide will ionize the phenol **A**, making it more polar and water soluble; it will be extracted from the organic layer into the water layer, while the alcohol will remain in the organic solvent. Separating these immiscible solvents will separate the original compounds. The alcohol can be retrieved by evaporating the organic solvent. The phenol can be isolated by acidifying the basic aqueous solution and filtering if the phenol is a solid, or separating the layers if the phenol is a liquid.

10-11 The Grignard reaction needs a solvent containing an ether functional group: (b), (f), (g), and (h) are possible solvents. Dimethyl ether, (b), is a gas at room temperature, however, so it would have to be liquefied at low temperature for it to be a useful solvent.

10-12

(a) CH_3CH_2MgBr
(b) [structure] Li + LiI
(c) F—[cyclohexane]—MgBr
(d) [structure with Li] + LiCl

195

10-13 Any of three halides—chloride, bromide, iodide, but not fluoride can be used. Ether is the solvent for Grignard reagents.

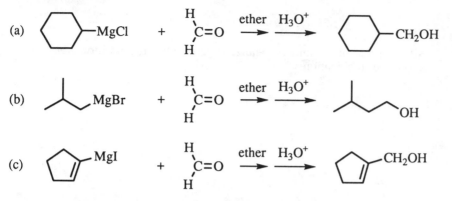

(a) cyclohexyl—MgCl + H₂C=O →(ether) →(H₃O⁺) cyclohexyl—CH₂OH

(b) isobutyl—MgBr + H₂C=O →(ether) →(H₃O⁺) isobutyl—CH₂CH₂OH

(c) cyclopentenyl—MgI + H₂C=O →(ether) →(H₃O⁺) cyclopentenyl—CH₂OH

Note: the alternative arrow symbolism could also be used, where the two steps are numbered around one arrow:

$$\xrightarrow[\text{2) } H_3O^+]{\text{1) ether}}$$

10-14 Any of three halides—chloride, bromide, iodide, but not fluoride—can be used. Grignard reactions are always performed in ether solvent; ether is not shown here.

(a) two methods

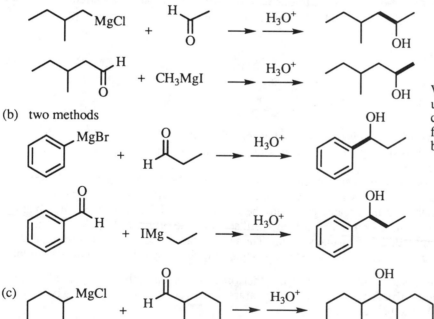

Where two methods can be used to form the target compound, the newly formed bond is shown in bold.

(b) two methods

(c)

10-15 Grignard reactions are always performed in ether. Here, the ether is not shown.

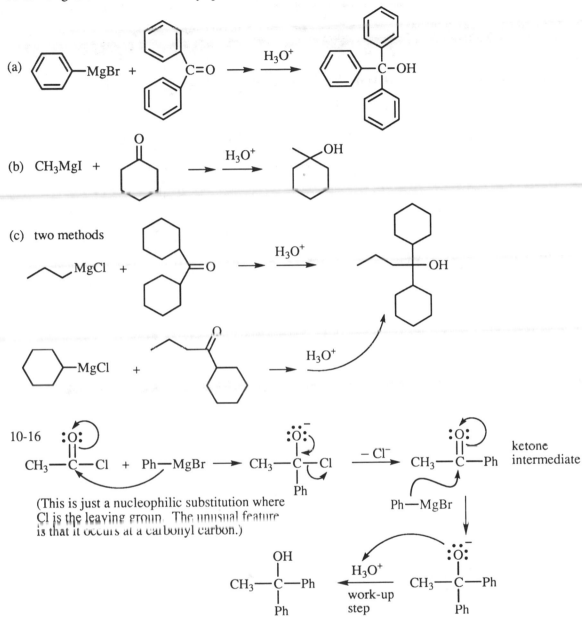

(a)

(b) CH_3MgI +

(c) two methods

10-16

(This is just a nucleophilic substitution where Cl is the leaving group. The unusual feature is that it occurs at a carbonyl carbon.)

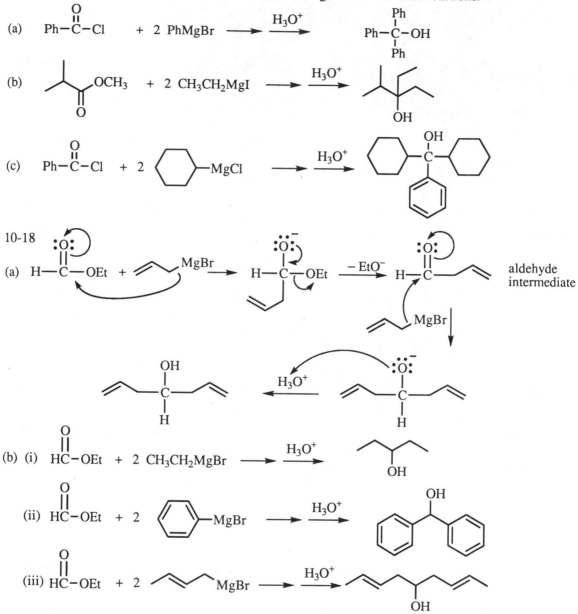

10-19 Ether is the solvent in all Grignard reactions.

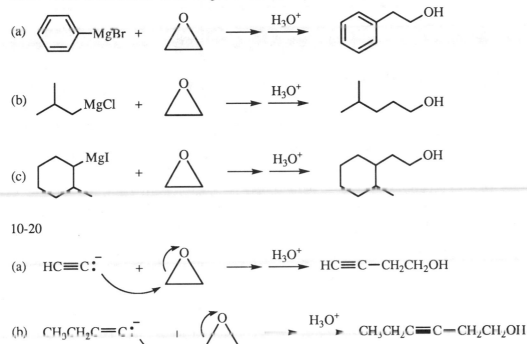

(a) [phenyl]—MgBr + [epoxide] $\xrightarrow{\quad}$ $\xrightarrow{H_3O^+}$ [2-phenylethanol with OH]

(b) [isobutyl]MgCl + [epoxide] $\xrightarrow{\quad}$ $\xrightarrow{H_3O^+}$ [product with OH]

(c) [methylcyclohexyl]MgI + [epoxide] $\xrightarrow{\quad}$ $\xrightarrow{H_3O^+}$ [product with OH]

10-20

(a) $HC\equiv C:^-$ + [epoxide] $\xrightarrow{\quad}$ $\xrightarrow{H_3O^+}$ $HC\equiv C-CH_2CH_2OH$

(h) $CH_3CH_2C\equiv C:^-$ + [epoxide] $\xrightarrow{\quad}$ $\xrightarrow{H_3O^+}$ $CH_3CH_2C\equiv C-CH_2CH_2OH$

10-21 There are several possible synthetic routes to each structure, of which these are representative. Your answers may be different and still be correct.

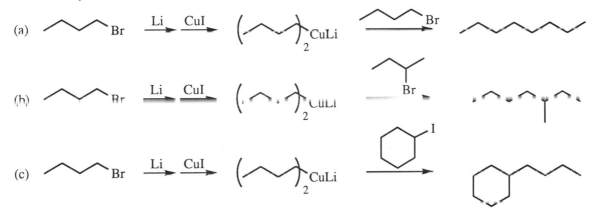

(a) [CH₂CH₂CH₂CH₂]Br $\xrightarrow{Li}$ $\xrightarrow{CuI}$ $\left(\ \sim\ \right)_2 CuLi$ $\xrightarrow{\quad Br}$ [product]

(h) [chain]Br $\xrightarrow{Li}$ $\xrightarrow{CuI}$ $\left(\ \sim\ \right)_2 CuLi$ $\xrightarrow{\quad Br}$ [product]

(c) [chain]Br $\xrightarrow{Li}$ $\xrightarrow{CuI}$ $\left(\ \sim\ \right)_2 CuLi$ $\xrightarrow{\quad I}$ [cyclohexyl product]

Alternatively, coupling lithium dicyclohexylcuprate with 1-bromobutane would also work. As this mechanism is not a typical S_N2, it is not as susceptible to steric hindrance as acetylide ion substitution or a similar S_N2 reaction.

(d) [chain]Br $\xrightarrow{Li}$ $\xrightarrow{CuI}$ $\left(\ \sim\ \right)_2 CuLi$ $\xrightarrow{\quad I}$ [product]

10-22 These reactions are acid-base reactions in which an acidic proton (or deuteron) is transferred to a basic carbon in either a Grignard reagent or an alkyllithium.

(a) CH_3D + $Mg(OD)I$

(b) $CH_3CH_2CH_2CH_3$ + $LiOCH_2CH_3$

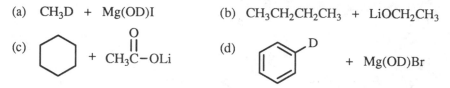

(c) ⬡ + $CH_3\overset{\overset{\text{O}}{\|}}{C}-OLi$

(d) ⬡—D + $Mg(OD)Br$

10-23 Grignard reagents are incompatible with acidic hydrogens and with electrophilic, polarized multiple bonds.

(a) As Grignard reagent is formed, it would instantaneously react with N—H present in other molecules of the same substance.

(b) As Grignard reagent is formed, it would immediately react with the ester functional group present in other molecules of the same substance.

(c) Care must be taken in how reagents are written above and below arrows. If reagents are numbered "1. ... 2. ... *etc.*", it means they are added sequentially, the same as writing reagents over separate arrows. If reagents written around an arrow are not numbered, it means they are added all at once in the same mixture. In this problem, the ketone is added in the presence of aqueous acid. The acid will immediately protonate and destroy the Grignard reagent before reaction with the ketone can occur.

(d) The ethyl Grignard reagent will be immediately protonated and consumed by the OH. This reaction *could* be made to work, however, by adding two equivalents of ethyl Grignard reagent—the first to consume the OH proton, the second to add across the ketone. Aqueous acid will then protonate both oxygens.

10-24 Sodium borohydride does not reduce esters.

(a) $CH_3(CH_2)_8CH_2OH$

(b) no reaction

(c) no reaction ($PhCOO^-$ before acid work-up)

(d) ⬡—OH (cyclohexanol)

(e) HO—⬡—C(=O)OCH₃ with OH substituent

(f) ester

10-25 Lithium aluminum hydride reduces esters as well as other carbonyl groups.

(a) $CH_3(CH_2)_8CH_2OH$

(b) $CH_3CH_2CH_2OH$ + $HOCH_3$

(c) $PhCH_2OH$

(d) ⬡—OH (cyclohexanol)

(e) HO—⬡—OH with OH substituent + $HOCH_3$

(f) HO—CH₂...—OH with HO substituent

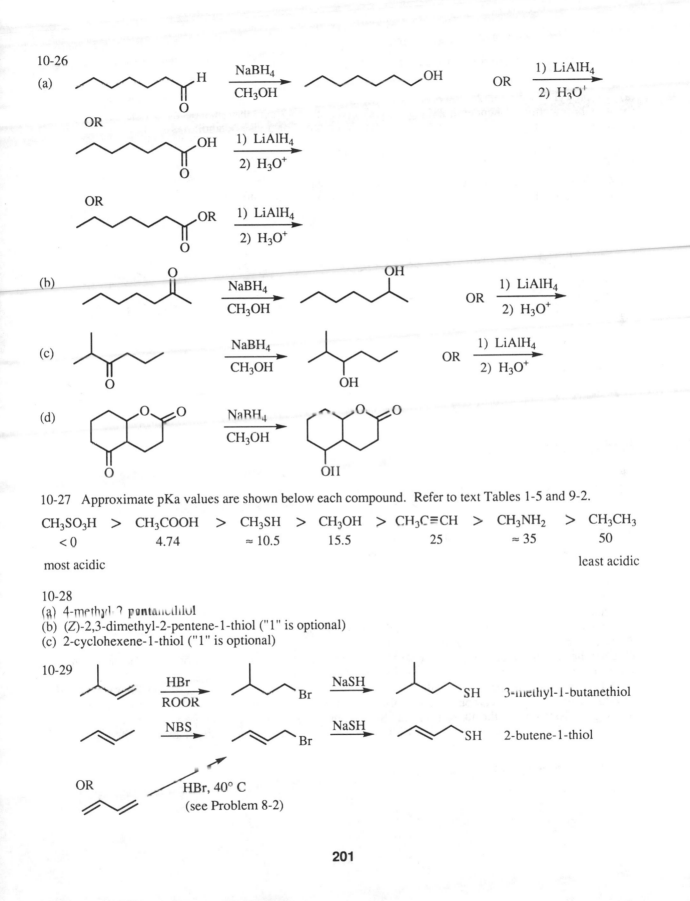

10-26

(a)

OR

OR

(b)

(c)

(d)

10-27 Approximate pKa values are shown below each compound. Refer to text Tables 1-5 and 9-2.

CH_3SO_3H > CH_3COOH > CH_3SH > CH_3OH > $CH_3C≡CH$ > CH_3NH_2 > CH_3CH_3

   < 0   4.74   ≈ 10.5   15.5   25   ≈ 35   50

most acidic   least acidic

10-28

(a) 4-methyl-2-pentanethiol

(b) (Z)-2,3-dimethyl-2-pentene-1-thiol ("1" is optional)

(c) 2-cyclohexene-1-thiol ("1" is optional)

10-29

3-methyl-1-butanethiol

2-butene-1-thiol

OR

HBr, 40° C

(see Problem 8-2)

201

10-30 Please refer to solution 1-19, page 12 of this Solutions Manual.

10-31

(a) 4-*n*-propyl-2-heptanol; 2°
(b) 4-(1-bromoethyl)-3-heptanol; 2°
(c) (*E*)-4,5-dimethyl-3-hexen-1-ol; 1°

(d) 3-bromo-3-cyclohexen-1-ol; 2° ("1" is optional)
(e) *cis*-4-chloro-2-cyclohexen-1-ol; 2° ("1" is optional)
(f) 6-chloro-3-phenyl-3-octanol; 3°

10-32

(a) 4-chloro-1-phenyl-1,5-hexanediol
(b) *trans*-1,2-cyclohexanediol

(c) 3-nitrophenol
(d) 4-bromo-2-chlorophenol

10-33

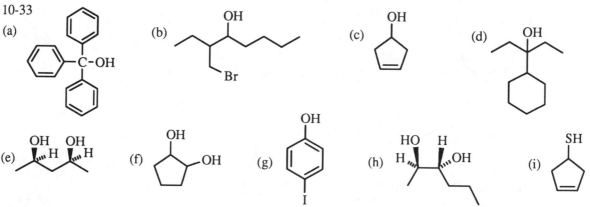

(j) CH₃S—SCH₃

10-34

(a) 1-Hexanol will boil at a higher temperature as it is less branched than 3,3-dimethyl-1-butanol.
(b) 2-Hexanol will boil at a higher temperature because its molecules hydrogen bond with each other, whereas molecules of 2-hexanone have no intermolecular hydrogen bonding.
(c) 1,5-Hexanediol will boil at a higher temperature as it has two OH groups for hydrogen bonding. 2-Hexanol has only one group for hydrogen bonding.

10-35

(a) 3-Chlorophenol is more acidic than cyclopentanol. In general, phenols are many orders of magnitude more acidic than alcohols.
(b) 2-Chlorocyclohexanol is slightly more acidic than cyclohexanol; the proximity of the electronegative chlorine to the OH increases its acidity.
(c) Cyclohexanecarboxylic acid is more acidic than cyclohexanol. In general, carboxylic acids are many orders of magnitude more acidic than alcohols.
(d) 1-Butanol is slightly more acidic than 2,2-dimethyl-1-butanol, for two reasons. In the latter compound, the two methyl groups on carbon-2 are slightly electron-donating and will increase the electron density on the oxygen, destabilizing the alkoxide anion. Second, in solution, the two methyl groups prevent stabilization of the alkoxide by solvation with molecules of solvent, lowering the stability of the anion, making the equilibrium less favorable.

10-36

(a) 2-Methyl-2-propanol is the most soluble as it is the most highly branched and therefore the least hydrophobic among these alcohols of identical molecular weight.

(b) 1,2-Cyclohexanediol is the most soluble as it has two OH groups for hydrogen bonding. Cyclohexanol has only one OH group; chlorocyclohexane cannot hydrogen bond and is the least soluble.

(c) Cyclohexanol is the most soluble as it can hydrogen bond. Chlorocyclohexane cannot hydrogen bond, and 4-methylcyclohexanol has the added hydrophobic methyl group, decreasing its water solubility.

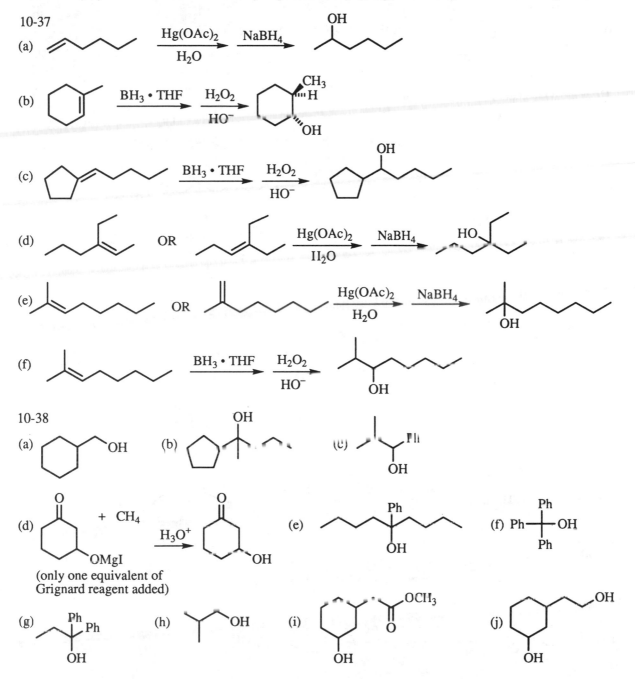

10-37

(a) [alkene] $\xrightarrow[H_2O]{Hg(OAc)_2}$ $\xrightarrow{NaBH_4}$ [alcohol]

(b) [methylcyclohexene] $\xrightarrow{BH_3 \cdot THF}$ $\xrightarrow[HO^-]{H_2O_2}$ [trans-2-methylcyclohexanol]

(c) [alkene] $\xrightarrow{BH_3 \cdot THF}$ $\xrightarrow[HO^-]{H_2O_2}$ [alcohol]

(d) [alkene] OR [alkene] $\xrightarrow[II_2O]{Hg(OAc)_2}$ $\xrightarrow{NaBH_4}$ [alcohol]

(e) [alkene] OR [alkene] $\xrightarrow[H_2O]{Hg(OAc)_2}$ $\xrightarrow{NaBH_4}$ [alcohol]

(f) [alkene] $\xrightarrow{BH_3 \cdot THF}$ $\xrightarrow[HO^-]{H_2O_2}$ [alcohol]

10-38

(a) [cyclohexylmethanol with OH]   (b) [alcohol with OH]   (c) [alcohol with OH]

(d) [cyclohexanone with CH$_4$ + OMgI] $\xrightarrow{H_3O^+}$ [product with OH]
(only one equivalent of Grignard reagent added)

(e) [Ph, OH structure]   (f) Ph—C(Ph)(Ph)—OH

(g) Ph, Ph, OH structure   (h) [isopropyl OH]   (i) [cyclohexyl OCH$_3$ ester with OH]   (j) [cyclohexyl with OH]

203

10-38 continued

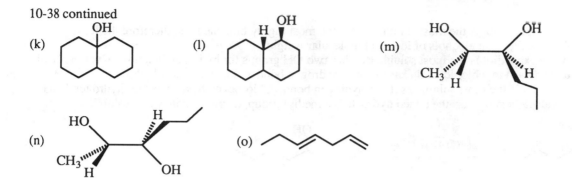

(k)

(l)

(m)

(n)

(o)

10-39   All Grignard reactions are run in ether solvent.

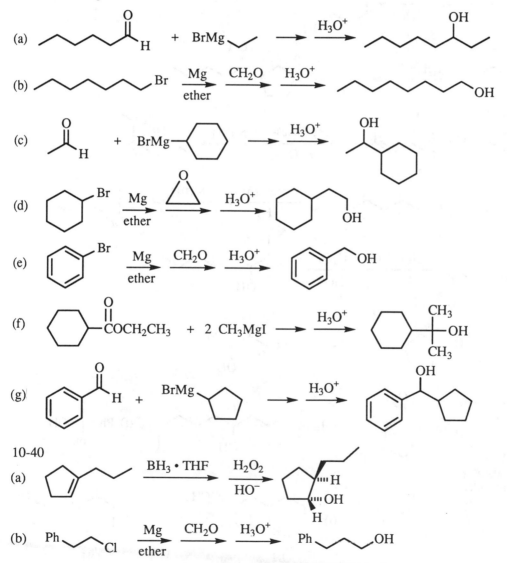

(a)

(b)

(c)

(d)

(e)

(f)

(g)

10-40

(a)

(b)

10-40 continued

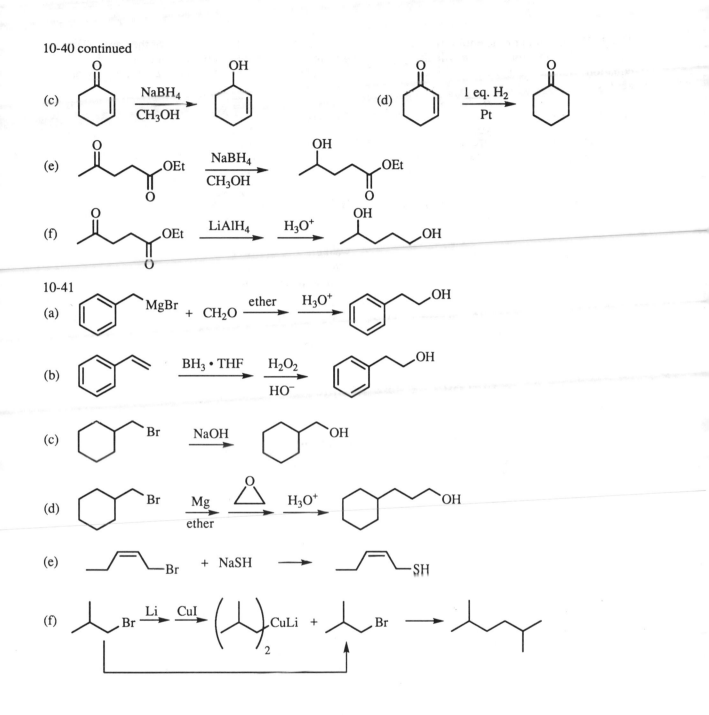

10-42   The position of the equilibrium can be determined by the strength of the acids or the bases. The stronger acid and stronger base will always react to give the weaker acid and base, so the side of the equation with the weaker acid and base will be favored at equilibrium. See Appendix 2 for a review of acidity.

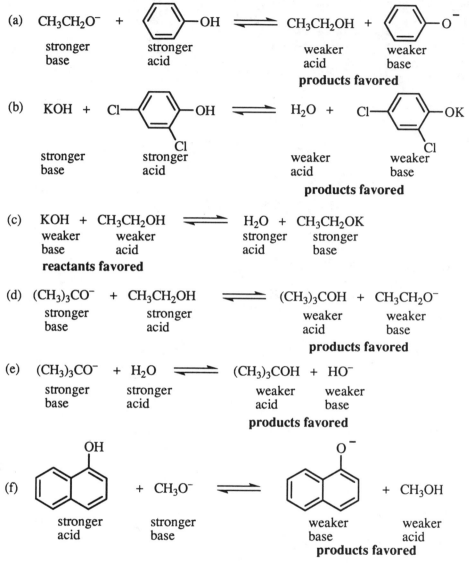

(a) $CH_3CH_2O^-$ +   ⬡—OH   ⇌ CH_3CH_2OH +   ⬡—O⁻

     stronger   stronger   weaker   weaker
     base   acid   acid   base

products favored

(b)   KOH   +   Cl—⬡—OH   ⇌ H_2O +   Cl—⬡—OK

                            Cl   Cl

     stronger   stronger   weaker   weaker
     base   acid   acid   base

products favored

(c)   KOH   + CH_3CH_2OH ⇌ H_2O + CH_3CH_2OK
    weaker   weaker   stronger   stronger
    base   acid   acid   base
reactants favored

(d) $(CH_3)_3CO^-$ + CH_3CH_2OH ⇌ $(CH_3)_3COH$ + $CH_3CH_2O^-$
    stronger   stronger   weaker   weaker
    base   acid   acid   base

products favored

(e) $(CH_3)_3CO^-$ + H_2O ⇌ $(CH_3)_3COH$ + HO^-
    stronger   stronger   weaker   weaker
    base   acid   acid   base
products favored

(f)   ⬡⬡—OH   + CH_3O^- ⇌   ⬡⬡—O⁻   + CH_3OH

    stronger   stronger   weaker   weaker
    acid   base   base   acid
products favored

10-43

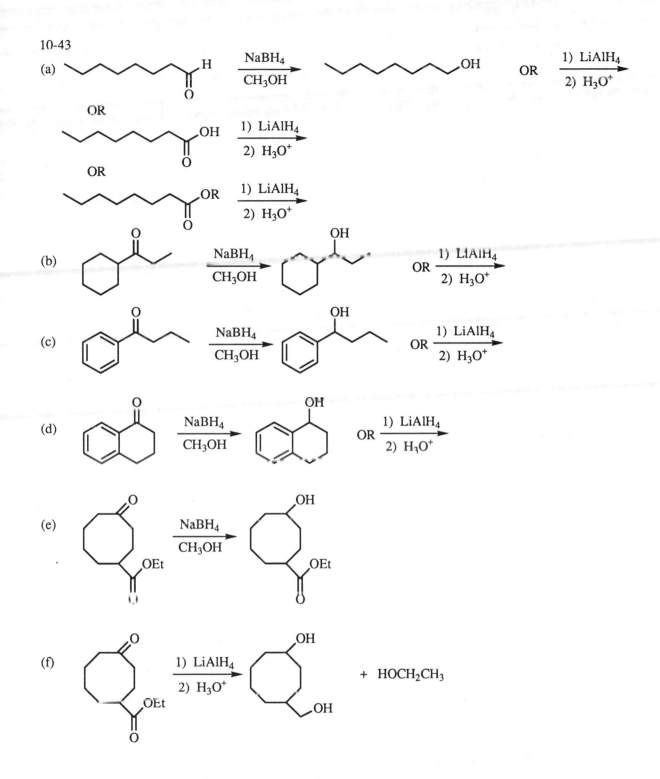

10-44 All steps are reversible.

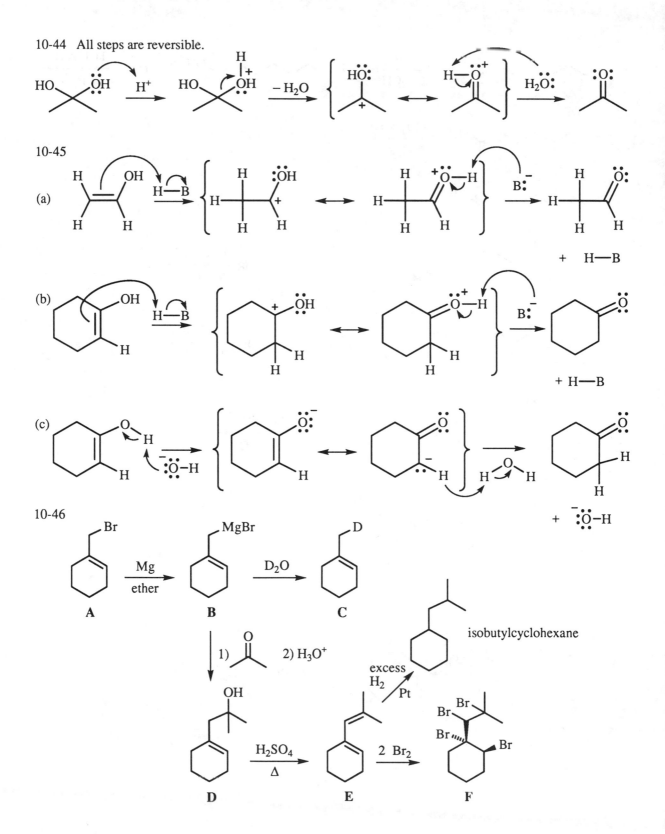

10-45

(a)

(b)

(c)

10-46

isobutylcyclohexane

10-47 This mechanism is similar to cleavage of the epoxide in ethylene oxide by Grignard reagents. The driving force for the reaction is relief of ring strain in the 4-membered cyclic ether.

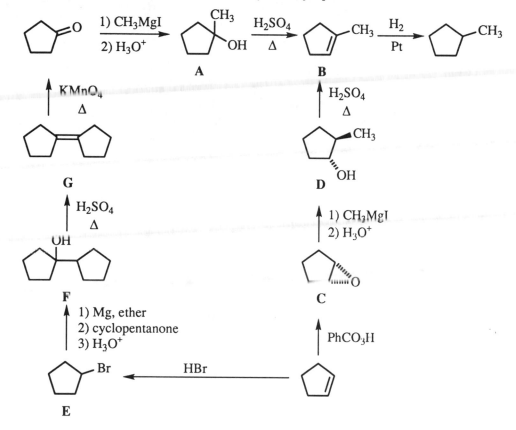

10-48 When mixtures of isomers can result, only the major product is shown.

CHAPTER 11—REACTIONS OF ALCOHOLS

11-1

(a) both are oxidations
(b) oxidation, oxidation, reduction, oxidation
(c) one carbon is oxidized and one carbon is reduced—no net change
(d) reduction
(e) neither oxidation nor reduction
(f) oxidation (addition of X_2)
(g) neither oxidation nor reduction (addition of HX)

11-2

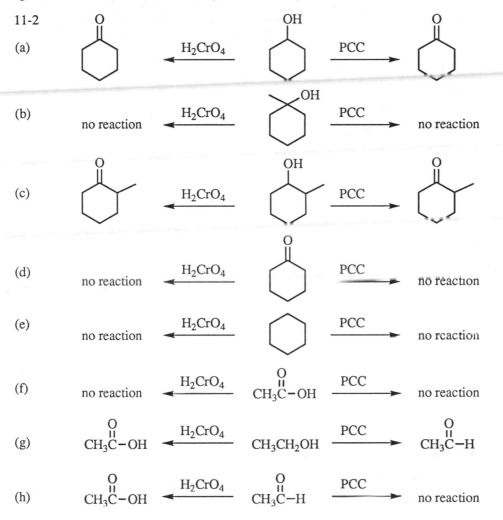

11-3

(a) The 1° alcohol loses two hydrogens when transformed to the aldehyde; the alcohol is oxidized. DMSO loses an oxygen from the sulfur; it is clearly reduced. If you got those two, you did the problem correctly. To be rigorous, oxalyl chloride undergoes a disproportionation reaction: one C is oxidized to CO_2 and the other C is reduced to CO; however, the net effect on the elements in oxalyl chloride is "no change".
(b) Text section 8-15 shows that dimethyl sulfide reduces an ozonide. In the process, dimethyl sulfide is oxidized to DMSO.

11-4

(a) Dehydrogenation does not occur at 25° C—either: 1) there is a high kinetic barrier (a high activation energy) for this reaction, or 2) it is thermodynamically unfavorable, with $\Delta G > 0$. The latter possibility is supported by the fact that the reverse reaction (catalytic hydrogenation of a carbonyl) *is* spontaneous at 25° C (see text section 10-11C) and therefore has $\Delta G < 0$. This makes the question of kinetics academic—a reaction that cannot proceed must be uselessly slow.

(b) and (c) Kinetics will improve with increasing temperature for virtually all reactions, so both the hydrogenation and dehydrogenation reactions will go faster. In this case, however, the question is how to favor the dehydrogenation reaction. The answer is that thermodynamics will favor this reaction as the temperature is raised. The key is the fundamental thermodynamic equation $\Delta G = \Delta H - T\Delta S$. We can estimate that $\Delta H > 0$ since the product ketone plus hydrogen is less stable than the starting alcohol. Also, $\Delta S > 0$ since one molecule is converted to two: therefore, $-T\Delta S < 0$. At low temperature (25° C), ΔH dominates because T is so small, so $\Delta G > 0$. At a high enough temperature, the $-T\Delta S$ term will begin to overwhelm ΔH, and ΔG will become negative. For the reaction in question, this must be the case at 300° C.

11-5

(a)

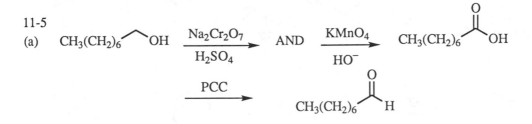

(b) all three reagents give the same ketone product with a secondary alcohol

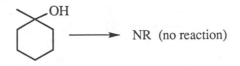

(c)

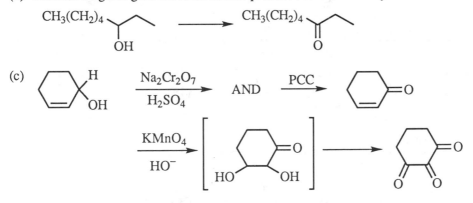

(d) all three reagents give no reaction with a tertiary alcohol

Note to the student: For simplicity, this book will use these standard laboratory methods of oxidation:

—PCC (pyridinium chlorochromate) to oxidize 1° alcohols to aldehydes;
—H_2CrO_4 (chromic acid) to oxidize 1° alcohols to carboxylic acids;
—CrO_3, H_2SO_4, acetone (Jones reagent) to oxidize 2° alcohols to ketones.

Understand that other choices are legitimate; for example, Collins reagent (CrO_3/pyridine) works about as well as PCC in the preparation of aldehydes, and Collins reagent or PCC will oxidize a 2° alcohol to a ketone as well as chromic acid. If you have a question about the appropriateness of a reagent you choose, consult the table in the text before Problem 11-2.

11-6

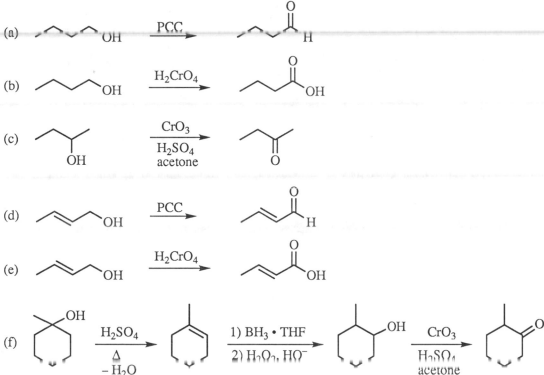

11-7 A chronic alcoholic has induced more ADH enzyme to be present to handle large amounts of imbibed ethanol, so requires more ethanol "antidote" molecules to act as a competitive inhibitor to "tie up" the extra enzyme molecules.

11-8

$$CH_3-\underset{\underset{OH}{|}}{CH}-\underset{\underset{OH}{|}}{CH_2} \xrightarrow{[O]} CH_3-\underset{\underset{O}{\|}}{C}-\underset{\underset{O}{\|}}{CH} \xrightarrow{[O]} CH_3-\underset{\underset{O}{\|}}{C}-\underset{\underset{O}{\|}}{COH}$$

pyruvaldehyde   pyruvic acid

pyruvic acid is a normal metabolite
in the breakdown of glucose ("blood sugar")

11-9

(a) CH_3CH_2-OTs + $KO-\overset{\underset{\displaystyle CH_3}{|}}{\underset{\underset{\displaystyle CH_3}{|}}{C}}-CH_3$ $\longrightarrow$ $CH_3CH_2O-\overset{\underset{\displaystyle CH_3}{|}}{\underset{\underset{\displaystyle CH_3}{|}}{C}}-CH_3$ + $KOTs$

(E2 is also possible with this hindered base; the product would be ethylene, $CH_2{=}CH_2$.)

(b) + NaI $\longrightarrow$ + NaOTs

(c) + NaCN $\longrightarrow$ + NaOTs   inversion—S$_N$2

(d) $\xrightarrow{\text{NH}_3}$ $\xrightarrow[\text{NH}_3]{\text{excess}}$ + $\overset{+}{N}H_4\ \overset{-}{O}Ts$

(e) + Na^+ $\overset{..}{:}C{\equiv}CH$ $\longrightarrow$ + NaOTs

11-10

(a) $\xrightarrow[\text{pyridine}]{\text{TsCl}}$ $\xrightarrow{\text{NaBr}}$

(b) $\xrightarrow[\text{pyridine}]{\text{TsCl}}$ $\xrightarrow[\text{NH}_3]{\text{excess}}$

(c) $\xrightarrow[\text{pyridine}]{\text{TsCl}}$ $\xrightarrow{\text{NaOCH}_2CH_3}$

(d) $\xrightarrow[\text{pyridine}]{\text{TsCl}}$ $\xrightarrow{\text{KCN}}$

11-11 From this problem on, "Ts" will refer to the "tosyl" or "p-toluenesulfonyl" group:

Ts $\Longrightarrow$

(a) $\xrightarrow[\text{pyridine}]{\text{TsCl}}$ OR

11-11 continued

(b)

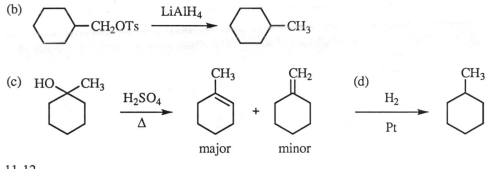

(c)

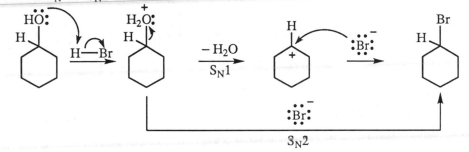

major minor

(d)

11-12

(a) either S_N1 or S_N2 on 2° alcohols

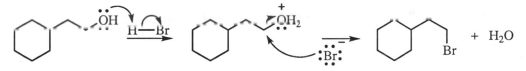

S_N2

(b) S_N2 on 1° alcohols

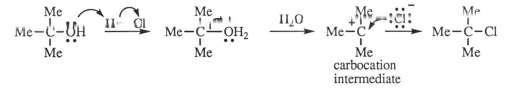

11-13 "Me" is the abbreviation for methyl.

Me—C̈—ÜH →H—Cl→ Me—C—ÖH₂ →H₂O→ Me—C⁺ ··· :C̈l:⁻ → Me—C—Cl
  | | | |
 Me   Me   Me   Me
                                          carbocation
                                          intermediate

11-14   The two standard qualitative tests are:

1) <u>chromic acid</u>—distinguishes 3° alcohol from either 1° or 2°

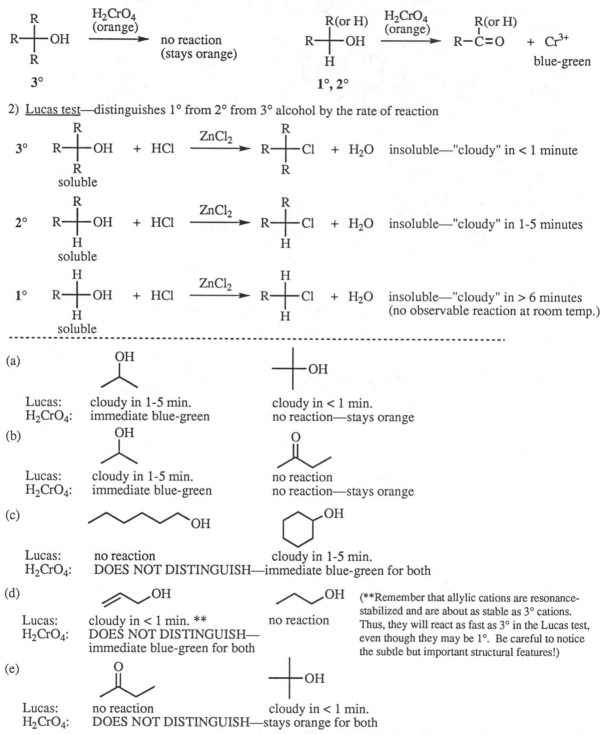

2) <u>Lucas test</u>—distinguishes 1° from 2° from 3° alcohol by the rate of reaction

(a)

Lucas: cloudy in 1-5 min.   cloudy in < 1 min.
H_2CrO_4:   immediate blue-green   no reaction—stays orange

(b)

Lucas: cloudy in 1-5 min.   no reaction
H_2CrO_4:   immediate blue-green   no reaction—stays orange

(c)

Lucas: no reaction   cloudy in 1-5 min.
H_2CrO_4:   DOES NOT DISTINGUISH—immediate blue-green for both

(d)

Lucas: cloudy in < 1 min. **   no reaction
H_2CrO_4:   DOES NOT DISTINGUISH—
            immediate blue-green for both

(**Remember that allylic cations are resonance-stabilized and are about as stable as 3° cations. Thus, they will react as fast as 3° in the Lucas test, even though they may be 1°. Be careful to notice the subtle but important structural features!)

(e)

Lucas: no reaction   cloudy in < 1 min.
H_2CrO_4:   DOES NOT DISTINGUISH—stays orange for both

216

11-15

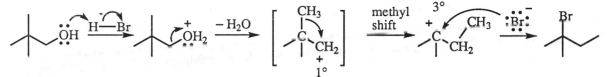

Even though 1°, the neopentyl carbon is hindered to backside attack, so S_N2 cannot occur easily. Instead, an S_N1 mechanism occurs, with rearrangement.

11-16

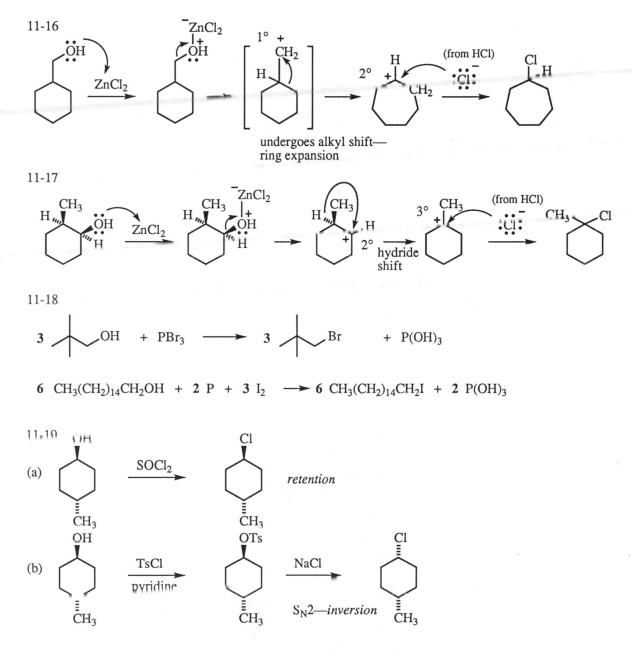

11-17

11-18

$$3 \quad \text{\Large$\curlywedge$}_{\text{OH}} \quad + \quad PBr_3 \quad \longrightarrow \quad 3 \quad \text{\Large$\curlywedge$}_{\text{Br}} \quad + \quad P(OH)_3$$

$$6 \ CH_3(CH_2)_{14}CH_2OH \ + \ 2\ P \ + \ 3\ I_2 \quad \longrightarrow \quad 6 \ CH_3(CH_2)_{14}CH_2I \ + \ 2\ P(OH)_3$$

11-19

(a) $\xrightarrow{\text{SOCl}_2}$ *retention*

(b) $\xrightarrow[\text{pyridine}]{\text{TsCl}}$ $\xrightarrow[\text{$S_N2$—inversion}]{\text{NaCl}}$

11-20

(a)

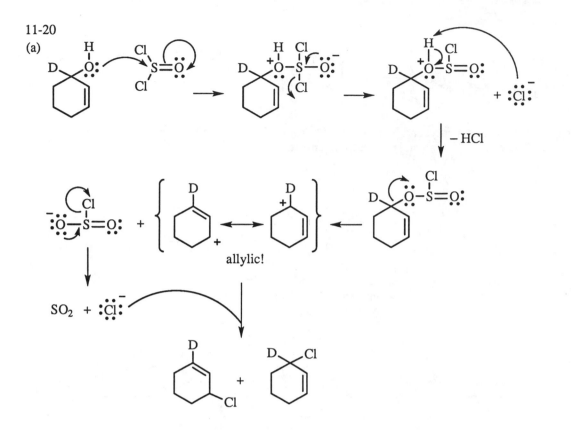

(b) The key is that the intermediate carbocation is allylic, very stable and relatively long-lived. It can therefore escape the ion pair and become a "free carbocation". The nucleophilic chloride can attack any carbon with positive charge, not just the one closest. Since two carbons have partial positive charge, two products result.

11-21

(a)

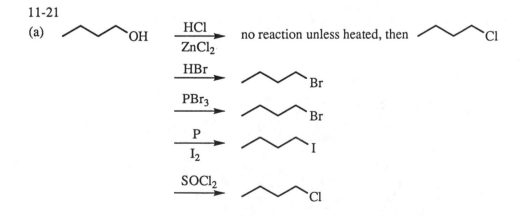

11-21 continued

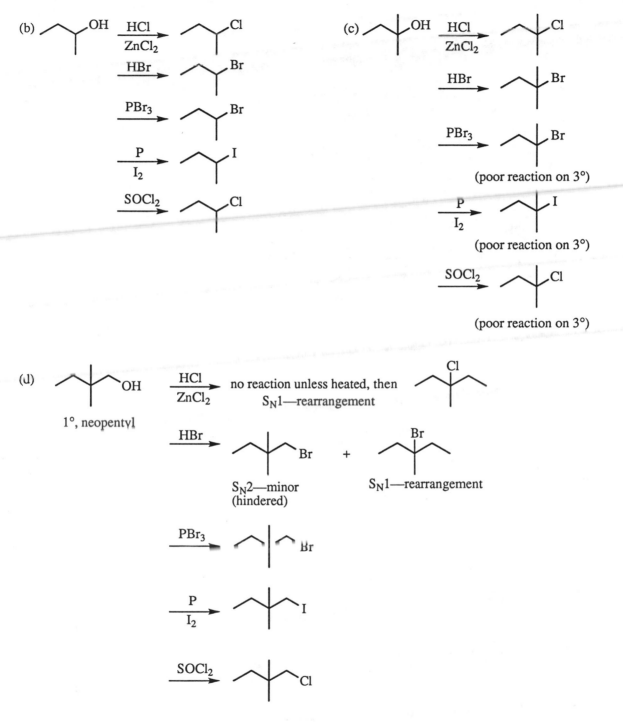

(b)   OH $\xrightarrow[\text{ZnCl}_2]{\text{HCl}}$ Cl

$\xrightarrow{\text{HBr}}$ Br

$\xrightarrow{\text{PBr}_3}$ Br

$\xrightarrow[\text{I}_2]{\text{P}}$ I

$\xrightarrow{\text{SOCl}_2}$ Cl

(c)   OH $\xrightarrow[\text{ZnCl}_2]{\text{HCl}}$ Cl

$\xrightarrow{\text{HBr}}$ Br

$\xrightarrow{\text{PBr}_3}$ Br

(poor reaction on 3°)

$\xrightarrow[\text{I}_2]{\text{P}}$ I

(poor reaction on 3°)

$\xrightarrow{\text{SOCl}_2}$ Cl

(poor reaction on 3°)

(d)   OH $\xrightarrow[\text{ZnCl}_2]{\text{HCl}}$ no reaction unless heated, then   Cl

S_N1—rearrangement

1°, neopentyl

$\xrightarrow{\text{HBr}}$ Br   +   Br

S_N2—minor S_N1—rearrangement
(hindered)

$\xrightarrow{\text{PBr}_3}$ Br

$\xrightarrow[\text{I}_2]{\text{P}}$ I

$\xrightarrow{\text{SOCl}_2}$ Cl

11-22

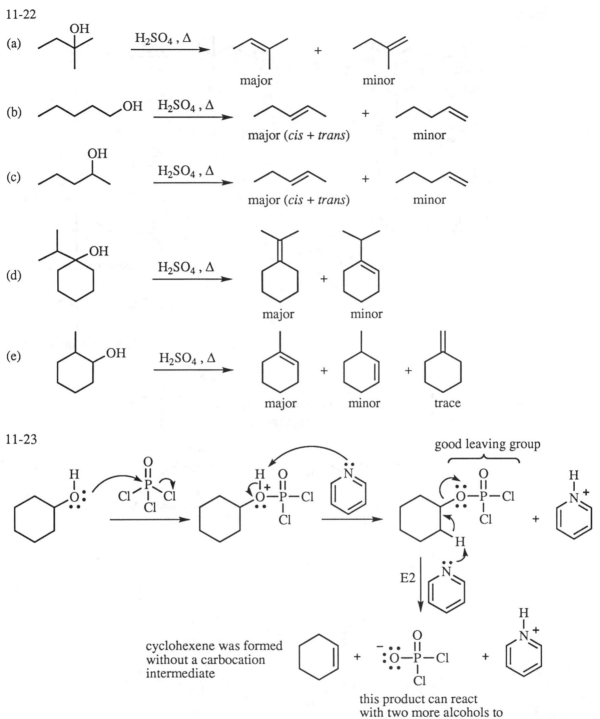

(a) H₂SO₄, Δ → major + minor

(b) H₂SO₄, Δ → major (*cis* + *trans*) + minor

(c) H₂SO₄, Δ → major (*cis* + *trans*) + minor

(d) H₂SO₄, Δ → major + minor

(e) H₂SO₄, Δ → major + minor + trace

11-23

good leaving group

E2

cyclohexene was formed
without a carbocation
intermediate

this product can react
with two more alcohols to
become leaving groups in
the E2 elimination

11-24

Both mechanisms begin with protonation of the oxygen.

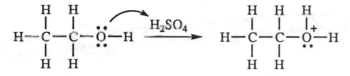

One mechanism involves another molecule of ethanol acting as a base, giving elimination.

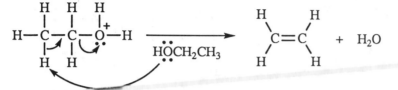

The other mechanism involves another molecule of ethanol acting as a nucleophile, giving substitution.

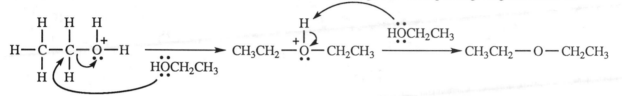

11-25 An equimolar mixture of methanol and ethanol would produce all three possible ethers. The difficulty in separating these compounds would preclude this method from being a practical route to any one of them. This method is practical only for symmetric ethers, that is, where both alkyl groups are identical.

$$CH_3CH_2OH + HOCH_3 \xrightarrow[\Delta]{H^+} H_2O + CH_3CH_2OCH_3 + CH_3OCH_3 + CH_3CH_2OCH_2CH_3$$

11-26

$$CH_3-\ddot{O}-H \xrightarrow{H_2SO_4} CH_3-\overset{H}{\underset{+}{\ddot{O}}}-H \xrightarrow[CH_3-\ddot{O}-H]{-H_2O} CH_3-\overset{H}{\underset{+}{\ddot{O}}}-CH_3 \xrightarrow{CH_3\ddot{O}H} CH_3-O-CH_3$$

11-27

(a) 2 $H_3C-\overset{\overset{O}{\|}}{C}-H$

(b) cyclopentanone $+$ $CH_2=O$

(c) acetophenone $+$ propanal

(d) bicyclic dialdehyde

11-28

(a)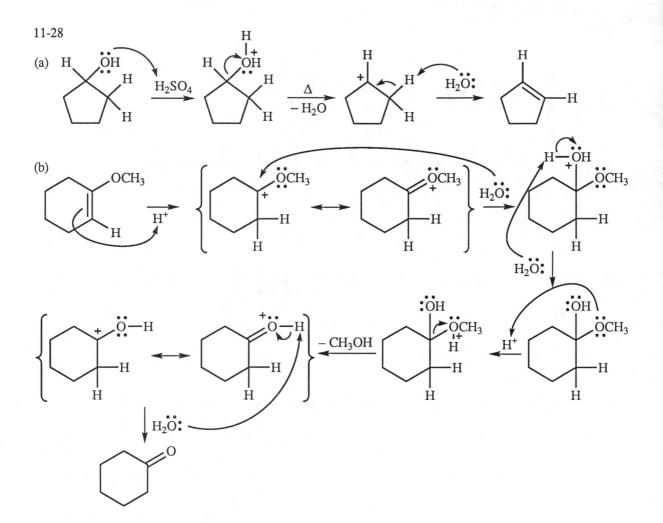

(b)

(c)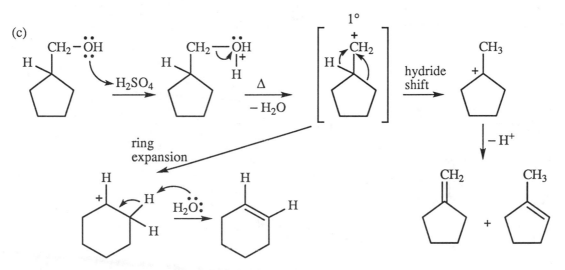

222

11-28 continued

(d)

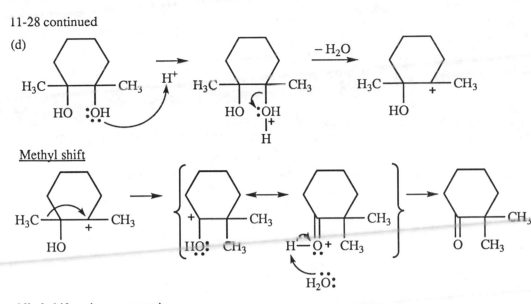

Methyl shift

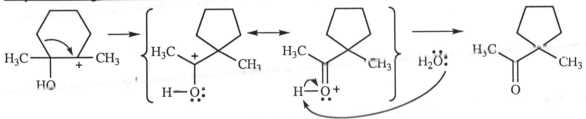

Alkyl shift—ring *contraction*

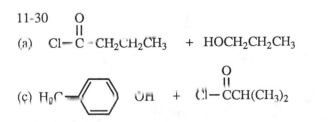

11-29

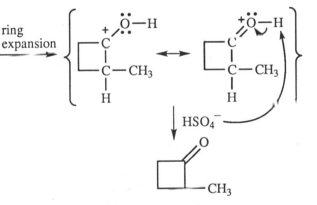

Similar to the pinacol rearrangement, this mechanism involves a carbocation next to an alcohol, with rearrangement to a protonated carbonyl. Relief of some ring strain in the cyclopropane is an added advantage of the rearrangement.

11-30

(a) $\text{Cl}-\overset{\text{O}}{\overset{\|}{\text{C}}}-\text{CH}_2\text{CH}_2\text{CH}_3$ + $\text{HOCH}_2\text{CH}_2\text{CH}_3$

(b) $\text{CH}_3(\text{CH}_2)_3\text{OH}$ + $\text{Cl}-\overset{\text{O}}{\overset{\|}{\text{C}}}\text{CH}_2\text{CH}_3$

(c) H_3C— 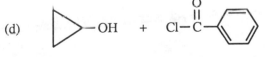 —OH + $\text{Cl}-\overset{\text{O}}{\overset{\|}{\text{C}}}\text{CH}(\text{CH}_3)_2$

(d) ▷—OH + $\text{Cl}-\overset{\text{O}}{\overset{\|}{\text{C}}}$—

11-31

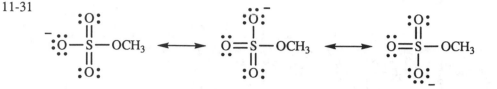

11-32 Proton transfer (acid-base) reactions are much faster than almost any other reaction. Methoxide will act as a base and remove a proton from the oxygen much faster than methoxide will act as a nucleophile and displace water.

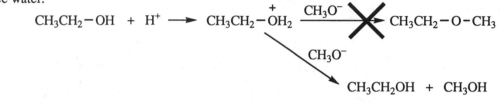

11-33

(a)

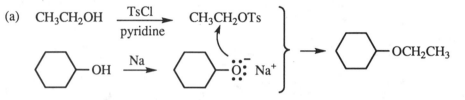

(b) There are two problems with this attempted bimolecular dehydration. First, all three possible ether combinations of cyclohexanol and ethanol would be produced. Second, hot sulfuric acid are the conditions for dehydrating secondary alcohols like cyclohexanol, so elimination would compete with substitution.

11-34

(a) What the student did:

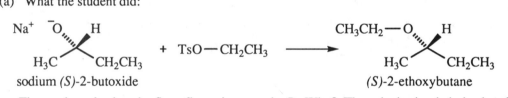

sodium (S)-2-butoxide   (S)-2-ethoxybutane

The product also has the S configuration, not the R. Why? The substitution is indeed an S_N2 reaction, but the *substitution did not take place at the chiral center*, so the configuration of the starting material is retained, not inverted.

(b) There are two ways to make (R)-2-ethoxybutane. Start with (R)-2-butanol, make the anion, and substitute on ethyl tosylate similar to part (a), or do an S_N2 inversion at the chiral center of (S)-2-butanol. S_N2 works better at 1° carbons so the former method would be preferred to the latter.

(c) This is not the optimum method because it requires S_N2 at a 2° carbon.

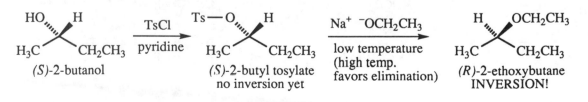

11-35

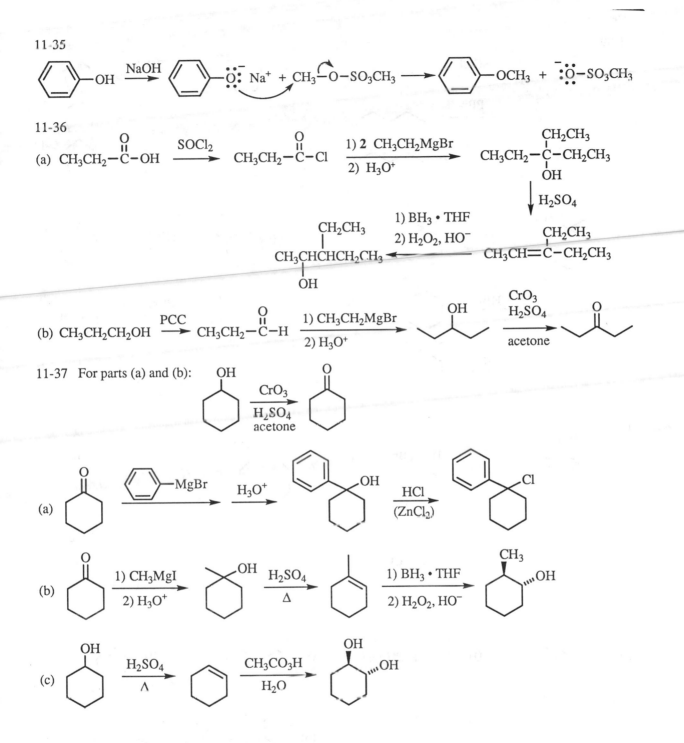

11-36

(a) $CH_3CH_2-\overset{O}{\overset{\|}{C}}-OH$ $\xrightarrow{SOCl_2}$ $CH_3CH_2-\overset{O}{\overset{\|}{C}}-Cl$ $\xrightarrow[\text{2) } H_3O^+]{\text{1) } 2 \ CH_3CH_2MgBr}$ $CH_3CH_2-\overset{CH_2CH_3}{\underset{OH}{\overset{|}{\underset{|}{C}}}}-CH_2CH_3$

$\downarrow H_2SO_4$

$CH_3\overset{CH_2CH_3}{\underset{OH}{\overset{|}{\underset{|}{CHCHCH_2CH_3}}}}$ $\xleftarrow[\text{2) } H_2O_2, HO^-]{\text{1) } BH_3 \cdot THF}$ $CH_3CH=\overset{CH_2CH_3}{\overset{|}{C}}-CH_2CH_3$

(b) $CH_3CH_2CH_2OH$ $\xrightarrow{PCC}$ $CH_3CH_2-\overset{O}{\overset{\|}{C}}-H$ $\xrightarrow[\text{2) } H_3O^+]{\text{1) } CH_3CH_2MgBr}$ [structure with OH] $\xrightarrow[\text{acetone}]{\overset{CrO_3}{H_2SO_4}}$ [ketone structure]

11-37 For parts (a) and (b):

[cyclohexanol] $\xrightarrow[\text{acetone}]{\overset{CrO_3}{H_2SO_4}}$ [cyclohexanone]

(a) [cyclohexanone] $\xrightarrow{\text{Ph-MgBr}}$ $\xrightarrow{H_3O^+}$ [1-phenylcyclohexanol] $\xrightarrow[(ZnCl_2)]{HCl}$ [1-phenyl-1-chlorocyclohexane]

(b) [cyclohexanone] $\xrightarrow[\text{2) } H_3O^+]{\text{1) } CH_3MgI}$ [1-methylcyclohexanol] $\xrightarrow[\Delta]{H_2SO_4}$ [1-methylcyclohexene] $\xrightarrow[\text{2) } H_2O_2, HO^-]{\text{1) } BH_3 \cdot THF}$ [2-methylcyclohexanol]

(c) [cyclohexanol] $\xrightarrow{\overset{H_2SO_4}{\Lambda}}$ [cyclohexene] $\xrightarrow[H_2O]{CH_3CO_3H}$ [trans-cyclohexane-1,2-diol]

11-38 Please refer to solution 1-19, page 12 of this Solutions Manual.

11-39

(a)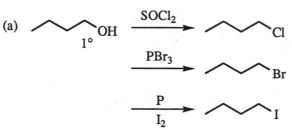

(b)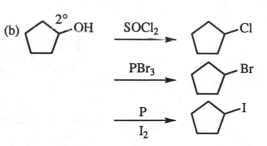

(c)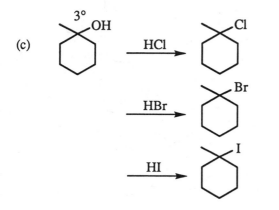

(d)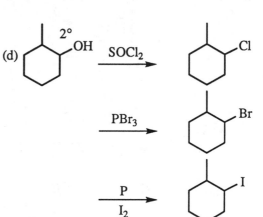

11-40

(a)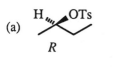
R

(b)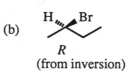
R
(from inversion)

(c)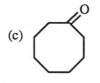

(d)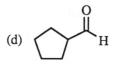

(e)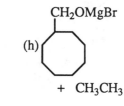

(f)

(g)

(h)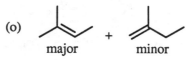
+ CH₃CH₃

(i)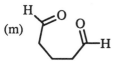

(j)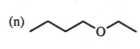
+ CH₃OH

(k)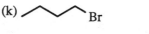

(l)

(m)

(n)

(o)
major   +   minor

11-41

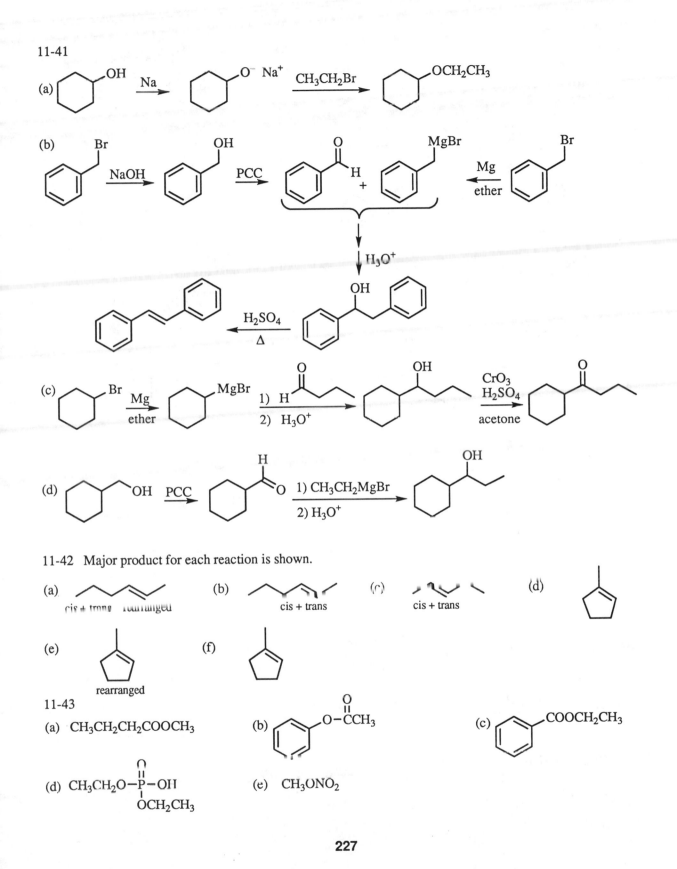

11-42 Major product for each reaction is shown.

(a) cis + trans rearranged

(b) cis + trans

(c) cis + trans

(d)

(e) rearranged

(f)

11-43

(a) CH₃CH₂CH₂COOCH₃

(b)

(c)

(d) CH₃CH₂O—P—OH

(e) CH₃ONO₂

227

11-44

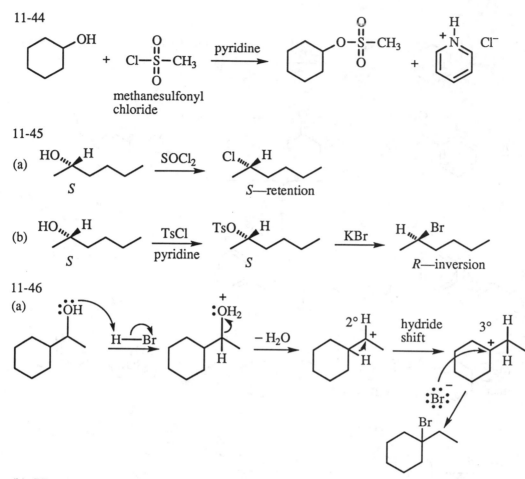

methanesulfonyl
chloride

11-45

(a)

HO,,,H → SOCl₂ → Cl,,,H
S   S—retention

(b)

HO,,,H → TsCl/pyridine → TsO,,,H → KBr → H,,,Br
S   S   R—inversion

11-46

(a)

2° H   hydride shift   3° H

(b) PBr₃ converts alcohols to bromides without rearrangement because no carbocation intermediate is produced.

OH → PBr₃ → Br

11-47

(a)   OH → PCC → H, O

(b)   OH → PBr₃ → Br

(c)   OH → Na → O⁻ Na⁺ → CH₃I → OCH₃

11-47 continued

(d)

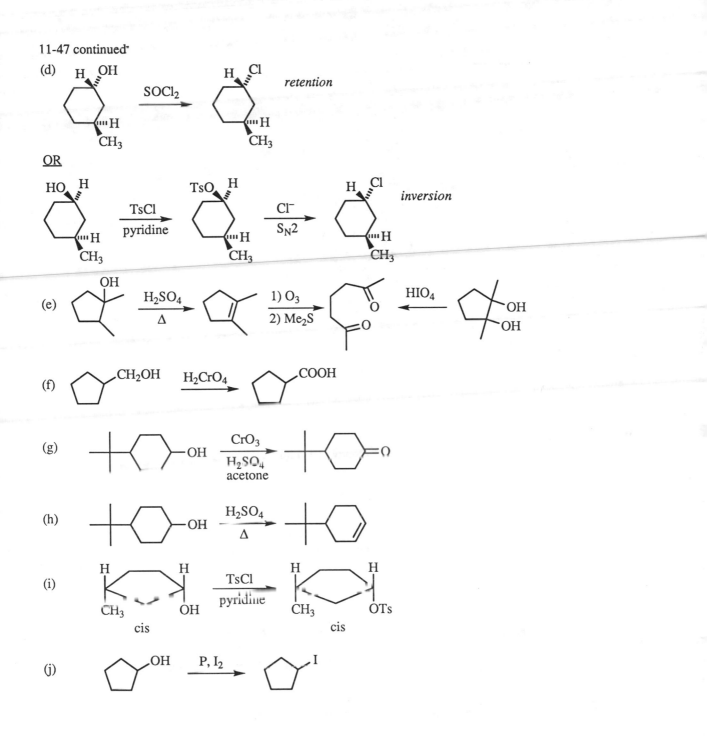

(e)

(f)

(g)

(h)

(i)

cis   cis

(j)

11-48

(a)

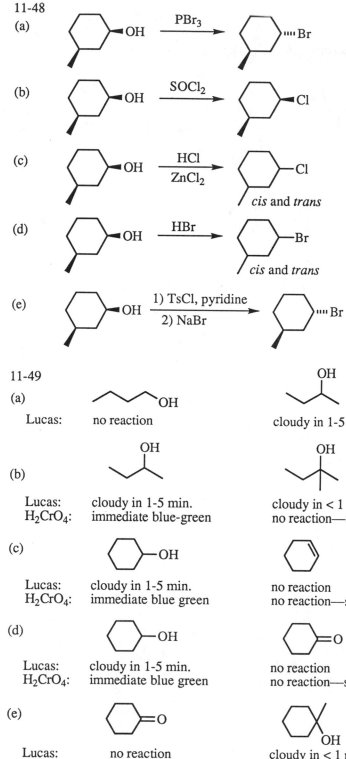

PBr₃

(b)

SOCl₂

(c)

HCl
ZnCl₂

cis and *trans*

(d)

HBr

cis and *trans*

(e)

1) TsCl, pyridine
2) NaBr

11-49

(a)

Lucas:　　　no reaction

cloudy in 1-5 min.

(b)

Lucas:　　　cloudy in 1-5 min.
H₂CrO₄:　　immediate blue-green

cloudy in < 1 min.
no reaction—stays orange

(c)

Lucas:　　　cloudy in 1-5 min.
H₂CrO₄:　　immediate blue green

no reaction
no reaction—stays orange

(d)

Lucas:　　　cloudy in 1-5 min.
H₂CrO₄:　　immediate blue green

no reaction
no reaction—stays orange

(e)

Lucas:　　　no reaction

cloudy in < 1 min.

11-50

(a)

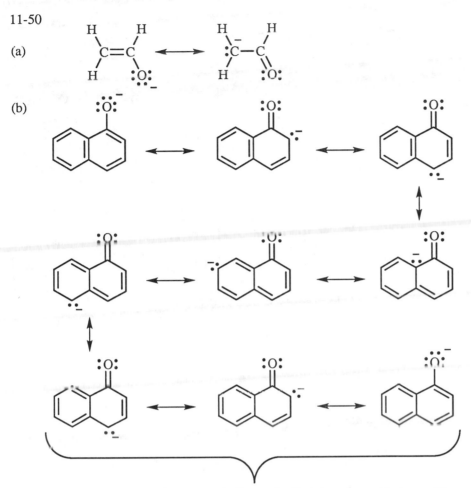

(b)

These last three resonance forms are similar to the first three but with the positions of the electrons in the left ring changed. To be rigorously correct, these three should be included, but most chemists would not write them since they do not reveal extra charge delocalization; understand that they would still be significant, even if not written with the others.

(c)

11-51

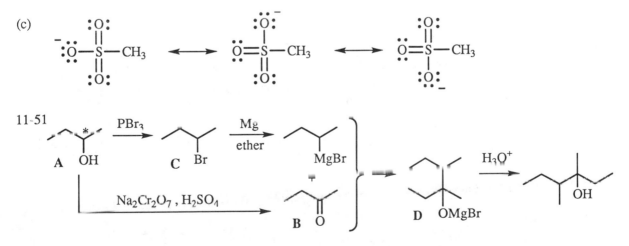

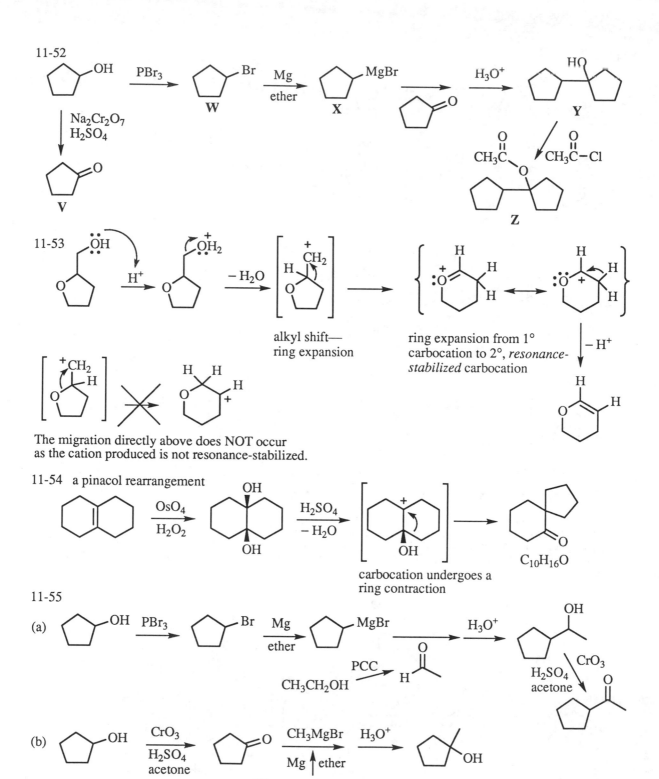

11-52

11-53

alkyl shift—
ring expansion

ring expansion from 1°
carbocation to 2°, *resonance-stabilized* carbocation

−H⁺

The migration directly above does NOT occur
as the cation produced is not resonance-stabilized.

11-54 a pinacol rearrangement

$C_{10}H_{16}O$

carbocation undergoes a
ring contraction

11-55

(a)

(b)

232

11-55 continued

(c)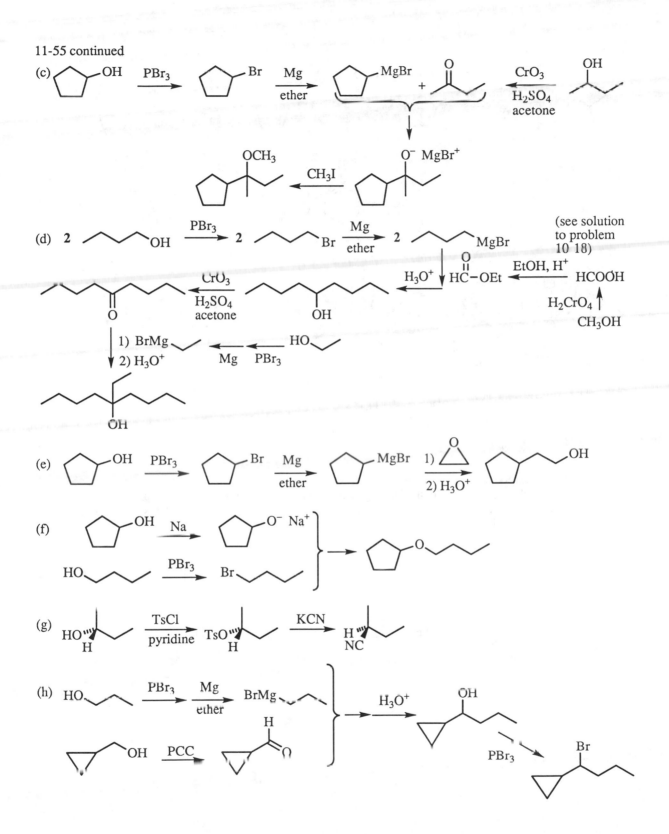

233

11-56

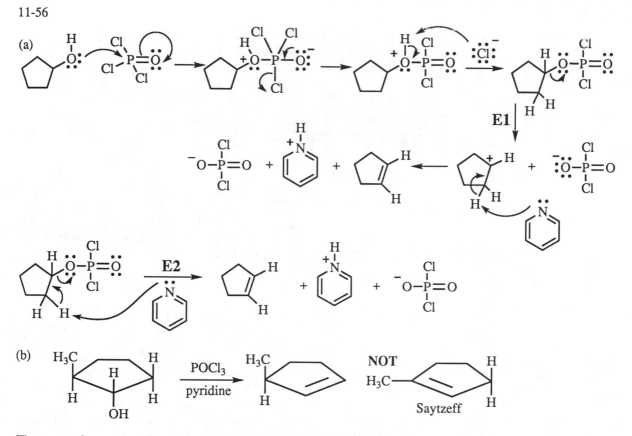

(b)

H₃C ... H / H / H / H / OH →(POCl₃ / pyridine)→ H₃C ... H / H   NOT   H₃C— ... H / H / Saytzeff

There must be a stereochemical requirement in this elimination. If the Saytzeff alkene is not produced because the methyl group is *trans* to the leaving group, then the H and the leaving group must be *trans* and the elimination must be anti—the characteristic stereochemistry of E2 elimination. This evidence differentiates between the two possibilities in part (a).

11-57

Compound X : —must be a 1° or 2° alcohol with an alkene; no reaction with Lucas leads to a 1° alcohol; can't be allylic as this would give a positive Lucas test

Compound Y : —must be a cyclic ether, not an alcohol and not an alkene; other isomers of cyclic ethers possible

11-58

(a)

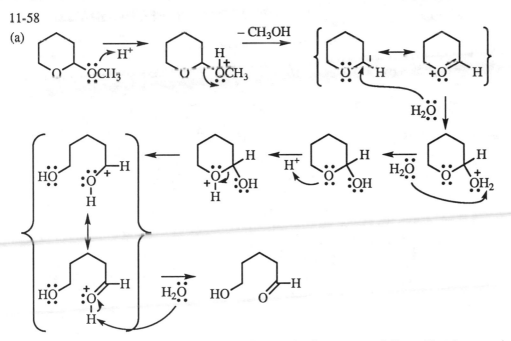

It is equally likely for protonation to occur first on the ring oxygen, followed by ring opening, then replacement of OCH₃ by water.

(b)

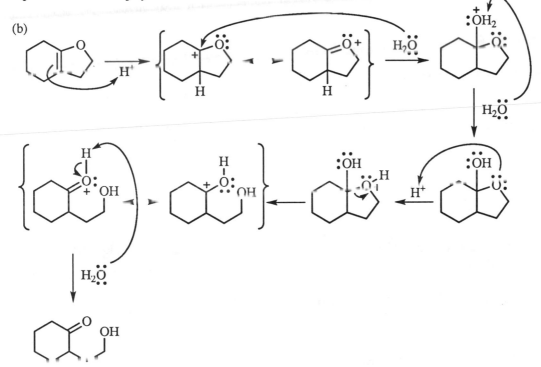

235

12-1 The table is completed by recognizing that: $(\overline{v})(\lambda) = 10{,}000$

$\overline{v}$ (cm^{-1})	4000	**3300**	**3003**	**2198**	1700	1640	1600	400
λ (μm)	2.50	3.03	3.33	4.55	**5.88**	**6.10**	**6.25**	25.0

12-2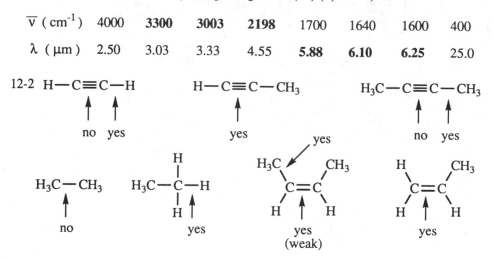

12-3
(a) alkene: C=C at 1640 cm^{-1} , =C—H at 3080 cm^{-1}
(b) alkane: no peaks indicating sp or sp^2 carbons present
(c) terminal alkyne: C≡C at 2100 cm^{-1} , ≡C—H at 3300 cm^{-1} ; =C—H at 3050 cm^{-1}, C=C at 1600 cm^{-1}
(perhaps benzene)

12-4
(a) 2° amine, R—NH—R: one peak at 3300 cm^{-1} indicates an N—H bond; this spectrum also shows a C=C at 1640 cm^{-1}
(b) carboxylic acid: the extremely broad absorption in the 2500-3500 cm^{-1} range, with a "shoulder" around 2500-2700 cm^{-1} , and a C=O at 1710 cm^{-1} , are compelling evidence for a carboxylic acid
(c) alcohol: strong, broad O—H at 3330 cm^{-1}

12-5
(a) conjugated ketone: the small peak at 3030 cm^{-1} suggests =C—H, and the strong peak at 1685 cm^{-1} is consistent with a ketone conjugated with the alkene. The C=C is indicated by a very small peak around 1600 cm^{-1} .
(b) ester: the C=O absorption at 1738 cm^{-1} (higher than the ketone's 1710 cm^{-1}), in conjunction with the strong C—O at 1200 cm^{-1}, points to an ester
(c) amide: the two peaks at 3160-3360 cm^{-1} are likely to be an NH_2 group; the strong peak at 1640 cm^{-1} is too strong for an alkene, so it must be a different type of C=X, in this case a C=O, so low because it is part of an amide

12-6

(a) The small peak at 1642 cm^{-1} indicates a C=C, consistent with the =C—H at 3080 cm^{-1}. This appears to be a simple alkene.

(b) The strong absorption at 1691 cm^{-1} is unmistakably a C=O. The smaller peak at 1626 cm^{-1} indicates a C=C, probably conjugated with the C=O. The two peaks at 2712 cm^{-1} and at 2814 cm^{-1} represent H—C=O confirming that this is an aldehyde.

(c) The strong peak at 1650 cm^{-1} is C=C, probably conjugated as it is unusually strong. The 1703 cm^{-1} peak appears to be a conjugated C=O, undeniably a carboxylic acid because of the strong, broad O—H absorption from 2400-3400 cm^{-1}.

(d) The C=O absorption at 1742 cm^{-1} coupled with C—O at 1220 cm^{-1} suggest an ester. The small peak at 1604 cm^{-1}, peaks above 3000 cm^{-1}, and peaks in the 600-800 cm^{-1} region indicate a benzene ring.

12-7

(a) The M and M+2 peaks of equal intensity identify the presence of bromine. The mass of M (156) minus the weight of the lighter isotope of bromine (79) gives the mass of the rest of the molecule: $156 - 79 = 77$. The C$_6$H$_5$ (phenyl) group weighs 77; this compound is bromobenzene, C$_6$H$_5$Br.

(b) The m/z 127 peak shows that iodine is present. The molecular ion minus iodine gives the remainder of the molecule: $156 - 127 = 29$. The C$_2$H$_5$ (ethyl) group weighs 29; this compound is iodoethane, C$_2$H$_5$I.

(c) The M and M+2 peaks have relative intensities of about 3:1, a sure sign of chlorine. The mass of M minus the mass of the lighter isotope of chlorine gives the mass of the remainder of the molecule: $90 - 35 = 55$. A fragment of mass 55 is not one of the common alkyl groups (15, 29, 43, 57, etc., increasing in increments of 14 mass units (CH$_2$)), so the presence of an atom like oxygen must be considered. In addition to the chlorine atom, mass 55 could be C$_4$H$_7$ or C$_3$H$_3$O. Possible molecular formulas are C$_4$H$_7$Cl or C$_3$H$_3$ClO.

(d) The odd-mass molecular ion indicates the presence of an odd number of nitrogen atoms (always begin by assuming *one* nitrogen). The rest of the molecule must be: $115 - 14 = 101$; this is most likely C$_7$H$_{17}$ which is a heptyl group, C$_7$H$_{15}$, plus two hydrogens on the nitrogen. The formula C$_7$H$_{17}$N is the correct formula of a molecule with no elements of unsaturation.

12-8 Recall that radicals are not detected in mass spectrometry; only positively-charged ions are detected.

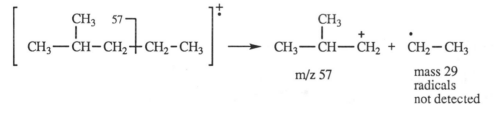

12-9

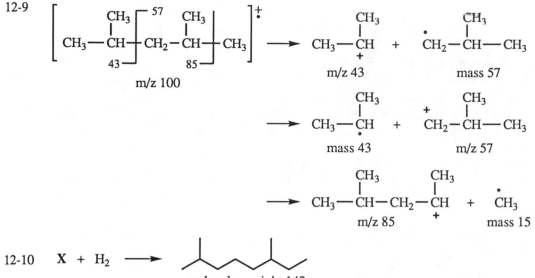

12-10   X + H$_2$ $\longrightarrow$

molecular weight 142

X has molecular weight 140 (from the mass spec), so X has one double bond—where?

A mass spec fragment of 57 is C$_4$H$_9$, a butyl group with no unsaturation; a mass spec fragment of 83 is C$_6$H$_{11}$, a six-carbon group with one unsaturation. If we presume that the cleavage is most likely allylic, then the alkene must be in one of only two possible positions. **X** must be one of these:

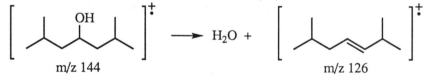

12-11   2,6-Dimethyl-4-heptanol, C$_9$H$_{20}$O, has molecular weight 144. The highest mass peak at 126 is *not* the molecular ion, but rather is the loss of water (18) from the molecular ion.

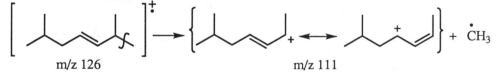

The peak at m/z 111 is loss of another 15 (CH$_3$) from the fragment of m/z 126. This is called allylic cleavage; it generates a 2°, allylic, resonance-stabilized carbocation.

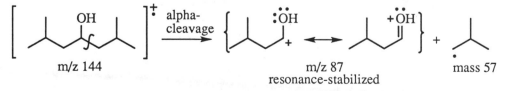

The peak at m/z 87 results from fragmentation on one side of the alcohol:

12-12 Please refer to solution 1-19, page 12 of this Solutions Manual.

12-13 Divide the numbers into 10,000 to arrive at the answer.

(a) 1603 cm^{-1} (b) 2959 cm^{-1} (c) 1709 cm^{-1} (d) 1739 cm^{-1} (e) 2212 cm^{-1} (f) 3300 cm^{-1}

12-14

(a)

1660 cm^{-1} 1710 cm^{-1}
                              stronger absorption—larger dipole

(b)

1660 cm^{-1} 1640 cm^{-1}
                              stronger absorption—larger dipole

(c)

1660 cm^{-1} 1660 cm^{-1}
stronger absorption—
larger dipole

(d)

1660 cm^{-1} 1660 cm^{-1}
(no dipole moment)   **stronger absorption**   larger dipole

12-15

(a)

1660 cm^{-1} 1660 cm^{-1}
weak or non-existent   moderate intensity

(b)

1620 cm^{-1} and 1645 cm^{-1}

conjugated   not conjugated

239

12-15 continued

(c) both carbonyls show strong absorptions around 1710 cm^{-1}

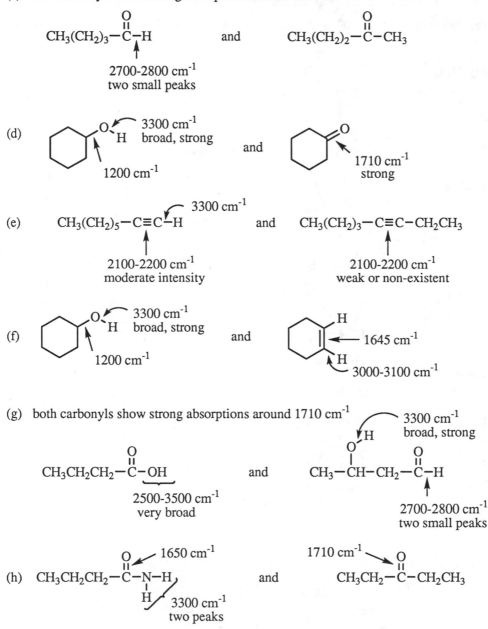

$$CH_3(CH_2)_3-\overset{\overset{\displaystyle O}{\|}}{C}-H \qquad and \qquad CH_3(CH_2)_2-\overset{\overset{\displaystyle O}{\|}}{C}-CH_3$$

2700-2800 cm^{-1}
two small peaks

(d)

3300 cm^{-1}
broad, strong

1200 cm^{-1}

and

1710 cm^{-1}
strong

(e) $CH_3(CH_2)_5-C\equiv C-H$

3300 cm^{-1}

2100-2200 cm^{-1}
moderate intensity

and $CH_3(CH_2)_3-C\equiv C-CH_2CH_3$

2100-2200 cm^{-1}
weak or non-existent

(f)

3300 cm^{-1}
broad, strong

1200 cm^{-1}

and

H

1645 cm^{-1}

H

3000-3100 cm^{-1}

(g) both carbonyls show strong absorptions around 1710 cm^{-1}

3300 cm^{-1}
broad, strong

$$CH_3CH_2CH_2-\overset{\overset{\displaystyle O}{\|}}{C}-OH$$

2500-3500 cm^{-1}
very broad

and

$$CH_3-\overset{\overset{\displaystyle O-H}{|}}{CH}-CH_2-\overset{\overset{\displaystyle O}{\|}}{C}-H$$

2700-2800 cm^{-1}
two small peaks

(h) $CH_3CH_2CH_2-\overset{\overset{\displaystyle O}{\|}}{C}-N-H$

1650 cm^{-1}

H

3300 cm^{-1}
two peaks

and

1710 cm^{-1}

$$CH_3CH_2-\overset{\overset{\displaystyle O}{\|}}{C}-CH_2CH_3$$

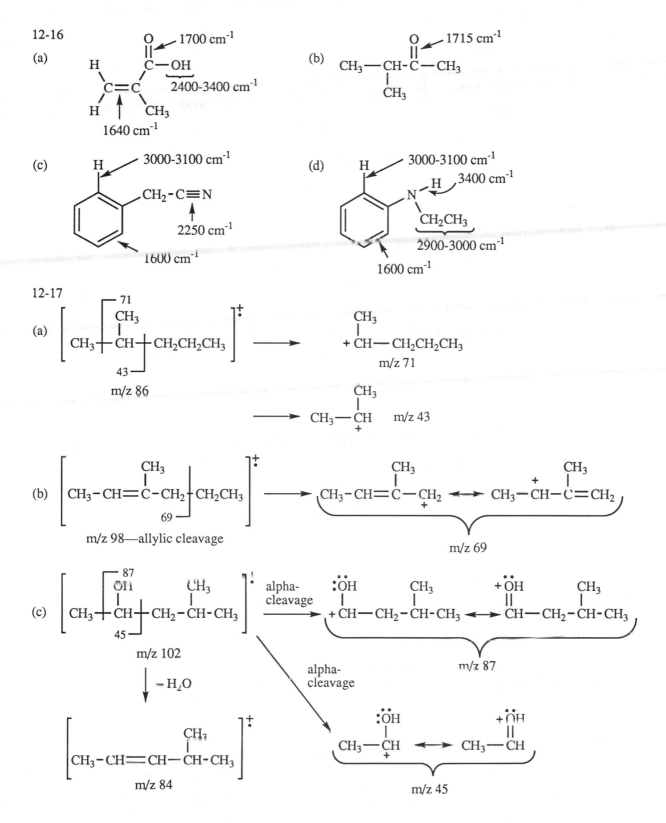

12-18

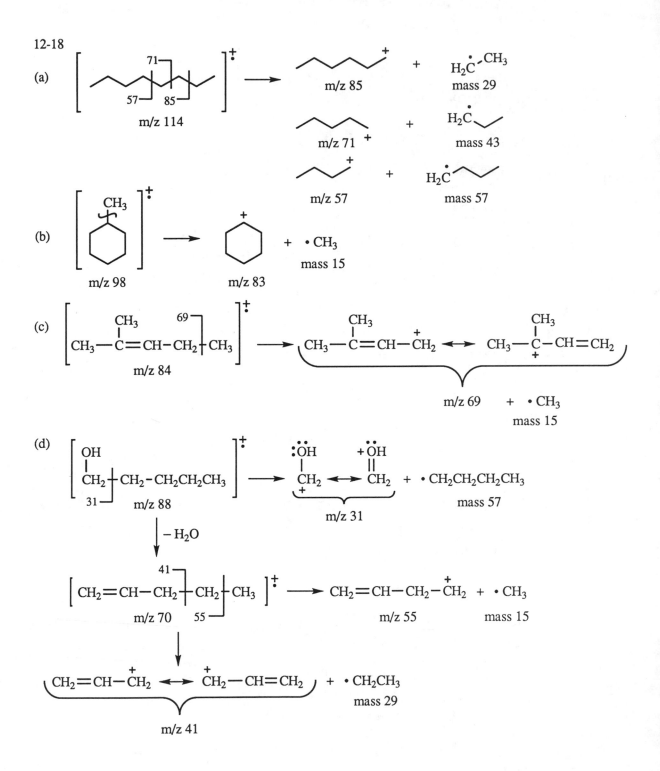

242

12-19

(a) The characteristic frequencies of the OH absorption and the C=C absorption will indicate the presence or absence of the groups. A spectrum with an absorption around 3300 cm^{-1} will have some cyclohexanol in it; if that same spectrum also has a peak at 1645 cm^{-1}, then the sample will also contain some cyclohexene. Pure samples will have peaks representative of only one of the compounds and not the other. Note that *quantitation* of the two compounds would be very difficult by IR because the strength of absorptions are very different. Usually, other methods are used in preference to IR for quantitative measurements.

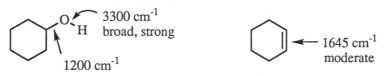

(b) Mass spectrometry can be misleading with alcohols. Usually, alcohols dehydrate in the inlet system of a mass spectrometer, and the characteristic peaks observed in the mass spectrum are those of the alkene, not of the parent alcohol. For this particular analysis, mass spectrometry would be unreliable and perhaps misleading.

12-20

(a) The "student prep" compound must be 1-bromobutane. The most obvious feature of the mass spectrum is the pair of peaks at M and M+2 of approximately equal heights, characteristic of a bromine atom. Loss of bromine (79) from the molecular ion at 136 gives a mass of 57, C_4H_9, a butyl group. Which of the four possible butyl groups? The peaks at 107 (loss of 29, C_2H_5) and 93 (loss of 43, C_3H_7) are consistent with a linear chain, not a branched chain.

(b) The base peak at 57 is so strong because the carbon-halogen bond is the weakest in the molecule. Typically, loss of halogen is the dominant fragmentation in alkyl halides.

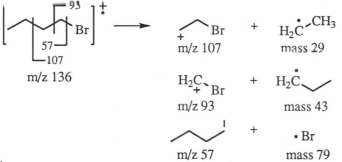

12-21

(a) Deuterium has twice the mass of hydrogen, but similar spring constant, k. Compare the frequency of C—D vibration to C—H vibration by setting up a ratio, changing only the mass (substitute 2m for m).

$$\frac{v_D}{v_H} = \frac{\sqrt{k/2m}}{\sqrt{k/m}} = \frac{\sqrt{1/2}\ \sqrt{k/m}}{\sqrt{k/m}} = \sqrt{1/2} = 0.707$$

$$v_D = 0.707\ v_H = 0.707\ (3000\ cm^{-1}) \approx 2100\ cm^{-1}$$

(b) The functional group most likely to be confused with a C—D stretch is the alkyne (carbon-carbon triple bond), which appears in the same region and is often very weak.

12-22

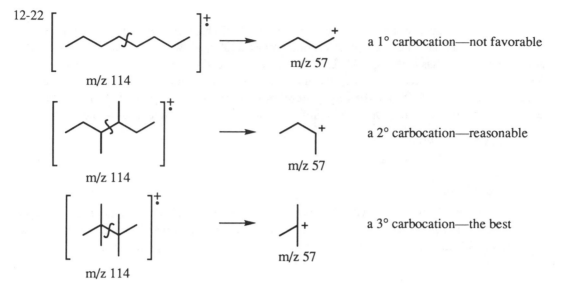

a 1° carbocation—not favorable

a 2° carbocation—reasonable

a 3° carbocation—the best

The most likely fragmentation of 2,2,3,3-tetramethylbutane will give a 3° carbocation, the most stable of the common alkyl cations. The molecular ion should be small or non-existent while m/z 57 is likely to be the base peak.

12-23

(a) The information that this mystery compound is a hydrocarbon makes interpreting the mass spectrum much easier. (It is relatively simple to tell if a compound has chlorine, bromine, or nitrogen by a mass spectrum, but oxygen is difficult to determine by mass spectrometry alone.) A hydrocarbon with molecular ion of 110 can have only 8 carbons (8 x 12 = 96) and 14 hydrogens. The formula C_8H_{14} has two elements of unsaturation.

(b) The IR will be useful in determining what the elements of unsaturation are. Cycloalkanes are generally not distinguishable in the IR. An alkene should have an absorption around 1600-1650 cm^{-1}; none is present in this IR. An alkyne should have a small, sharp peak around 2200 cm^{-1}—PRESENT AT 2120 cm^{-1}! Also, a sharp peak around 3300 cm^{-1} indicates a hydrogen on an alkyne, so the alkyne is at one end of the molecule. Both elements of unsaturation are accounted for by the alkyne.

(c) The only question is how are the other carbons arranged. The mass spectrum shows a progression of peaks from the molecular ion at 110 to 95 (loss of CH_3), to 81 (loss of C_2H_5), to 67 (loss of C_3H_7). The mass spectrum suggests it is a linear chain. The extra evidence that hydrogenation of the mystery compound gives *n*-octane verifies that the chain is linear. The original compound must be 1-octyne.

(d) The base peak is so strong because the ion produced is stabilized by resonance.

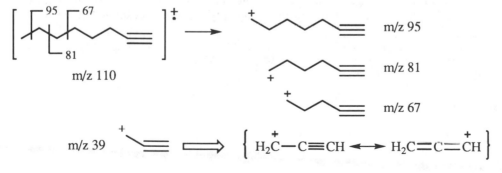

244

12-24

(a) and (b) The mass spec is consistent with the formula of the alkyne, C_8H_{14}, mass 110. The IR is not consistent with the alkyne, however. Often, symmetrically substituted alkynes have a miniscule $C{\equiv}C$ peak, so the fact that the IR does not show this peak does not prove that the alkyne is absent. The important evidence in the IR is the significant peak at 1620 cm^{-1}; this absorption is characteristic of a conjugated diene. Instead of the alkyne being formed, the reaction must have been a double elimination to the diene.

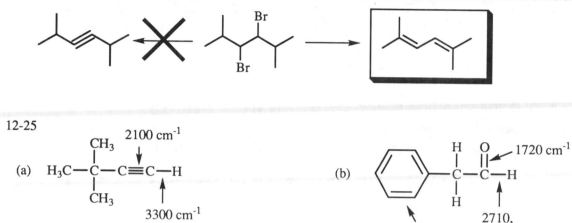

12-25

(a)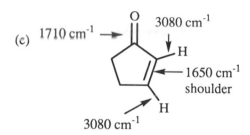

(b)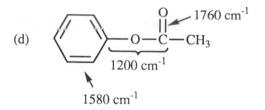

(c)

(d)

A benzene ring can be written with three alternating double bonds or with a circle in the ring. All of the carbons and hydrogens in an unsubstituted benzene ring are equivalent, regardless of which symbolism is used.

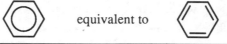

equivalent to

13-1

(a) $\dfrac{130 \text{ Hz}}{60 \times 10^6 \text{ Hz}} = 2.17 \times 10^{-6} = 2.17$ ppm downfield from TMS

(b) difference in magnetic field $= 14{,}092$ gauss $\times$ $(2.17 \times 10^{-6}) = 0.031$ gauss

(c) The chemical shift does not change with field strength: $\delta\,2.17$ at both 60 MHz and 300 MHz.

(d) $(2.17 \text{ ppm}) \times (300 \text{ MHz}) = (2.17 \times 10^{-6}) \times (300 \times 10^6 \text{ Hz}) = 651$ Hz

13-2 Numbers are chemical shift values, in ppm, derived from Table 13-3 and the Appendix in the text. *Your predictions should be in the given range, or within 0.5 ppm of the given value.*

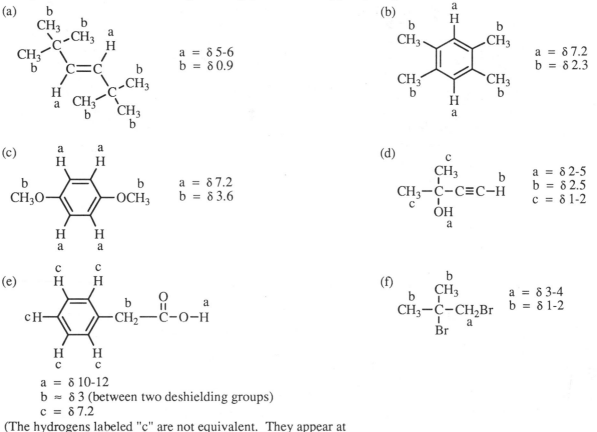

(a)
$a = \delta\,5\text{-}6$
$b = \delta\,0.9$

(b)
$a = \delta\,7.2$
$b = \delta\,2.3$

(c)
$a = \delta\,7.2$
$b = \delta\,3.6$

(d)
$a = \delta\,2\text{-}5$
$b = \delta\,2.5$
$c = \delta\,1\text{-}2$

(e)
$a = \delta\,10\text{-}12$
$b \approx \delta\,3$ (between two deshielding groups)
$c = \delta\,7.2$

(f)
$a = \delta\,3\text{-}4$
$b = \delta\,1\text{-}2$

(The hydrogens labeled "c" are not equivalent. They appear at roughly the same chemical shift because the substituent is neither strongly electron-donating nor withdrawing.)

13-3

(a) CH₃CH₂CH₂Cl
     c   b   a

three types of H

(b) CH₃CHCH₃
     b   a b
          |
          Cl

two types of H

(c)
            CH₃   a
  a | b   c
CH₃—C—CH₂CH₃
           |
          CH₃
           a

three types of H

(d)

five types of H

Note: NMR spectra drawn in this Solutions Manual will resemble 60 MHz spectra.

13-4

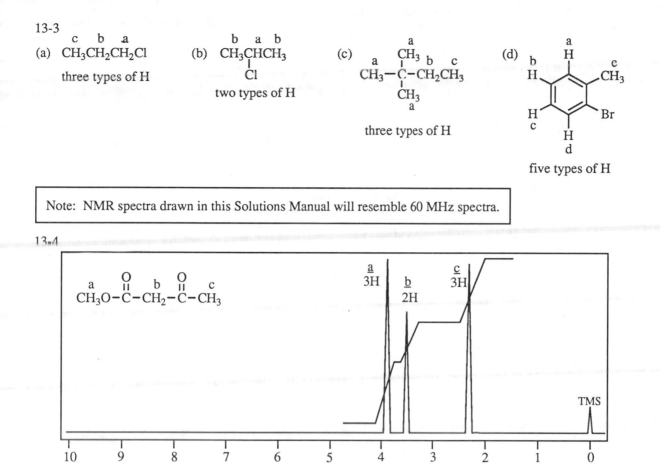

$$CH_3O-\overset{O}{\overset{\|}{\underset{a}{C}}}-\underset{b}{CH_2}-\overset{O}{\overset{\|}{\underset{}{C}}}-\underset{c}{CH_3}$$

13-5 The three spectra are identified with their structures. Data are given as chemical shift values, with the integration ratios of each peak given in parentheses.

Spectrum (a)

$$\underset{c}{CH_3}\overset{c}{}$$
$$CH_3-\underset{|}{\overset{|}{C}}-C\equiv C-H\quad a$$
$$\underset{OH}{}$$
$$b$$

a = δ 2.5 (1) (1H)
b = δ 2.3 (1) (1H)
c = δ 1.5 (6) (6H)

Spectrum (b)

a = δ 6.7 (2) (4H)
b = δ 3.7 (3) (6H)

Spectrum (c)

$$CH_3-\underset{|}{\overset{|}{C}}-CH_2Br$$
$$\underset{Br}{}$$

a = δ 3.9 (1) (2H)
b = δ 1.8 (3) (6H)

(This is the compound in Problem 13-2(f). You may wish to check your answer to that question against the spectrum.)

13-6 Chemical shift values are approximate and may vary slightly from yours. The splitting and integration values should match exactly, however.

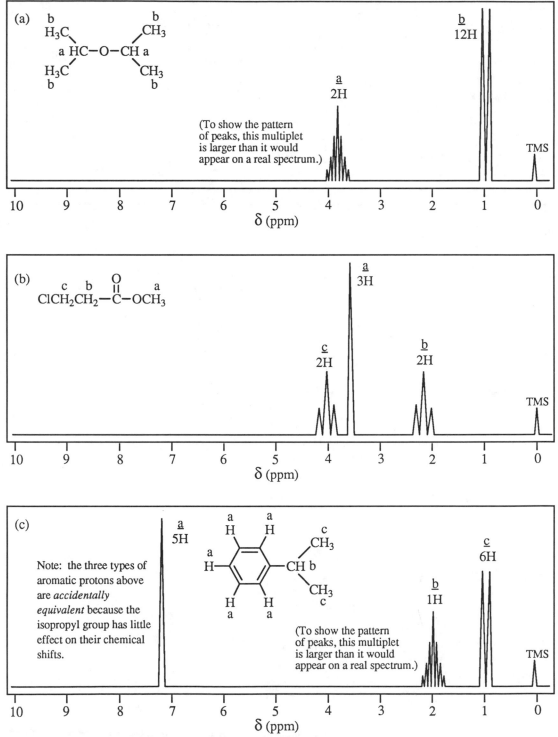

(a)

b
H_3C
a HC—O—CH a
H_3C CH_3
b   b

b
CH_3

b
12H

a
2H

(To show the pattern
of peaks, this multiplet
is larger than it would
appear on a real spectrum.)

TMS

δ (ppm)

(b)

$\underset{\text{c}}{\text{ClCH}_2}\underset{\text{b}}{\text{CH}_2}-\underset{\underset{\text{O}}{\|}}{\text{C}}-\underset{\text{a}}{\text{OCH}_3}$

a
3H

c
2H

b
2H

TMS

δ (ppm)

(c)

a
5H

Note: the three types of
aromatic protons above
are *accidentally*
equivalent because the
isopropyl group has little
effect on their chemical
shifts.

a   a
H   H

a
H—

c
CH_3

CH b

CH_3
c

H   H
a   a

(To show the pattern
of peaks, this multiplet
is larger than it would
appear on a real spectrum.)

b
1H

c
6H

TMS

δ (ppm)

248

13-6 continued

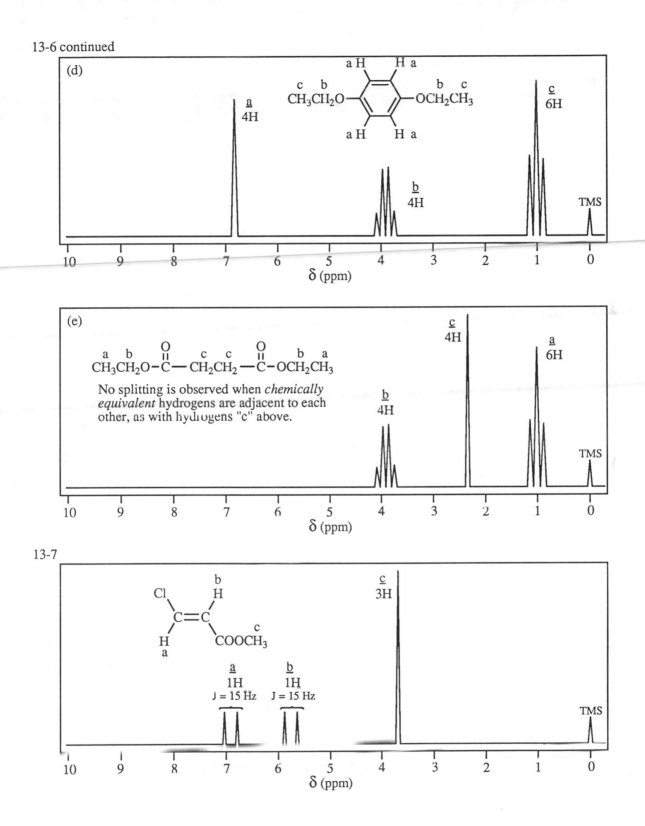

13-8

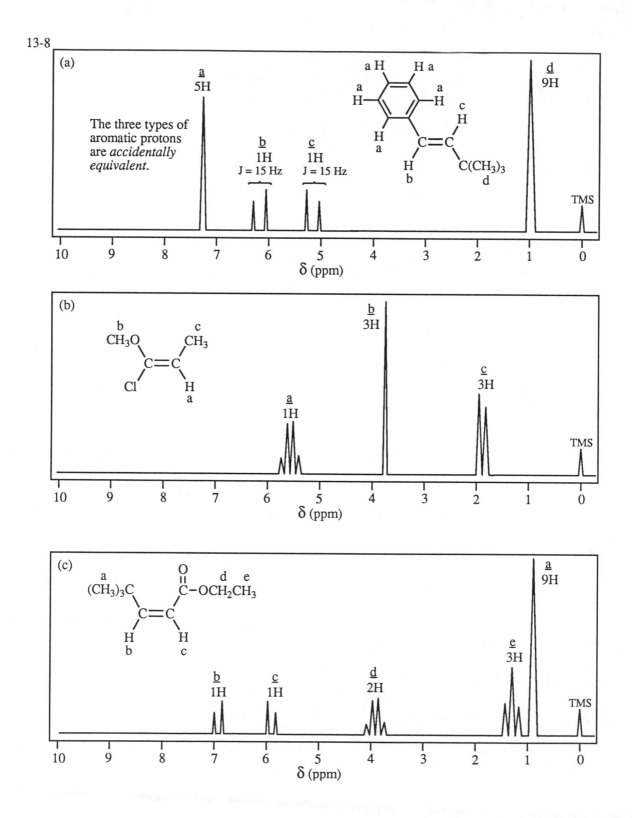

(d)

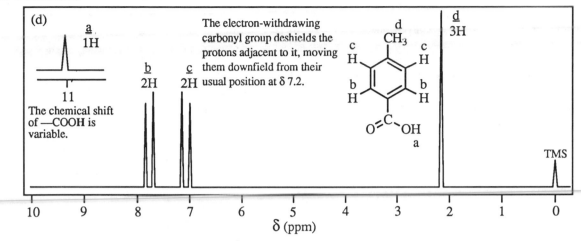

The electron-withdrawing carbonyl group deshields the protons adjacent to it, moving them downfield from their usual position at δ 7.2.

11
The chemical shift of —COOH is variable.

TMS

δ (ppm)

13-9 The formula C_3H_2NCl has three elements of unsaturation. The IR peak at 1650 cm⁻¹ indicates an alkene, while the absorption at 2200 cm⁻¹ must be from a nitrile (not enough carbons left for an alkyne). These two groups account for the three elements of unsaturation. So far, we have:

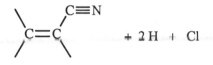

The NMR gives the coupling constant for the two protons as 14 Hz. This large J value shows the two protons as *trans* (*cis*, J = 10 Hz; geminal, J = 2 Hz). The structure must be:

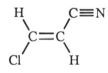

13-10

(a) C_3H_7Cl—no elements of unsaturation; 3 types of protons in the ratio of 2 : 2 : 3 .

$$\overset{a\quad b\quad c}{Cl—CH_2CH_2CH_3}$$

a = δ 3.5 (triplet, 2H); b = δ 1.8 (multiplet, 2H); c = δ 1.1 (triplet, 3H)

13-10 continued

(b) $C_9H_{10}O_2$—5 elements of unsaturation; four protons in the aromatic region of the NMR indicate a disubstituted benzene; the pair of doublets with $J = 8$ Hz indicate the substituents are on opposite sides of the ring (*para*).

$+ 3C + 6H + 2O + 1$ element of unsaturation

The other NMR signals are two 3H singlets, two CH_3 groups. One at δ 3.9 must be a CH_3O group. The other at δ 2.4 is most likely a CH_3 group on the benzene ring.

CH_3—⟨benzene⟩— $+ OCH_3 + C + O + 1$ element of unsaturation

One way to assemble these pieces consistent with the NMR is:

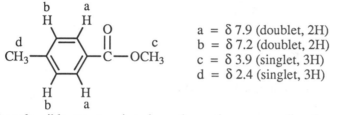

a = δ 7.9 (doublet, 2H)
b = δ 7.2 (doublet, 2H)
c = δ 3.9 (singlet, 3H)
d = δ 2.4 (singlet, 3H)

(Another plausible structure is to have the methoxy group directly on the ring and to put the carbonyl between the ring and the methyl. This does not fit the chemical shift values quite as well as the above structure, as the methyl would appear around δ 2.1 or 2.2 instead of 2.4.)

13-11

δ 3.4
triplet

δ 1.5
sextet

Br—CH_2—CH_2—CH_2—CH_3

δ 1.8
pentet

δ 0.9
triplet

(The two CH_2 groups in the middle of the chain are coupled with protons on either side. Even though the neighbors on either side are not chemically equivalent, the coupling constants of protons in an aliphatic chain are usually of the same magnitude, so the neighbors couple as if they were equivalent, and the **N+1** rule applies.)

13-12 H_c, δ 5.1

δ 5.1

$J_{ac} = 11$ Hz

$J_{bc} = 1.4$ Hz

$J_{bc} = 1.4$ Hz

13-13

(a)

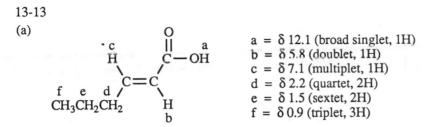

a = δ 12.1 (broad singlet, 1H)
b = δ 5.8 (doublet, 1H)
c = δ 7.1 (multiplet, 1H)
d = δ 2.2 (quartet, 2H)
e = δ 1.5 (sextet, 2H)
f = δ 0.9 (triplet, 3H)

(b) The vinyl proton at δ 7.1 is H_c; it is coupled with H_b and H_d, with two different coupling constants, J_{bc} and J_{cd}, respectively. The value of J_{bc} can be measured most precisely from the signal for H_b at δ 5.8; the two peaks are separated by about 15 Hz, corresponding to 0.05 ppm in a 300 MHz spectrum. The value of J_{cd} appears to be about the standard value 8 Hz, judging from the signal at δ 7.1. The splitting tree would thus appear:

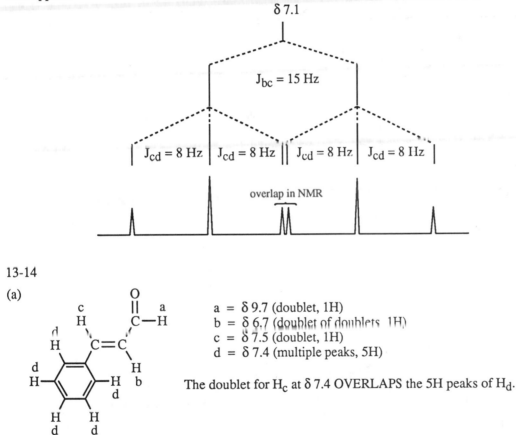

13-14

(a)

a = δ 9.7 (doublet, 1H)
b = δ 6.7 (doublet of doublets, 1H)
c = δ 7.5 (doublet, 1H)
d = δ 7.4 (multiple peaks, 5H)

The doublet for H_c at δ 7.4 OVERLAPS the 5H peaks of H_d.

(b) J_{ab} can be determined most accurately from H_a at δ 9.7: $J_{ab} \approx 8$ Hz, about the same as "normal" alkyl coupling.

J_{bc} can be measured from H_b at δ 6.7, as the distance between either the first and third peaks or the second and fourth peaks (see diagram on the next page): $J_{bc} \approx 18$ Hz, about double the "normal" alkyl coupling.

13-14 continued

(c)

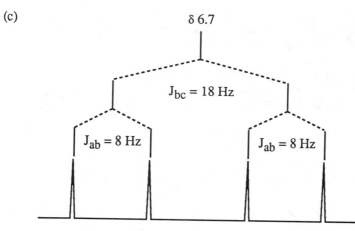

13-15

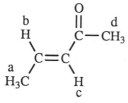

(a) a = δ 1.7   (b) a = doublet
     b ≈ δ 6.8   b = multiplet (two overlapping quartets—see part (c))
     c = δ 5-6   c = doublet
     d = δ 2.1   d = singlet

(c) using J_{bc} = 15 Hz and J_{ab} = 7 Hz:

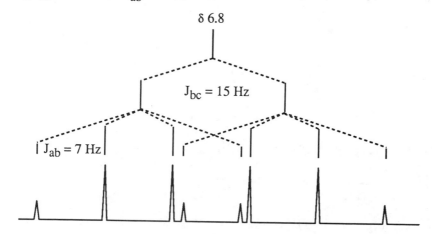

13-16

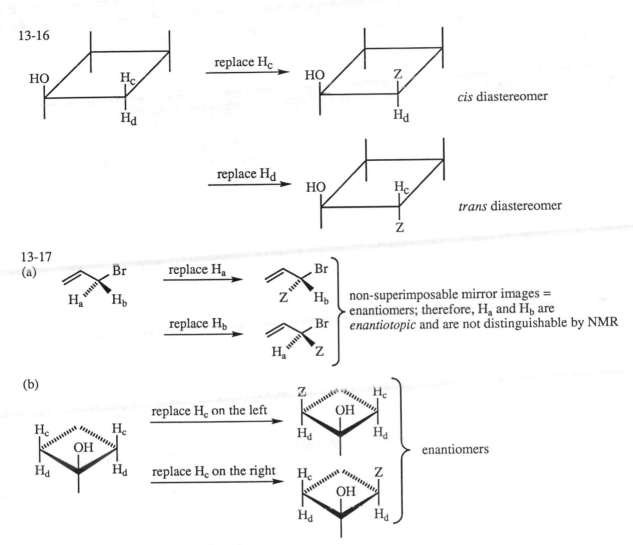

13-17

(a)

non-superimposable mirror images = enantiomers; therefore, H_a and H_b are *enantiotopic* and are not distinguishable by NMR

(b)

enantiomers

(c) The H_d protons are also enantiotopic.

13-18

(a)

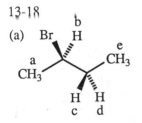

This compound has five types of protons. H_c and H_d are diastereotopic. a = δ 1.5; b = δ 3.6; c,d = δ 1.7; e = δ 1.0

(b)

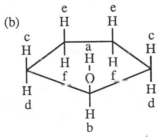

This compound has six types of protons. H_c and H_d are diastereotopic, as are H_e and H_f. a = δ 2-5; b = δ 3.9; c,d = δ 1.6; e,f = δ 1.3

13-18 continued

(c)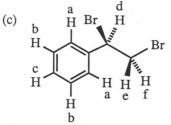

This compound has six types of protons.
H_e and H_f are diastereotopic. a,b,c = δ 7.2;
d ≈ δ 5.0; e,f = δ 3.6

(d)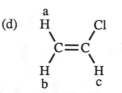

This compound has three types of
protons. H_a and H_b are diastereotopic.
a,b = δ 5-6; c = δ 7-8

13-19

(a)

$$CH_3CH_2-\overset{..}{\underset{..}{O}}-H \; + \; H-B \longrightarrow CH_3CH_2-\overset{H}{\underset{..}{O^+}}-H \; + :B^- \longrightarrow CH_3CH_2-\overset{H}{\underset{..}{O}:} \; + \; H-B$$

(b)

$$CH_3CH_2-\overset{..}{\underset{..}{O}}-H \; + :B^- \longrightarrow CH_3CH_2-\overset{..}{\underset{..}{O}}:^- \; + \; H-B \longrightarrow CH_3CH_2-\overset{H}{\underset{..}{O}:} \; + \; :B^-$$

13-20 The protons from the OH in ethanol exchange with the deuteriums in D_2O. Thus, the OH in ethanol is replaced with OD which does not absorb in the NMR. What happens to the H? It becomes HOD which can usually be seen as a broad singlet around δ 5.25. (If the solvent is $CDCl_3$, the HOD will float on top of the solvent, out of the spectrometer beam, and its signal will be missing.)

$$ROH \; + \; D_2O \longrightarrow ROD \; + \; HOD$$

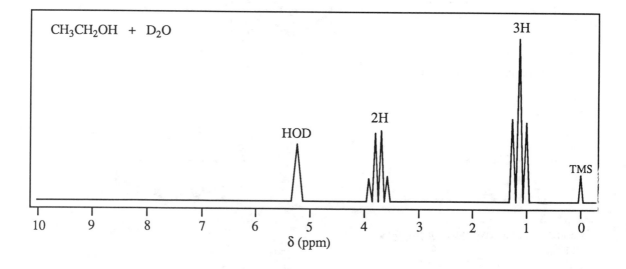

256

13-21

(a) The formula $C_4H_{10}O_2$ has no elements of unsaturation, so the oxygens must either be alcohol or ether functional groups. The doublet at δ 1.1 represents 3H and must be a CH_3 next to a CH. The peaks centered at δ 1.6 integrating to 2H appear to be an uneven quartet and signify a CH_2 between two sets of non-equivalent protons. The broad, 2H singlet at δ 3.3 has the earmark of OH protons, so apparently the molecule is a di-alcohol. The 3H multiplet centered around δ 3.8 is very complex; while the splitting is hard to interpret, the chemical shift suggests that these 3 hydrogens are on carbons bonded to oxygen. To put the pieces together:

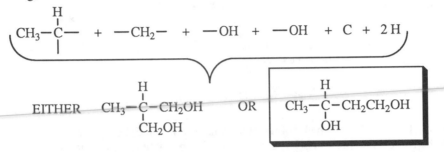

The structure on the left does not fit the NMR; it would show a 4H doublet about δ 3.8. The structure on the right must be correct. The multiplet at δ 3.8 are the overlapping signals of the CH and the CH_2 bonded to the OH's.

(b) The formula C_2H_7NO has no elements of unsaturation. The N must be an amine and the O must be an alcohol or ether. Two triplets, each with 2H, are certain to be —CH_2CH_2— . Since there are no carbons left, the N and O, with enough hydrogens to fill their valences, must go on the ends of this chain. The rapidly exchanging OH and NH_2 protons appear as a broad, 3H singlet at δ 2.4.

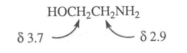

HOCH$_2$CH$_2$NH$_2$

δ 3.7 ⟶   ⟵ δ 2.9

13-22

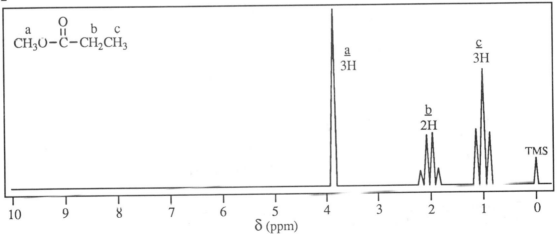

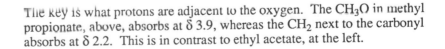

$$CH_3-\overset{\overset{\displaystyle O}{\|}}{C}-OCH_2CH_3$$
δ 2.2   δ 4.1

The key is what protons are adjacent to the oxygen. The CH_3O in methyl propionate, above, absorbs at δ 3.9, whereas the CH_2 next to the carbonyl absorbs at δ 2.2. This is in contrast to ethyl acetate, at the left.

257

13-23

(a) $H_a = \delta\, 4.0$ (singlet, 1H)
 $H_b = \delta\, 3.4$ (doublet, 2H)
 $H_c = \delta\, 1.8$ (multiplet, 1H)
 $H_d = \delta\, 0.9$ (doublet, 6H)

(b) The methine (CH) absorption is split into 7 peaks by 6 equivalent protons on the adjacent methyl groups. The signal is also split into triplets by the adjacent methylene (CH_2). Theoretically, each of seven peaks is split into a triplet, for a total of 21 peaks. You would look ragged too!

13-24

(a) The formula $C_4H_8O_2$ has one element of unsaturation. The 1H singlet at $\delta\, 12.1$ indicates carboxylic acid. The 1H multiplet and the 6H doublet scream isopropyl group.

$$\begin{array}{c} CH_3 \\ \diagdown \\ \qquad CH-\overset{\displaystyle O}{\overset{\|}{C}}-OH \\ \diagup \\ CH_3 \end{array}$$

(b) The formula C_8H_8O has five elements of unsaturation. The broad 5H singlet at $\delta\, 7.2$ indicates monosubstituted benzene. The peak at $\delta\, 9.7$ is unmistakably an aldehyde, trying to be a triplet because it is weakly coupled to an adjacent CH_2, whose own signal at $\delta\, 3.7$ appears to be a weakly coupled doublet.

$$\text{Ph}-CH_2-\overset{\displaystyle O}{\overset{\|}{C}}-H$$

(c) The formula $C_5H_8O_2$ has two elements of unsaturation. A 3H singlet at $\delta\, 2.3$ is probably a CH_3 next to carbonyl. The 2H quartet and the 3H triplet are certain to be ethyl; with the CH_2 at $\delta\, 2.7$, this also appears to be next to carbonyl.

$$CH_3CH_2-\overset{\displaystyle O}{\overset{\|}{C}}-\overset{\displaystyle O}{\overset{\|}{C}}-CH_3$$

(d) The formula C_4H_8O has one element of unsaturation, and the signals from $\delta\, 5.0\text{-}6.0$ indicate a vinyl pattern ($CH_2{=}CH-$). The complex quartet for 1H at $\delta\, 4.3$ is a CH bonded to an alcohol, next to CH_3. The OH appears as a 1H singlet at $\delta\, 2.5$, and the CH_3 next to CH is a doublet at $\delta\, 1.3$. Put together:

$$\begin{array}{c} \qquad\quad OH \\ \qquad\quad | \\ CH_2{=}CH-CH-CH_3 \end{array}$$

(e) The formula $C_6H_{10}O$ has two elements of unsaturation. No splitting means that none of the CH_x groups have neighboring hydrogens. (Sometimes this simplifies the problem, sometimes it makes it harder.) The 6H singlet at $\delta\, 1.4$ must be two CH_3 groups in identical environments; a reasonable first guess is that they are on the same carbon: $(CH_3)_2C$. The 3H singlet at $\delta\, 3.3$ is probably a CH_3O, although it is slightly upfield; it must be in a shielded environment. So far:

$$CH_3O-\quad+\quad CH_3-\overset{\displaystyle |}{\underset{\displaystyle CH_3}{C}}-\quad+\quad 2\,C\quad+\quad H\quad+\quad 2\text{ elements of unsaturation}$$

How to fit two elements of unsaturation in the remaining two carbons? An alkyne! The lone H must be on the end of the alkyne, at $\delta\, 2.4$ in the NMR. The puzzle solved:

$$CH_3O-\overset{\displaystyle CH_3}{\underset{\displaystyle CH_3}{C}}-C{\equiv}C-H$$

258

13-25 Chemical shift values are estimates from Figure 13-43, except in (d), where the values are exact.

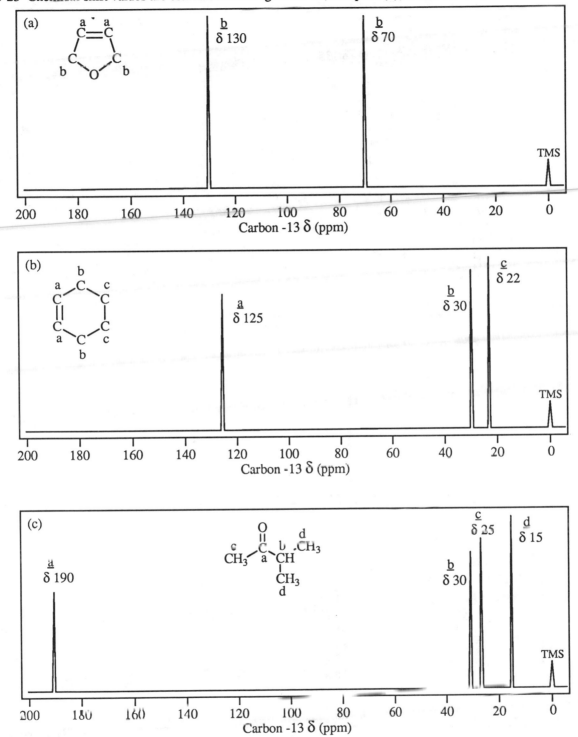

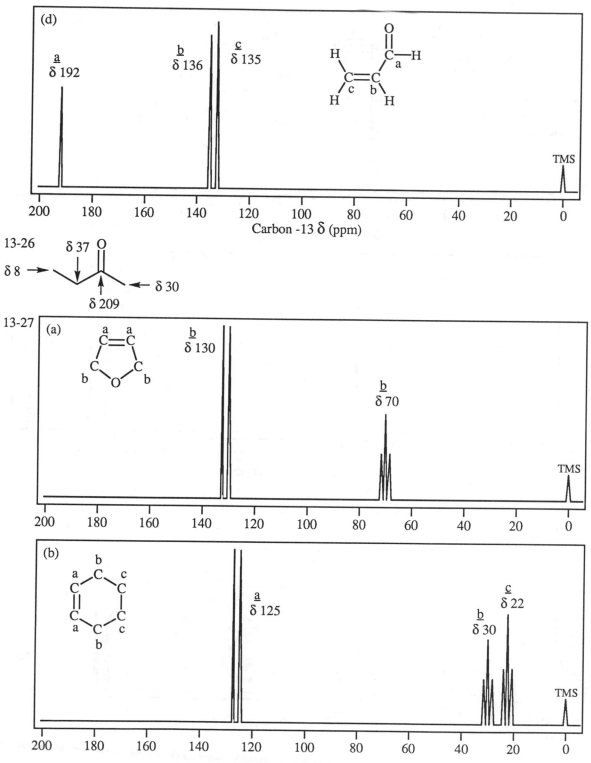

(d)

a
δ 192

b
δ 136

c
δ 135

TMS

Carbon -13 δ (ppm)

13-26

δ 37   O

δ 8 →

← δ 30

δ 209

13-27   (a)

b
δ 130

b
δ 70

TMS

(b)

a
δ 125

c
δ 22

b
δ 30

TMS

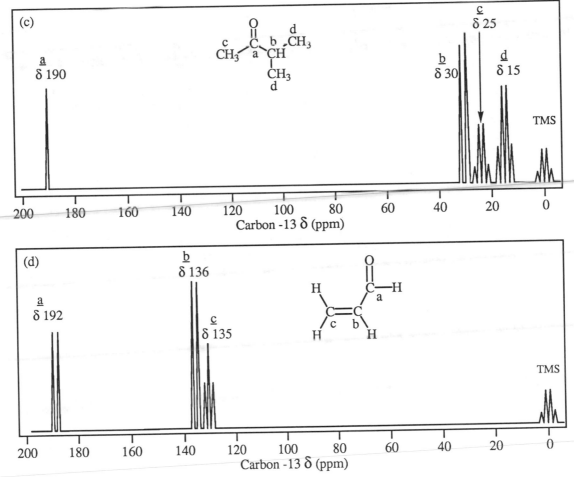

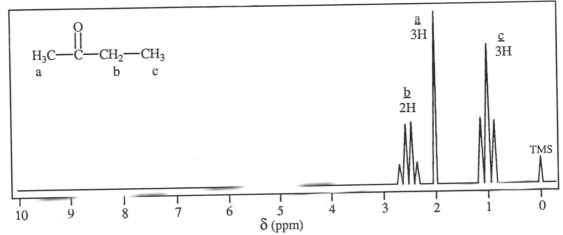

261

13-28 continued

Chemical shift values for 2-butanone:

	C-1	C-3	C-4
proton	2.0	2.5	1.0
carbon	30	37	8
factor	15	14.8	8

The "15 to 20 times as large" rule works well for protons and carbons near deshielding groups, less well for simple aliphatic H and C.

13-29

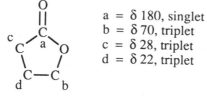

$$H_2C=CH-CH_2OH$$

δ 115 triplet   δ 138 doublet   δ 63 triplet

Allyl bromide is easily hydrolyzed by water.

$$H_2C=CH-CH_2Br \xrightarrow{H_2O} H_2C=CH-CH_2OH \ + \ HBr$$

13-30

a = δ 180, singlet
b = δ 70, triplet
c = δ 28, triplet
d = δ 22, triplet

13-31

a = δ 128, doublet
b = δ 25, triplet
c = δ 23, triplet

Using PBr$_3$ instead of H$_2$SO$_4$/NaBr would give a higher yield of bromocyclohexane.

13-32   Compound 2

Mass spectrum: the molecular ion at m/z 136 shows a peak at 138 of about equal height, indicating a bromine atom is present: 136 – 79 = 57. The fragment at m/z 57 is the base peak; this fragment is most likely a butyl group, C$_4$H$_9$, so a likely molecular formula is C$_4$H$_9$Br.
Infrared spectrum: Notable for the absence of functional groups: no O—H, no N—H, no =C—H, no C=C, no C=O ⇒ most likely an alkyl bromide.
NMR spectrum: The 6H doublet at δ 1.0 suggests two CH$_3$'s split by an adjacent H—an isopropyl group. The 2H doublet at δ 3.2 is a CH$_2$ between a CH and the Br.

Putting the pieces together gives isobutyl bromide.

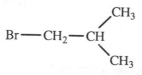

13-33

The formula $C_9H_{11}Br$ indicates four elements of unsaturation, just enough for a benzene ring.

Here is the most accurate method for determining the number of protons per signal from integration values *when the total number of protons is known.* Add the integration heights: 4.4 cm + 13.0 cm + 6.7 cm = 24.1 cm. Divide by the total number of hydrogens: 24.1 cm ÷ 11H = 2.2 cm/H. Each 2.2 cm of integration height = 1H, so the ratio of hydrogens is 2 : 6 : 3.

The 2H singlet at δ 7.1 means that only two hydrogens remain on the benzene ring, that is, it has 4 substituents. The 6H singlet at δ 2.3 must be two CH_3's on the benzene ring in identical environments. The 3H singlet at δ 2.2 is another CH_3 in a slightly different environment from the first two. Substitution of the three CH_3's and the Br in the most symmetric way leads to:

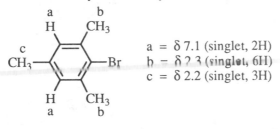

a = δ 7.1 (singlet, 2H)
b = δ 2.3 (singlet, 6H)
c = δ 2.2 (singlet, 3H)

13-34 The numbers in italics indicate the number of peaks in each signal.

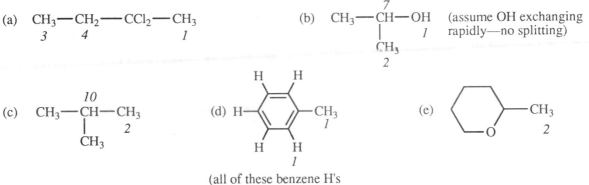

(a) CH_3—CH_2—CCl_2—CH_3
        3   *4*   *1*

(b) CH_3—$\overset{7}{CH}$—OH   (assume OH exchanging
                 | *1*   rapidly—no splitting)
              CH_3
                2

(c) CH_3—$\overset{10}{CH}$—CH_3
              | *2*
            CH_3

(d) structure with benzene ring, H's and CH_3 (*1*)
(all of these benzene H's are accidentally equivalent and do not split each other)

(e) ring with O and —CH_3 *2*

13-35 Consult Appendix 1 in the text for chemical shift values. *Your predictions should be in the given range, or within 0.5 ppm of the given value.*

(a) benzene—H   all at δ 7.2

(b) cyclohexane with H, H   all at δ 1.3

(c)   δ 1.6
    CH_3—O—CH_2—CH_2—$CHCl_2$
      δ 3.4   δ 3.8   δ 5.5

(d) CH_3—CH_2—$C\equiv C$—H
       δ 1.2   δ 2.2   δ 2.5

(e)   O
                ‖
    CH_3—CH_2—C—CH_3
      δ 1.0   δ 2.5   δ 2.0

(f) CH_3 δ 1.5   δ 2-5
          \
           CH—CH_2—CH_2—OH
          /
      CH_3 δ 1.4   δ 3.8
       δ 0.9

13-35 continued

(g)

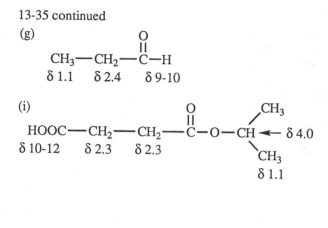

$$CH_3-CH_2-\overset{\overset{\textstyle O}{\|}}{C}-H$$
δ 1.1   δ 2.4   δ 9-10

(h)

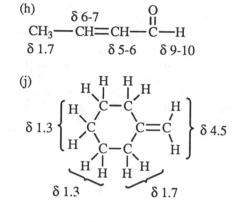

δ 6-7
$$CH_3-CH=CH-\overset{\overset{\textstyle O}{\|}}{C}-H$$
δ 1.7   δ 5-6   δ 9-10

(i)

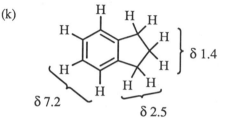

$$HOOC-CH_2-CH_2-\overset{\overset{\textstyle O}{\|}}{C}-O-CH \overset{\textstyle CH_3}{\underset{\textstyle CH_3}{\big<}}$$
δ 10-12   δ 2.3   δ 2.3   ← δ 4.0
δ 1.1

(j)

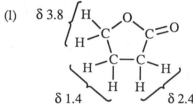

δ 1.3 {   } δ 4.5

δ 1.3   δ 1.7

(k)

δ 1.4

δ 7.2

δ 2.5

(l)

δ 3.8

δ 1.4   δ 2.4

13-36

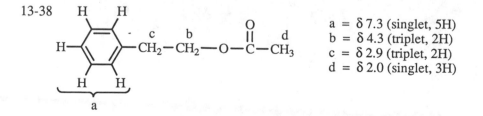

c   a   c
$$CH_3-CH-CH_3$$
       |
       OH
       b

a = δ 4.0 (septet, 1H)
b = δ 2.2 (singlet, 1H) (rapidly exchanging)
c = δ 1.2 (doublet, 6H)

13-37

(a) δ 4.00 = 4.00 ppm = $(4.00 \times 10^{-6}) \times (60 \times 10^6\,Hz)$ = 240 Hz

The signal is 240 Hz downfield from TMS.

(b) The chemical shift *in ppm* would not change: δ 4.00. At 300 MHz:

$(4.00 \times 10^{-6}) \times (300 \times 10^6\,Hz)$ = 1200 Hz

The signal is 1200 Hz downfield from TMS.

(c) Coupling constants do not change with field strength: J = 7 Hz, regardless of field strength.

13-38

c   b   d
$$CH_2-CH_2-O-\overset{\overset{\textstyle O}{\|}}{C}-CH_3$$

a = δ 7.3 (singlet, 5H)
b = δ 4.3 (triplet, 2H)
c = δ 2.9 (triplet, 2H)
d = δ 2.0 (singlet, 3H)

a

264

13-39

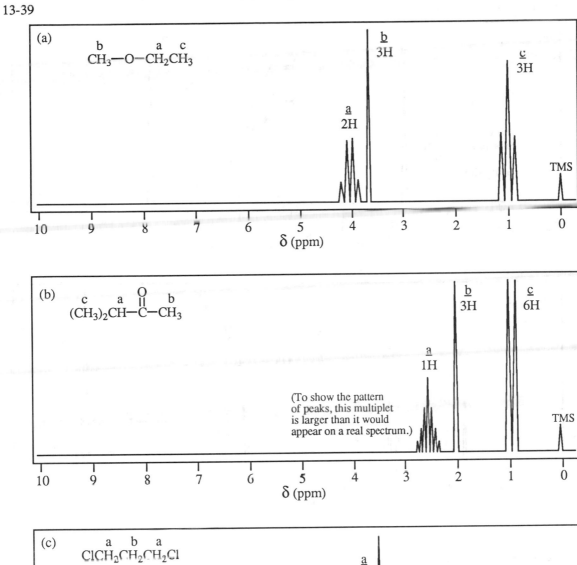

(a)

b   a   c
CH₃—O—CH₂CH₃

b
3H

c
3H

a
2H

TMS

10   9   8   7   6   5   4   3   2   1   0
δ (ppm)

(b)

c   a   O   b
(CH₃)₂CH—C—CH₃

b
3H

c
6H

a
1H

(To show the pattern
of peaks, this multiplet
is larger than it would
appear on a real spectrum.)

TMS

10   9   8   7   6   5   4   3   2   1   0
δ (ppm)

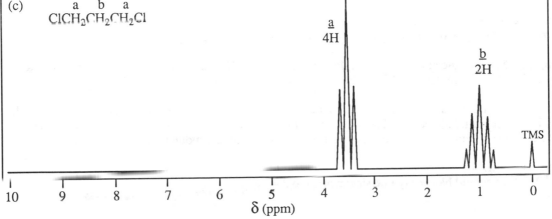

(c)

a   b   a
ClCH₂CH₂CH₂Cl

a
4H

b
2H

TMS

10   9   8   7   6   5   4   3   2   1   0
δ (ppm)

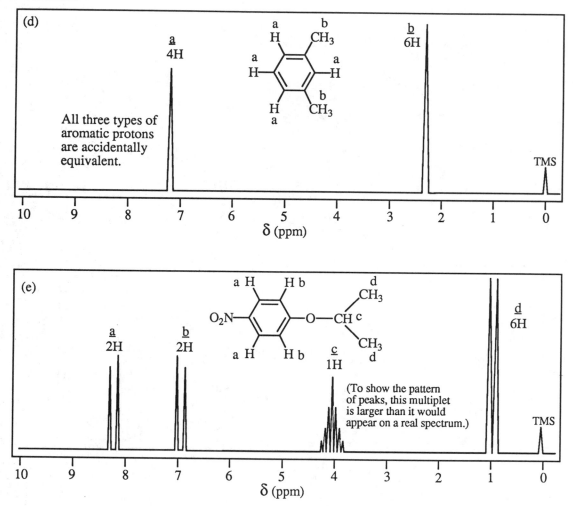

(d)

a
4H

All three types of aromatic protons are accidentally equivalent.

b
6H

TMS

10   9   8   7   6   5   4   3   2   1   0
δ (ppm)

(e)

a
2H

b
2H

c
1H

(To show the pattern of peaks, this multiplet is larger than it would appear on a real spectrum.)

d
6H

TMS

10   9   8   7   6   5   4   3   2   1   0
δ (ppm)

13-40

(a) The NMR of 1-bromopropane would have three sets of signals, whereas the NMR of 2-bromopropane would have only two sets (a septet and a doublet, the typical isopropyl pattern).

(b) The NMR of the aldehyde would show a triplet at δ 9-10 and an ethyl pattern, whereas the ketone would exhibit only one singlet at δ 2.0.

(c) The most obvious difference is the chemical shift of the CH_3 singlet. In the compound on the left, the CH_3 singlet would appear at δ 2.1, while the compound on the right would show the CH_3 singlet at δ 3.8. Refer to the solution of 13-22 for the spectrum of the second compound.

(d) The NMR of 1-butyne would show three signals: an ethyl pattern and a 1H singlet at δ 2.5. The NMR of 2-butyne would have only one singlet about δ 2.4.

13-41   The numbers below are chemical shift values in the carbon NMR spectra. The letters are the multiplicity: s = singlet; d = doublet; t = triplet; q = quartet.

(a)

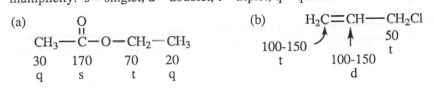

$$CH_3\overset{O}{\overset{\|}{-}C-O-CH_2-CH_3}$$
30   170   70   20
q   s   t   q

(b) $H_2C{=}CH-CH_2Cl$

100-150   50
    t   t
          100-150
             d

13-42   The multiplicity of the peaks in this off-resonance decoupled spectrum show two different CH's and a CH₃. There is only one way to assemble these pieces with three chlorines.

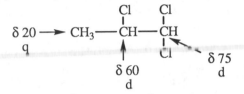

δ 20 ⟶ $CH_3-\overset{Cl}{\overset{|}{CH}}-\overset{Cl}{\overset{|}{\underset{|}{CH}}}$
     q   Cl   δ 75
               δ 60   d
                d

13-43   There is no evidence for vinyl hydrogens, so the double bond is gone. Integration gives eight hydrogens, so the formula must be $C_4H_8Br_2$, and the four carbons must be in a straight chain because the starting material was 2-butene. From the integration, the four carbons must be present as one CH₃, one CH, and two CH₂ groups. The methyl is split into a doublet, so it must be adjacent to the CH. The two CH₂ groups must follow in succession, with two bromine atoms filling the remaining valences.

```
   H   Br  H   Br
   |   |   |   |
H-C - C - C - C-H
   |   |   |   |
   H   H   H   H
‿‿‿‿‿         ‿‿‿‿‿
  d    a   c    b
```

a = δ 4.3 (sextet, 1H)
b = δ 3.5 (triplet, 2H)
c = δ 2.2 (multiplet, 2H)
d = δ 1.7 (doublet, 3H)

13-44   There is no evidence for vinyl hydrogens, so the compound must be a small, saturated, oxygen-containing molecule. Starting upfield (toward TMS), the first signal is a 3H triplet; this must be a CH₃ next to a CH₂. The CH₂ could be the signal at δ 1.5, but it has six peaks: it must have five neighboring hydrogens, a CH₃ on one side and a CH₂ on the other side. The third carbon must therefore be a CH₂; its signal is a triplet at δ 3.6. To be so far downfield, the final CH₂ must be bonded to oxygen. The remaining 1H signal must be from an OH. The compound must be 1-propanol.

b   a   c   d
$HOCH_2CH_2CH_3$

a = δ 3.6 (triplet, 2H)
b = δ 3.1 (singlet, 1H)
c = δ 1.5 (6 peaks, 2H)
d = δ 0.9 (triplet, 3H)

13-45

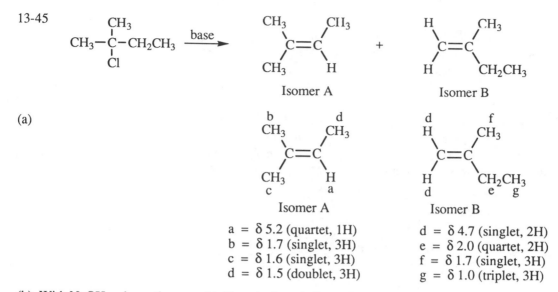

(a)

a = δ 5.2 (quartet, 1H)
b = δ 1.7 (singlet, 3H)
c = δ 1.6 (singlet, 3H)
d = δ 1.5 (doublet, 3H)

d = δ 4.7 (singlet, 2H)
e = δ 2.0 (quartet, 2H)
f = δ 1.7 (singlet, 3H)
g = δ 1.0 (triplet, 3H)

(b) With NaOH as base, the more highly substituted alkene, Isomer A, would be expected to predominate—the Saytzeff Rule. With KO-*t*-Bu as a hindered, bulky base, the less substituted alkene, Isomer B, would predominate (the Hofmann product).

13-46 "Nuclear waste" is comprised of radioactive products from either nuclear reactions, for example, from electrical generating stations powered by nuclear reactors, or residue from medical or scientific studies using radioactive nuclides as therapeutic agents (like iodine for thyroid treatment) or as molecular tracers (carbon-14, tritium H-3, phosphorus-32, nitrogen-15, and many others). The physical technique of *nuclear* magnetic resonance neither uses nor generates any radioactive elements, and does not generate "nuclear waste". (Some people assume that the medical application of NMR, medical resonance imaging or MRI, purposely dropped the word "nuclear" from the technique to avoid the confusion between "nuclear" and "radioactive".)

13-47

Mass spectrum: The molecular ion of m/z 117 suggests the presence of an odd number of nitrogens.
Infrared spectrum: No NH or OH appears. Hydrogens bonded to both sp^2 and sp^3 carbon are indicated around 3000 cm^{-1}. The characteristic C≡N peak appears at 2250 cm^{-1} and aromatic C=C is suggested by the peak at 1600 cm^{-1}.
NMR spectrum: Five aromatic protons are shown in the NMR at δ 7.3. A CH_2 singlet appears at δ 3.7. Assemble the pieces:

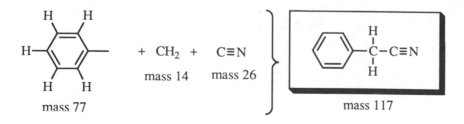

13-48 This is a challenging problem, despite the molecule being relatively small.

Mass spectrum: The molecular ion at 96 suggests no Cl, Br, or N. The molecule must have seven carbons or fewer.

Infrared spectrum: The dominant functional group peak is at 1685 cm^{-1}, a carbonyl that is conjugated with C=C (lower wavenumber than normal, very intense peak). The presence of an oxygen and a molecular ion of 96 lead to a formula of C_6H_8O, with three elements of unsaturation, a C=O and one or two C=C.

Carbon NMR spectrum: The six peaks show, by chemical shift, one carbonyl carbon (196), two alkene carbons (129, 151), and three aliphatic carbons (23, 26, 36). By off-resonance decoupling multiplicity, the groups are: three CH_2 groups, two alkene CH groups, and carbonyl.

Since the structure has one carbonyl and only two alkene carbons, the third element of unsaturation must be a ring.

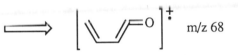

CH_2 + CH_2 + CH_2 + 1 ring

Since the structure has no methyl group, and no $H_2C=$, all of the carbons must be included in the ring. The only way these pieces can fit together is in 2-cyclohexenone. Notice that the proton NMR was unnecessary to determine the structure, fortunately, since the HNMR was not easily interpreted except for the two alkene hydrogens.

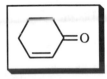

2-cyclohexenone

$\Longrightarrow$ $\left[\begin{array}{c} \end{array} \right]^{+\cdot}$ m/z 68

The mystery mass spec peak at m/z 68 comes from a fragmentation that will be discussed later; it is called a retro-Diels-Alder fragmentation.

CHAPTER 14—ETHERS, EPOXIDES AND SULFIDES

14-1

The four solvents decrease in polarity in this order: water, ethanol, ethyl ether, and dichloromethane. The three solutes decrease in polarity in this order: sodium acetate, 2-naphthol, and naphthalene. The guiding principle in determining solubility is, "Like dissolves like." Compounds of similar polarity will dissolve (in) each other. Thus, sodium acetate will dissolve in water, will dissolve only slightly in ethanol, and will be virtually insoluble in ethyl ether and dichloromethane. 2-Naphthol will be insoluble in water, somewhat soluble in ethanol, and soluble in ether and dichloromethane. Naphthalene will be insoluble in water, partially soluble in ethanol, and soluble in ethyl ether and dichloromethane. (Actual solubilities are difficult to predict, but you should be able to predict *trends*.)

14-2

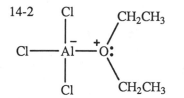

Oxygen shares one of its electron pairs with aluminum; oxygen is the Lewis base, and aluminum is the Lewis acid. An oxygen atom with three bonds and one unshared pair has a positive formal charge. An aluminum atom with four bonds has a negative formal charge.

14-3

The crown ether has two effects on $KMnO_4$: first, it makes $KMnO_4$ much more soluble in benzene; second, it holds the potassium ion tightly, making the permanganate more available for reaction. Chemists call this a "naked anion" because it is not complexed with solvent molecules.

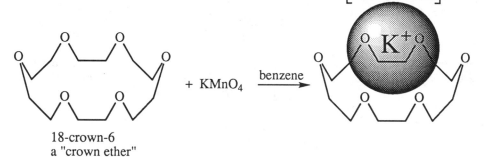

18-crown-6
a "crown ether"

14-4 IUPAC name first; then common name (see Appendix 1 in this Solutions Manual for a summary of IUPAC nomenclature)

(a) methoxycyclopropane; cyclopropyl methyl ether
(b) 2-ethoxypropane; ethyl isopropyl ether
(c) 1-chloro-2-methoxyethane; 2-chloroethyl methyl ether
(d) 2-ethoxy-2,3-dimethylpentane; no common name
(e) 2-*t*-butoxybutane; *sec*-butyl *t*-butyl ether
(f) *trans*-2-methoxy-1-cyclohexanol; no common name

14-5

(a)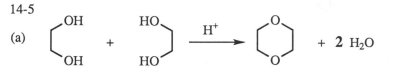

The alcohol is 1,2-ethanediol; the common name is ethylene glycol.

14-5 continued

(b)

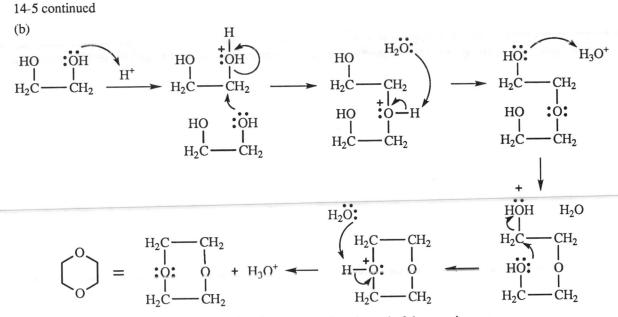

The mechanism shows that the acid catalyst is regenerated at the end of the reaction.

14-6
(a) dihydropyran
(b) 2-chloro-1,4-dioxane
(c) 3-isopropylpyran
(d) *trans*-2,3-diethyloxirane; *trans*-3,4-epoxyhexane; *trans*-3-hexene oxide
(e) 3-bromo-2-ethoxyfuran
(f) 3-bromo-2,2-dimethyloxetane

14-7

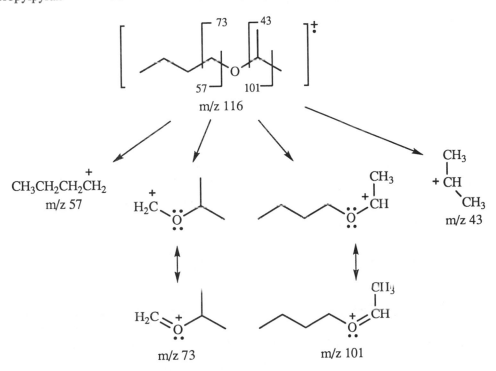

14-8 S_N2 reactions, including the Williamson ether synthesis, work best when the nucleophile attacks a 1° or methyl carbon. Instead of attempting to form the bond from oxygen to the 2° carbon on the ring, form the bond from oxygen to the 1° carbon of the butyl group.

The OH must first be transformed into a good leaving group: either a tosylate, or one of the halides (not fluoride).

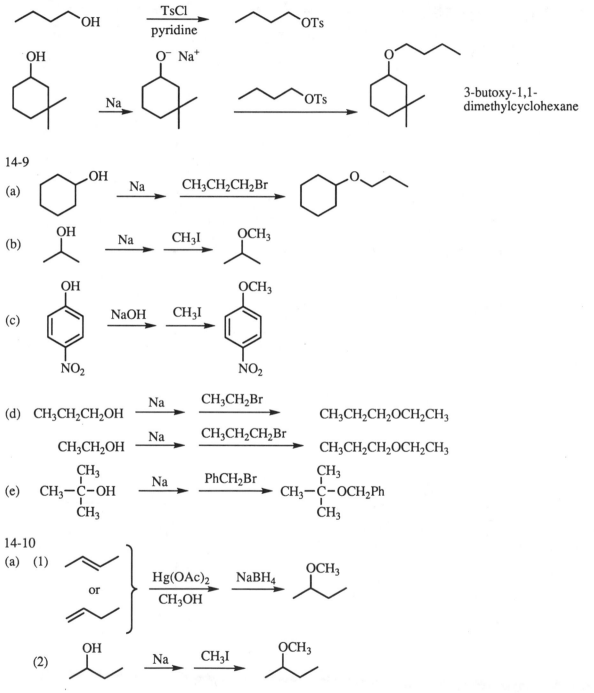

3-butoxy-1,1-dimethylcyclohexane

14-10 continued

(b) (1)

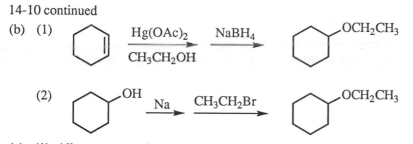

(2)

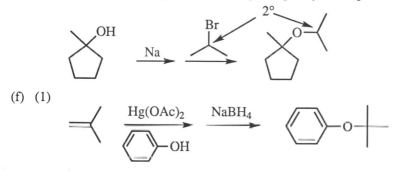

(c) (1) Alkoxymercuration is not practical here; the product does not have Markovnikov orientation.

(2)

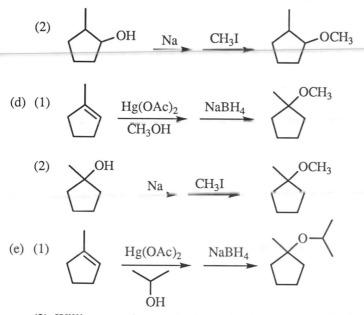

(d) (1)

(2)

(e) (1)

(2) Williamson ether synthesis would give a poor yield of product as the halide is on a 2° carbon.

(f) (1)

(2) Williamson ether synthesis is not feasible here. S_N2 does not work on either a benzene or a 3° halide.

14-11 An important principle of synthesis is to avoid mixtures of isomers wherever possible; minimizing separations increases recovery of products. Bimolecular dehydration is a random process. Heating a mixture of ethanol and methanol with acid will produce all possible combinations: dimethyl ether, ethyl methyl ether, and diethyl ether. This mixture would be troublesome to separate.

273

14-12

Ether formation

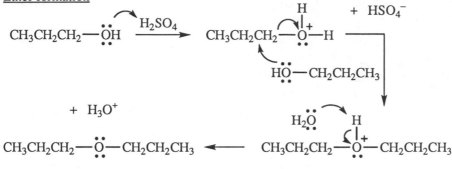

Dehydration

CH$_3$CH$_2$CH$_2$—ÖH $\xrightarrow{\text{H}_2\text{SO}_4}$ CH$_3$CHCH$_2$—Ö$^+$—H $\longrightarrow$ CH$_3$CH$=$CH$_2$

+ H$_2$O

Remember $\Delta G = \Delta H - T\Delta S$? Thermodynamics of a reaction depend on the sign and magnitude of ΔG. As temperature increases, the entropy term grows in importance. In ether formation, the ΔS is small because two molecules of alcohol give one molecule of ether plus one molecule of water—no net change in the number of molecules. In dehydration, however, one molecule of alcohol generates one molecule of alkene plus one molecule of water—a large increase in entropy. So $T\Delta S$ is more important for dehydration than for ether formation. As temperature increases, the competition will shift toward more dehydration.

14-13

(a) This symmetrical ether at 1° carbons could be produced in good yield by bimolecular dehydration.

(b) This unsymmetrical ether could not be produced in high yield by bimolecular dehydration. Williamson synthesis would be preferred.

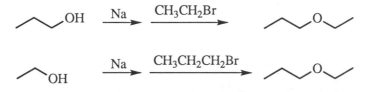

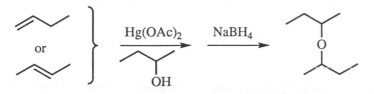

(c) Even though this ether is symmetrical, both carbons are 2°, so bimolecular dehydration would give low yields. Unimolecular dehydration to give alkenes would be the dominant pathway. Alkoxymercuration-demercuration is the preferred route.

274

14-14

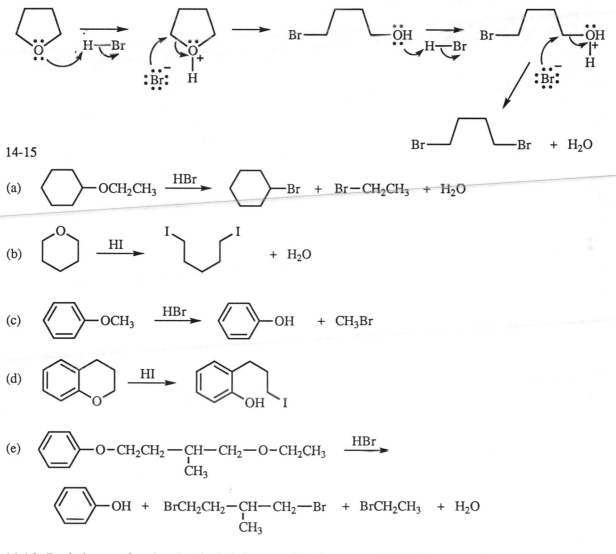

14-15

(a) [cyclohexyl]—OCH$_2$CH$_3$ $\xrightarrow{\text{HBr}}$ [cyclohexyl]—Br + Br—CH$_2$CH$_3$ + H$_2$O

(b) [tetrahydropyran] $\xrightarrow{\text{HI}}$ I—CH$_2$CH$_2$CH$_2$CH$_2$CH$_2$—I + H$_2$O

(c) [phenyl]—OCH$_3$ $\xrightarrow{\text{HBr}}$ [phenyl]—OH + CH$_3$Br

(d) [chroman] $\xrightarrow{\text{HI}}$ [aryl]—OH ...I

(e) [phenyl]—O–CH$_2$CH$_2$—CH—CH$_2$—O–CH$_2$CH$_3$ $\xrightarrow{\text{HBr}}$
                                    |
                                    CH$_3$

[phenyl]—OH + BrCH$_2$CH$_2$—CH—CH$_2$—Br + BrCH$_2$CH$_3$ + H$_2$O
                            |
                            CH$_3$

14-16 Begin by transforming the alcohols into good leaving groups like halides or tosylates:

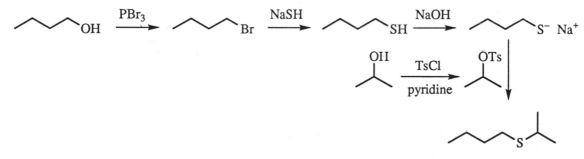

14-17 The sulfur at the center of mustard gas is an excellent nucleophile, and chloride is a decent leaving group. Sulfur can do in *internal* nucleophilic subsitution to make a reactive sulfonium salt and the sulfur equivalent of an epoxide.

(a)

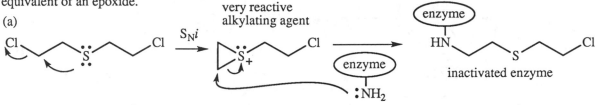

(b) NaOCl is a powerful oxidizing agent. It oxidizes sulfur to a sulfoxide or more likely a sulfone, either of which is no longer nucleophilic, preventing formation of the cyclic sulfonium salt.

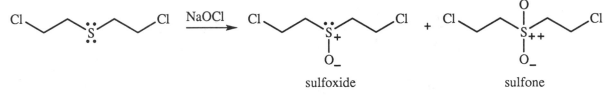

14-18

Generally, chemists prefer the peroxyacid method of epoxide formation to the halohydrin method. Reactions (a) and (b) show the peroxyacid method, but the halohydrin method could also be used.

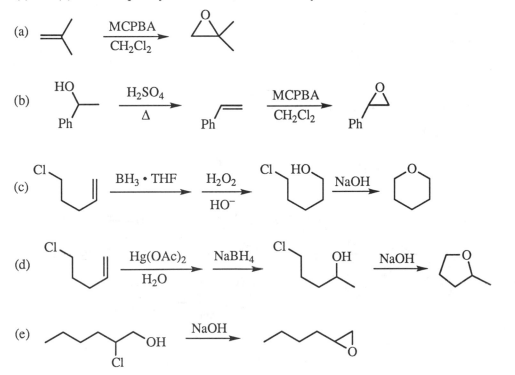

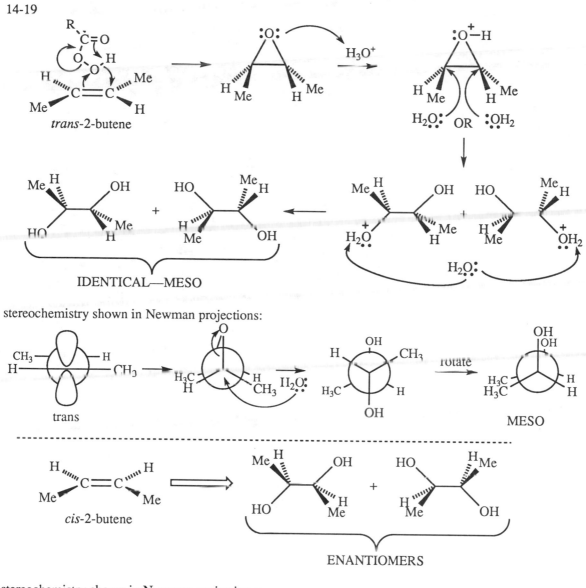

stereochemistry shown in Newman projections:

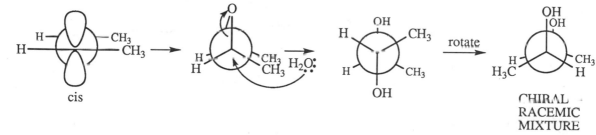

14-20

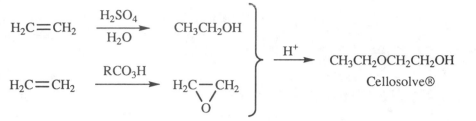

14-21

anhydrous HBr—only Br⁻ nucleophiles present

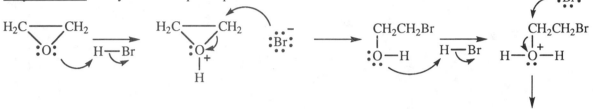

aqueous HBr—many more H_2O nucleophiles than Br⁻ nucleophiles

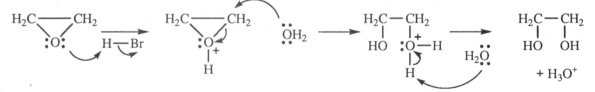

14-22 The cyclization of squalene via the epoxide is an excellent (and extraordinary) example of how Nature uses organic chemistry to its advantage. In one enzymatic step, Nature forms four rings and eight chiral centers! Out of 256 possible stereoisomers, only one is formed!

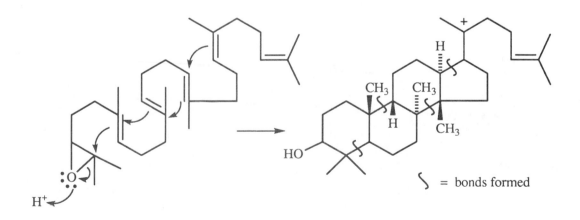

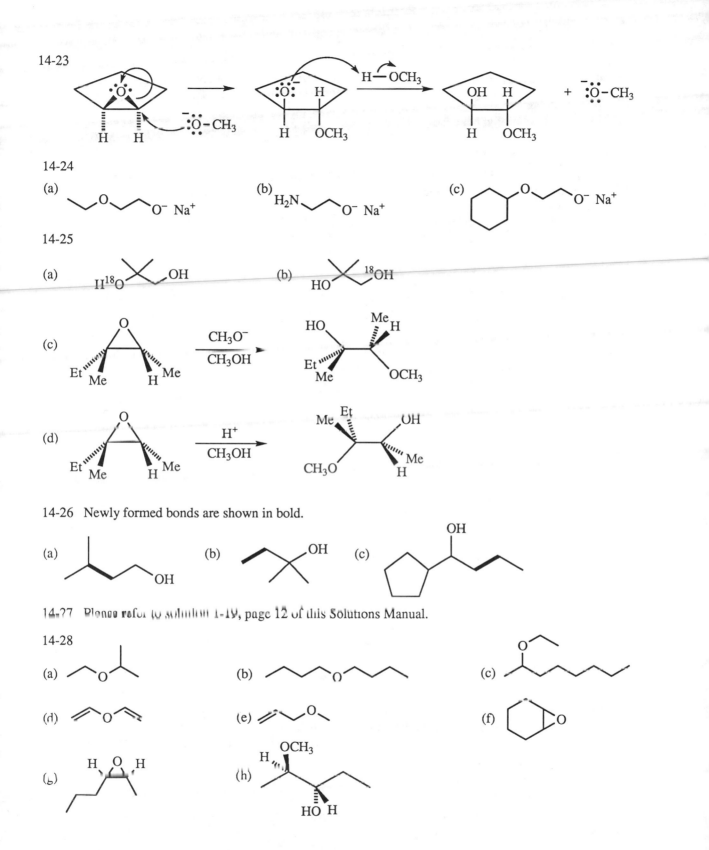

14-23

14-24

(a)

(b)

(c)

14-25

(a)

(b)

(c) $\xrightarrow[\text{CH}_3\text{OH}]{\text{CH}_3\text{O}^-}$

(d) $\xrightarrow[\text{CH}_3\text{OH}]{\text{H}^+}$

14-26 Newly formed bonds are shown in bold.

(a)

(b)

(c)

14-27 Please refer to solution 1-19, page 12 of this Solutions Manual.

14-28

(a)

(b)

(c)

(d)

(e)

(f)

(g)

(h)

279

14-29

(a) *sec*-butyl isopropyl ether
(b) *t*-butyl isobutyl ether
(c) ethyl phenyl ether
(d) chloromethyl *n*-propyl ether

(e) *trans*-cyclohexene glycol
(f) cyclopentyl methyl ether
(g) propylene oxide
(h) cyclopentene oxide

14-30

(a) 2-methoxy-1-propanol
(b) ethoxybenzene or phenoxyethane
(c) methoxycyclopentane
(d) 2,2-dimethoxy-1-cyclopentanol
(e) *trans*-1-methoxy-2-methylcyclohexane

(f) *trans*-3-chloro-1,2-epoxycycloheptane
(g) *trans*-1-methoxy-1,2-epoxybutane; or,
    trans-2-ethyl-3-methoxyoxirane
(h) 3-bromooxetane
(i) 1,3-dioxane

14-31

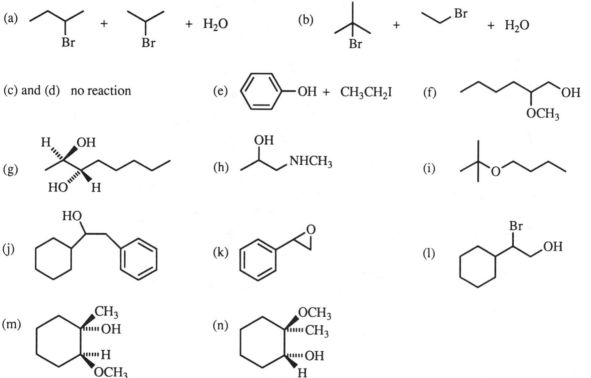

(c) and (d)   no reaction

14-32

(a) On long-term exposure to air, ethers form peroxides. Peroxides are explosive when concentrated or heated. (For exactly this reason, ethers should *never* be distilled to dryness.)
(b) Peroxide formation can be prevented by excluding oxygen. Ethers can be checked for the presence of peroxides, and peroxides can be destroyed safely by treatment with reducing agents.

14-33

(a) Beginning with *(R)*-2-butanol and producing the *(R)* sulfide requires two inversions of configuration.

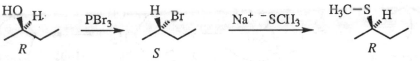

An alternative approach would be to make the tosylate, displace with chloride or bromide (S_N2 with inversion), then do a second inversion with $NaSCH_3$.

(b) Synthesis of the *(S)* isomer directly requires only one inversion.

14-34

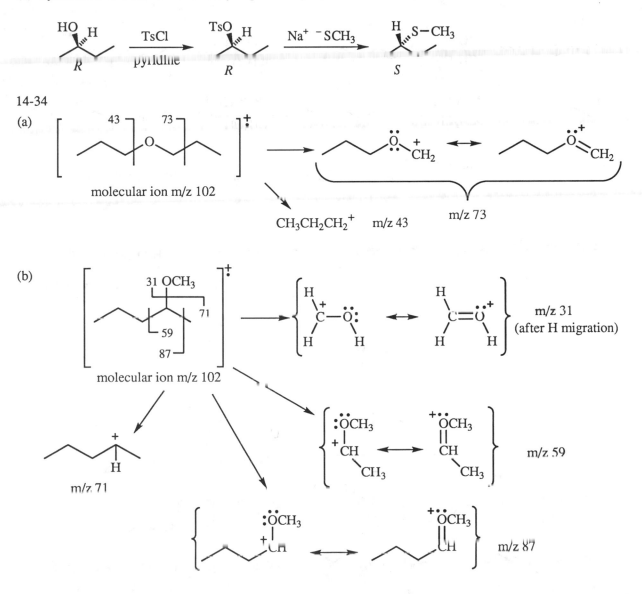

14-35

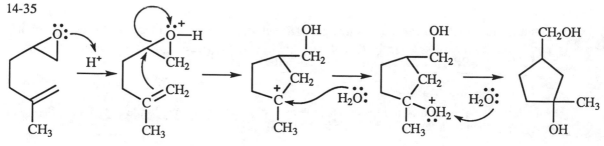

14-36

(a)

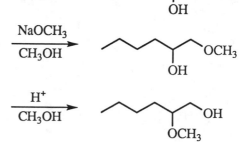

(b)

(c)

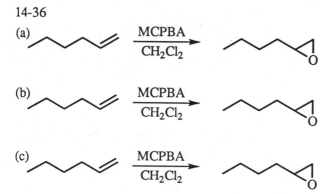

14-37

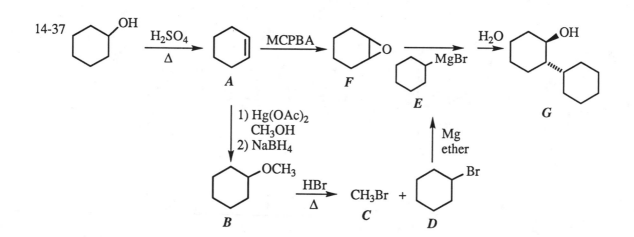

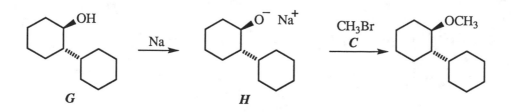

14-38 The student turned in the wrong product! Three pieces of information are consistent with the desired product: molecular formula $C_4H_{10}O$; O—H stretch in the IR at 3300 cm^{-1} (although it should be strong, not weak); and mass spectrum fragment at m/z 59 (loss of CH_3). The NMR of the product should have a 9H singlet at δ 1.0 and a 1H singlet between δ 2 and δ 5. Instead, the NMR shows CH_3CH_2 bonded to oxygen. The student isolated diethyl ether, *the typical solvent used in Grignard reactions*.

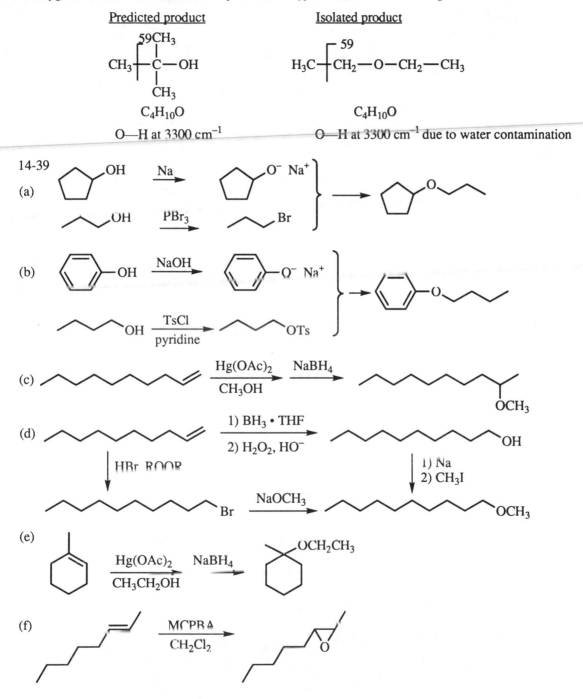

Predicted product

Isolated product

$C_4H_{10}O$

$C_4H_{10}O$

O—H at 3300 cm^{-1}

O—H at 3300 cm^{-1} due to water contamination

14-39

(a)

(b)

(c)

(d)

(e)

(f)

14-40 In the first sequence, no bond is broken to the chiral center, so the configuration of the product is the same as the configuration of the starting material.

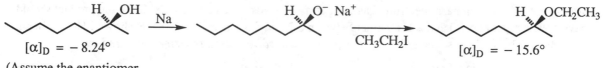

$[\alpha]_D = -8.24°$

(Assume the enantiomer shown is levorotatory.)

$[\alpha]_D = -15.6°$

In the second reaction sequence, however, bonds to the chiral carbon are broken twice, so the stereochemistry of each process must be considered.

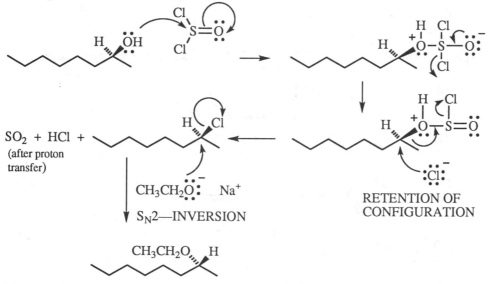

SO_2 + HCl +
(after proton transfer)

$CH_3CH_2\overset{..}{\underset{..}{O}}:$ Na^+

S_N2—INVERSION

RETENTION OF CONFIGURATION

The second sequence involves *retention* followed by *inversion*, thereby producing the *enantiomer* of the 2-ethoxyoctane generated by the first sequence. The optical rotation of the final product will have equal magnitude but opposite sign, $[\alpha]_D = +15.6°$.

14-41

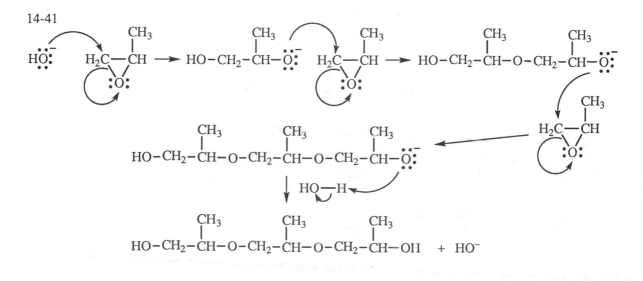

284

14-42 The text uses HA to indicate an acid; A⁻ is the conjugate base.

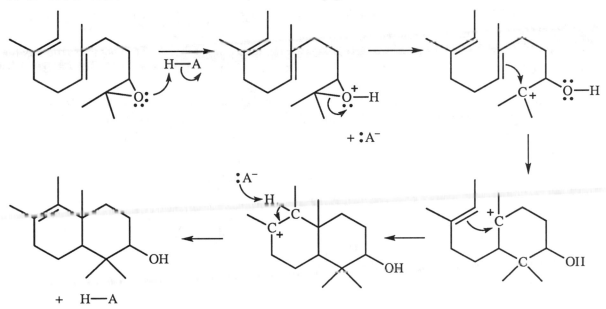

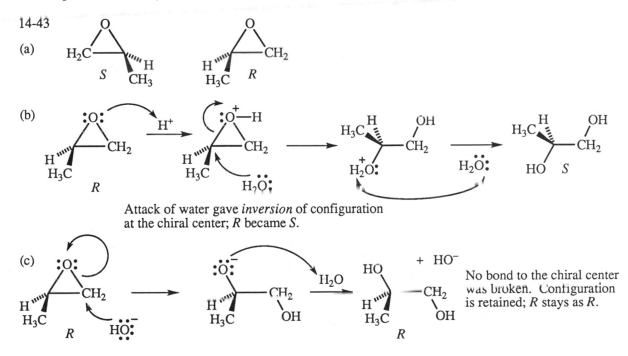

This process resembles the cyclization of squalene oxide to lanosterol. (See the solution to problem 14-22.) In fact, pharmaceutical synthesis of steroids uses the same type of reaction called a "biomimetic cyclization".

14-43

(a) *(structures shown)*

(b) *(mechanism shown)*

Attack of water gave *inversion* of configuration at the chiral center; *R* became *S*.

(c) *(mechanism shown)*

+ HO⁻

No bond to the chiral center was broken. Configuration is retained; *R* stays as *R*.

(d) The difference in these mechanisms lies in where the nucleophile attacks. Attack at the chiral carbon gives inversion; attack at the achiral carbon retains the configuration at the chiral carbon. These products are enantiomers and must necessarily have optical rotations of opposite sign.

14-44   methyl cellosolve $CH_3OCH_2CH_2OH$

To begin, what can be said about methyl cellosolve? Its molecular weight is 76; its IR would show C—O in the 1000-1200 cm^{-1} region and a strong O—H around 3300 cm^{-1}; and its NMR would show three sets of signals in the ratio of 3 : 4 : 1. (The two CH$_2$ groups are not identical but their environments are roughly the same, so that their NMR signals can be expected to be accidentally equivalent.)

The unknown has molecular weight 134; this is double the weight of methyl cellosolve, minus 18 (water). The IR shows no OH, only ether C—O. The NMR shows no OH, only H—C—O in the ratio of 3 : 2 : 2. Apparently, two molecules of methyl cellosolve have combined in an acid-catalyzed, bimolecular dehydration.

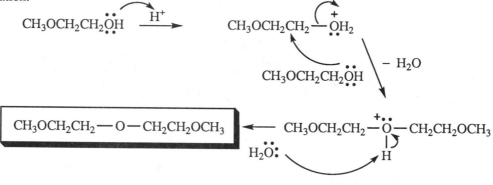

14-45

The formula C$_8$H$_{10}$O has five elements of unsaturation (enough (4) for a benzene ring). The IR is useful for what is does *not* show. There is neither OH nor C=O, so the oxygen must be an ether functional group.

The NMR shows a 5H signal at δ 7.2, a monosubstituted benzene. No peaks in the δ 4.5-6.0 range indicate the absence of an alkene, so the remaining element of unsaturation must be a ring. The three protons are non-equivalent, with complex splitting.

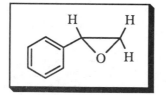
            + 2 C + O + 3 H + ring
                ether

These pieces can be assembled in only one manner consistent with the data.

(Note that the CH$_2$ hydrogens are not equivalent (one is *cis* and one is *trans* to the phenyl) and therefore have distinct chemical shifts.)

15-1   First, look for the number of double bonds to be hydrogenated, then for conjugation, then for degree of substitution of the alkenes.

(a)

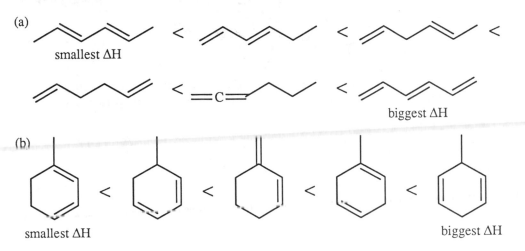

(b)

15-2   Reminder: H—B is used to symbolize the general form for an acid, that is, a protonated base.

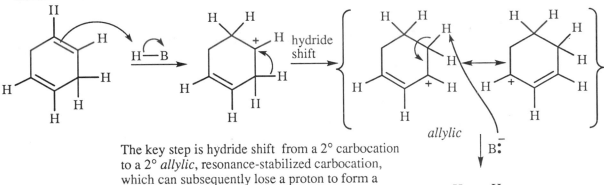

The key step is hydride shift from a 2° carbocation to a 2° *allylic*, resonance-stabilized carbocation, which can subsequently lose a proton to form a *conjugated* diene.

15-3 (You may wish to refer to problem 2-6.)

(a)

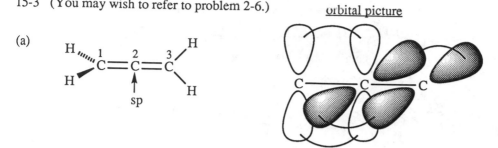

orbital picture

sp

The central carbon atom makes two π bonds with two p orbitals. These p orbitals must necessarily be perpendicular to each other, thereby forcing the groups on the ends of the allene system perpendicular.

(b)

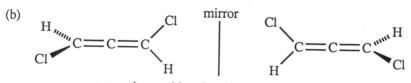

mirror

non-superimposable mirror images = enantiomers

15-4

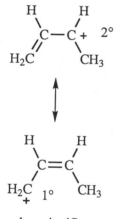

less significant
contributor

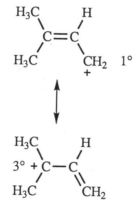

more significant
contributor

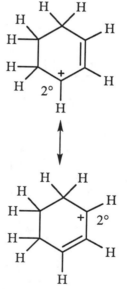

equivalent to the
first resonance form

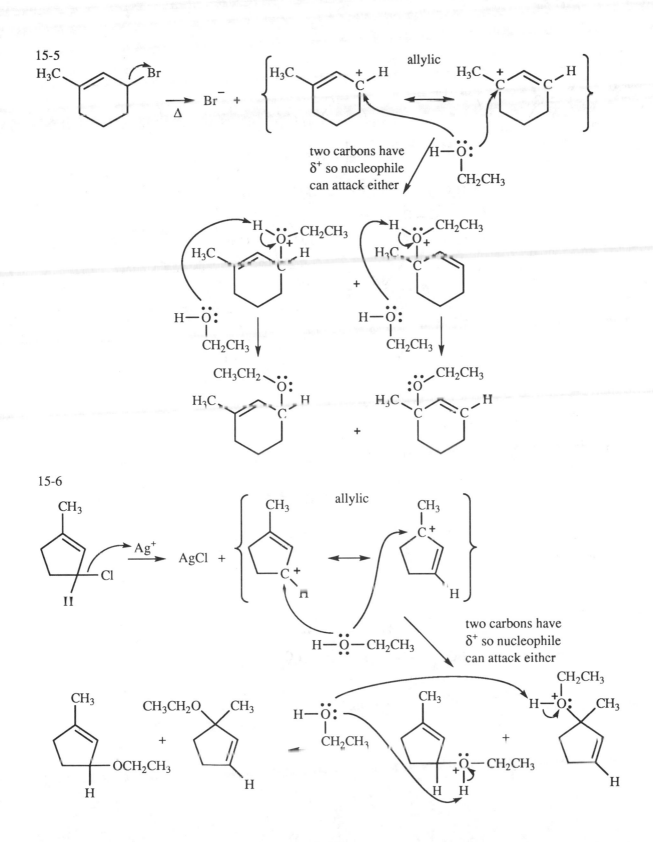

15-5

15-6

two carbons have δ⁺ so nucleophile can attack either

allylic

two carbons have δ⁺ so nucleophile can attack either

15-7

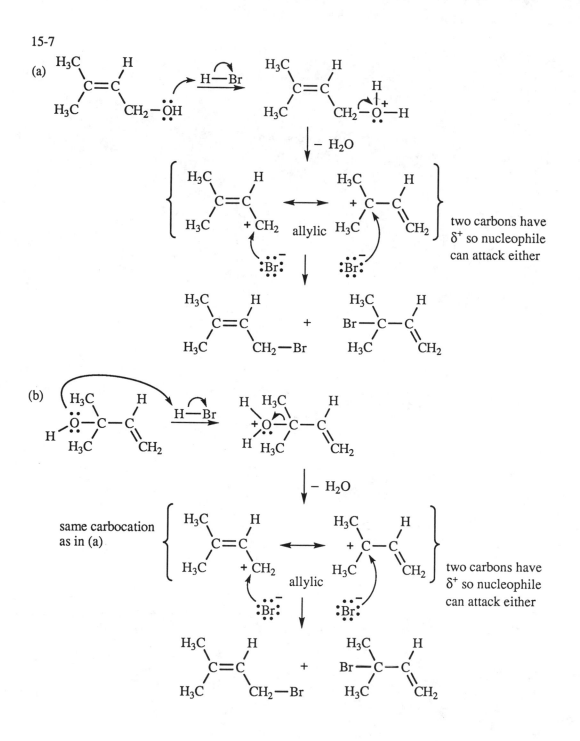

(a) two carbons have δ⁺ so nucleophile can attack either

(b) same carbocation as in (a)

two carbons have δ⁺ so nucleophile can attack either

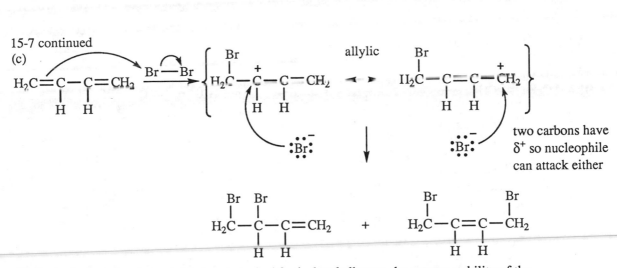

While the bromonium ion mechanism is typical for isolated alkenes, the greater stability of the resonance-stabilized carbocation will make it the lower energy intermediate for conjugated systems.

(d)

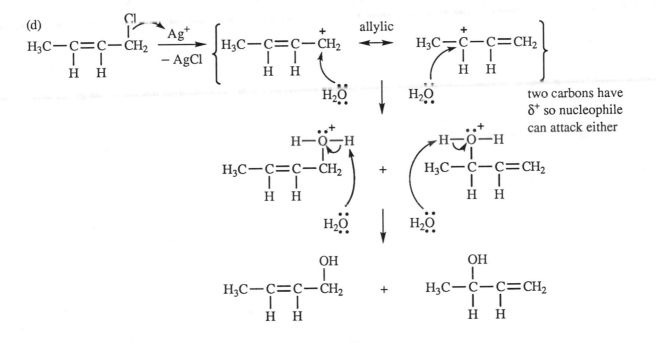

15-7 continued

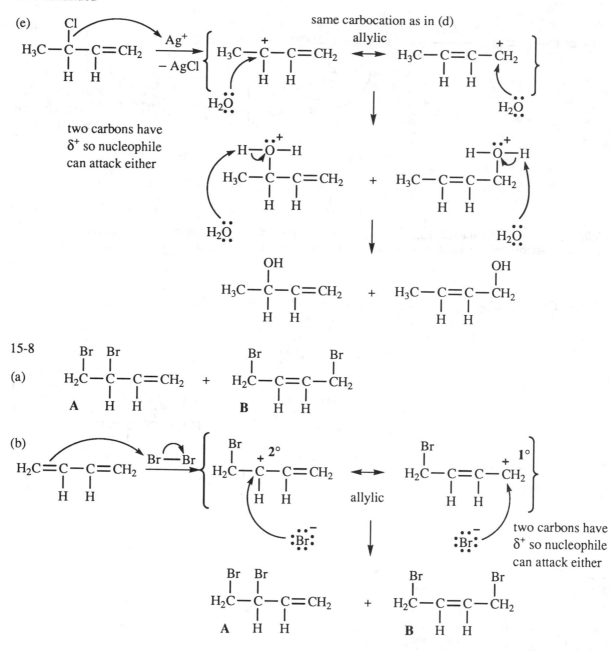

(e)

same carbocation as in (d)

allylic

two carbons have
δ^+ so nucleophile
can attack either

15-8

(a)

A   B

(b)

allylic

two carbons have
δ^+ so nucleophile
can attack either

A   B

(c) The resonance form **A**$^+$, which eventually leads to product **A**, has positive charge on a 2° carbon and is a more significant resonance contributor than structure **B**$^+$. With greater positive charge on the 2° carbon than on the 1° carbon, we would expect bromide ion attack on the 2° carbon to have lower activation energy. Therefore, **A** must be the *kinetic* product. At higher temperature, however, the last step becomes reversible, and the stability of the products becomes the dominant factor in determining product ratios. As **B** has a disubstituted alkene whereas **A** is only monosubstituted, it is reasonable that **B** is the major, *thermodynamic* product at 60° C.

15-8 continued

(d) At 60° C, ionization of **A** would lead to the same allylic carbocation as shown in (b), which would give the same product ratio as formation of **A** and **B** from butadiene.

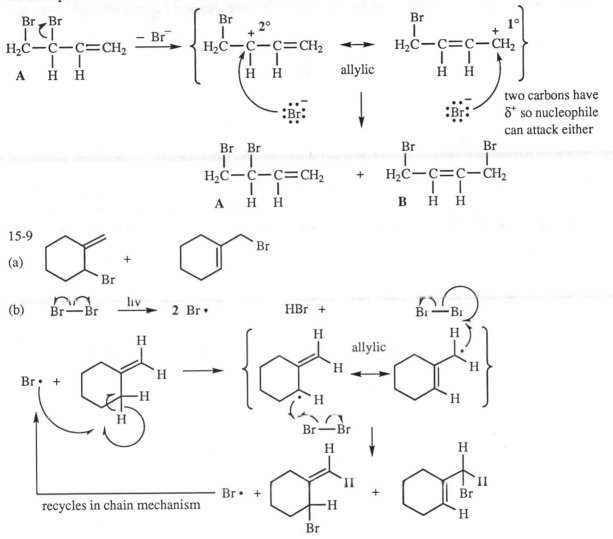

15-9

(a)

(b)

recycles in chain mechanism

15-10　("Pr" is the abbreviation for *n*-propyl, used below.)

<u>NBS generates a low concentration of Br$_2$</u>

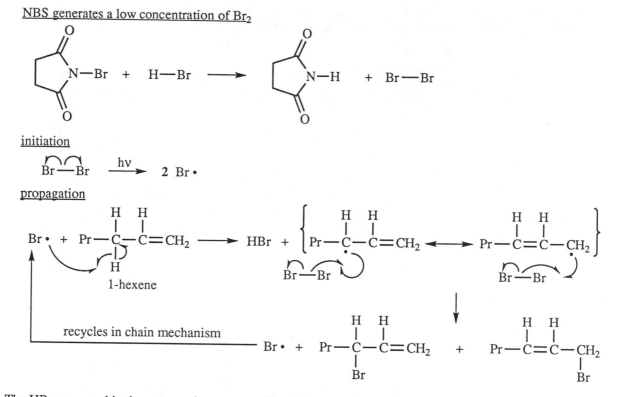

<u>initiation</u>

$$Br\overset{\frown}{-}Br \xrightarrow{\ h\nu\ } 2\ Br\cdot$$

<u>propagation</u>

The HBr generated in the propagation step combines with NBS to produce more Br$_2$, continuing the chain mechanism.

15-11

(a)

(b)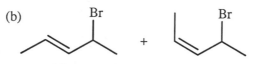

These are the major products from abstraction of a 2° allylic H.

(c)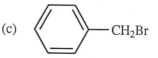

benzylic radicals are even more stable than allylic

15-12 Both halides generate the same allylic carbanion.

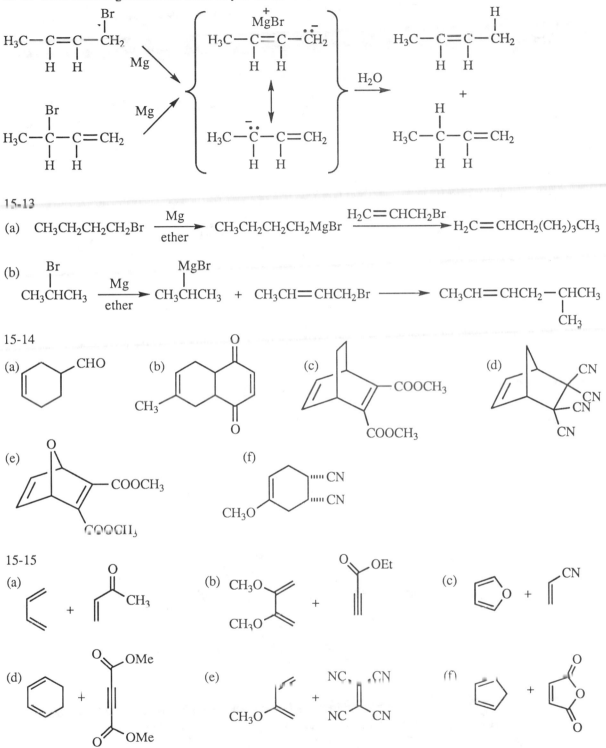

15-13

(a) $CH_3CH_2CH_2CH_2Br \xrightarrow[\text{ether}]{\text{Mg}} CH_3CH_2CH_2CH_2MgBr \xrightarrow{H_2C=CHCH_2Br} H_2C=CHCH_2(CH_2)_3CH_3$

(b)
$$\underset{\substack{| \\ Br}}{CH_3CHCH_3} \xrightarrow[\text{ether}]{\text{Mg}} \underset{\substack{| \\ MgBr}}{CH_3CHCH_3} + CH_3CH=CHCH_2Br \longrightarrow CH_3CH=CHCH_2-\underset{\substack{| \\ CH_3}}{CHCH_3}$$

15-14

(a) [structure: cyclohexene with CHO]
(b) [structure: bicyclic diketone with CH₃]
(c) [structure: bicyclo with two COOCH₃]
(d) [structure: bicyclic with four CN]

(e) [structure: oxabicyclic with COOCH₃ and COOCH₃]
(f) [structure: cyclohexene with CH₃O and two CN]

15-15

(a) [diene] + [methyl vinyl ketone with CH₃]

(b) CH_3O [structure] + [alkyne with OEt]

(c) [furan O] + [acrylonitrile CN]

(d) [cyclohexadiene] + [dimethyl acetylenedicarboxylate with OMe, OMe]

(e) [CH₃O structure] + [NC, CN, NC, CN tetracyanoethylene]

(f) [cyclopentadiene] + [maleic anhydride]

295

15-16

(a)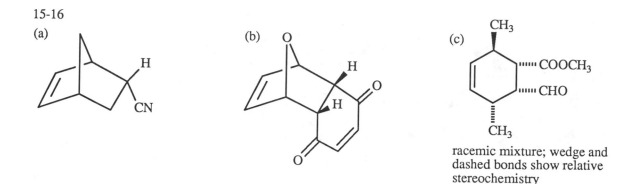

(b)

(c)

racemic mixture; wedge and
dashed bonds show relative
stereochemistry

15-17 These structures are imaginary; they are not intermediates because the Diels-Alder reaction follows a concerted, one-step mechanism.

(a)

(b)

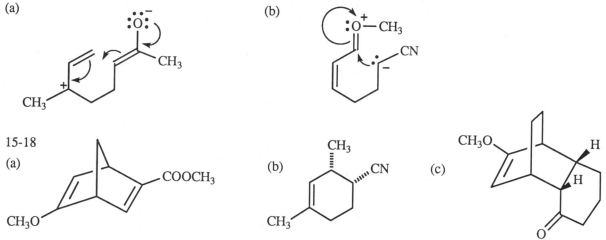

15-18

(a)

(b)

(c)

15-19 For a photochemically *allowed* process, one molecule must use an excited state in which an electron has been promoted to the first antibonding orbital. All orbital interactions between the excited molecule's HOMO* and the other molecule's LUMO must be bonding for the interaction to be allowed; otherwise, it is a forbidden process.

HOMO*
excited state of
diene HOMO

π_3^*

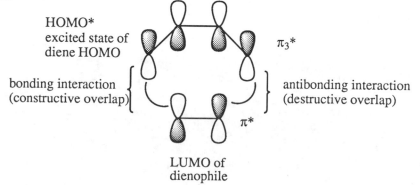

bonding interaction
(constructive overlap)

antibonding interaction
(destructive overlap)

π^*

LUMO of
dienophile

In the Diels-Alder cycloaddition, the LUMO of the dienophile and the excited state of the HOMO of the diene (labeled HOMO*) produce one bonding interaction and one antibonding interaction. Thus, this is a photochemically forbidden process.

15-20 For a [4 + 4] cycloaddition:

(a)

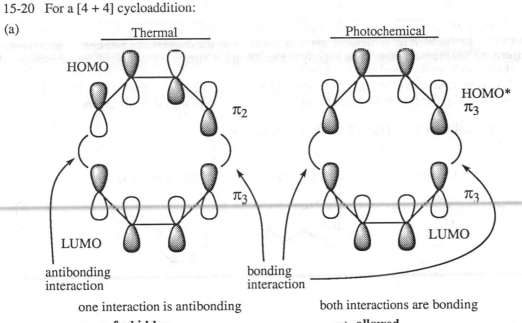

(b) A [4 + 4] cycloaddition is not thermally allowed, but a [4 + 2] (Diels-Alder) is!

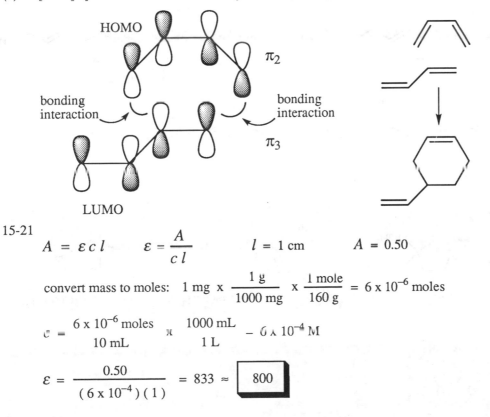

15-21

$$A = \varepsilon\, c\, l \qquad \varepsilon = \frac{A}{c\, l} \qquad l = 1 \text{ cm} \qquad A = 0.50$$

convert mass to moles: $1 \text{ mg} \times \dfrac{1 \text{ g}}{1000 \text{ mg}} \times \dfrac{1 \text{ mole}}{160 \text{ g}} = 6 \times 10^{-6} \text{ moles}$

$$c = \frac{6 \times 10^{-6} \text{ moles}}{10 \text{ mL}} \times \frac{1000 \text{ mL}}{1 \text{ L}} = 6 \times 10^{-4} \text{ M}$$

$$\varepsilon = \frac{0.50}{(6 \times 10^{-4})(1)} = 833 \approx \boxed{800}$$

297

15-22

(a) 353 nm: a conjugated tetraene—must have highest absorption maximum among these compounds;
(b) 313 nm: closest to the bicyclic conjugated triene in Table 15-2; the diene is in a more substituted ring, so it is not surprising for the maximum to be slightly higher than 304 nm;
(c) 232 nm: similar to 3-methylenecyclohexene in Table 15-2;
(d) 273 nm: 1,3 cyclohexadiene (256 nm) + 2 alkyl substituents (2 x 5 nm) = predicted value of 266 nm;
(e) 237 nm: like 3-methylenecyclohexene (232 nm) + 1 alkyl group (5 nm) = predicted value of 237 nm

15-23 Please refer to solution 1-19, page 12 of this Solutions Manual.

15-24

(a) isolated　　(b) conjugated　　(c) cumulated　　(d) conjugated and isolated (at C-6)　　(e) conjugated

15-25

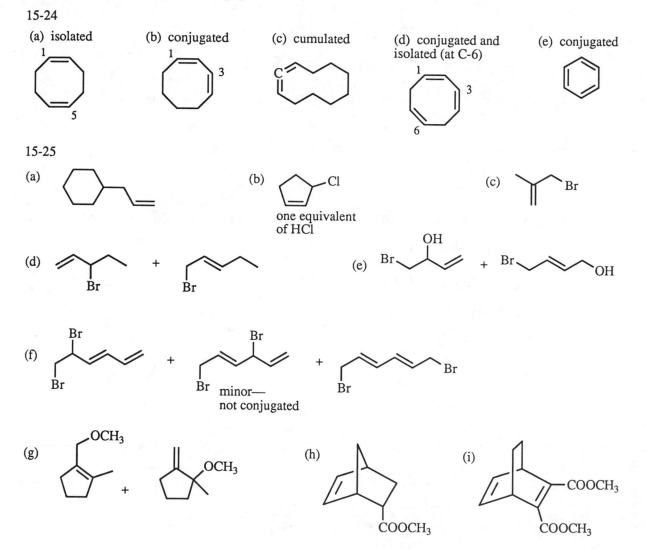

(b) one equivalent of HCl

(f) minor— not conjugated

298

15-26

(a)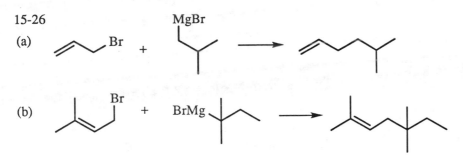

(b)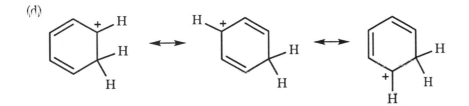

15-27

(a)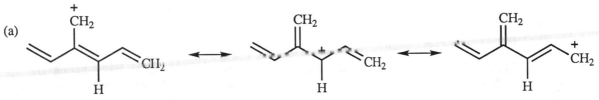

(b)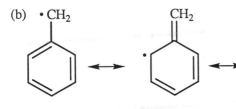

(c)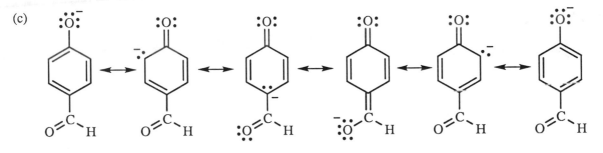

(d)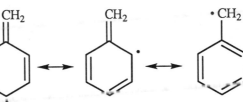

15-28

(a) $A = \varepsilon\, c\, l$ $\varepsilon = \dfrac{A}{c\, l}$ $l = 1$ cm $A = 0.74$

convert mass to moles: 0.0010 g $\times\ \dfrac{1\ \text{mole}}{255\ \text{g}} = 3.9 \times 10^{-6}$ moles

$c = \dfrac{3.9 \times 10^{-6}\ \text{moles}}{100\ \text{mL}}\ \times\ \dfrac{1000\ \text{mL}}{1\ \text{L}} = 3.9 \times 10^{-5}$ M

$\varepsilon = \dfrac{0.74}{(3.9 \times 10^{-5})\,(1)}\ \approx\ \boxed{19{,}000}$

(b) This large value of ε could only come from a conjugated system, eliminating the first structure. The absorption maximum at 235 nm is most likely a diene rather than a triene. The most reasonable structure is:

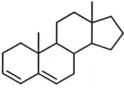

compare with:

λ_{max} = 235 nm

Solved Problem 15-3

λ_{max} = 232 nm

Table 15-2

15-29

(a)

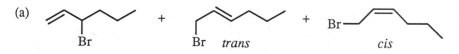

+   +

Br   trans   cis

(b)   ("Pr" is the abbreviation for *n*-propyl, used below.)

NBS generates a low concentration of Br_2

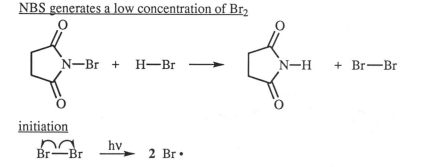

N—Br + H—Br ⟶ N—H + Br—Br

initiation

Br—Br $\xrightarrow{h\nu}$ 2 Br •

(see next page for propagation steps)

15-29 continued

propagation

first step is abstraction of allylic hydrogen to generate allylic radical

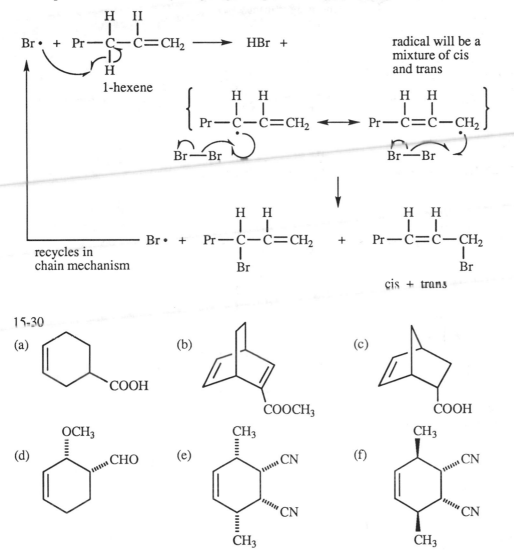

radical will be a
mixture of cis
and trans

15-30

(a), (b), (c), (d), (e), (f)

(Note: Yield of the product in (f) would be small or zero due to severe steric interaction in the *s-cis* conformation of the diene. See Figure 15-16.)

15-31

(a)

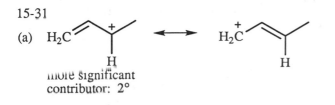

more significant
contributor: 2°

301

15-31 continued

(b)
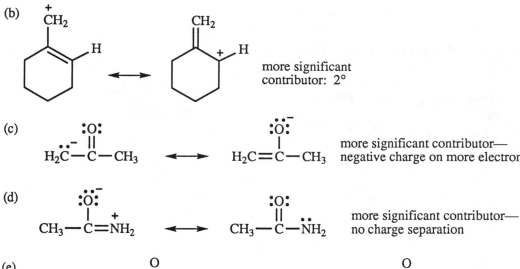

more significant
contributor: 2°

(c)
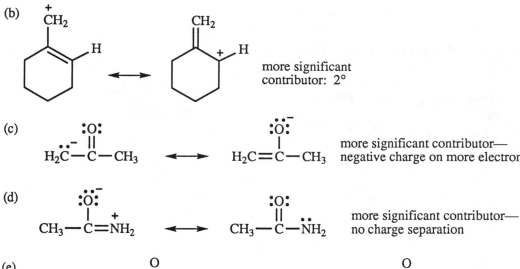

more significant contributor—
negative charge on more electronegative atom

(d)

$$:\overset{..}{\underset{..}{O}}:^-$$
$$CH_3-\overset{|}{\underset{+}{C}}=NH_2$$
$\longleftrightarrow$
$$:\overset{..}{O}:$$
$$CH_3-\overset{||}{C}-\overset{..}{N}H_2$$

more significant contributor—
no charge separation

(e)
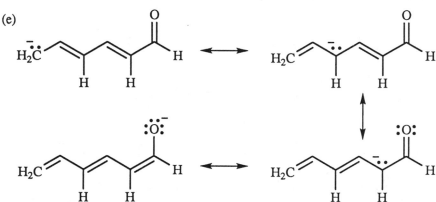

most significant contributor—
negative charge on more electronegative atom

(f)
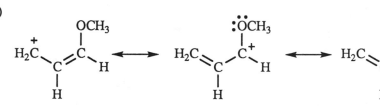

most significant
contributor—all atoms
have full octets

(g)
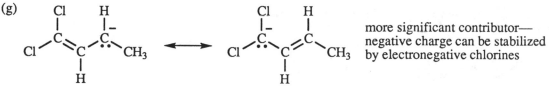

more significant contributor—
negative charge can be stabilized
by electronegative chlorines

302

15-32

(a) The absorption at 1630 cm^{-1} suggests a conjugated alkene. The higher temperature allowed for migration of the double bond.

(b)

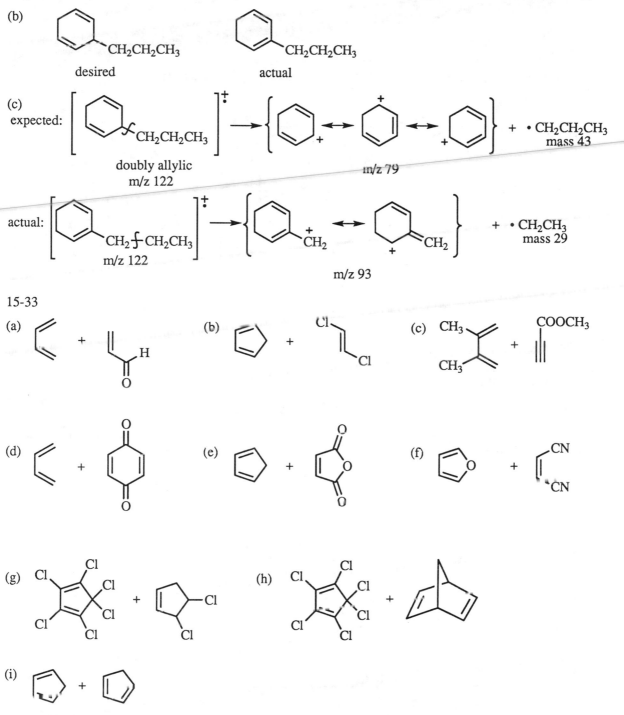

15-33

303

15-34

(a)

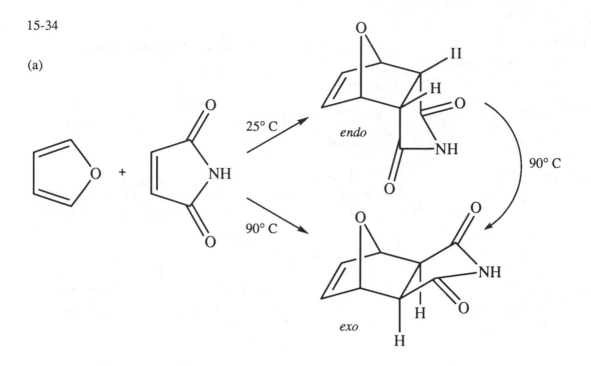

endo

exo

(b)   The *endo* isomer is usually preferred because of secondary p orbital overlap of C=O with the diene in the transition state.

(c)   The reasoning in (b) applies to stabilization of the transition state of the reaction, not the stability of the product. Arguments based on transition state stability apply to the rate of reaction, inferring that the *endo* product is the kinetic product.

(d)   At 25° C, the reaction cannot easily reverse, or at least not very rapidly. The *endo* product is formed faster and is the major product because its transition state is lower in energy—the reaction is under kinetic control. At 90° C, the reverse reaction is not as slow and equilibrium is achieved. The *exo* product is less crowded and therefore more stable—equilibrium control gives the *exo* as the major product.

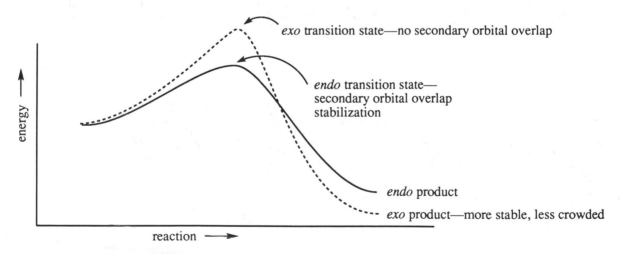

15.35 Nodes are represented by dashed lines.

(a)

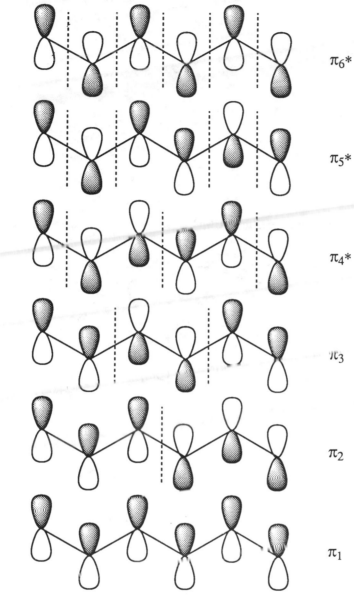

π_6^*

π_5^*

π_4^*

π_3

π_2

π_1

15-35 continued   (b)

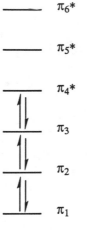

π_6^*

π_5^*

π_4^*

π_3

π_2

π_1

(c)

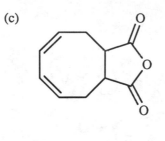

(d) Whether the triene is the HOMO and the alkene is the LUMO, or *vice versa*, the answer will be the same.

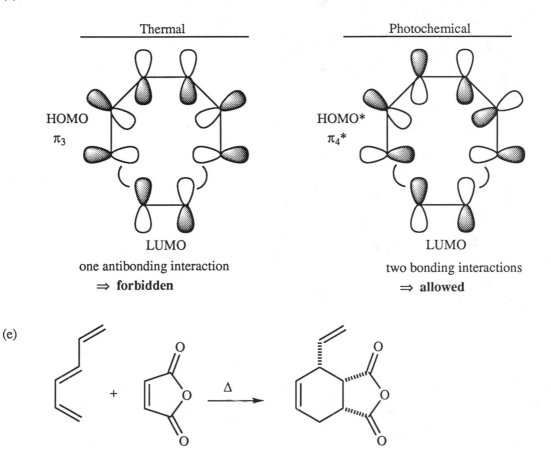

Thermal

HOMO
π_3

LUMO

one antibonding interaction
⇒ **forbidden**

Photochemical

HOMO*
π_4^*

LUMO

two bonding interactions
⇒ **allowed**

(e)

+ $\xrightarrow{\Delta}$

15-36

(a)   H$_2$C=C—C=C—ĊH$_2$ ⟷ H$_2$C=C—Ċ—C=CH$_2$ ⟷ H$_2$Ċ—C=C—C=CH$_2$
          | | | | | | | | |
          H   H   H   H   H   H   H   H   H

(b) Five atomic orbitals will generate five molecular orbitals.

(c) The lowest energy molecular orbital has no nodes. Each higher molecular orbital will have one more node, so the fifth molecular orbital will have four nodes.

(d) Nodes are represented by dashed lines.

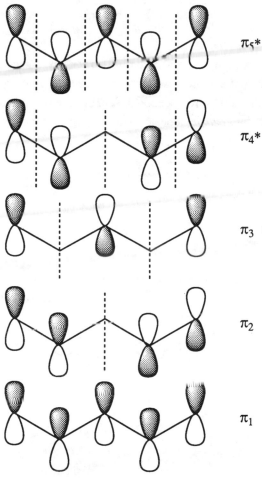

π_5^*

π_4^*

π_3 (nonbonding)

π_2

π_1

(e)

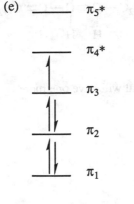

π₅*

π₄*

π₃

π₂

π₁

(f) The HOMO, π_3, contains an unpaired electron giving this species its radical character. The HOMO is a non-bonding orbital with lobes only on carbons 1, 3, and 5, consistent with the resonance picture.

(g)

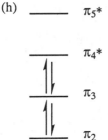

π₅*

π₄*

π₃

π₂

π₁

Again, it is π_3 that determines the character of this species. When the single electron in π_3 of the neutral radical is removed, positive charge appears only in the position(s) which that electron occupied. That is, the positive charge depends on the now *empty* π_3, with *empty* lobes (positive charge) on carbons 1, 3, and 5, consistent with the resonance description.

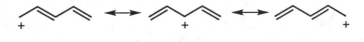

(h)

π₅*

π₄*

π₃

π₂

π₁

Again, it is π_3 that determines the character of this species. The negative charge depends on the *filled* π_3, with lobes (negative charge) on carbons 1, 3, and 5, consistent with the resonance description.

15-37

(a)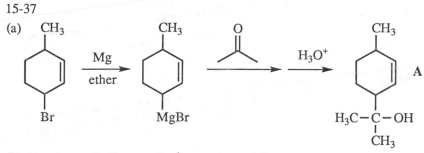

(b) Use Appendix 3 to predict λ_{max} values. Alkyl substituents are circled.

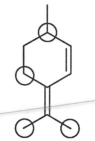 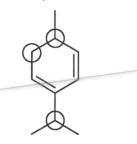 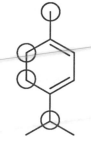

transoid cyclic diene = 217 nm
4 alkyl groups = 20 nm
exocyclic alkene = 5 nm
TOTAL = 242 nm

  desired product

cisoid cyclic diene = 253 nm
3 alkyl groups = 15 nm
TOTAL = 268 nm

cisoid cyclic diene = 253 nm
4 alkyl groups = 20 nm
TOTAL = 273 nm—AHA!

    actual product **B**

(c)

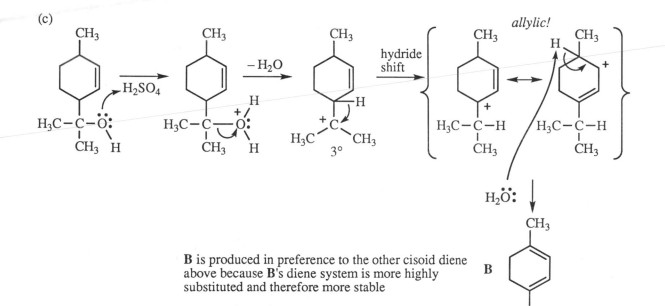

B is produced in preference to the other cisoid diene
above because **B**'s diene system is more highly
substituted and therefore more stable

Note: The representation of benzene with a circle to represent the π system is fine for questions of nomenclature, properties, isomers, and reactions. For questions of mechanism or reactivity, however, the representation with three alternating double bonds (the Kekulé picture) is more informative. For clarity and consistency, this Solutions Manual will use the Kekulé form exclusively.

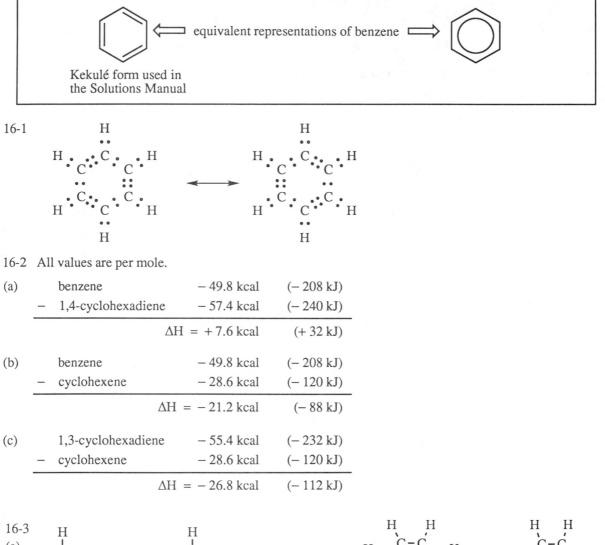

16-1

16-2   All values are per mole.

(a)   benzene   − 49.8 kcal   (− 208 kJ)
   −   1,4-cyclohexadiene   − 57.4 kcal   (− 240 kJ)
   ───
                   ΔH = + 7.6 kcal   (+ 32 kJ)

(b)   benzene   − 49.8 kcal   (− 208 kJ)
   −   cyclohexene   − 28.6 kcal   (− 120 kJ)
   ───
                   ΔH = − 21.2 kcal   (− 88 kJ)

(c)   1,3-cyclohexadiene   − 55.4 kcal   (− 232 kJ)
   −   cyclohexene   − 28.6 kcal   (− 120 kJ)
   ───
                   ΔH = − 26.8 kcal   (− 112 kJ)

16-3
(a)

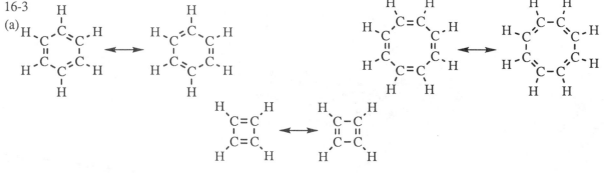

310

16-3 continued

(b)

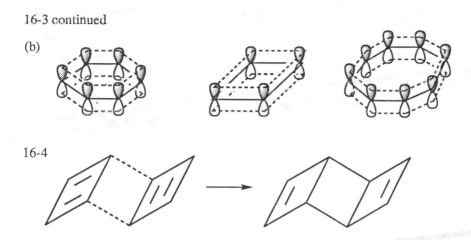

16-4

16-5 Figure 16-8 shows that the first 3 pairs of electrons are in three bonding molecular orbitals of cyclooctatetraene. Electrons 7 and 8, however, are located in two different nonbonding orbitals. As in cyclobutadiene, a planar cyclooctatetraene is predicted to be a diradical, a particularly unstable electron configuration.

16-6

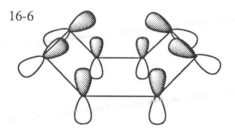

Models show that the angles between p orbitals on adjacent π bonds approach 90°.

16-7

(a) nonaromatic: internal hydrogens prevent planarity
(b) nonaromatic: not all atoms in the ring have a p orbital
(c) aromatic: [14]annulene
(d) aromatic: also a [14]annulene in the outer ring: the internal alkene is not part of the aromatic system

16-8 Azulene satisfies all the criteria for aromaticity, and it has a Huckel number of π electrons: 10. Both heptalene (12 π electrons) and pentalene (8 π electrons) are antiaromatic.

16-9

(a)

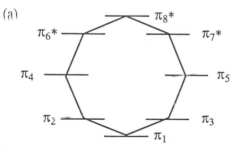

nonbonding energy

(b)

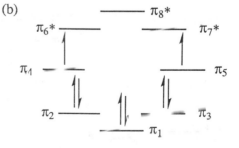

This electronic configuration is antiaromatic.

311

16-9 continued

(c)

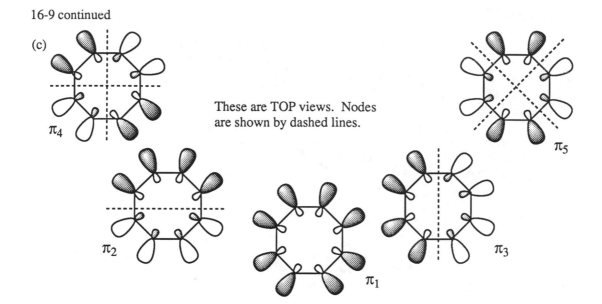

These are TOP views. Nodes are shown by dashed lines.

π_4

π_5

π_2

π_1

π_3

16-10

(a)

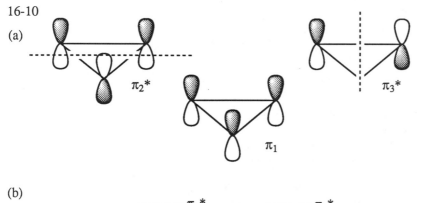

π_2^*

π_1

π_3^*

(b)

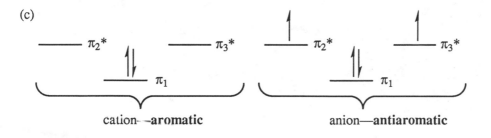

—— π_2^* —— π_3^*

--- nonbonding

—— π_1

π_1 is bonding; π_2^* and π_3^* are antibonding.

(c)

—— π_2^* —— π_3^* ↑ π_2^* ↑ π_3^*

⇅ —— π_1 ⇅ —— π_1

cation—**aromatic**   anion—**antiaromatic**

16-11

(a)

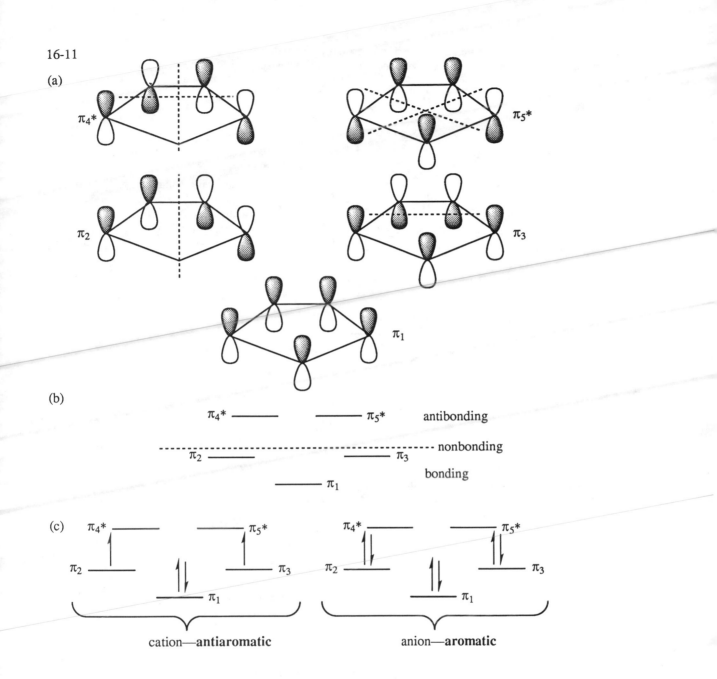

(b)

π_4^* ———   ——— π_5^* antibonding

- nonbonding
π_2 ———   ——— π_3

          ——— π_1 bonding

(c)

cation—**antiaromatic**   anion—**aromatic**

16-12

(a) antiaromatic: 8 π electrons, not a Huckel number
(b) aromatic: 10 π electrons, a Huckel number
(c) aromatic: 18 π electrons, a Huckel number
(d) antiaromatic: 20 π electrons, not a Huckel number
(e) nonaromatic: no cyclic π system
(f) aromatic: 18 π electrons, a Huckel number

16-13 The reason for the dipole can be seen in a resonance form distributing the electrons to give each ring 6 π electrons. This resonance picture gives one ring a negative charge and the other ring a positive charge.

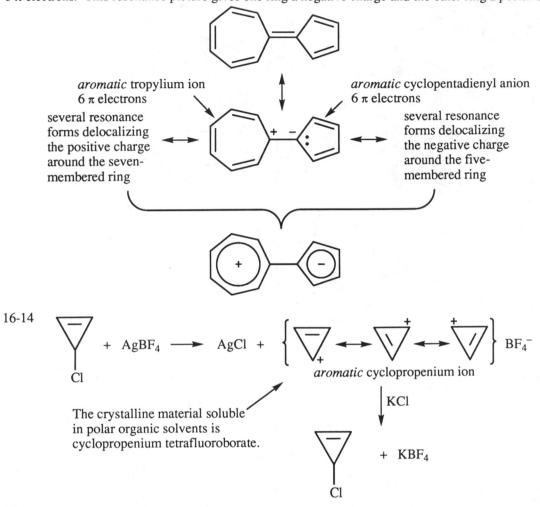

aromatic tropylium ion
6 π electrons

several resonance forms delocalizing the positive charge around the seven-membered ring

aromatic cyclopentadienyl anion
6 π electrons

several resonance forms delocalizing the negative charge around the five-membered ring

16-14

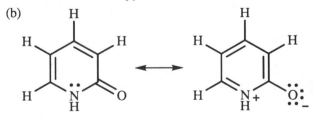

The crystalline material soluble in polar organic solvents is cyclopropenium tetrafluoroborate.

aromatic cyclopropenium ion

16-15

(a) The proton NMR of benzene shows a single peak at δ 7.2; alkene hydrogens absorb at δ 4.5-6. The chemical shifts of 2-pyridone are more similar to benzene's absorptions than they are to alkenes. It would be correct to infer that 2-pyridone is aromatic.

(b)

The lone pair of electrons on the nitrogen in the first resonance form is part of the cyclic pi system. The second resonance form shows three alternating double bonds with six electrons in the cyclic pi system, consistent with an aromatic electronic system.

16-15 continued

(b) continued

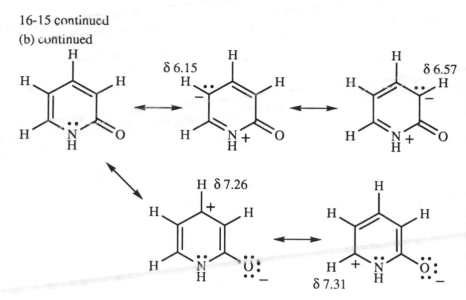

δ 6.15   δ 6.57

These resonance forms show that the hydrogens at positions with greater electron density are shielded, decreasing chemical shift.

H δ 7.26

δ 7.31

These resonance forms show that the hydrogens at positions with greater positive charge are deshielded, increasing chemical shift.

(c)

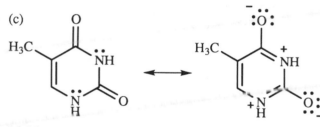

This resonance form of thymine shows a cyclic pi system with 6 pi electrons, consistent with an aromatic system. Four of these electrons came from the two lone pairs of electrons on nitrogens in the first resonance form.

16-16

(a) aromatic: one electron pair from oxygen joins the two π bonds to make a 6 π electron system

(b) nonaromatic: no cyclic π system

(c) aromatic: cation on carbon-4 indicates an empty p orbital; two π bonds plus a pair of electrons from oxygen makes a 6 π electron system

(d) nonaromatic: no cyclic π system

(e) aromatic: one electron pair from sulfur joins the two π bonds to make a 6 π electron system

16-17

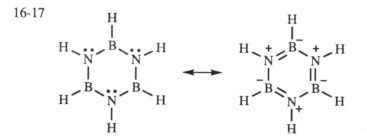

Borazole is a non-carbon equivalent of benzene. Each boron is hybridized in its normal sp^2. Each nitrogen is also sp^2 with its pair of electrons in its p orbital. The system has six π electrons in 6 p orbitals—aromatic!

16-18

(a)

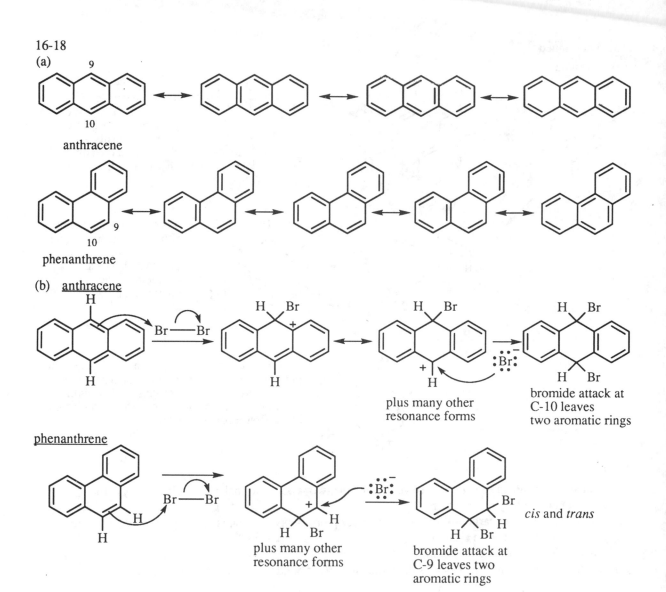

anthracene

phenanthrene

(b) <u>anthracene</u>

plus many other
resonance forms

bromide attack at
C-10 leaves
two aromatic rings

<u>phenanthrene</u>

plus many other
resonance forms

bromide attack at
C-9 leaves two
aromatic rings

cis and trans

(c) A typical addition of bromine occurs with a bromonium ion intermediate which can give only anti addition. Addition of bromine to phenanthrene, however, generates a free carbocation because the carbocation is benzylic, stabilized by resonance over two rings. In the second step of the mechanism, bromide nucleophile can attack either side of the carbocation giving a mixture of cis and trans products.

(d)

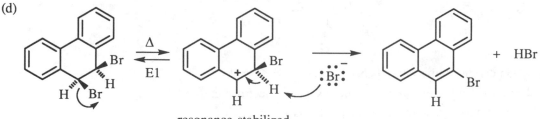

resonance-stabilized

+ HBr

16-19

chlorobenzene

o-dichlorobenzene
(1,2-dichlorobenzene)

m-dichlorobenzene
(1,3-dichlorobenzene)

p-dichlorobenzene
(1,4-dichlorobenzene)

1,2,3-trichlorobenzene

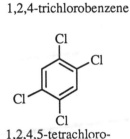

1,2,4-trichlorobenzene

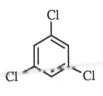

1,3,5-trichlorobenzene

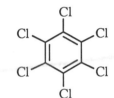

1,2,3,4-tetrachlorobenzene

1,2,3,5-tetrachloro-
benzene

1,2,4,5-tetrachloro-
benzene

1,2,3,4,5-pentachloro-
benzene

1,2,3,4,5,6-hexachloro-
benzene

16-20

(a) fluorobenzene
(b) 4-phenyl-1-butyne
(c) m-methylphenol, or
3-methylphenol
(common name: m-cresol)

(d) o-nitrostyrene
(e) p-bromobenzoic acid, or
4-bromobenzoic acid
(f) isopropoxybenzene, or
isopropyl phenyl ether

(g) 3,4-dinitrophenol
(h) benzyl ethyl ether, or
benzoxyethane, or
(ethoxymethyl)benzene, or
α-ethoxytoluene

16-21 These examples are representative. Your examples may be different from these and still be correct.

(a)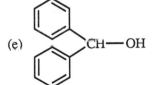

methyl phenyl ether
or methoxybenzene
or anisole

(b) 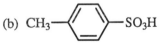 CH$_3$—⟨⟩—SO$_3$H

p-toluenesulfonic acid

(c) ⟨⟩—Li

phenyllithium

(d) O$_2$N—⟨⟩—OH

These are *phenols*.
This is 4-nitrophenol.

(e) 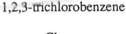 CH—OH

diphenylmethanol

(f)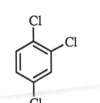

1,3-dibromo-2-phenylbenzene

(g)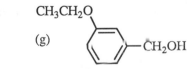

3-ethoxybenzyl alcohol
or 3-ethoxyphenylmethanol

317

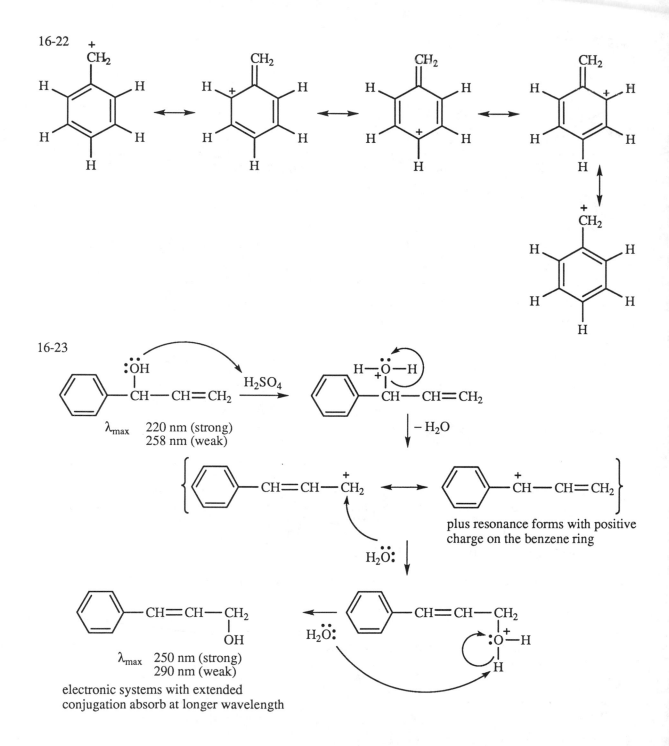

16-24 Please refer to solution 1-19, page 12 of this Solutions Manual.

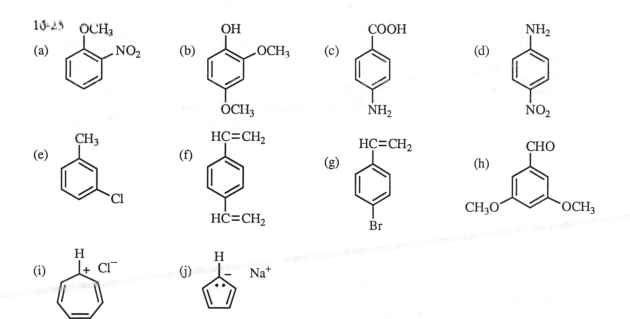

16-25

(a) [structure: OCH₃, NO₂]
(b) [structure: OH, OCH₃, OCH₃]
(c) [structure: COOH, NH₂]
(d) [structure: NH₂, NO₂]
(e) [structure: CH₃, Cl]
(f) [structure: HC=CH₂, HC=CH₂]
(g) [structure: HC=CH₂, Br]
(h) [structure: CHO, CH₃O, OCH₃]
(i) + Cl⁻
(j) − Na⁺

16-26

(a) 1,2-dichlorobenzene (*ortho*)
(b) 4-nitroanisole (*para*)
(c) 2,3-dibromobenzoic acid
(d) 2,7-dimethoxynaphthalene
(e) 3-chlorobenzoic acid (*meta*)
(f) 2,4,6-trichlorophenol
(g) 2-*sec*-butylbenzaldehyde (*ortho*)
(h) cyclopropenium tetrafluoroborate

16-27

toluene   *o*-xylene   *m*-xylene   *p*-xylene

1,2,3-trimethylbenzene   1,2,4-trimethylbenzene   1,3,5-trimethylbenzene (common name: mesitylene)

16-28 Aromaticity is one of the strongest stabilizing forces in organic molecules. The cyclopentadienyl system is stabilized in the anion form where it has 6 π electrons, a Huckel number. The question then becomes: which of the four structures can lose a proton to become aromatic?

While the first, third, and fourth structures can lose protons from sp^3 carbons to give resonance-stabilized anions, only structure two can make a cyclopentadienide anion. It will lose a proton most easily of these four structures which, by definition, means it is the strongest acid.

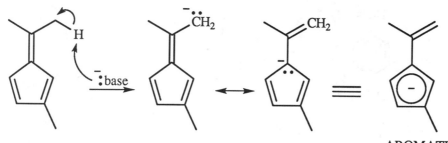

AROMATIC

16-29 Draw resonance forms showing the carbonyl polarization, leaving a positive charge on the carbonyl carbon.

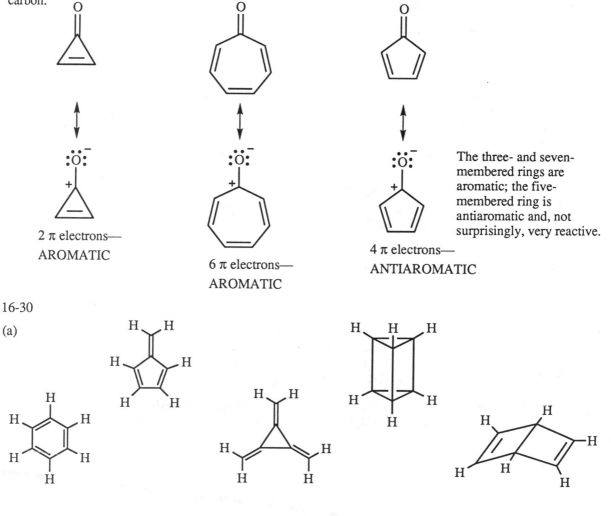

2 π electrons—
AROMATIC

6 π electrons—
AROMATIC

4 π electrons—
ANTIAROMATIC

The three- and seven-membered rings are aromatic; the five-membered ring is antiaromatic and, not surprisingly, very reactive.

16-30

(a)

320

16-30 continued

(b)

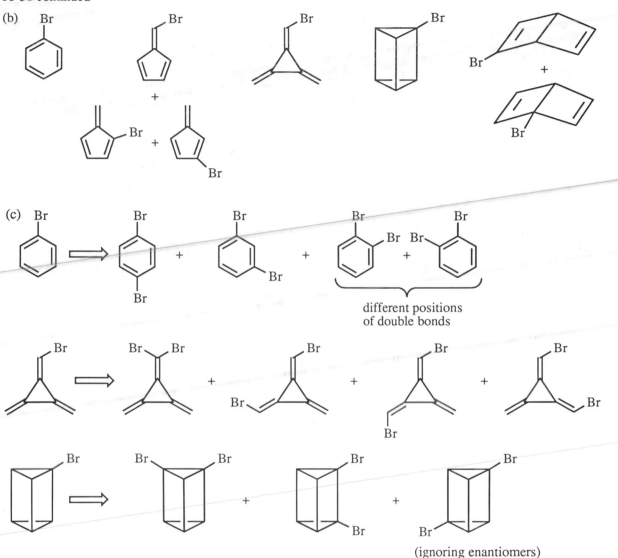

(c)

different positions
of double bonds

(ignoring enantiomers)

(d) The only structure consistent with three isomers of dibromobenzene is the prism structure, called Ladenburg benzene. It also gives no test for alkenes, consistent with the behavior of benzene. (Kekulé defended his structure by claiming that the "two" structures of *ortho*-dibromobenzene were rapidly interconverted, equilibrating so quickly that they could never be separated.)

(e) We now know that three- and four-membered rings are the least stable, but this fact was unknown to chemists during the mid-1800s when the benzene controversy was raging. Ladenburg benzene has two three-membered rings and three four-membered rings (of which only four of the rings are independent), which we would predict to be unstable. (In fact, the structure has been synthesized. Called *prismane*, it is NOT aromatic, but rather, is very reactive toward addition reactions.)

16-31

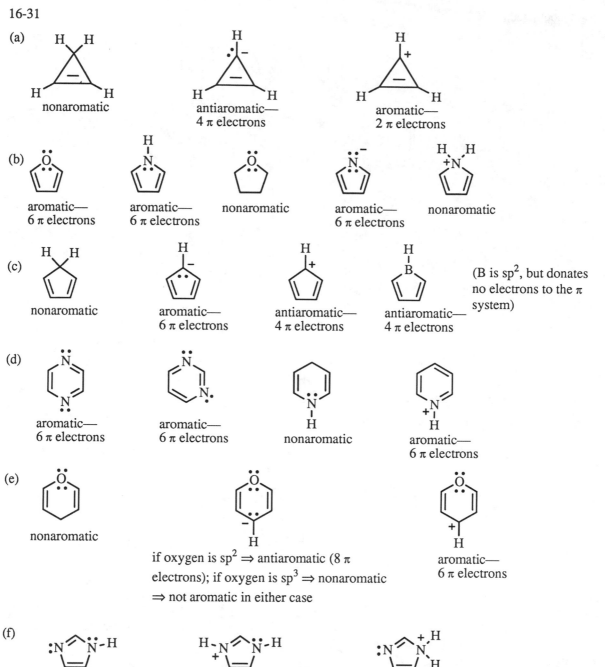

(a)

H H

nonaromatic

antiaromatic—
4 π electrons

aromatic—
2 π electrons

(b)

aromatic—
6 π electrons

aromatic—
6 π electrons

nonaromatic

aromatic—
6 π electrons

nonaromatic

(c)

H H

nonaromatic

aromatic—
6 π electrons

antiaromatic—
4 π electrons

antiaromatic—
4 π electrons

(B is sp², but donates
no electrons to the π
system)

(d)

aromatic—
6 π electrons

aromatic—
6 π electrons

nonaromatic

aromatic—
6 π electrons

(e)

nonaromatic

if oxygen is sp² ⇒ antiaromatic (8 π
electrons); if oxygen is sp³ ⇒ nonaromatic
⇒ not aromatic in either case

aromatic—
6 π electrons

(f)

aromatic—
6 π electrons

aromatic—
6 π electrons

nonaromatic

16-31 continued

(g)

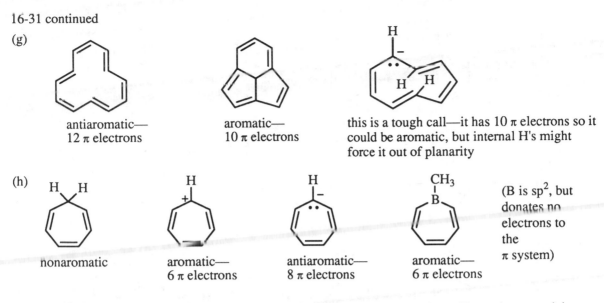

antiaromatic—
12 π electrons

aromatic—
10 π electrons

this is a tough call—it has 10 π electrons so it could be aromatic, but internal H's might force it out of planarity

(h)

nonaromatic

aromatic—
6 π electrons

antiaromatic—
8 π electrons

aromatic—
6 π electrons

(B is sp², but donates no electrons to the π system)

16-32 The clue to azulene is recognition of the five- and seven-membered rings. To attain aromaticity, a seven-membered carbon ring must have a positive charge; a five-membered carbon ring must have a negative charge. Drawing a resonance form of azulene shows this:

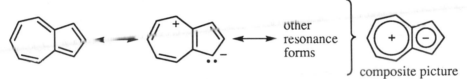

other resonance forms

composite picture

The composite picture shows that the negative charge is concentrated in the five-membered ring, giving rise to the dipole.

16-33 Whether a nitrogen is strongly basic or weakly basic depends on the location of its electron pair. If the electron pair is needed for an aromatic π system, the nitrogen will not be basic (shown here as "weak base"). If the electron pair is in either an sp² or sp³ orbital, it is available for bonding, and the nitrogen is a "strong base".

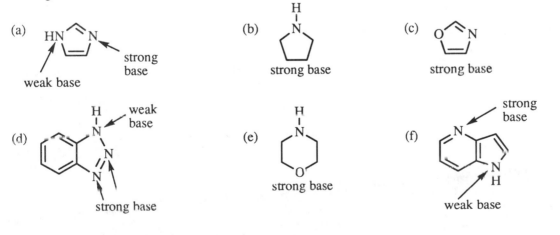

(a)

strong base

weak base

(b)

strong base

(c)

strong base

(d)

weak base

strong base

(e)

strong base

(f)

strong base

weak base

16-34

(a)

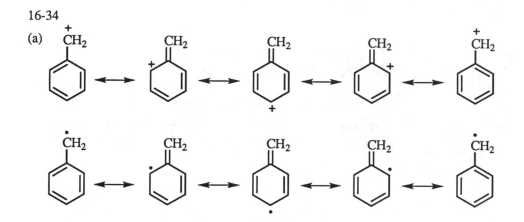

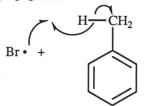

(b) <u>initiation</u>

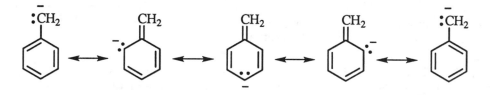

<u>propagation</u>

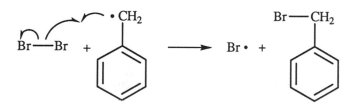

resonance-stabilized

16-34 continued

(c) Both reactions are S_N2 on primary carbons, but the one at the benzylic carbon occurs faster. In the transition state of S_N2, as the nucleophile is approaching the carbon and the leaving group is departing, the electron density resembles that of a p orbital. As such, it can be stabilized through overlap with the π system of the benzene ring.

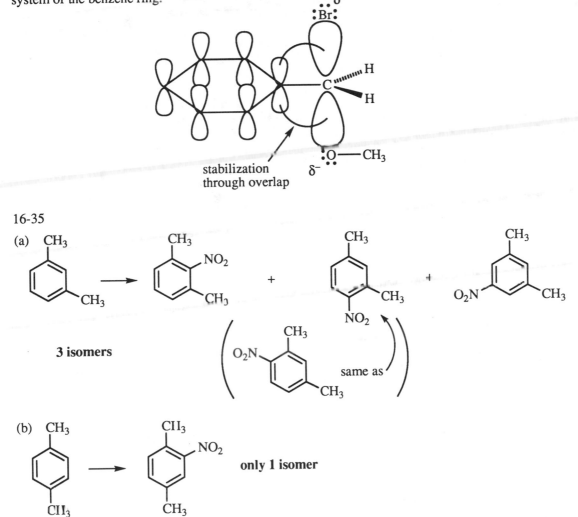

stabilization
through overlap

16-35

(a)

3 isomers

same as

(b)

only 1 isomer

(c) The original compound had to have been *meta*-dibromobenzene as this is the only dibromo isomer that gives three mononitrated products.

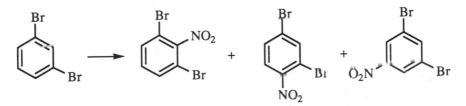

16-36

(a) The formula C_8H_7OCl has five elements of unsaturation, probably a benzene ring (4) plus either a double bond or a ring. The IR suggests a carbonyl at 1690 cm^{-1} and an aromatic ring at 1602 cm^{-1}. The NMR shows a total of five aromatic protons, indicating a monosubstituted benzene. A 2H singlet at δ 4.7 is a deshielded methylene.

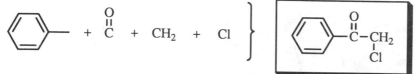

(b) The mass spectral evidence of molecular ion peaks of 1 : 1 intensity at 184 and 186 shows the presence of a bromine atom. The m/z 184 minus 79 for bromine gives a mass of 105 for the rest of the molecule, which is about a benzene ring plus two carbons and a few hydrogens. The NMR shows four aromatic hydrogens in a typical *para* pattern (two doublets), indicating a *para*-disubstituted benzene. The 2H quartet and 3H triplet are characteristic of an ethyl group.

16-37
(a)

like the ends of a
conjugated diene

(b)

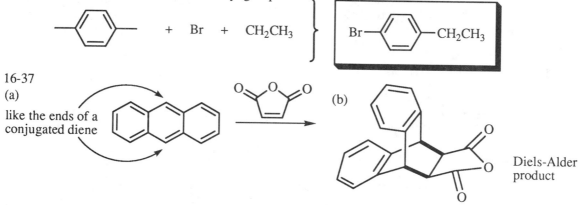

Diels-Alder
product

new sigma bonds shown in bold

16-38
(a) No, biphenyl is not fused. The rings must share two atoms to be labeled "fused".
(b) There are 12 π electrons in biphenyl compared with 10 for naphthalene.
(c) Biphenyl has 6 "double bonds". An isolated alkene releases 28.6 kcal/mole upon hydrogenation.

  predicted: 6 x 28.6 kcal/mole (120 kJ/mole) ≈ 172 kcal/mole (720 kJ/mole)
  observed: 100 kcal/mole (418 kJ/mole)
  ───
    resonance energy: 72 kcal/mole (302 kJ/mole)

(d) On a "per ring" basis, biphenyl is 72 ÷ 2 = 36 kcal/mole, identical to the value for benzene. Naphthalene's resonance energy is 60 kcal/mole (252 kJ/mole); on a "per ring" basis, naphthalene has only 30 kcal/mole of stabilization per ring. This is consistent with the greater reactivity of naphthalene compared with benzene. In fact, the more fused rings, the lower the resonance energy per ring, and the more reactive the compound. (Refer to Problem 16-18.)
(Here is an unconventional view of resonance stabilization energy. Are we being fair to naphthalene? Naphthalene does not have two separate rings like biphenyl, and it does not have two independent aromatic systems. What is the stabilization energy *per pair of π electrons*? For biphenyl: 72 kcal ÷ 6 pairs = 12 kcal per pair of π electrons; for benzene: 36 kcal ÷ 3 pairs = 12 kcal per pair of π electrons; for naphthalene: 60 kcal ÷ 5 pairs = 12 kcal per pair of π electrons! All these compounds have identical stabilization energies per pair of π electrons!)

16-39 Two protons are removed from sp³ carbons to make sp² carbons and to generate a π system with 10 π electrons.

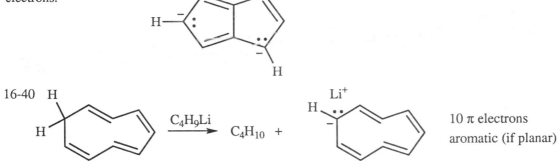

16-40

10 π electrons
aromatic (if planar)

16-41 These four bases can be aromatic, partially aromatic, or aromatic in a tautomeric form. In other words, aromaticity plays an important role in the chemistry of all four structures. (Only electron pairs involved in the important resonance are shown.)

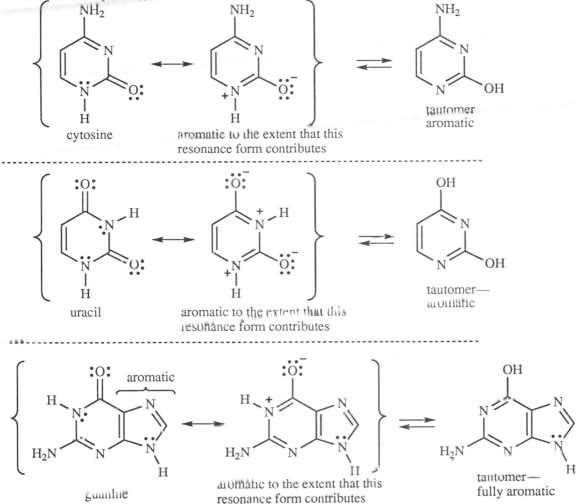

cytosine

aromatic to the extent that this resonance form contributes

tautomer aromatic

uracil

aromatic to the extent that this resonance form contributes

tautomer— aromatic

guanine

aromatic to the extent that this resonance form contributes

tautomer— fully aromatic

16-41 continued

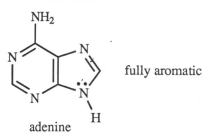

adenine

fully aromatic

16-42

(a) Antiaromatic—only 4 π electrons.

(b) This molecule is electronically equivalent to cyclobutadiene. Cyclobutadiene is unstable and undergoes a Diels-Alder reaction with another molecule of itself. The *t*-butyl groups prevent dimerization by blocking approach of any other molecule.

(c) Yes, the nitrogen should be basic. The pair of electrons on the nitrogen is in an sp^2 orbital and is not part of the π system.

(d)

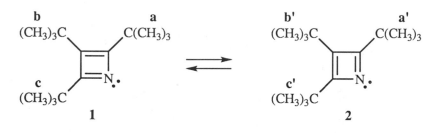

Analysis of structure **1** shows the three *t*-butyl groups in unique environments in relation to the nitrogen. We would expect three different signals in the NMR, as is observed at –110° C. Why do signals coalesce as the temperature is increased? Two of the *t*-butyl groups become equivalent—which two? Most likely, they are **a** and **c** that become equivalent as they are symmetric around the nitrogen. But they are *not* equivalent in structure **1**—what is happening here?

What must happen is an equilibration between structures **1** and **2**, very slow at –110° C, but very fast at room temperature, faster than the NMR can differentiate. So the signal which has coalesced is an average of **a** and **a'** and **c** and **c'**. (This type of low temperature NMR experiment is also used to differentiate axial and equatorial hydrogens on a cyclohexane.)

The NMR data prove that **1** is not aromatic, and that **1** and **2** are isomers, not resonance forms. If **1** were aromatic, then **a** and **c** would have identical NMR signals at all temperatures.

16-43

Mass spectrum: Molecular ion at 150; base peak at 135, M − 15, is loss of methyl.

Infrared spectrum: The broad peak at 3500 cm^{-1} is OH; thymol must be an alcohol. The peak at 1620 cm^{-1} suggests an aromatic compound.

NMR spectrum: The singlet at δ 4.8 is OH; it disappears upon shaking with D_2O. The 6H doublet at δ 1.2 and the 1H multiplet at δ 3.2 are an isopropyl group, apparently on the benzene ring. A 3H singlet at δ 2.3 is a methyl group, also on the benzene ring.

   Analysis of the aromatic protons suggests the substitution pattern. The three aromatic hydrogens confirm that there are three substituents. The singlet at δ 6.5 is a proton between two substituents (no neighboring H's). The doublets at δ 6.75 and δ 7.1 are ortho hydrogens, splitting each other.

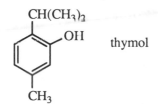   + CH_3 +   OH   + $CH(CH_3)_2$

Several isomeric combinations are consistent with the spectra (although the single H giving δ 6.5 suggests that either Y or Z is the OH group—an OH on a benzene ring shields hydrogens ortho to it, moving them upfield). The structure of thymol is:

CH(CH₃)₂
OH   thymol
CH₃

The final question is how the molecule fragments in the mass spectrometer:

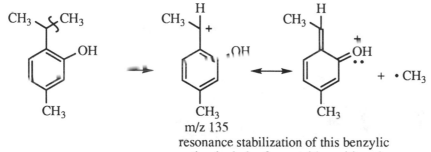

m/z 135
resonance stabilization of this benzylic
cation includes forms with positive charge
on three ring carbons and on oxygen (shown)

16-44

Mass spectrum: Molecular ion at 170; two prominent peaks are M − 15 (loss of methyl) and M − 43 (as we shall see, most likely the loss of acetyl, CH_3CO).

Infrared spectrum: The two most significant peaks are at 1680 cm^{-1} (conjugated carbonyl) and 1600 cm^{-1} (aromatic C=C).

NMR spectrum: A 3H singlet at δ 2.7 is methyl next to a carbonyl, shifted slightly downfield by an aromatic ring. The other signals are seven aromatic protons. The 1H at d 8.7 is a deshielded proton next to a carbonyl. Since there is only one, the carbonyl can have only one neighboring hydrogen.

Conclusions:

$$CH_3 - \overset{\overset{\displaystyle O}{\|}}{C} - \quad + \text{ m/z 127 including 7H} \Rightarrow \text{ mass 120 for carbons} \Rightarrow \text{ 10 C}$$

The fragment $C_{10}H_7$ is almost certainly a naphthalene. Which isomer is correct is indicated by the NMR.

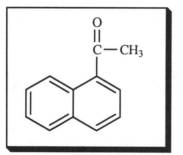

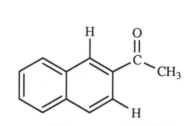

would have two deshielded protons in the NMR

16-45

Although all carbons in hexahelicene are sp^2, the molecule is not flat. Because of the curvature of the ring system, one end of the molecule has to sit on top of the other end—the carbons and the hydrogens would bump into each other if they tried to occupy the same plane. In other words, the molecule is the beginning of a spiral. An "upward" spiral is the nonsuperimposable mirror image of a "downward" spiral, so the molecule is chiral and therefore optically active.

The magnitude of the optical rotation is extraordinary: it is one of the largest rotations ever recorded. In general, alkanes have small rotations and aromatic compounds have large rotations, so it is reasonable to expect that it is the interaction of plane-polarized light (electromagnetic radiation) with the electrons in the twisted pi system (which can also be considered as having wave properties) that causes this enormous rotation.

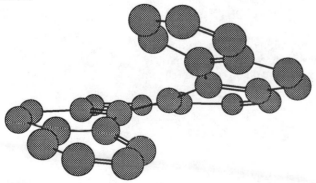

three-dimensional picture of hexahelicene showing the twist in the system of six rings

The representation of benzene with a circle to represent the π system is fine for questions of nomenclature, properties, isomers, and reactions. For questions of mechanism or reactivity, however, the representation with three alternating double bonds (the Kekulé picture) is more informative. For clarity and consistency, this Solutions Manual will use the Kekulé form exclusively.

17-1

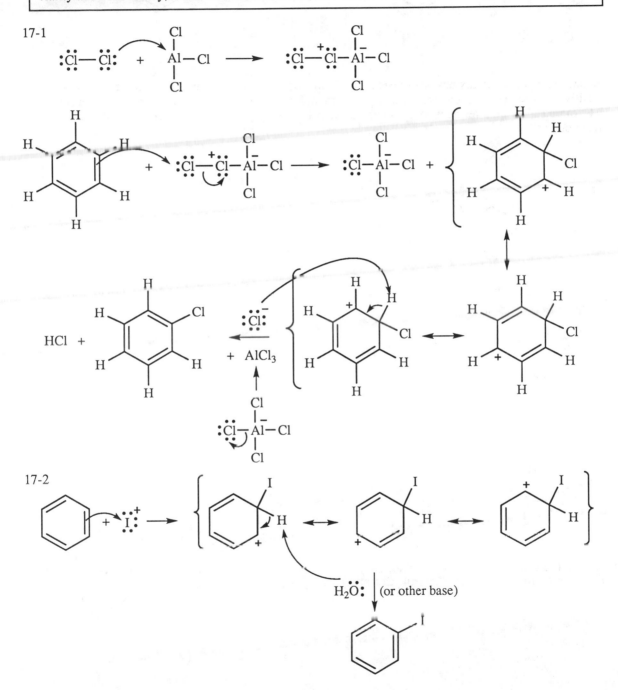

17-2

17-3

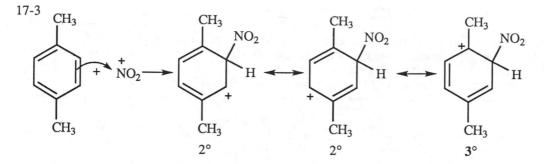

2°   2°   3°

Benzene's sigma complex has positive charge on three 2° carbons. The sigma complex above shows positive charge in one resonance form on a 3° carbon, lending greater stabilization to this sigma complex. The more stable the intermediate, the lower the activation energy required to reach it, and the faster the reaction will be.

17-4

<u>delocalization of the positive charge on the ring</u>

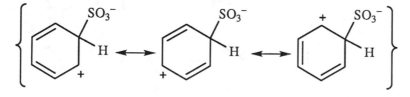

<u>delocalization of the negative charge on the sulfonate group</u>

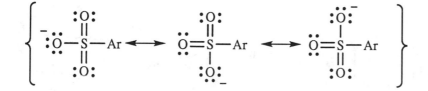

("Ar" is the general abbreviation for an "aromatic" or "aryl" group, in this case, benzene; "R" is the general abbreviation for an "aliphatic" or "alkyl" group.)

17-5

(a) The key to electrophilic aromatic substitution lies in the stability of the sigma complex. When the electrophile bonds at ortho or para positions of ethylbenzene, the positive charge is shared by the 3° carbon with the ethyl group. Bonding of the electrophile at the meta position lends no particular advantage because the positive charge in the sigma complex is never adjacent to, and therefore never stabilized by, the ethyl group.

ortho

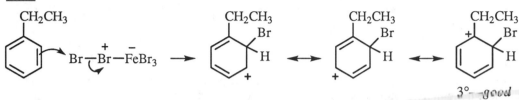

3°—good

meta

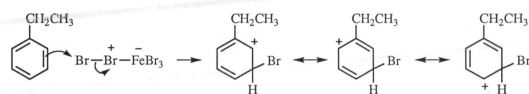

para

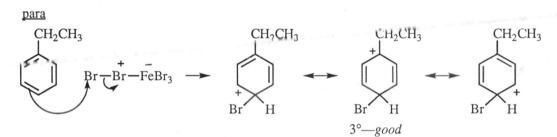

3°—good

(b) Electrophilic attack on *p*-xylene gives an intermediate in which only one of the three resonance forms is stabilized by a substituent (see the solution to Problem 17-3). *m*-Xylene, however, is stabilized in two of its three resonance forms. A more stable intermediate gives a faster reaction.

m-xylene

17-6 For ortho and para attack, the positive charge in the sigma complex can be shared by resonance with the vinyl group. This cannot happen with meta attack because the positive charge is never adjacent to the vinyl group. (Ortho attack is shown; para attack gives an intermediate with positive charge on the same carbons.)

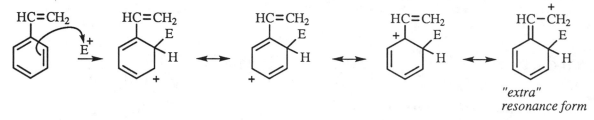

17-7 Attack at only ortho and para positions (not meta) places the positive charge on the carbon with the ethoxy group, where the ethoxy group can stabilize the positive charge by resonance donation of a lone pair of electrons. (Ortho attack is shown; para attack gives a similar intermediate.)

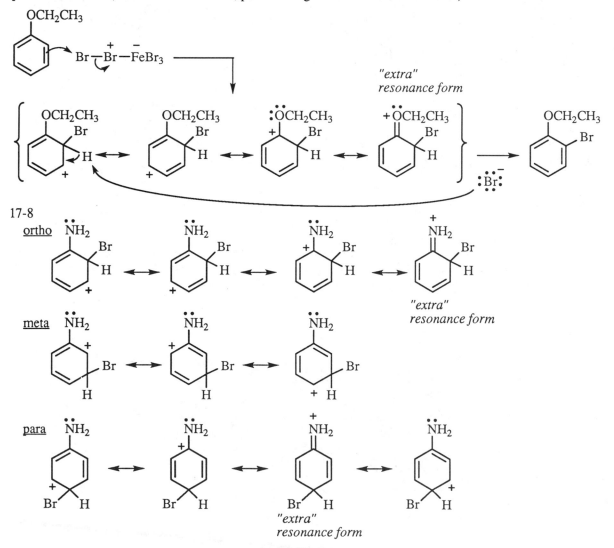

17-9

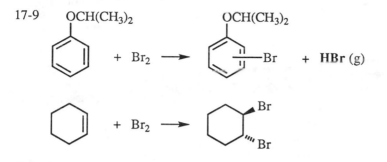

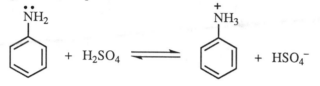

Substitution generates HBr whereas the addition does not. If the reaction is performed in an organic solvent, bubbles of HBr can be observed, and HBr gas escaping into moist air will generate a cloud. If the reaction is performed in water, then adding moist litmus paper to test for acid will differentiate the results of the two compounds.

17-10

(a) Nitration is performed with nitric acid and a sulfuric acid catalyst. In strong acid, amines in general, including aniline, are protonated.

(b) The NH_2 group is a strongly activating ortho,para-director. In acid, however, it exists as the protonated ammonium ion—a strongly **deactivating meta-director**. The strongly acidic nitrating mixture itself forces the reaction to be slower.

(c) The acetyl group removes some of the electron density from the nitrogen, making it much less basic; the nitrogen of this amide is not protonated under the reaction conditions. The N retains enough electron density to share with the benzene ring, so the $NHCOCH_3$ group is still an activating ortho,para-director, though weaker than NH_2.

$$Ph-\underset{\underset{H}{|}}{N}-\overset{\overset{\ddot{O}:}{\|}}{C}-CH_3 \longleftrightarrow Ph-\underset{\underset{H}{|}}{\overset{+}{N}}=\overset{\overset{:\ddot{O}:^-}{|}}{C}-CH_3$$

17-11 Nitronium ion attack at the ortho and para (see next page) positions places positive charge on the carbon adjacent to the bromine, allowing resonance stabilization by an unshared electron pair from the bromine. Meta attack does not give a stabilized intermediate.

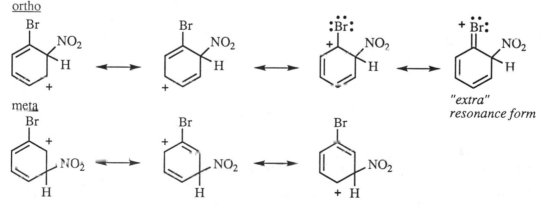

ortho

meta

"extra"
resonance form

17-11 continued

<u>para</u>

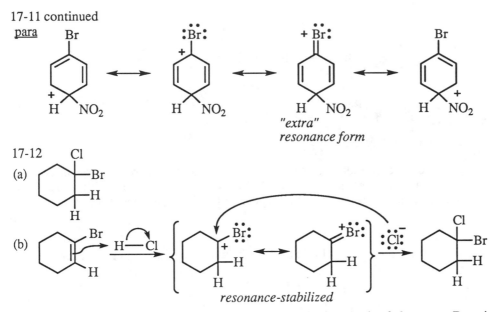

"extra"
resonance form

17-12

(a)

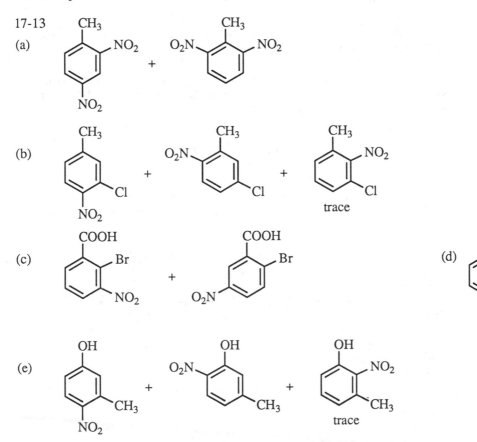

(b)

resonance-stabilized

(c) A bromine atom can stabilize positive charge by sharing a pair of electrons. Bromine can do this in any cationic species, whether from electrophilic addition or electrophilic aromatic substitution.

17-13

(a)

(b)

trace

(c)

(d)

(e)

trace

17-14

(a)

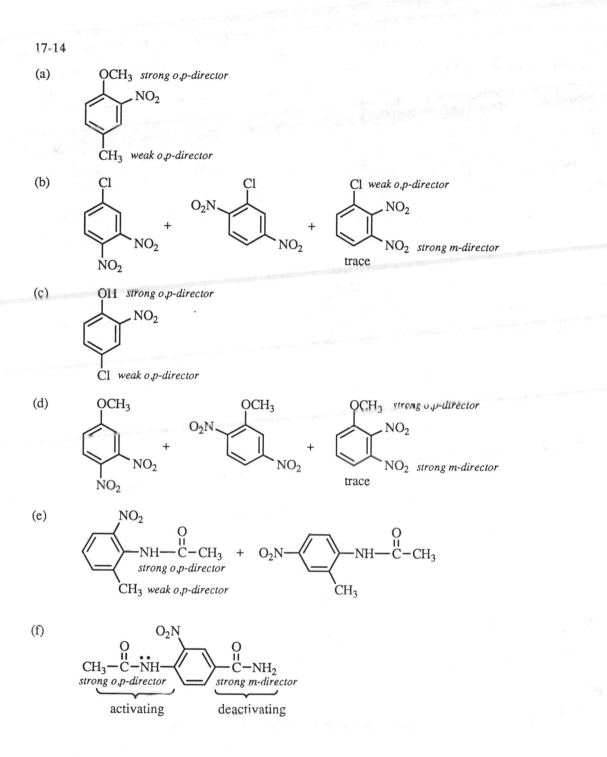

OCH₃ *strong o,p-director*

NO₂

CH₃ *weak o,p-director*

(b)

Cl

NO₂

NO₂

+

Cl

O₂N

NO₂

+

Cl *weak o,p-director*

NO₂

NO₂ *strong m-director*

trace

(c)

OH *strong o,p-director*

NO₂

Cl *weak o,p-director*

(d)

OCH₃

NO₂

NO₂

+

OCH₃

O₂N

NO₂

+

OCH₃ *strong o,p-director*

NO₂

NO₂ *strong m-director*

trace

(e)

NO₂

—NH—C—CH₃
     ‖
     O

strong o,p-director

CH₃ *weak o,p-director*

+

O₂N

—NH—C—CH₃
     ‖
     O

CH₃

(f)

O₂N

CH₃—C—NH—
    ‖
    O

strong o,p-director

C—NH₂
‖
O

strong m-director

activating   deactivating

337

17-15

(a) Sigma complex of ortho attack—benzene stabilizes positive charge by resonance:

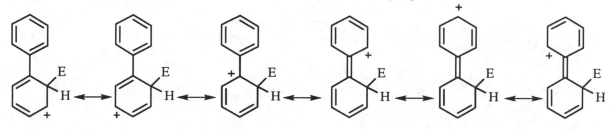

Para attack gives similar stabilization. Meta attack does not permit delocalization of the positive charge on the phenyl substituent.

(b)

(i)

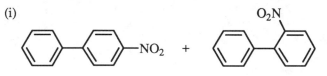

(ii)

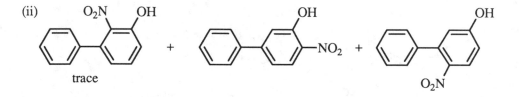

trace

(iii)

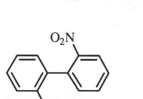

(iv)

(v)

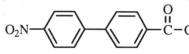

major (minor amounts of nitration on the outer rings)

17-16

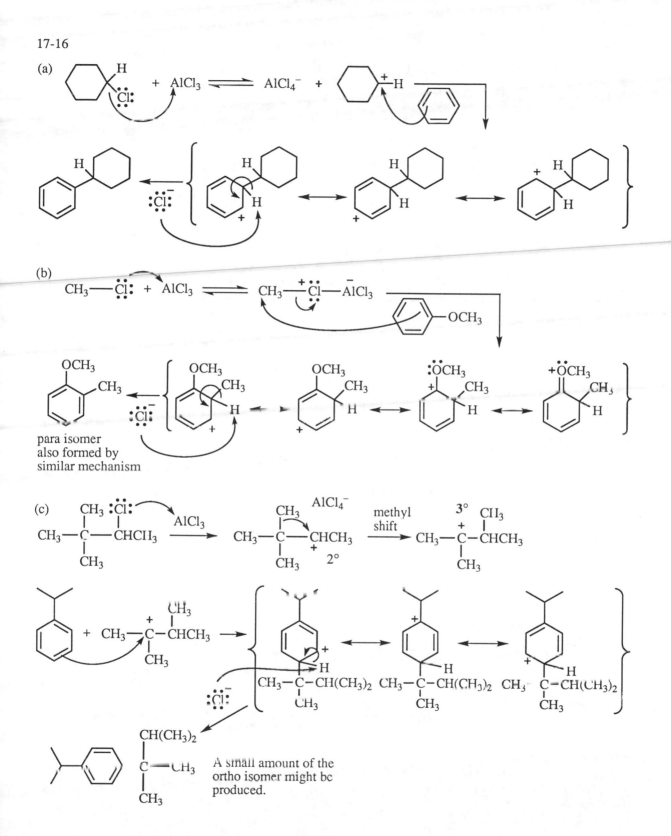

(a)

(b)

para isomer
also formed by
similar mechanism

(c)

methyl
shift

A small amount of the
ortho isomer might be
produced.

17-17

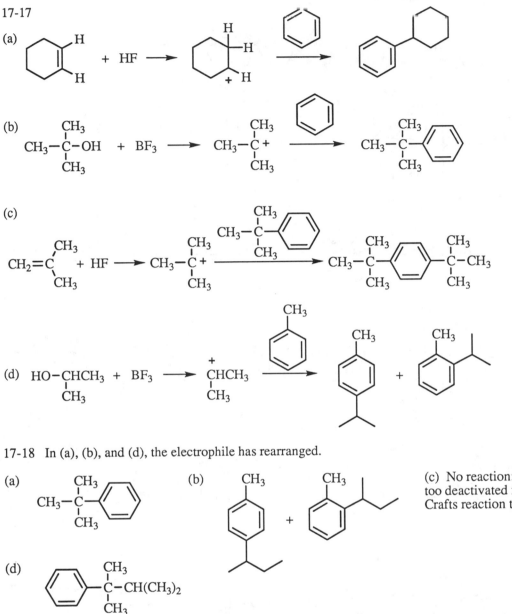

17-18 In (a), (b), and (d), the electrophile has rearranged.

(a)

CH₃—C(CH₃)₂—C₆H₅

(b)

(c) No reaction: nitrobenzene is too deactivated for the Friedel-Crafts reaction to succeed.

(d)

17-19

(a)

benzene + CH₃CH₂CH₂CH₂Br —AlCl₃→ plus over-alkylation products

(b) gives desired product
(c) gives desired product, plus ortho isomer; use excess bromobenzene to avoid overalkylation
(d) gives desired product, plus ortho isomer
(e) gives desired product

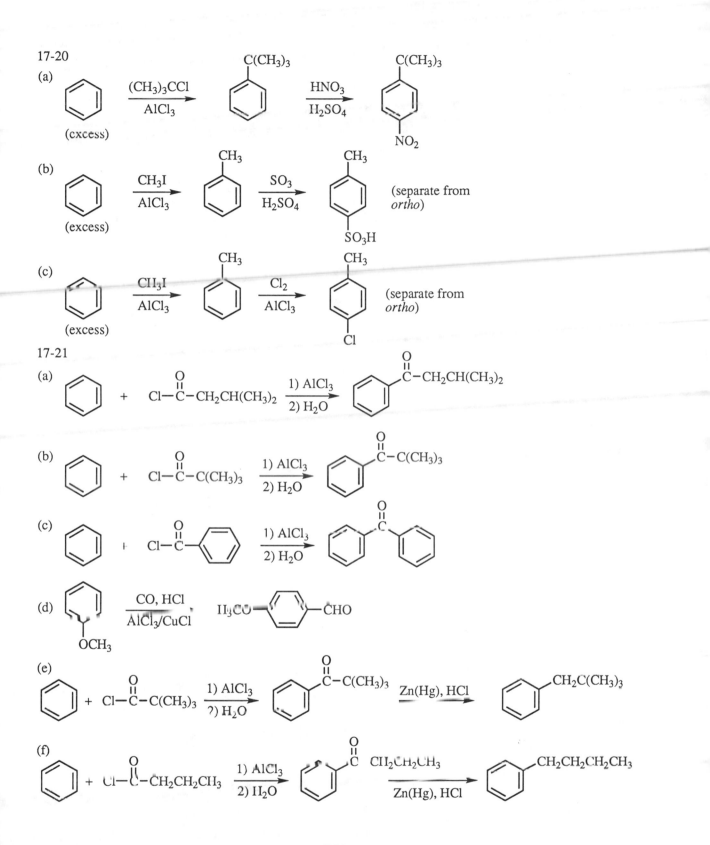

17-20

(a)

(b)

(c)

17-21

(a)

(b)

(c)

(d)

(e)

(f)

17-22 Another way of asking this question is this: Why is fluoride ion a good leaving group from **A** but not from **B** (either by S_N1 or S_N2)?

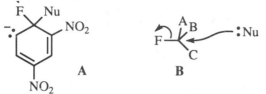

A   B

Formation of the anionic sigma complex **A** is the rate-determining (slow) step in nucleophilic aromatic substitution. The loss of fluoride ion occurs in a subsequent fast step where the nature of the leaving group does not affect the overall reaction rate. In the S_N1 or S_N2 mechanisms, however, the carbon-fluorine bond is breaking in the rate-determining step, so the poor leaving group ability of fluoride does indeed affect the rate.

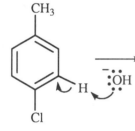

nucleophilic aromatic
substitution

S_N1

S_N2

17-23

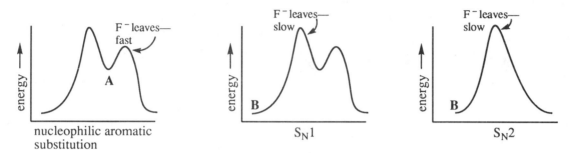

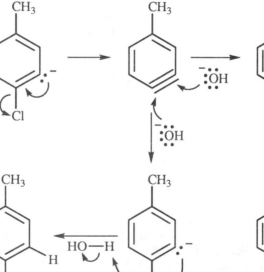

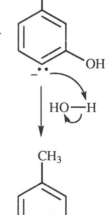

17-24

(a)

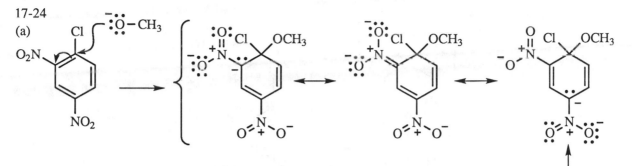

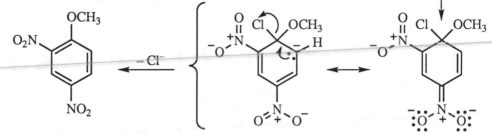

(b)

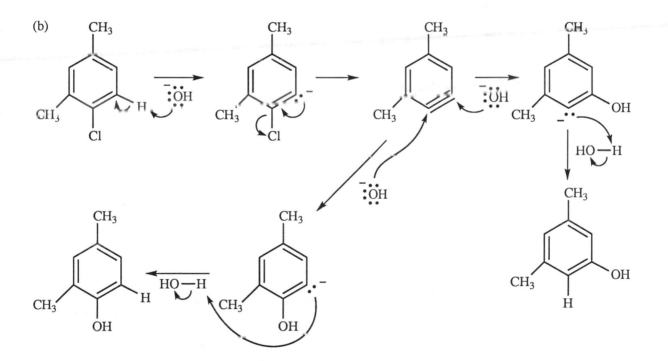

17-24 continued

(c)

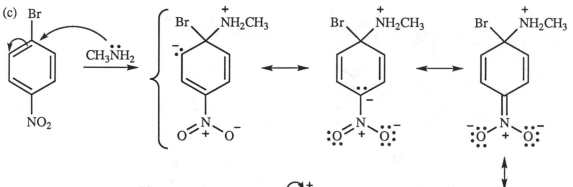

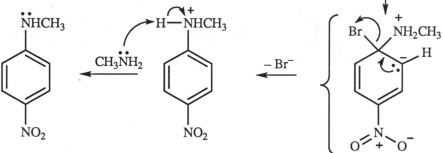

(d)

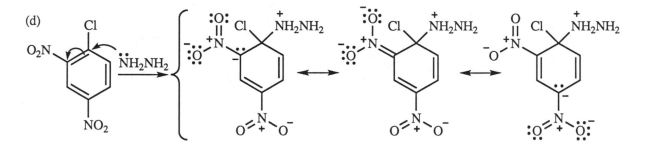

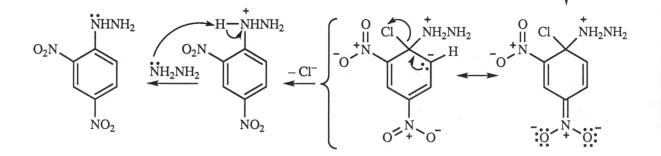

344

17-25

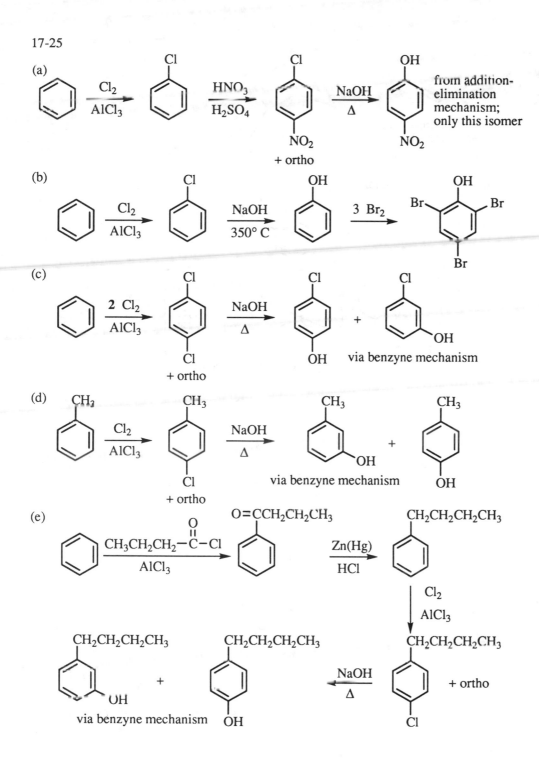

17-26

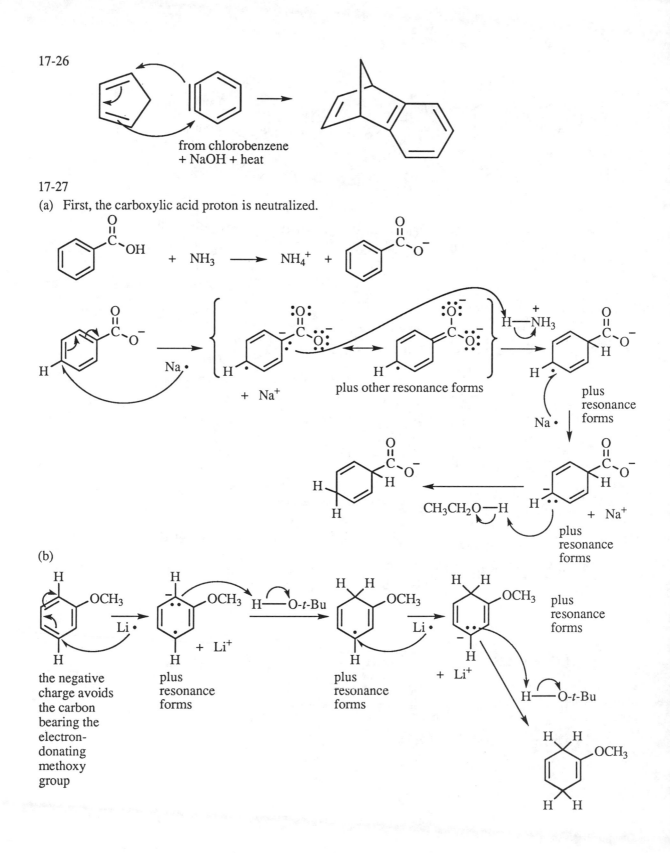

from chlorobenzene
+ NaOH + heat

17-27

(a) First, the carboxylic acid proton is neutralized.

plus other resonance forms

plus resonance forms

plus resonance forms

plus resonance forms

(b)

the negative
charge avoids
the carbon
bearing the
electron-
donating
methoxy
group

plus
resonance
forms

plus
resonance
forms

plus
resonance
forms

17-28

(a) 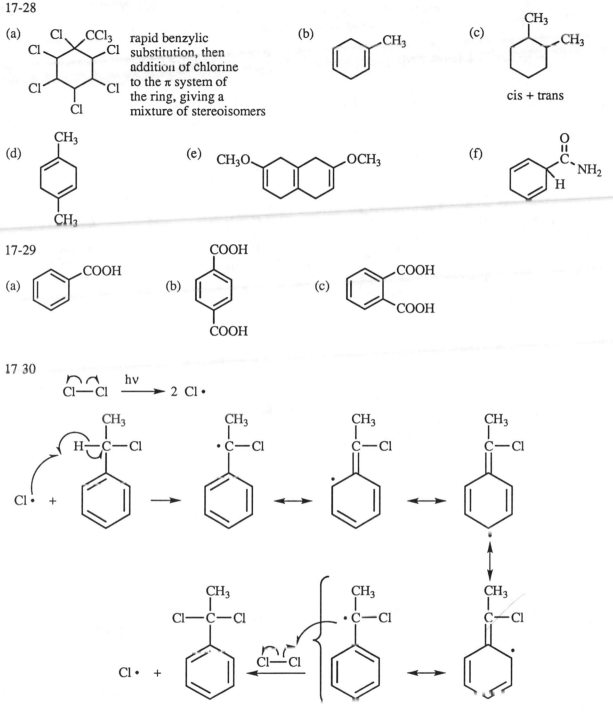 Cl, CCl₃, Cl, Cl, Cl, Cl, Cl   rapid benzylic substitution, then addition of chlorine to the π system of the ring, giving a mixture of stereoisomers

(b) CH₃

(c) CH₃ CH₃ cis + trans

(d) CH₃ ... CH₃

(e) CH₃O ... OCH₃

(f) O, C, NH₂, H

17-29

(a) COOH

(b) COOH ... COOH

(c) COOH ... COOH

17 30

Cl—Cl —hv→ 2 Cl·

347

17-31 A statistical mixture would give 2 : 3 or 40% : 60% α to β. To calculate the relative reactivities, the percents must be corrected for the numbers of each type of hydrogen.

α: $\dfrac{56\%}{2H}$ = 28 relative reactivity   β: $\dfrac{44\%}{3H}$ = 14.7 relative reactivity

The reactivity of α to β is $\dfrac{28}{14.7}$ = 1.9 to 1

17-32

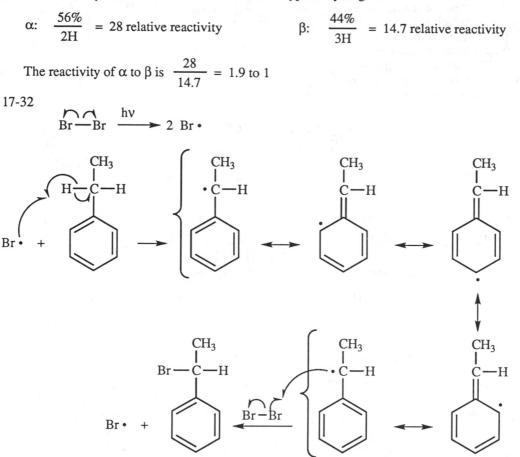

17-33 Replacement of aliphatic hydrogens with bromine can be done under free radical substitution conditions, but reaction at aromatic carbons is unfavorable because of the very high energy of the aryl radical.

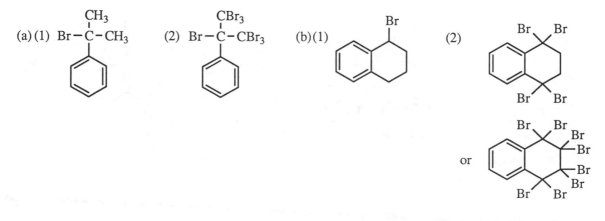

17-34

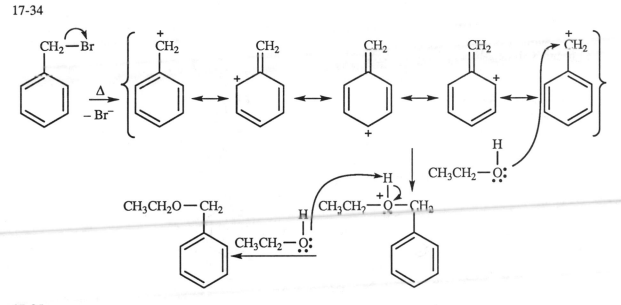

17-35

(a) Benzylic cations are stabilized by resonance and are much more stable than regular alkyl cations. The product is 1-bromo-1-phenylpropane.

(b)

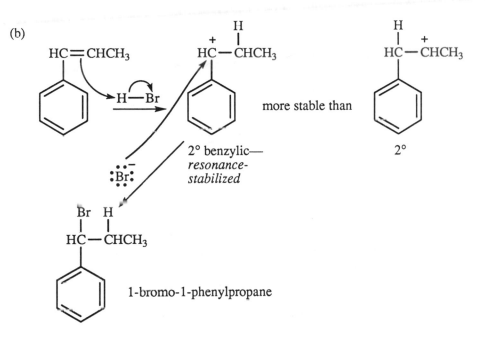

17-37

(a) The combination of HBr with a free-radical initiator generates bromine radicals and leads to anti-Markovnikov orientation. (Recall that whatever species adds *first* to an alkene determines orientation.) The product will be 2-bromo-1-phenylpropane.

(b) Assume the free-radical initiator is a peroxide.

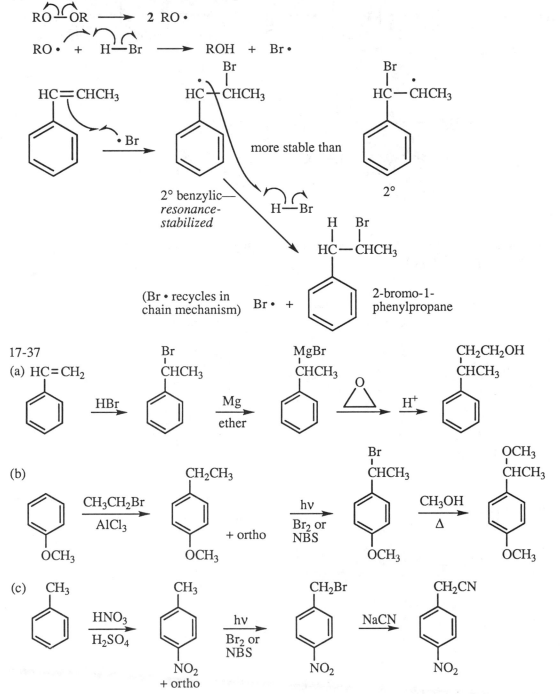

17-37

(a)

(b)

(c)

17-38

(a)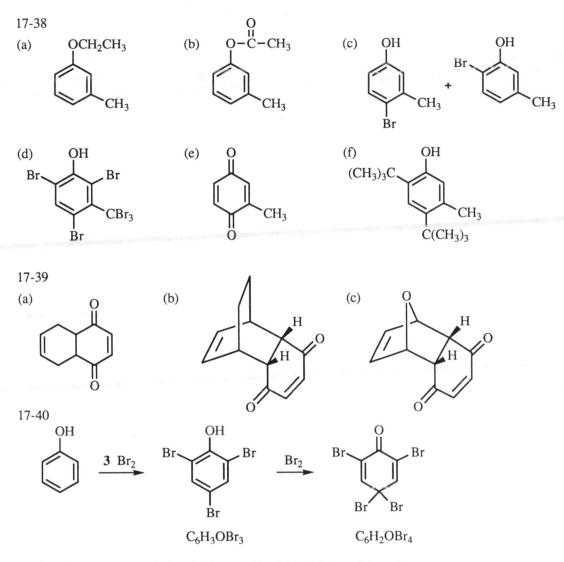

17-39

17-40

$C_6H_3OBr_3$

$C_6H_2OBr_4$

17-41 Please refer to solution 1-19, page 12 of this Solutions Manual.

17-42

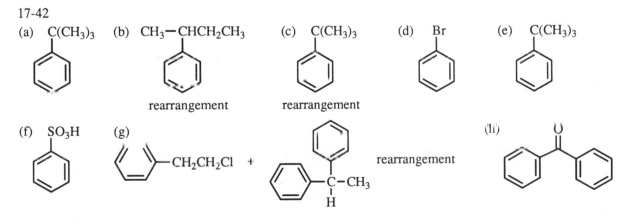

(a) $C(CH_3)_3$ (b) $CH_3-CHCH_2CH_3$ (c) $C(CH_3)_3$ (d) Br (e) $C(CH_3)_3$

rearrangement   rearrangement

(f) SO_3H (g)   (h)

$-CH_2CH_2Cl$ +   rearrangement

$\overset{|}{\underset{H}{C}}-CH_3$

17-42 continued

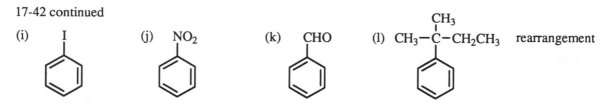

(i) I

(j) NO₂

(k) CHO

(l) CH₃—C(CH₃)—CH₂CH₃ rearrangement

17-43 Products from substitution at the ortho position will be minor because of the steric bulk of the isopropyl group.

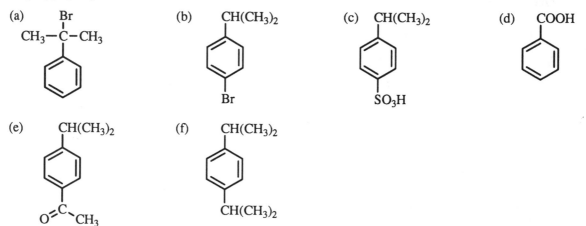

(a) Br CH₃—C(Br)—CH₃

(b) CH(CH₃)₂ ... Br

(c) CH(CH₃)₂ ... SO₃H

(d) COOH

(e) CH(CH₃)₂ ... C(=O)CH₃

(f) CH(CH₃)₂ ... CH(CH₃)₂

17-44

(a)

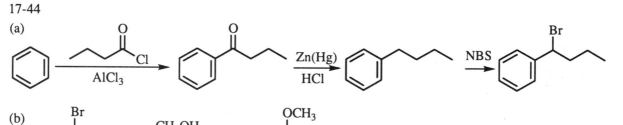

(b)

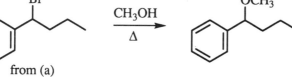

from (a)   CH₃OH / Δ   OCH₃

(c)

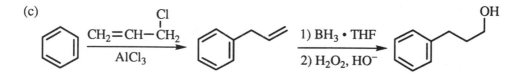

OR

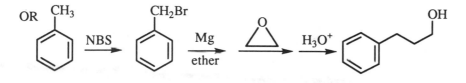

17-44 continued

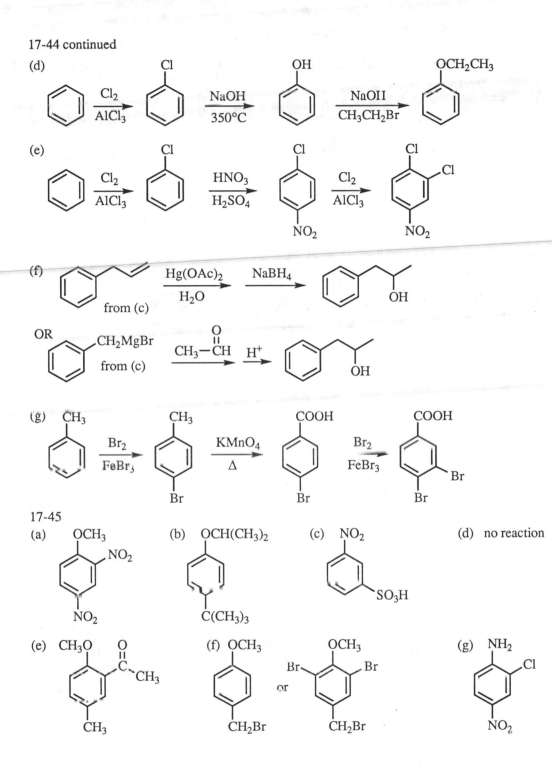

(d)

(e)

(f)

from (c)

OR

CH₂MgBr
from (c)

(g)

17-45

(a) OCH₃
NO₂
NO₂

(b) OCH(CH₃)₂
C(CH₃)₃

(c) NO₂
SO₃H

(d) no reaction

(e) CH₃O O
‖
C—CH₃
CH₃

(f) OCH₃
CH₂Br
or
OCH₃
Br   Br
CH₂Br

(g) NH₂
Cl
NO₂

353

17-45 continued

(h) 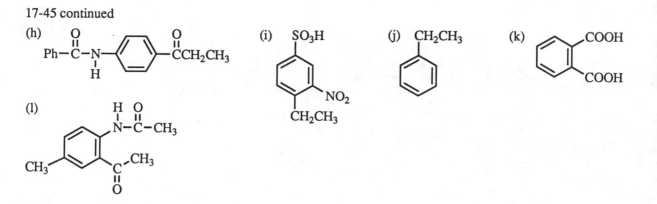 Ph–C(=O)–N(H)–⟨benzene⟩–C(=O)CH₂CH₃

(i) SO₃H / NO₂ / CH₂CH₃

(j) CH₂CH₃

(k) COOH / COOH

(l) N(H)–C(=O)–CH₃ / CH₃ / C(=O)–CH₃

17-46   Major products are shown. Other isomers are possible.

(a)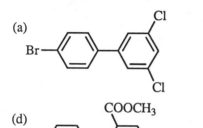

(b)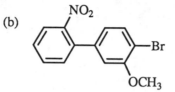

(c)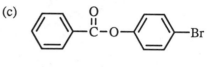

(d)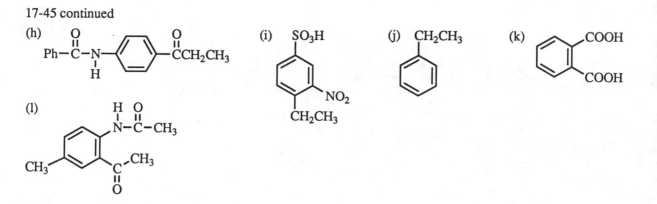

17-47

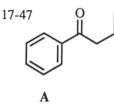

A

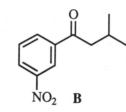

NO₂ **B**

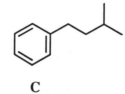

C

D

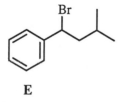

E

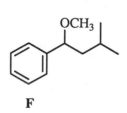

F

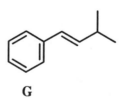

G

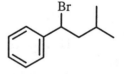

H (same as **E**)

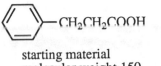

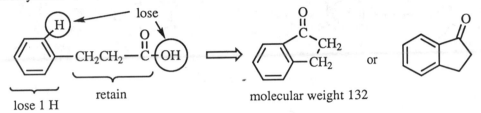

starting material
molecular weight 150

product

molecular weight 132 =
loss of 18 = loss of H_2O

IR spectrum: The dominant peak is the carbonyl at 1710 cm^{-1}. No COOH stretch.

NMR spectrum: The splitting is complicated but the integration is helpful. In the region of δ 2.6-3.2, there are two signals each with integration value of 2H; these must be the two adjacent methylenes, CH_2CH_2. The aromatic region from δ 7.3-7.8 has integration of 4H, so the ring must be disubstituted.

Carbon NMR: Four of the aromatic signals are tall, indicating C—H; two are short, with no attached hydrogens, also showing a disubstituted benzene. Also indicated are carbons in a carbonyl and two methylenes.

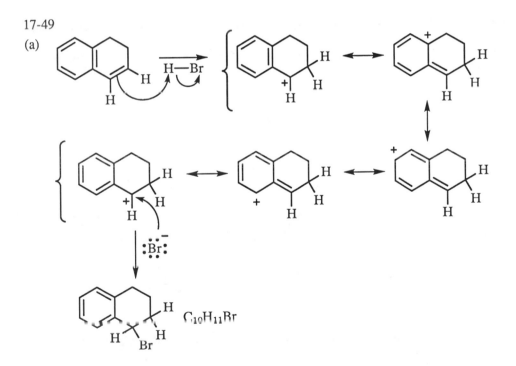

lose 1 H   retain   molecular weight 132

The product must be the cyclized ketone, formed in an intramolecular Friedel-Crafts acylation.

17-49

(a)

$C_{10}H_{11}Br$

17-49 continued

(b) Assume the free-radical initiator is a peroxide.

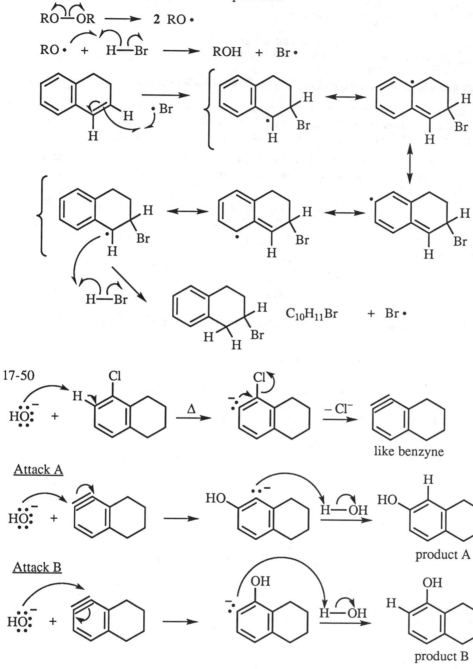

17-50

like benzyne

Attack A

product A

Attack B

product B

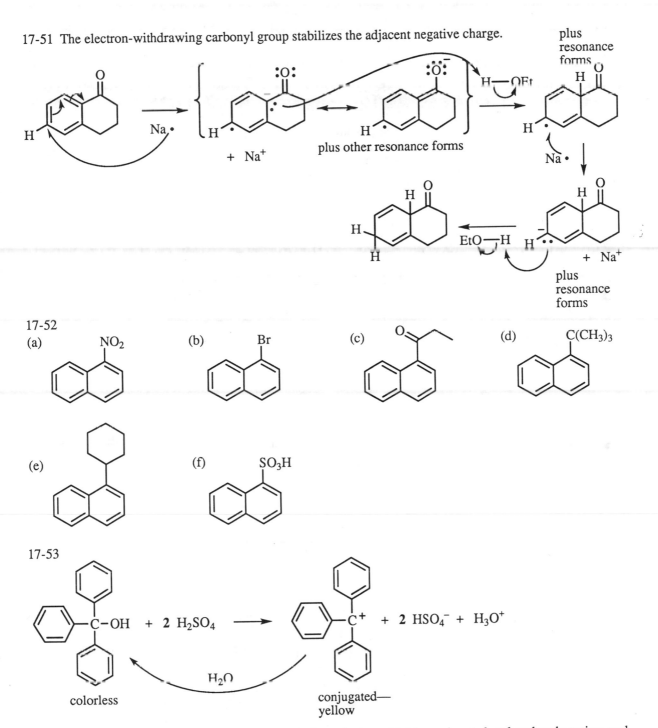

17-51 The electron-withdrawing carbonyl group stabilizes the adjacent negative charge.

+ Na⁺

plus other resonance forms

plus
resonance
forms

plus
resonance
forms

17-52

(a) NO₂

(b) Br

(c) O

(d) C(CH₃)₃

(e)

(f) SO₃H

17-53

$$C-OH + 2\ H_2SO_4 \longrightarrow C^+ + 2\ HSO_4^- + H_3O^+$$

colorless

H₂O

conjugated—
yellow

Concentrated sulfuric acid "dehydrates" the alcohol, producing a highly conjugated, colored carbocation, and protonates the water to prevent the reverse reaction. Upon adding more water, however, there are too many water molecules for the acid to protonate, and triphenylmethanol is regenerated.

17-54

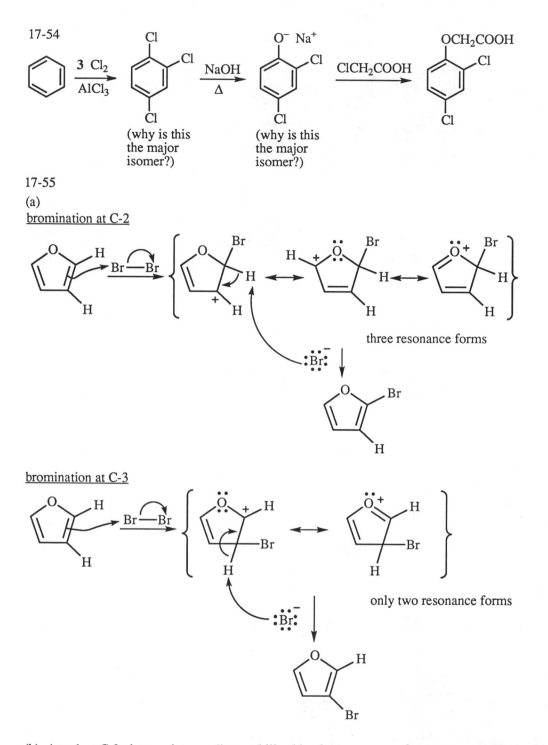

(why is this the major isomer?)

(why is this the major isomer?)

17-55

(a)

bromination at C-2

three resonance forms

bromination at C-3

only two resonance forms

(b) Attack at C-2 gives an intermediate stabilized by three resonance forms, as opposed to only two resonance forms stabilizing attack at C-3. Bromination at C-2 will occur more readily.

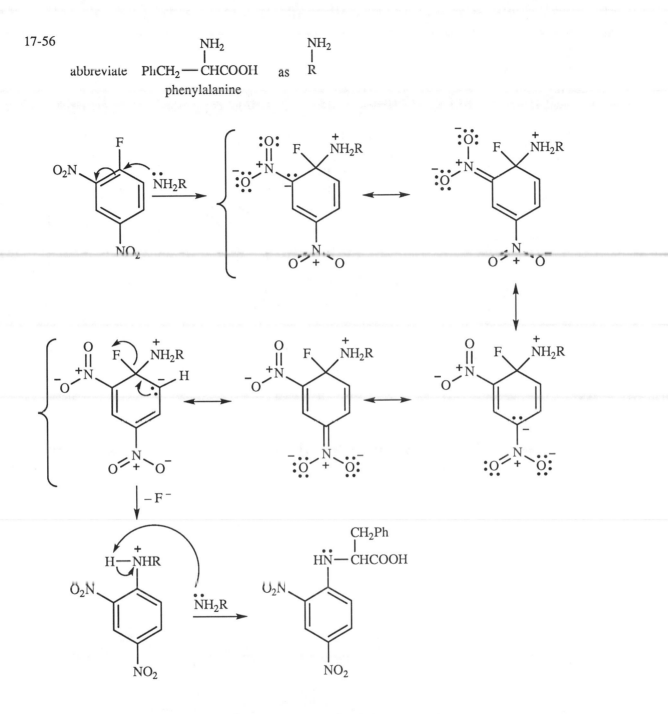

17-57

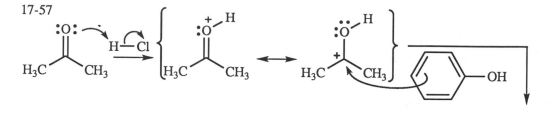

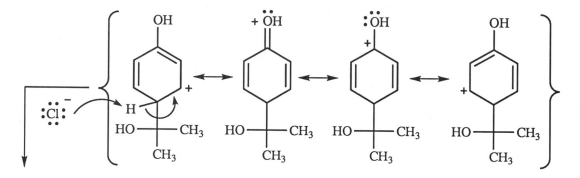

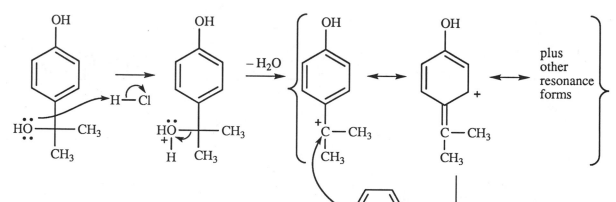

plus
other
resonance
forms

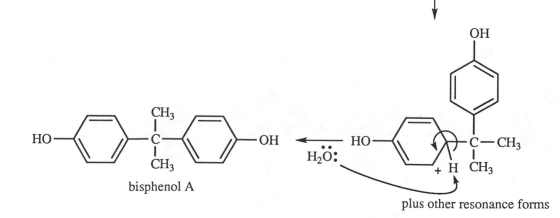

bisphenol A

plus other resonance forms

360

17-58
(a) This is an example of kinetic versus thermodynamic control of a reaction. At low temperature, the kinetic product predominates: in this case, almost a 1 : 1 mixture of ortho and para. These two isomers must be formed at approximately equal rates at 0° C. At 100° C, however, enough energy is provided for the *desulfonation* to occur rapidly; the large excess of the para isomer indicates the para is more stable, even though it is formed initially at the same rate as the ortho.

(b) The product from the 0° C reaction will equilibrate as it is warmed, and at 100° C will produce the same ratio of products as the reaction which was run initially at 100° C.

17-59

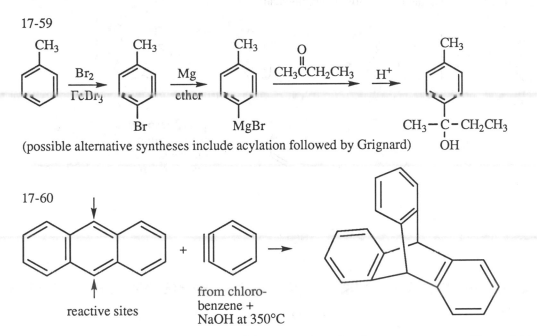

(possible alternative syntheses include acylation followed by Grignard)

17-60

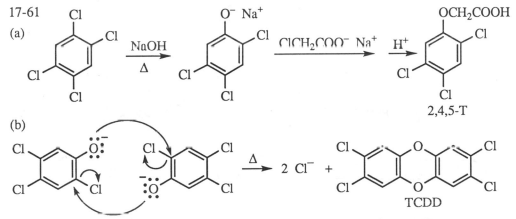

reactive sites

from chloro-
benzene +
NaOH at 350°C

As we saw in Chapter 16, the carbons of the center ring of anthracene are susceptible to electrophilic addition, leaving two isolated benzene rings on the ends. Benzyne is such a reactive dienophile that the reluctant anthracene is forced into a Diels-Alder reaction.

17-61
(a)

2,4,5-T

(b)

2 Cl⁻ +

TCDD

Two nucleophilic aromatic substitutions form a new six-membered ring. (Though not shown here, this reaction would follow the standard addition-elimination mechanism.)

17-61 continued

(c) To minimize formation of TCDD during synthesis: 1) keep the solutions dilute; 2) avoid high temperature; 3) replace chloroacetate with a more reactive molecule like bromoacetate or iodoacetate; 4) add an excess of the haloacetate.

To separate TCDD from 2,4,5-T at the end of the synthesis, take advantage of the acidic properties of 2,4,5-T. The 2,4,5-T will dissolve in an aqueous solution of a weak base like NaHCO$_3$. The TCDD will remain insoluble and can be filtered or extracted into an organic solvent like ether or dichloromethane. The 2,4,5-T can be precipitated from aqueous solution by adding acid.

17-62

(a)

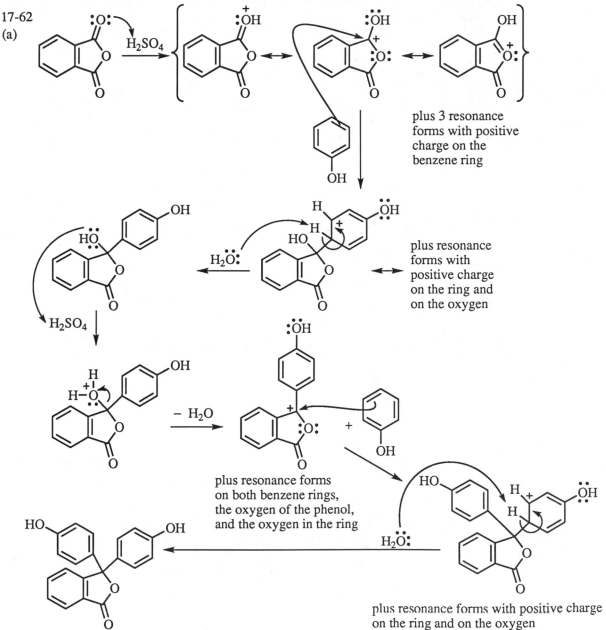

17-62 continued

(b)

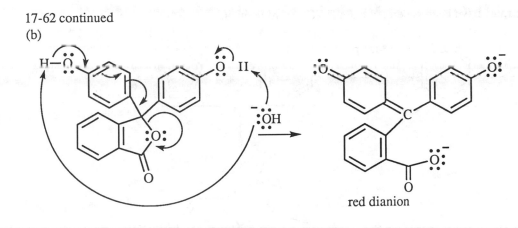

red dianion

(c)

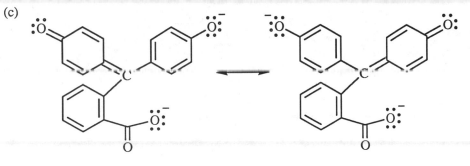

17-63

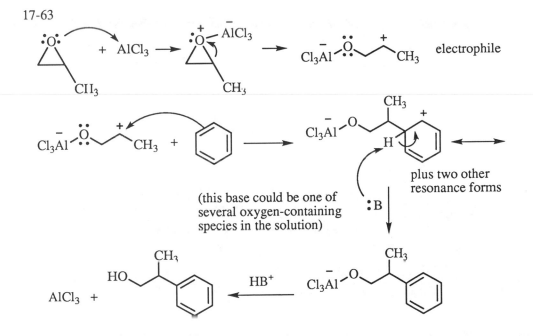

(this base could be one of
several oxygen-containing
species in the solution)

plus two other
resonance forms

363

17-64 A benzyne must have been generated from the Grignard reagent.

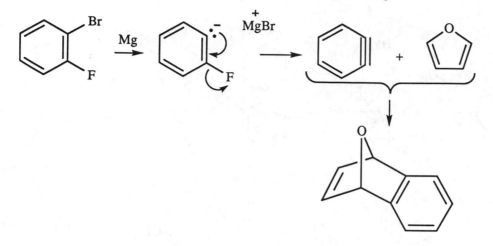

18-1

(a) 5-hydroxy-3-hexanone; ethyl β-hydroxypropyl ketone

(b) 3-phenylbutanal; β-phenylbutyraldehyde

(c) *trans*-2-methoxycyclohexanecarbaldehyde (or *(R,R)* if you named this enantiomer; no common name

(d) 6,6-dimethyl-2,4-cyclohexadienone; no common name

18-2

(a) $C_9H_{10}O$ ⇒ 5 elements of unsaturation

   1H doublet (very small coupling constant) at δ 9.7 ⇒ aldehyde hydrogen, next to CH

   5H multiple peaks at δ 7.2-7.4 ⇒ monosubstituted benzene

   1H multiplet at δ 3.6 and 3H doublet at δ 1.4 ⇒ CHCH₃

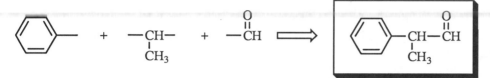

The splitting of the hydrogen on carbon-2, next to the aldehyde, is worth examining. In its overall shape, it looks like a quartet due to the splitting from the adjacent CH₃. A closer examination of the peaks shows that each peak of the quartet is split into two peaks: this is due to the splitting from the aldehyde hydrogen. The aldehyde hydrogen and the methyl hydrogens are not equivalent, so it is to be expected that the coupling constants will not be equal. If a hydrogen is coupled to different neighboring hydrogens by different coupling constants, they must be considered separately, just as you would by drawing a splitting tree for each type of adjacent hydrogen.

(b) C_8H_8O → 5 elements of unsaturation

   cluster of 4 peaks at δ 128-145 ⇒ mono- or para-substituted benzene ring

   peak at δ 197 ⇒ carbonyl carbon (the small peak height suggests a ketone rather than an aldehyde)

   peak at δ 26 → methyl next to carbonyl or benzene

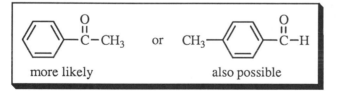

18-3 A compound has to have a hydrogen on a γ carbon (or other atom) in order for the McLafferty rearrangement to occur. 2-Butanone has no γ hydrogen.

$$\overset{\gamma}{\underset{\alpha \qquad \alpha \quad \beta}{CH_3-\overset{\overset{\displaystyle O}{\|}}{C}-CH_2-CH_3}}$$

18-4

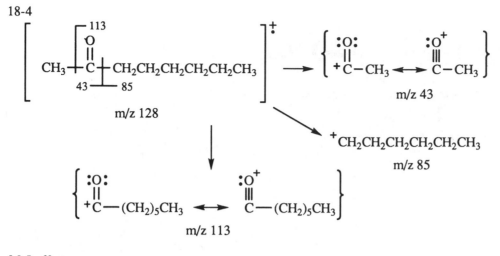

m/z 128

m/z 43

m/z 85

m/z 113

McLafferty rearrangement

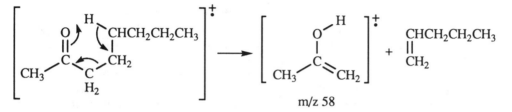

m/z 58

18-5
The first value is the π to π^*; the second value is the n to π^*. The values are approximate.

(a) < 200 nm; 280 nm; this simple ketone should have values similar to acetone

(b) 230 nm; 310 nm; conjugated system (210) plus 2 alkyl groups (20) = 230; the value of 310 nm is similar to Figure 18-7: ketone (280 nm base value) plus 30 nm for the conjugated double bond = 310 nm

(c) 280 nm; 360 nm; conjugated system (210) plus 1 extra double bond (30) plus 4 alkyl groups (40) = 280; similar reasoning for the other transition, starting with an average base value of 290

(d) 270 nm; 350 nm; same as in (c) except only 3 alkyl groups instead of 4

REMINDERS ABOUT SYNTHESIS PROBLEMS:
1. There may be more than one legitimate approach to a synthesis, especially as the list of reactions gets longer.
2. Begin your analysis by comparing the target to the starting material. If the product has more carbons than the reactant, you will need to use one of the small number of reactions that form carbon-carbon bonds.
3. Where possible, work backwards from the target back to the starting material.
4. KNOW THE REACTIONS. There is no better test of whether you know the reactions than attempting synthesis problems.

18-6 All three target molecules in this problem have more than six carbons, so all answers will include carbon-carbon-bond-forming reactions. So far, there are three types of reactions that form carbon-carbon bonds: the Grignard reaction, S_N2 substitution by an acetylide ion, and the Friedel-Crafts reactions (alkylation and acylation) on benzene.

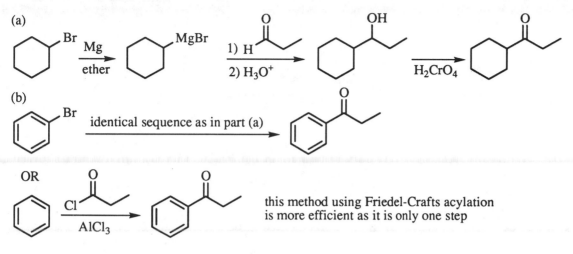

(b)

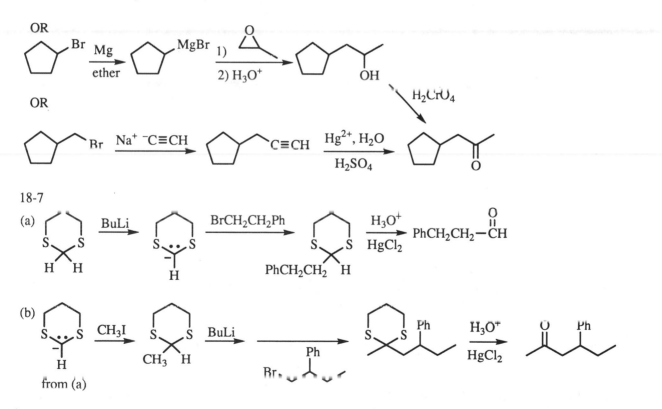

18-7

18-7

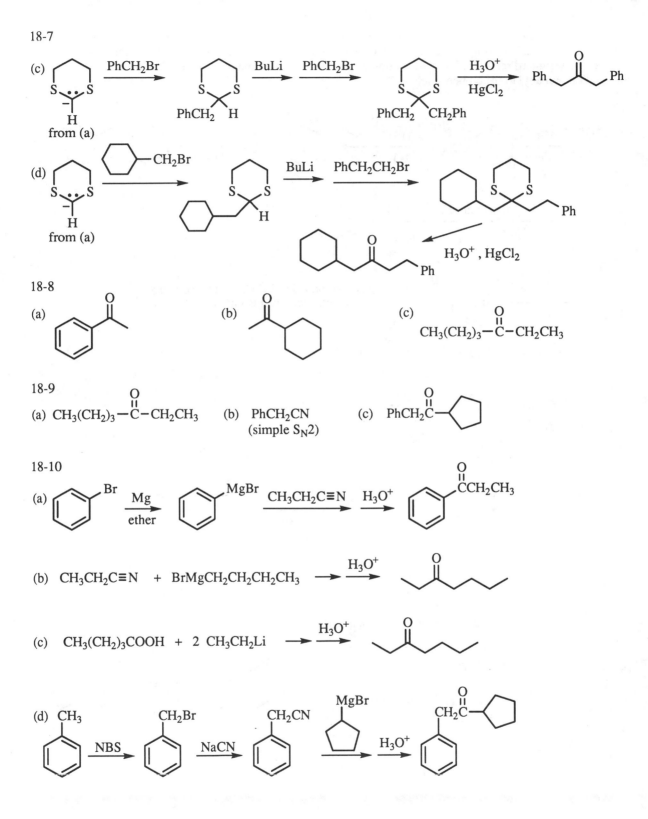

(c) from (a) → PhCH₂Br → PhCH₂ H → BuLi, PhCH₂Br → PhCH₂ CH₂Ph → H₃O⁺, HgCl₂ → Ph–CO–Ph

(d) from (a)

18-8

(a)

(b)

(c) CH₃(CH₂)₃—C(=O)—CH₂CH₃

18-9

(a) CH₃(CH₂)₃—C(=O)—CH₂CH₃

(b) PhCH₂CN (simple S_N2)

(c) PhCH₂C(=O)-cyclopentyl

18-10

(a)

(b) CH₃CH₂C≡N + BrMgCH₂CH₂CH₂CH₃ → H₃O⁺ →

(c) CH₃(CH₂)₃COOH + 2 CH₃CH₂Li → H₃O⁺ →

(d)

18-11

(a) 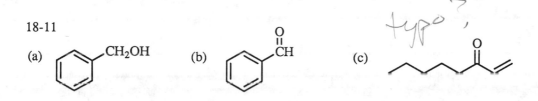 **CH₂OH** (b) **CH** with O   (c) *typo?*

18-12 Review the reminders on p. 366. There are often more than one correct way to do syntheses, but a more direct route with fewer steps is usually better.

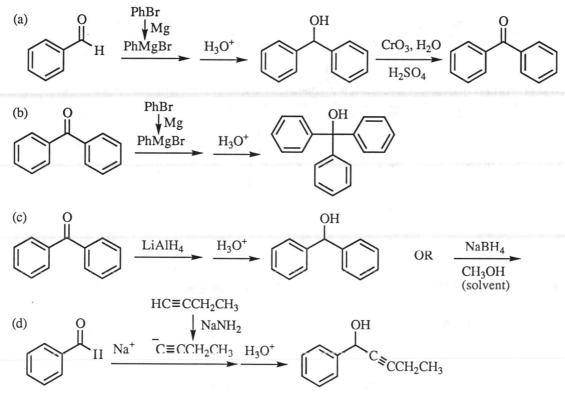

(a)
$$\xrightarrow[\text{PhMgBr}]{\overset{\text{PhBr}}{\downarrow \text{Mg}}} \quad \xrightarrow{H_3O^+} \quad \xrightarrow[H_2SO_4]{CrO_3, H_2O}$$

(b)
$$\xrightarrow[\text{PhMgBr}]{\overset{\text{PhBr}}{\downarrow \text{Mg}}} \quad \xrightarrow{H_3O^+}$$

(c)
$$\xrightarrow{LiAlH_4} \quad \xrightarrow{H_3O^+} \qquad OR \qquad \xrightarrow[\substack{CH_3OH \\ (\text{solvent})}]{NaBH_4}$$

(d)
$$\xrightarrow[Na^+ \;\; ^-C\equiv CCH_2CH_3]{\overset{HC\equiv CCH_2CH_3}{\downarrow NaNH_2}} \quad \xrightarrow{H_3O^+}$$

18-13 Trimethylphosphine has α-hydrogens that could be removed by butyllithium, generating undesired ylides.

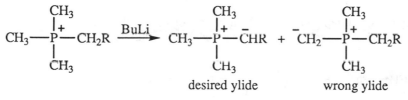

$$CH_3-\overset{\overset{\displaystyle CH_3}{|}}{\underset{\underset{\displaystyle CH_3}{|}}{P}}{}^+-CH_2R \xrightarrow{BuLi} CH_3-\overset{\overset{\displaystyle CH_3}{|}}{\underset{\underset{\displaystyle CH_3}{|}}{P}}{}^+-\overset{-}{C}HR + {}^-CH_2-\overset{\overset{\displaystyle CH_3}{|}}{\underset{\underset{\displaystyle CH_3}{|}}{P}}{}^+-CH_2R$$

desired ylide   wrong ylide

18-14

(a)

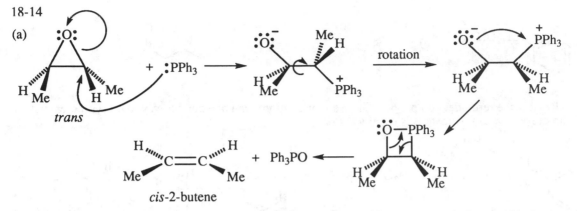

cis-2-butene

The stereochemistry is inverted. The nucleophile triphenylphosphine must attack the epoxide in an anti fashion, yet the triphenylphosphine oxide must eliminate with syn geometry.

(b)

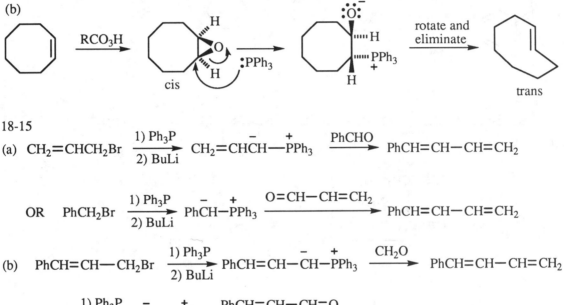

18-15

(a) $CH_2=CHCH_2Br$ $\xrightarrow[\text{2) BuLi}]{\text{1) Ph}_3\text{P}}$ $CH_2=CHCH{-}\overset{+}{P}Ph_3$ $\xrightarrow{\text{PhCHO}}$ $PhCH=CH{-}CH=CH_2$

OR $PhCH_2Br$ $\xrightarrow[\text{2) BuLi}]{\text{1) Ph}_3\text{P}}$ $PhCH{-}\overset{+}{P}Ph_3$ $\xrightarrow{O=CH{-}CH=CH_2}$ $PhCH=CH{-}CH=CH_2$

(b) $PhCH=CH{-}CH_2Br$ $\xrightarrow[\text{2) BuLi}]{\text{1) Ph}_3\text{P}}$ $PhCH=CH{-}CH{-}\overset{+}{P}Ph_3$ $\xrightarrow{CH_2O}$ $PhCH=CH{-}CH=CH_2$

OR CH_3I $\xrightarrow[\text{2) BuLi}]{\text{1) Ph}_3\text{P}}$ $CH_2{-}\overset{+}{P}Ph_3$ $\xrightarrow{PhCH=CH{-}CH=O}$ $PhCH=CH{-}CH=CH_2$

18-16   Many alkenes can be synthesized by two different Wittig reactions (as in the previous problem). The ones shown here form the phosphonium salt from the less hindered alkyl halide.

(a) $PhCH_2Br$ $\xrightarrow[\text{2) BuLi}]{\text{1) Ph}_3\text{P}}$ $PhCH{-}\overset{+}{P}Ph_3$ $+$ (acetone) $\longrightarrow$ $PhCH=C(CH_3)_2$

(b) CH_3I $\xrightarrow[\text{2) BuLi}]{\text{1) Ph}_3\text{P}}$ $CH_2{-}\overset{+}{P}Ph_3$ $+$ (acetophenone) $\longrightarrow$ (isopropenylbenzene)

18-16 continued

(c)

$$PhCH_2Br \xrightarrow[\text{2) BuLi}]{\text{1) Ph}_3\text{P}} Ph\overset{-}{C}H\overset{+}{-}PPh_3 \xrightarrow{PhCH=CH-CH=O} PhCH=CH-CH=CHPh$$

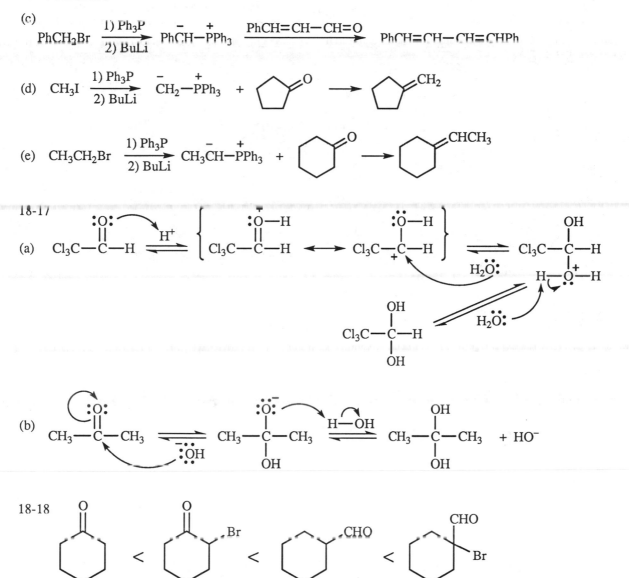

(d) $CH_3I \xrightarrow[\text{2) BuLi}]{\text{1) Ph}_3\text{P}} \overset{-}{C}H_2\overset{+}{-}PPh_3 +$ [cyclopentanone] $\longrightarrow$ [methylenecyclopentane]

(e) $CH_3CH_2Br \xrightarrow[\text{2) BuLi}]{\text{1) Ph}_3\text{P}} CH_3\overset{-}{C}H\overset{+}{-}PPh_3 +$ [cyclohexanone] $\longrightarrow$ [ethylidenecyclohexane]

18-17

(a) $Cl_3C-\underset{\underset{}{\overset{\|}{:O:}}}{C}-H \xrightarrow{H^+} \left\{ Cl_3C-\underset{\overset{\|}{\overset{}{}}}{C}-H \longleftrightarrow Cl_3C-\overset{+}{C}-H \right\} \underset{H_2\overset{..}{O}:}{\rightleftharpoons} Cl_3C-\underset{\underset{H-\overset{+}{\underset{..}{O}}-H}{|}}{\overset{OH}{C}}-H$

$Cl_3C-\underset{\underset{OH}{|}}{\overset{OH}{C}}-H \underset{H_2\overset{..}{O}:}{\rightleftharpoons}$

(b) $CH_3-\underset{\underset{}{\overset{\|}{:O:}}}{C}-CH_3 \underset{:\overset{..}{O}H}{\rightleftharpoons} CH_3-\underset{\underset{OH}{|}}{\overset{\overset{-}{:\overset{..}{O}:}}{C}}-CH_3 \xrightarrow{H-OH} CH_3-\underset{\underset{OH}{|}}{\overset{OH}{C}}-CH_3 + HO^-$

18-18

[cyclohexanone] < [2-bromocyclohexanone] < [cyclohexanecarbaldehyde with Br] < [1-bromocyclohexanecarbaldehyde CHO]

least amount   greatest amount
of hydrate   of hydrate

18-19

(a)

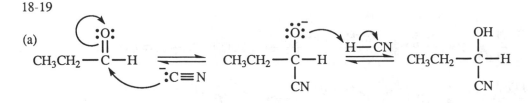

(b)

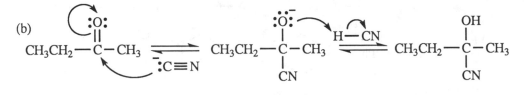

(c)

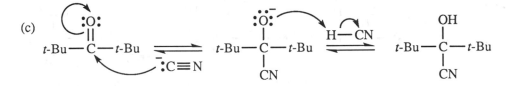

18-20

(a)

(b)

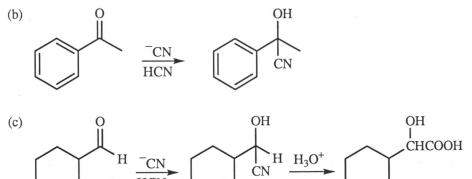

(c)

Note: Mechanisms of nucleophilic attack at carbonyl carbon frequently include species with both positive and negative charges. These species are very short-lived as each charge is quickly neutralized by a rapid proton transfer; in fact, these steps are probably the fastest of the whole mechanism. In most cases, this Solutions Manual will show these *two* steps as occurring at the same time, even though you have been admonished to show *all* steps of a mechanism separately. The practice of showing these proton transfers in one step is legitimate as long as it is understood that these are *two* fast steps.

18-21

(a)

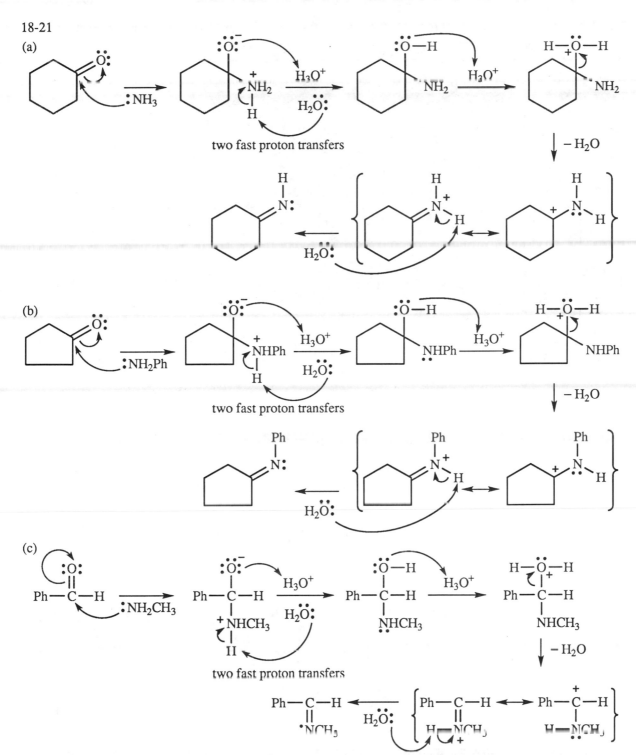

two fast proton transfers

− H₂O

(b)

two fast proton transfers

− H₂O

(c)

two fast proton transfers

− H₂O

plus three resonance forms with positive charge on the benzene ring

18-22 Whenever a double bond is formed, stereochemistry must be considered. The two compounds are the *Z* and *E* isomers.

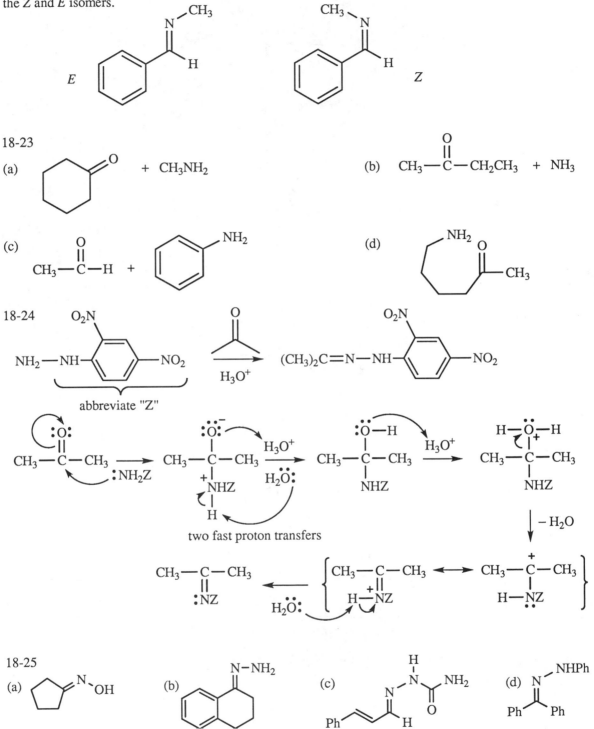

18-23

(a) [cyclohexanone] + CH₃NH₂

(b) $CH_3-\overset{\overset{O}{\|}}{C}-CH_2CH_3$ + NH₃

(c) $CH_3-\overset{\overset{O}{\|}}{C}-H$ + [aniline with NH₂]

(d) [structure with NH₂ and C=O with CH₃]

18-24

O_2N ... NH_2-NH-[benzene ring]$-NO_2$ (abbreviate "Z") + [acetone] → $\xrightarrow{H_3O^+}$ $(CH_3)_2C=N-NH-$[benzene ring with O_2N]$-NO_2$

$CH_3-\overset{\overset{:O:}{\|}}{C}-CH_3$:NH₂Z → $CH_3-\overset{\overset{:O:^-}{|}}{\underset{\overset{+}{N}HZ\ \ \underset{H}{|}}{C}}-CH_3$ $\xrightarrow[H_2\overset{..}{O}:]{H_3O^+}$ $CH_3-\overset{\overset{:O-H}{|}}{\underset{NHZ}{C}}-CH_3$ $\xrightarrow[]{H_3O^+}$ $CH_3-\overset{\overset{H-\overset{..}{O}-H}{|^+}}{\underset{NHZ}{C}}-CH_3$

two fast proton transfers

$\Big\downarrow -H_2O$

$CH_3-\overset{\overset{\|}{C}}{\underset{:NZ}{}}-CH_3$ ← $\underset{H_2\overset{..}{O}:}{}$ $\Big\{ CH_3-\overset{\overset{+\|}{C}}{\underset{H-NZ}{}}-CH_3 \longleftrightarrow CH_3-\overset{\overset{+}{C}}{\underset{H-NZ}{}}-CH_3 \Big\}$

18-25

(a) [cyclopentanone oxime] N–OH

(b) [tetralone] N–NH₂

(c) Ph–CH=CH–CH=N–N(H)–C(=O)–NH₂

(d) Ph₂C=N–NHPh

374

18-26

(a) PhCHO + H₂NNH—C(=O)—NH₂

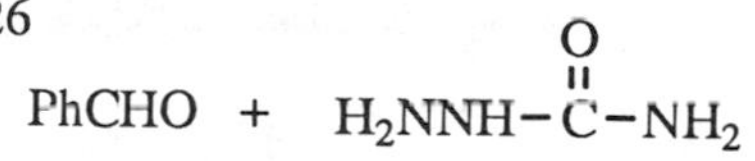

$$\text{(a)} \quad \text{PhCHO} + \text{H}_2\text{NNH}-\overset{\displaystyle O}{\overset{\|}{\text{C}}}-\text{NH}_2$$

(b) + H₂NOH

(c) + H₂N—NHPh

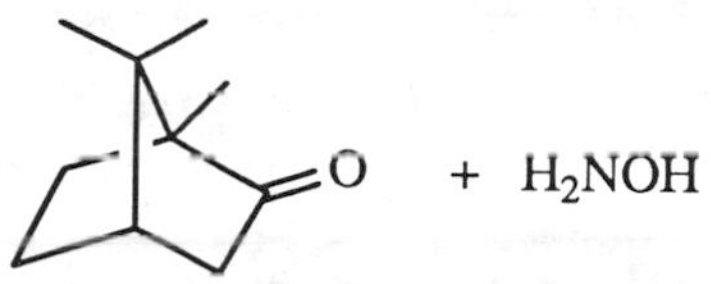

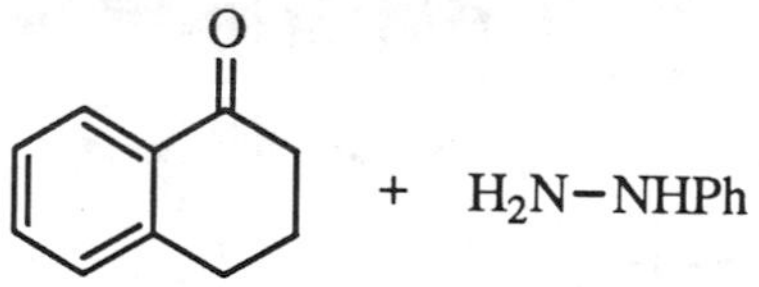

(d) + NH₂—NH— (2,4-dinitrophenyl)

$$\text{(d)} \quad + \quad \text{NH}_2-\text{NH}-\text{C}_6\text{H}_3(\text{O}_2\text{N})(\text{NO}_2)$$

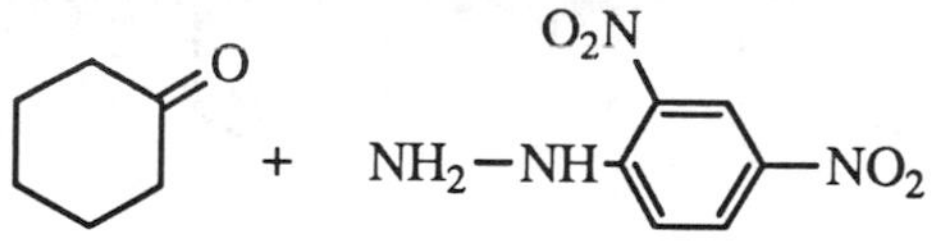

(e)

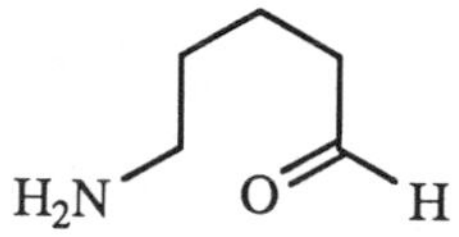

(f)

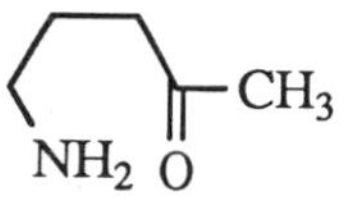

18-27

hemiacetal

acetal

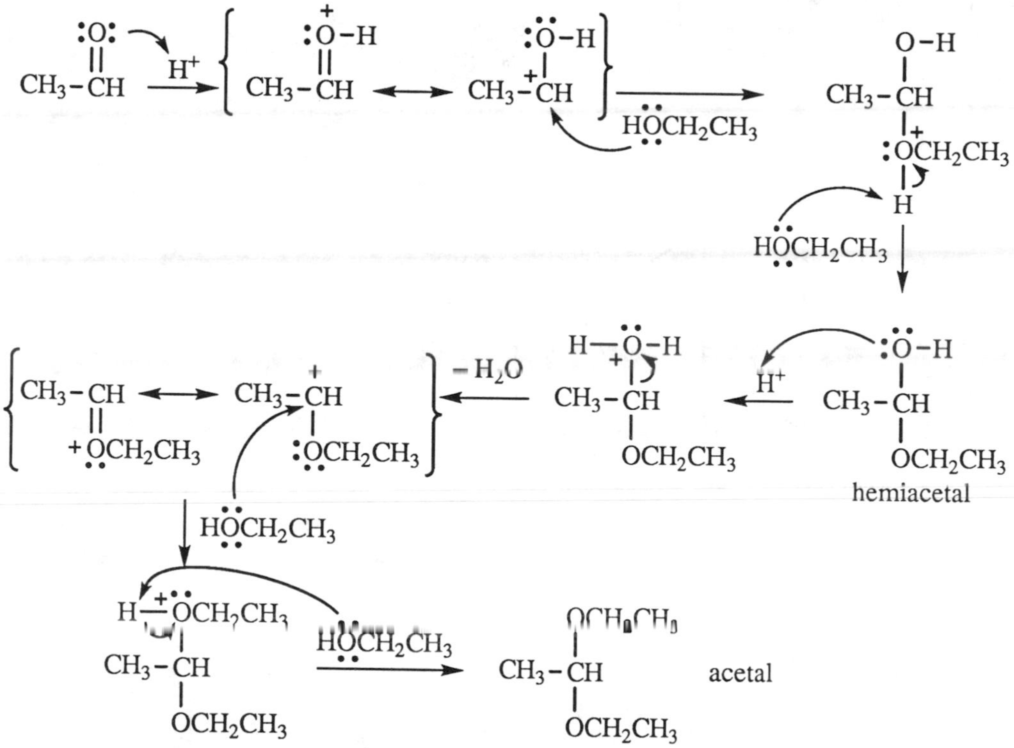

18-28

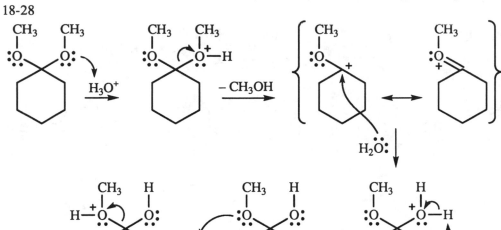

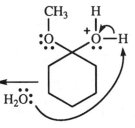

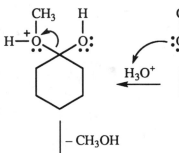

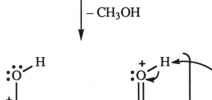

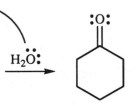

18-29

(a)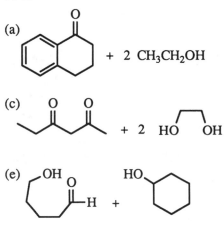

(b)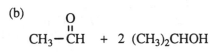

(c)

(d)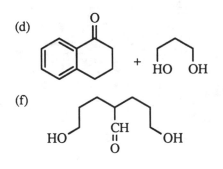

(e)

(f)

18-30

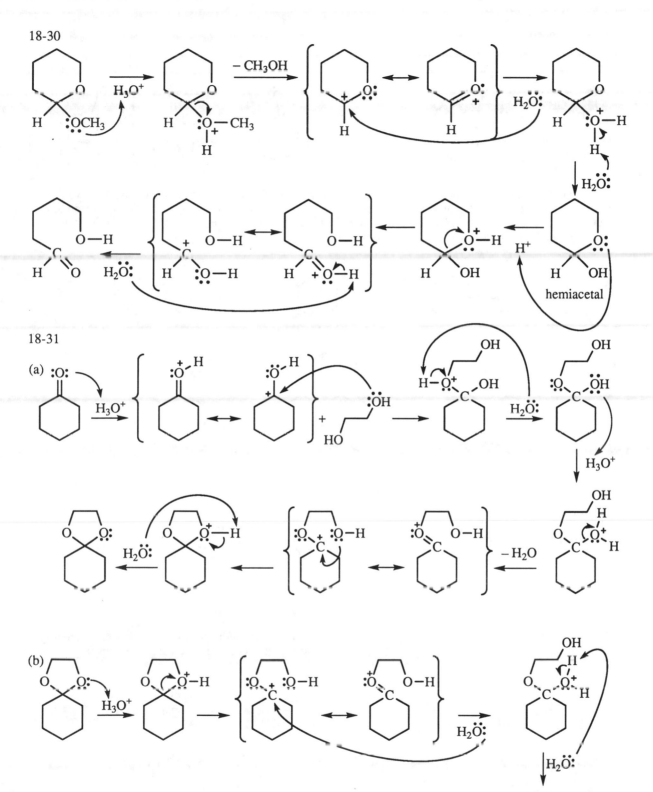

18-31

(a)

(b)

hydrolysis mechanism
continued on next page

18-31 continued

(b) hydrolysis mechanism continued

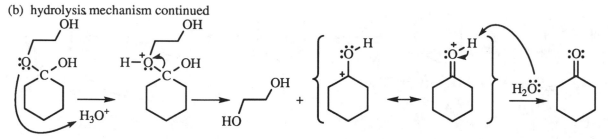

(c) The mechanisms of formation of an acetal and hydrolysis of an acetal are identical, just the reverse order. This has to be true because this process is an equilibrium: if the forward steps follow a minimum energy path, then the reverse steps have to follow the identical minimum energy path. This is the famous Principle of Microscopic Reversibility, text section 8-4A.

(d)

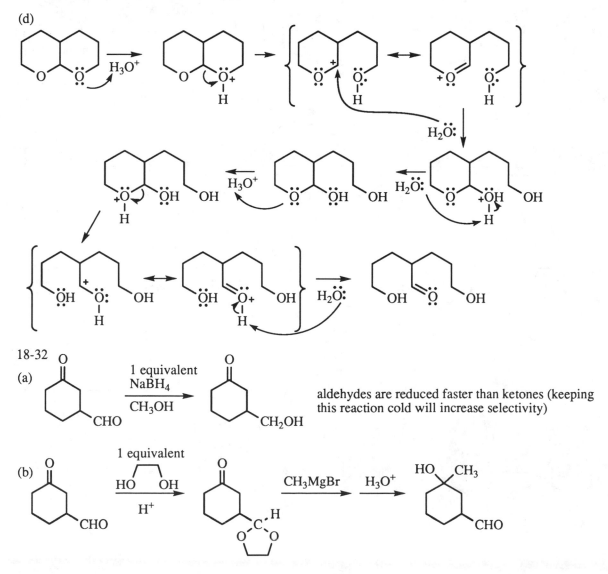

18-32

(a) aldehydes are reduced faster than ketones (keeping this reaction cold will increase selectivity)

(b)

378

18-32 continued

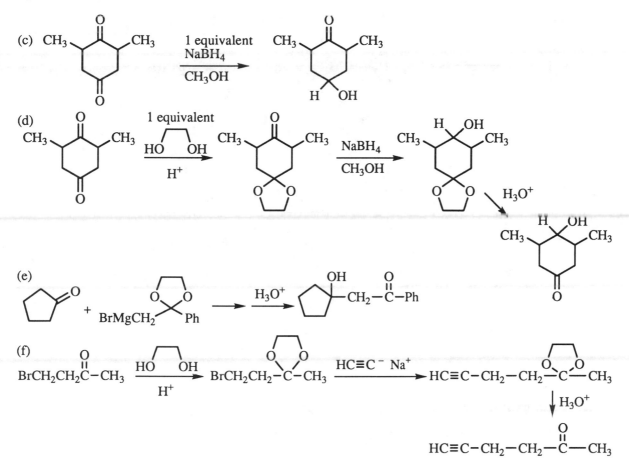

(c)

(d)

(e)

(f)

18-33

(a)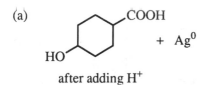

+ Ag⁰

after adding H⁺

(b)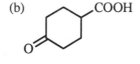

(c)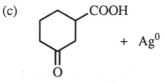

+ Ag⁰

after adding II⁺

(d)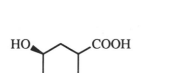

18-34

hydrazone formation

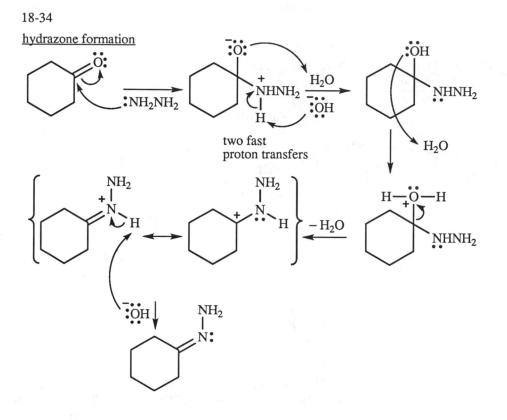

two fast
proton transfers

reduction of the hydrazone

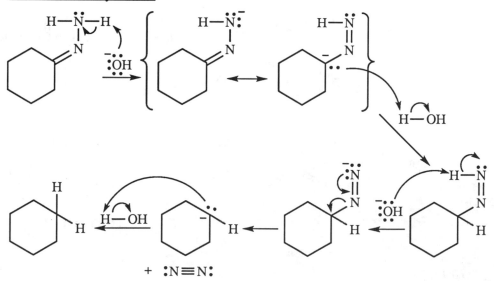

18-35

(a)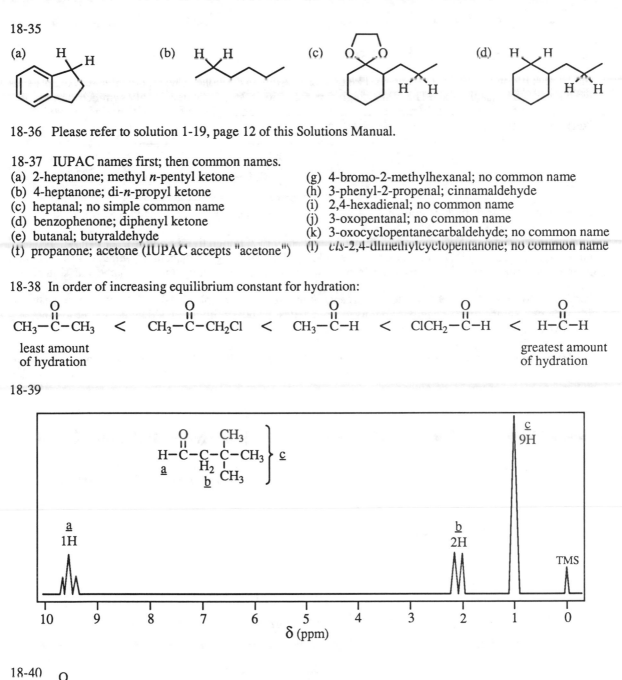
(b) H H
(c)
(d)

18-36 Please refer to solution 1-19, page 12 of this Solutions Manual.

18-37 IUPAC names first; then common names.
(a) 2-heptanone; methyl *n*-pentyl ketone
(b) 4-heptanone; di-*n*-propyl ketone
(c) heptanal; no simple common name
(d) benzophenone; diphenyl ketone
(e) butanal; butyraldehyde
(f) propanone; acetone (IUPAC accepts "acetone")
(g) 4-bromo-2-methylhexanal; no common name
(h) 3-phenyl-2-propenal; cinnamaldehyde
(i) 2,4-hexadienal; no common name
(j) 3-oxopentanal; no common name
(k) 3-oxocyclopentanecarbaldehyde; no common name
(l) *cis*-2,4-dimethylcyclopentanone; no common name

18-38 In order of increasing equilibrium constant for hydration:

$$CH_3-\overset{O}{\overset{||}{C}}-CH_3 \quad < \quad CH_3-\overset{O}{\overset{||}{C}}-CH_2Cl \quad < \quad CH_3-\overset{O}{\overset{||}{C}}-H \quad < \quad ClCH_2-\overset{O}{\overset{||}{C}}-H \quad < \quad H-\overset{O}{\overset{||}{C}}-H$$

least amount
of hydration
          greatest amount
of hydration

18-39

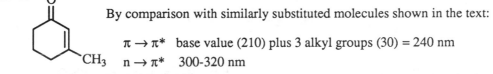

18-40

By comparison with similarly substituted molecules shown in the text:

$\pi \rightarrow \pi^*$ base value (210) plus 3 alkyl groups (30) = 240 nm
$n \rightarrow \pi^*$ 300-320 nm

18-41

$C_6H_{10}O_2$ indicates two elements of unsaturation.

The IR absorption at 1708 cm^{-1} suggests a ketone, or possibly two ketones since there are two oxygens and two elements of unsaturation. The NMR singlets in the ratio of 2 : 3 indicate a highly symmetric molecule. The singlet at δ 2.15 is probably methyl next to carbonyl, and the singlet at δ 2.67 integrating to two is likely to be CH_2 on the other side of the carbonyl.

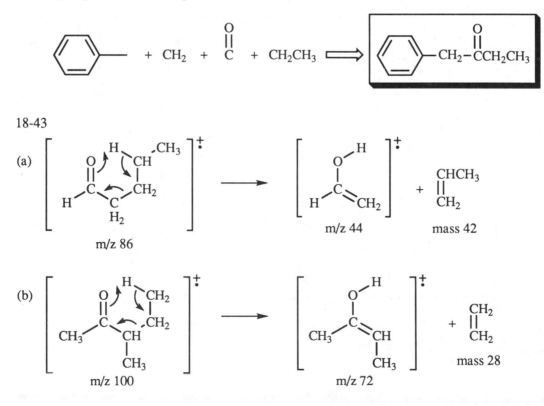

Since the molecular formula is double this fragment, the molecule must be twice the fragment.

Two questions arise. Why is the integration 2 : 3 and not 4 : 6? Integration provides a *ratio*, not absolute numbers, of hydrogens. Why don't the two methylenes show splitting? Adjacent, *identical* hydrogens, with identical chemical shifts, do not split each other; the signals for ethane or cyclohexane appear as singlets.

18-42 The formula $C_{10}H_{12}O$ indicates 5 elements of unsaturation. A solid 2,4-DNP derivative suggests an aldehyde or a ketone, but a negative Tollens test precludes the possibility of an aldehyde; therefore, the unknown must be a ketone.

The NMR shows the typical ethyl pattern at δ 1.0 (3H, triplet) and δ 2.5 (2H, quartet), and a monosubstituted benzene at δ 7.3 (5H, multiplet). The singlet at δ 3.7 is a CH_2, but quite far downfield, apparently deshielded by two groups. Assemble the pieces:

18-43

18-43

(c)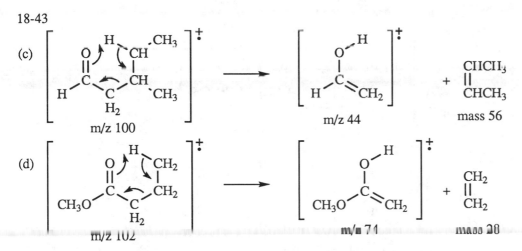

m/z 100 → m/z 44 + mass 56

(d)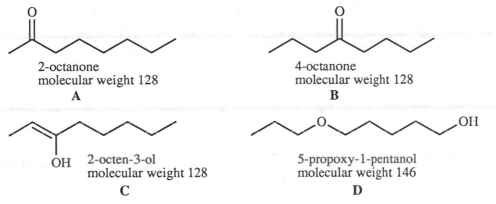

m/z 102 → m/z 71 + mass 28

18-44 A solid 2,4-DNP indicates an aldehyde or a ketone. Possible structures:

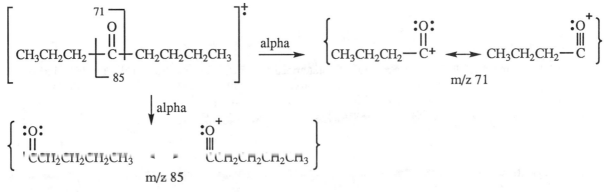

2-octanone
molecular weight 128
A

4-octanone
molecular weight 128
B

2-octen-3-ol
molecular weight 128
C

5-propoxy-1-pentanol
molecular weight 146
D

The positive 2,4-DNP test eliminates **D** as a possibility as **D** has no carbonyl. (**C** is the enol form of 3-octanone; it could give a positive 2,4-DNP test.) The mass spectrum of **A** would show a strong M – 15, loss of methyl, not present in the data; thus, the unknown cannot be **A**.

What are the MS fragmentations of **B**? Two α-cleavages and *two* McLafferty rearrangements:

$$\left[CH_3CH_2CH_2 \overset{71}{\underset{85}{|}} \overset{O}{\underset{}{\overset{||}{C}}} CH_2CH_2CH_2CH_3 \right]^{+} \xrightarrow{alpha} \left\{ CH_3CH_2CH_2-\overset{:O:}{\underset{}{\overset{||}{C^+}}} \longleftrightarrow CH_3CH_2CH_2-\overset{:O:^+}{\underset{}{\overset{|||}{C}}} \right\}$$

m/z 71

↓ alpha

$$\left\{ \overset{:O:}{\underset{}{\overset{||}{C}}}CH_2CH_2CH_2CH_3 \longleftrightarrow \overset{:O^+}{\underset{}{\overset{|||}{C}}}CH_2CH_2CH_2CH_3 \right\}$$

m/z 85

More MS fragmentations on the next page.

383

18-44 continued

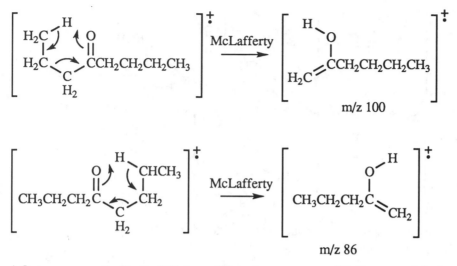

m/z 100

m/z 86

4-Octanone exactly fits the MS data. What about structure **C**? Because of the location of the π bond (in either the enol or keto forms), **C** could only do one McLafferty rearrangement and could not give rise to two even-mass peaks.

Therefore, the unknown must be 4-octanone, **B**.

18-45
A molecular ion of m/z 70 means a fairly small molecule. A solid semicarbazone derivative and a negative Tollens test indicate a ketone. The carbonyl (CO) has mass 28, so $70 - 28 = 42$, enough mass for only 3 more carbons. The molecular formula is probably C_4H_6O (mass 70); with two elements of unsaturation, we can infer the presence of a double bond or a ring in addition to the carbonyl.

The IR shows a strong peak at 1790 cm^{-1}, indicative of a ketone in a small ring. No peak in the 1600-1650 cm^{-1} region shows the absence of an alkene. The only possibilities for a small ring ketone containing four carbons are these:

The NMR can distinguish these. No methyl doublet appears in the NMR spectrum, ruling out **B**. The NMR does show a 4H triplet at δ 3.1; this signal comes from the two methylenes (C-2 and C-4) adjacent to the carbonyl, split by the two hydrogens on C-3. The signal for the methylene at C-3 appears at δ 2.0, roughly a quintet because of splitting by four neighboring protons.

The unknown is cyclobutanone, **A**. The IR absorption of the carbonyl at 1790 cm^{-1} is characteristic of small ring ketones; ring strain strengthens the carbon-oxygen double bond, increasing its frequency of vibration. (See Section 12-9 in the text.)

18-46

(a) The conjugated diene has a maximum at 235 nm (see Solved Problem 15-3) and the ketone has a maximum at about 237 nm, so the π to π* transition cannot be used to differentiate the compounds.

(b) The ketone has an n to π* transition around 315 nm that the diene cannot have.

18-47

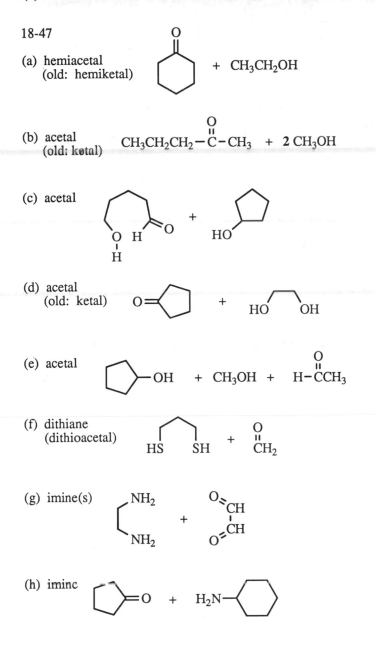

(a) hemiacetal (old: hemiketal)

(b) acetal (old: ketal)

(c) acetal

(d) acetal (old: ketal)

(e) acetal

(f) dithiane (dithioacetal)

(g) imine(s)

(h) imine

18-48

(a)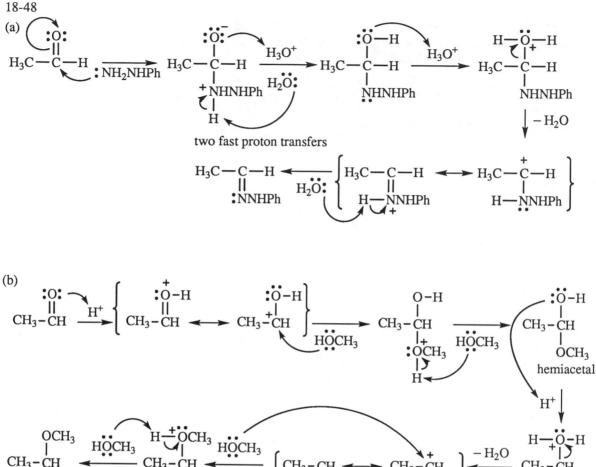

two fast proton transfers

(b)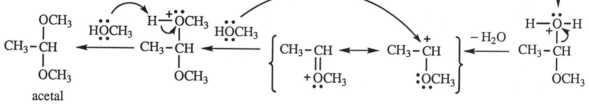

hemiacetal

acetal

(c)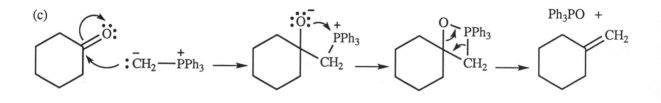

(d)

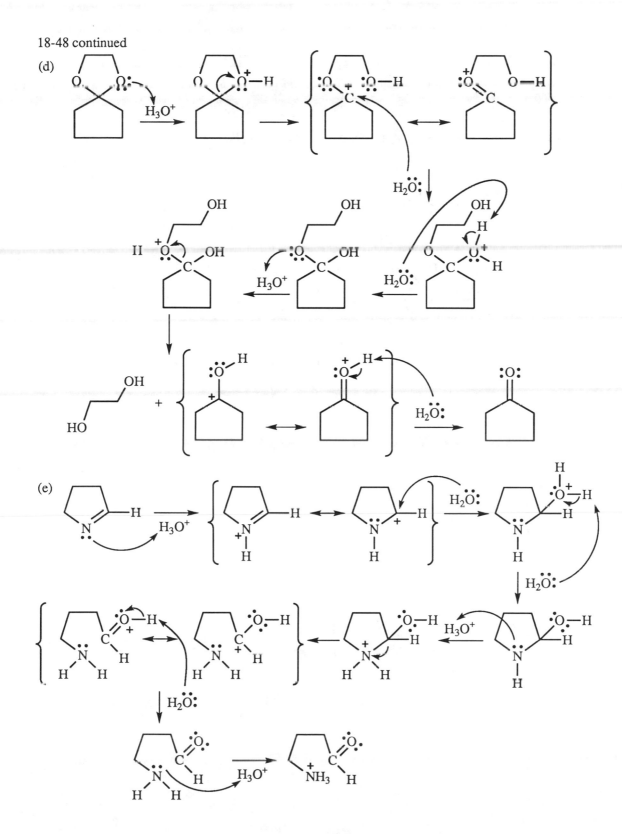

(e)

18-49

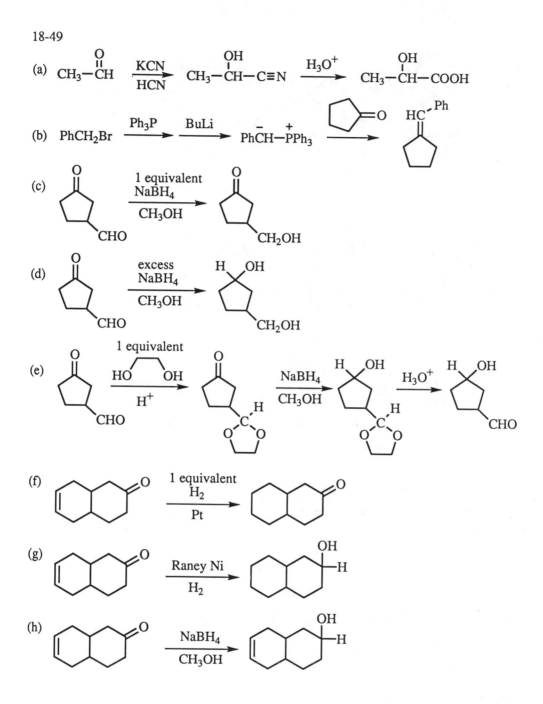

18-50 All of these reactions would be acid-catalyzed.

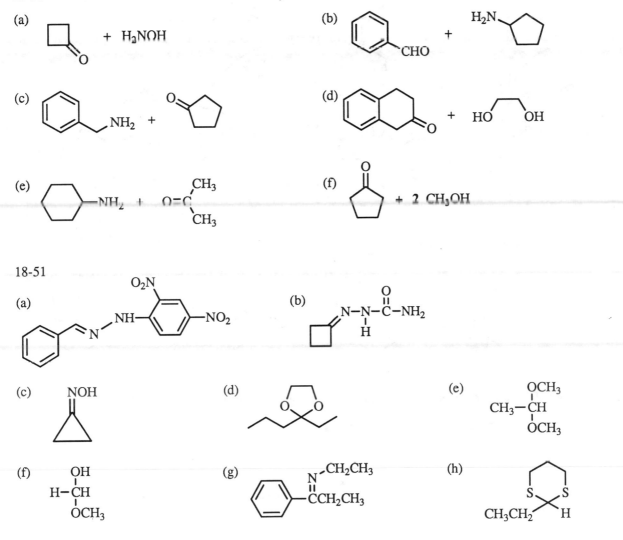

(a) + H₂NOH

(b) + H₂N

(c) +

(d) + HO OH

(e) +

(f) + 2 CH₃OH

18-51

(a)

(b)

(c)

(d)

(e)

(f)

(g)

(h)

18-52

(a)

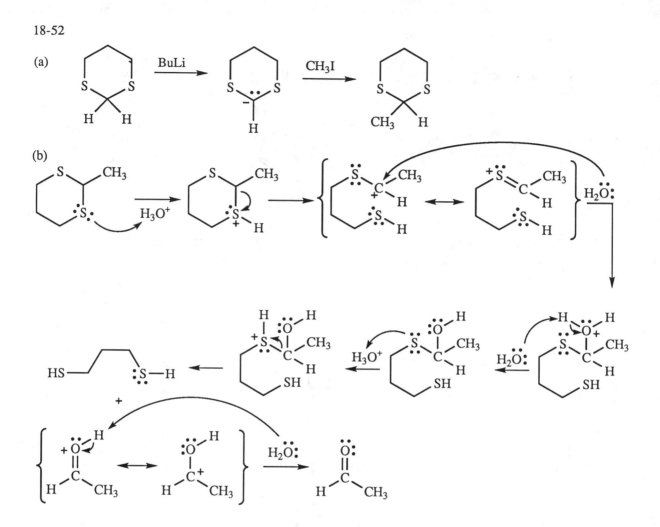

(b)

(c) Mercuric ion, Hg^{2+}, assists the hydrolysis in two ways. First, mercuric ion is a Lewis acid of moderate strength, performing the same function as a proton from a protic acid.

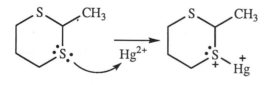

The effectiveness of Hg^{2+} as a Lewis acid is partly due to its charge: even complexed with a sulfur, the mercury atom still has a positive charge, attracting the sulfur's electrons.

A second explanation for mercuric ion's effectiveness in hydrolysis of thioacetals lies in the complex formed between the ion and the two sulfur atoms. This stable complex effectively removes $HSCH_2CH_2SH$ from the equilibrium, shifting the equilibrium to product. (An example of whose principle? His initials are "Le Châtelier".)

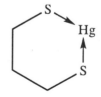

390

18-53 The key to this problem is understanding that the relative proximity of the two oxygens can dramatically affect their chemistry.

1,2-dioxane

1,3-dioxane

1,4-dioxane

The "second" isomer described: two oxygens connected by a sigma bond are a peroxide. The O—O bond is easily cleaved to give radicals. In the presence of organic compounds, radical reactions can be explosive.

The "third" isomer described: two oxygens bonded to the same sp^3 carbon constitute an acetal which is hydrolyzed in aqueous acid. See the mechanism below.

The "first" isomer described: an excellent solvent (although toxic), these oxygens are far enough apart to act independently. It is a simple ether.

Mechanism of acetal hydrolysis

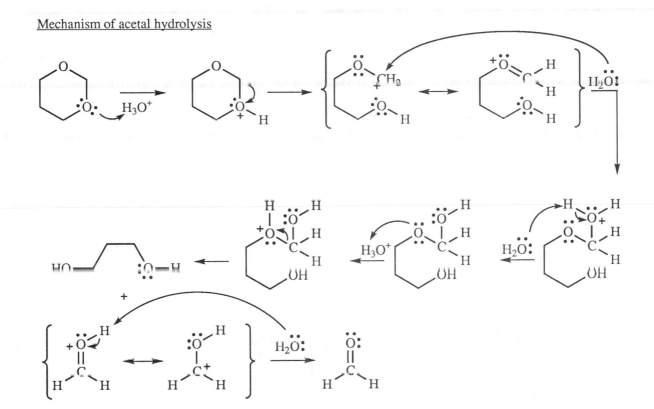

391

18-54

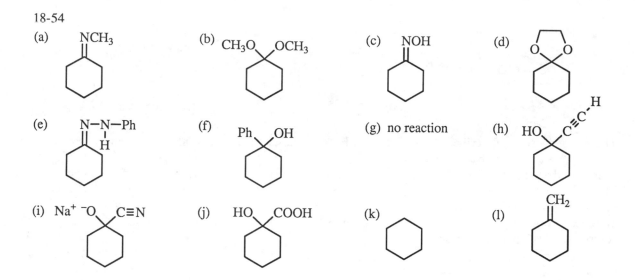

(a) NCH₃

(b) CH₃O OCH₃

(c) NOH

(d)

(e) N—N—Ph H

(f) Ph OH

(g) no reaction

(h) HO C≡C-H

(i) Na⁺ ⁻O C≡N

(j) HO COOH

(k)

(l) CH₂

18-55 The new bond to carbon comes from the NaBH₄ or NaBD₄. The new bond to oxygen comes from the protic solvent.

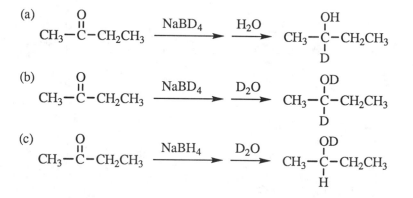

(a)

$$CH_3-\overset{O}{\underset{||}{C}}-CH_2CH_3 \xrightarrow{NaBD_4} \xrightarrow{H_2O} CH_3-\overset{OH}{\underset{D}{\underset{|}{C}}}-CH_2CH_3$$

(b)

$$CH_3-\overset{O}{\underset{||}{C}}-CH_2CH_3 \xrightarrow{NaBD_4} \xrightarrow{D_2O} CH_3-\overset{OD}{\underset{D}{\underset{|}{C}}}-CH_2CH_3$$

(c)

$$CH_3-\overset{O}{\underset{||}{C}}-CH_2CH_3 \xrightarrow{NaBH_4} \xrightarrow{D_2O} CH_3-\overset{OD}{\underset{H}{\underset{|}{C}}}-CH_2CH_3$$

18-56 While hydride is a small group, the actual chemical species supplying it, AlH_4^-, is fairly large, so it prefers to approach from the less hindered side of the molecule, that is, the side opposite the methyl. This forces the oxygen to go to the same side as the methyl, producing the *cis* isomer as the major product.

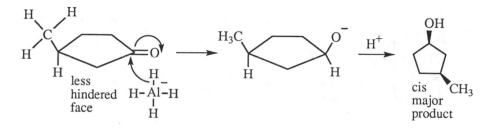

18-57 A comment about the first step: this step is similar to an S_N2 reaction, but we saw earlier that RO^- is not a leaving group in nucleophilic substitution. Why does this work? What is different here?

Two factors facilitate this step. First, the Mg, a mild Lewis acid, complexes with the oxygen as it leaves, making the oxygen more neutral than ionic (similar to protonation to make a neutral leaving group). Second, the other two oxygens assist in two ways: a) they withdraw electrons by induction, making the carbon more positive and therefore more susceptible to attack by the carbanion; and b) they donate electrons by resonance, stabilizing any positive charge that develops on this carbon in the transition state.

No evidence is presented to suggest whether substitution by the Grignard reagent is one step or more; it will be shown here as one step, for simplicity.

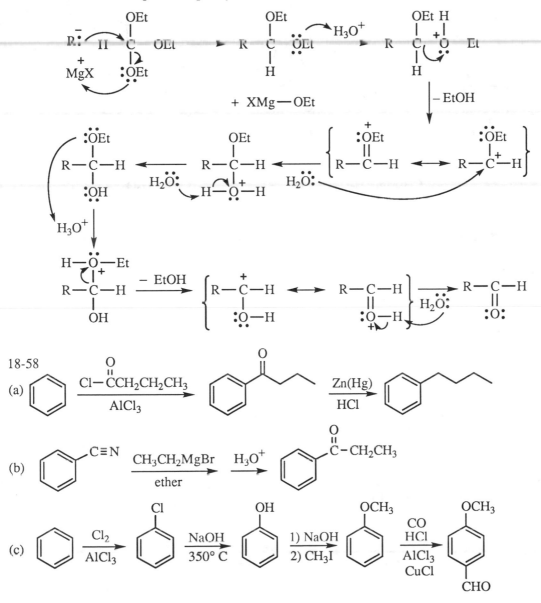

18-58

(d)

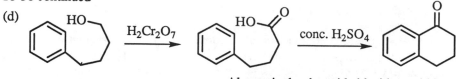

Alternatively, the acid chloride could be made
with SOCl₂, then cyclized by Friedel-Crafts
acylation with AlCl₃.

18-59

(a)

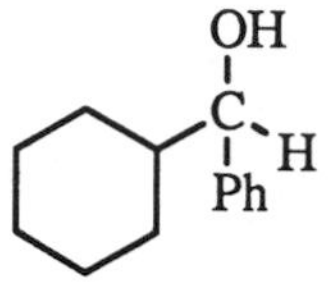

OH
|
C—H
|
Ph

(b)

O
‖
C—O⁻

+ Ag⁰

(c)

H O
| ‖
N—N—C—NH₂
‖
C—H

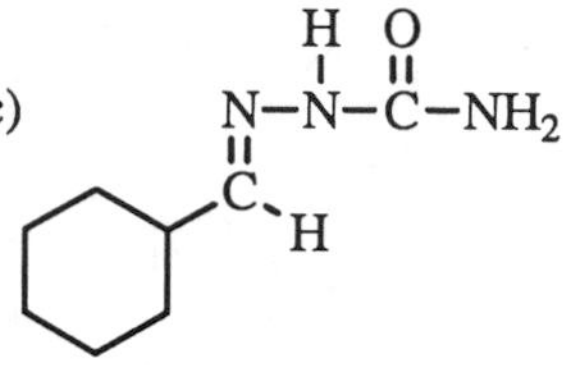

(d)

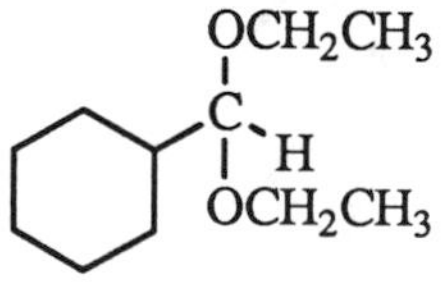

OCH₂CH₃
|
C—H
|
OCH₂CH₃

(e)

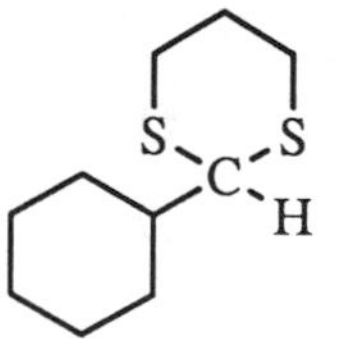

S   S
\ /
C—H

(f)

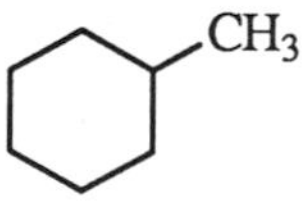

CH₃

18-60

(a) $\xrightarrow[\text{ether}]{\text{CH}_3\text{MgI}\quad\text{H}_3\text{O}^+}$ $\xrightarrow[\text{H}_2\text{SO}_4]{\text{CrO}_3\,,\,\text{H}_2\text{O}}$

(b) C≡CH $\xrightarrow[\substack{\text{H}_2\text{O}\\\text{H}_2\text{SO}_4}]{\text{HgSO}_4}$

(c) $\xrightarrow[\text{2) CH}_3\text{I}]{\text{1) BuLi}}$ $\xrightarrow[\text{2) CH}_3(\text{CH}_2)_5\text{Br}]{\text{1) BuLi}}$ $\xrightarrow[\text{HgCl}_2]{\text{H}_3\text{O}^+}$

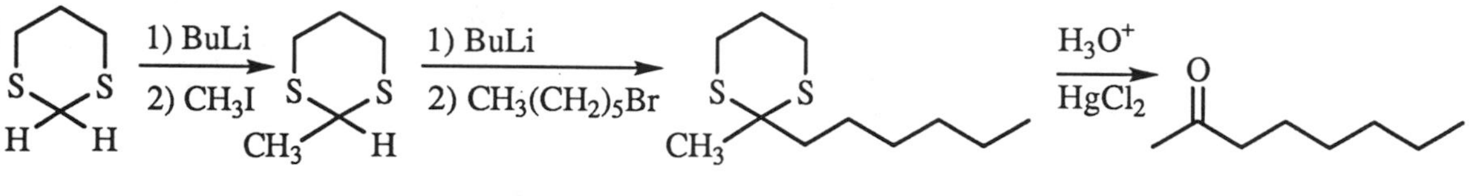

(d) $\xrightarrow[\text{H}_2\text{SO}_4]{\text{CrO}_3\,,\,\text{H}_2\text{O}}$

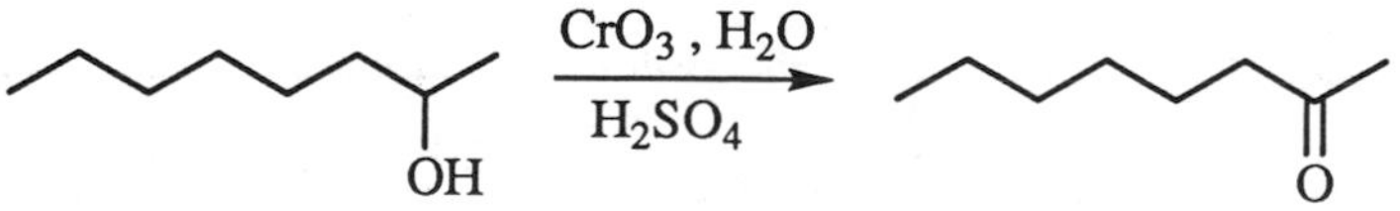

(e) $\xrightarrow[\text{2) H}_3\text{O}^+]{\text{1) excess CH}_3\text{Li}}$

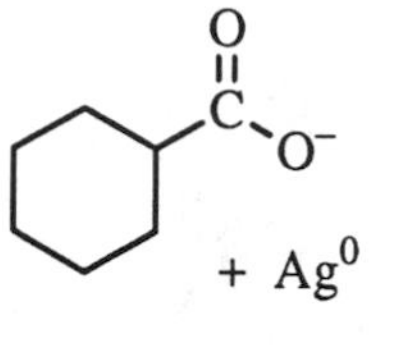

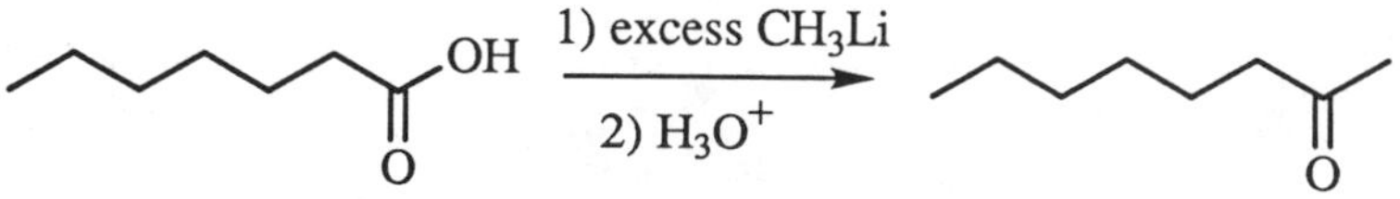

(f) C≡N $\xrightarrow[\text{ether}]{\text{CH}_3\text{MgI}\quad\text{H}_3\text{O}^+}$

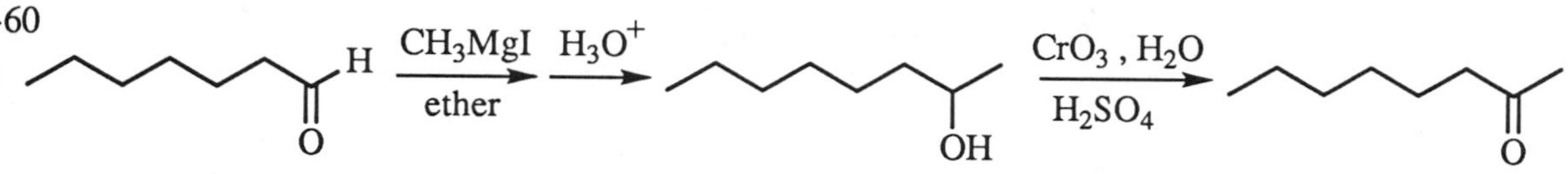

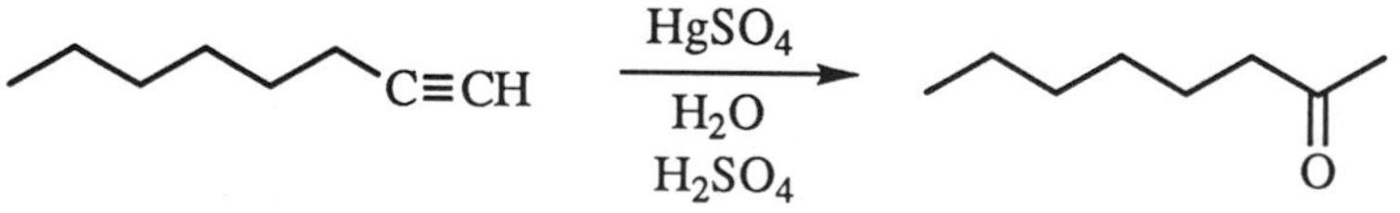

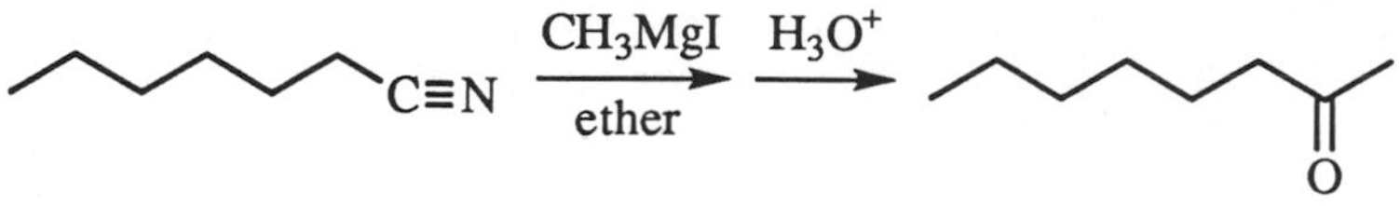

18-60 continued

(g)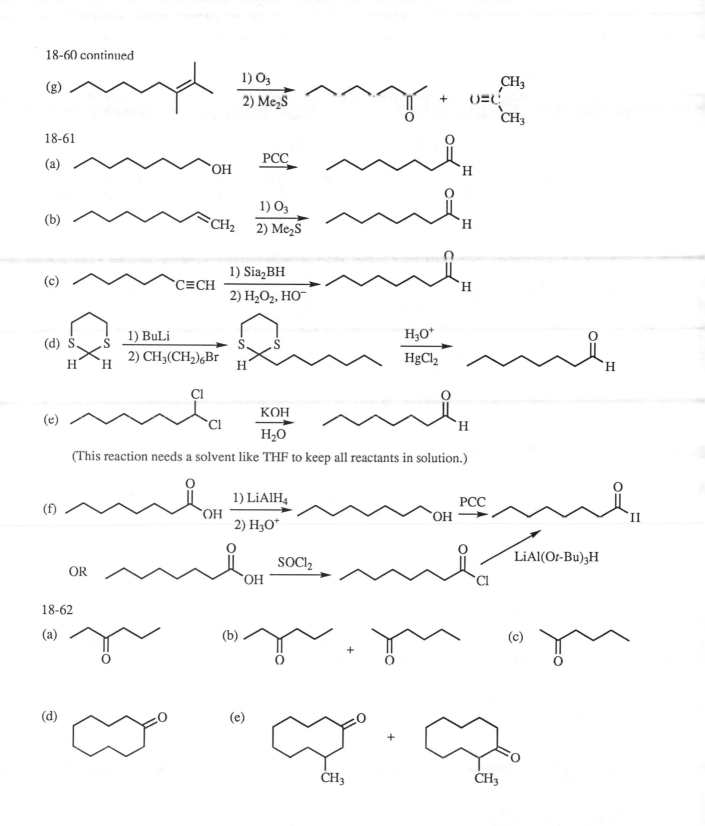

18-61

(a)

(b)

(c)

(d)

(e)

(This reaction needs a solvent like THF to keep all reactants in solution.)

(f)

OR

18-62

(a) (b) + (c)

(d) (e) +

18-63

(a) ketone: no reaction
(b) aldehyde: positive
(c) enol of an aldehyde—tautomerizes to aldehyde in base: positive
(d) hemiacetal of an aldehyde in equilibrium with the aldehyde in base: positive
(e) acetal—stable in base: no reaction
(f) hemiacetal of an aldehyde in equilibrium with the aldehyde in base: positive

18-64 The structure of **A** can be deduced from its reaction with **J** and **K**. What is common to both products of these reactions is the 2-heptanol part; the reactions must be Grignard reactions with 2-heptanone, so **A** must be 2-heptanone.

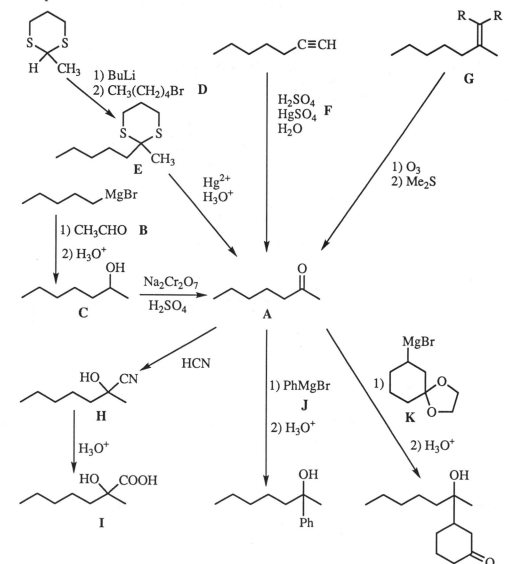

18-65 The very strong π to π* absorption at 225 nm in the UV spectrum suggests a conjugated ketone or aldehyde. The IR confirms this: strong, conjugated carbonyl at 1690 cm^{-1} and small alkene at 1610 cm^{-1}. The absence of peaks at 2700-2800 cm^{-1} shows that the unknown is not an aldehyde.

The molecular ion at 96 leads to the molecular formula:

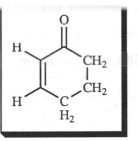

$$\begin{array}{r} 96 \\ -64 \\ \hline 32 \text{ mass units} \Rightarrow \text{ add 2 carbons and 8 hydrogens} \end{array}$$

molecular formula = C_6H_8O = 3 elements of unsaturation

Two elements of unsaturation are accounted for in the enone. The other one is likely a ring.

The NMR shows two vinyl hydrogens. The doublet at δ 6.0 says that the two hydrogens are on neighboring carbons (two peaks = one neighboring H).

doublet: 1 neighboring H

$$\underset{\text{C}-\text{C}=\text{C}-\text{C}-\text{C}}{\overset{\text{H}\quad\text{H}\quad\text{O}}{}} \quad + \quad 1\,C \ + \ 6\,H \ + \ 1 \text{ ring}$$

No methyls are apparent in the NMR, so the 6H group of peaks at δ 2.0-2.4 is most likely 3 CH_2 groups. Combining the pieces:

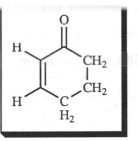

The mass spectral fragmentation can be explained by a "retro" or reverse Diels-Alder fragmentation:

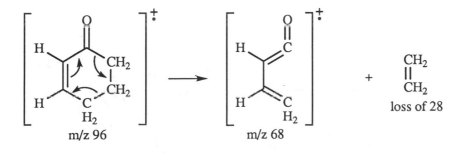

18-66 Building a model will help visualize this problem.

(a)

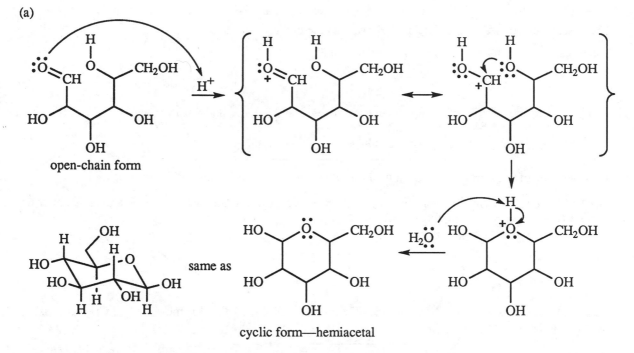

cyclic form—hemiacetal

(b) Yes, the cyclic form of glucose will give a positive Tollens test. In the basic solution of the Tollens test, the hemiacetal is in equilibrium with the open-chain aldehyde. It is the open-chain aldehyde that reacts with silver ion. As more of the open-chain form is oxidized by silver ion, more cyclic form will open to replace the consumed open-chain form. Le Châtelier's Principle strikes again!

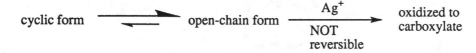

398

18-67

(a)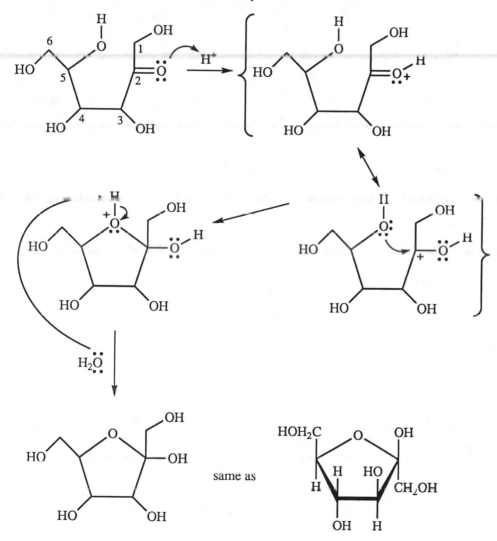

Any carbon with two oxygens bonded to it with single bonds belongs to the acetal family. If one of the oxygen groups is an OH, then the functional group is a hemiacetal. (The old name for this group is hemiketal as it came from a ketone.)

(b) Models will help. Ignore stereochemistry for the mechanism.

same as

18-68 Recall that "dilute acid" means an aqueous solution, and aqueous acid will remove acetals.

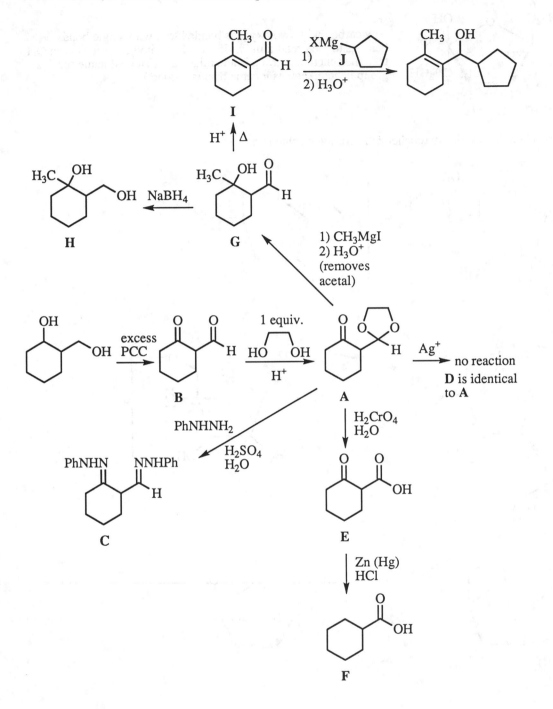

400

18-69

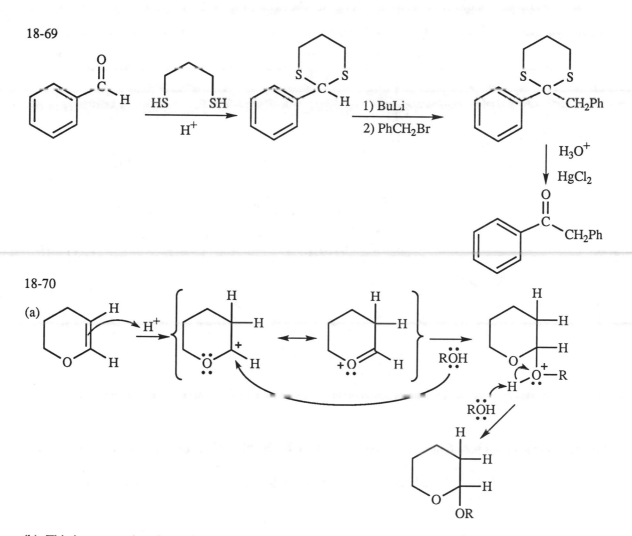

18-70

(b) This is not an ether, but rather an acetal, stable to base but reactive with aqueous acid.

18-70 continued

(c)

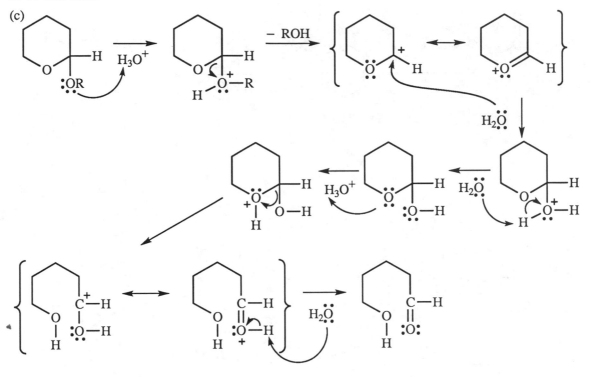

18-71

(a) First, deduce what functional groups are present in **A** and **B**. The IR of **A** shows no alkene and no carbonyl: the strongest peak is at 1060 cm^{-1}, possible a C—O bond. After acid hydrolysis of **A**, the IR of **B** shows a carbonyl at 1715 cm^{-1}: a ketone. (If it were an aldehyde, it would have aldehyde C—H around 2700-2800 cm^{-1}, absent in the spectrum of **B**.) What functional group has C—O bonds and is hydrolyzed to a ketone? An acetal (ketal)!

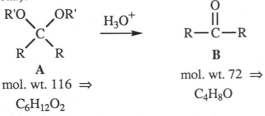

There is only one ketone of formula C$_4$H$_8$O: 2-butanone.

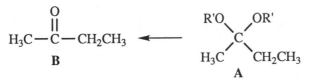

continued on next page

A must have the same alkyl groups as **B**. **A** has one element of unsaturation and is missing only C_2H_4 from the partial structure above. The most likely structure is the ethylene ketal. Is this consistent with the NMR?

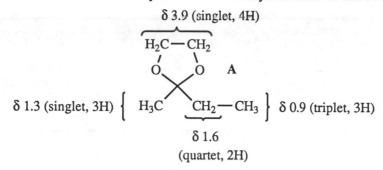

δ 3.9 (singlet, 4H)

δ 1.3 (singlet, 3H) { H_3C CH_2—CH_3 } δ 0.9 (triplet, 3H)

δ 1.6
(quartet, 2H)

What about the peak in the MS at m/z 87? This is the loss of 29 from the molecular ion at 116.

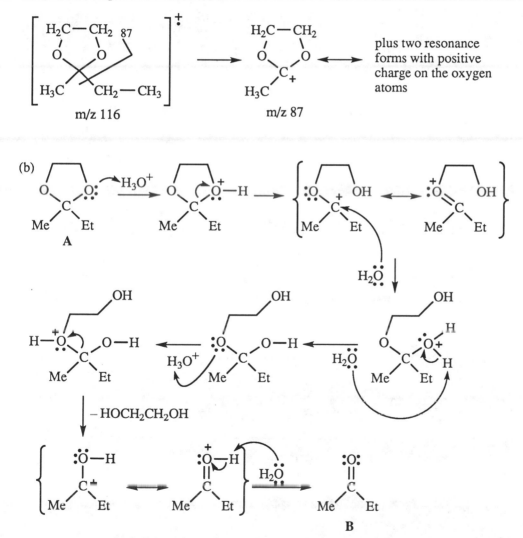

plus two resonance
forms with positive
charge on the oxygen
atoms

m/z 116   m/z 87

(b)

– HOCH₂CH₂OH

18-72 The strong UV absorption at 220 nm indicates a conjugated aldehyde or ketone. The IR shows a strong carbonyl at 1690 cm^{-1}, alkene at 1625 cm^{-1}, and two peaks at 2720 cm^{-1} and 2810 cm^{-1} —aldehyde!

$$C=C-\overset{\overset{\displaystyle O}{\|}}{C}-H$$

The NMR shows the aldehyde proton at δ 9.5 split into a doublet, so it has one neighboring H. There are only two vinyl protons, so there must be an alkyl group coming off the β carbon:

$$R-\overset{\overset{\displaystyle H}{|}}{C}=\overset{\overset{\displaystyle H}{|}}{C}-\overset{\overset{\displaystyle O}{\|}}{C}-H$$

The only other NMR signal is a 3H doublet: R must be methyl.

$$CH_3-\overset{\overset{\displaystyle H}{|}}{C}=\overset{\overset{\displaystyle H}{|}}{C}-\overset{\overset{\displaystyle O}{\|}}{C}-H$$

"crotonaldehyde"

18-73

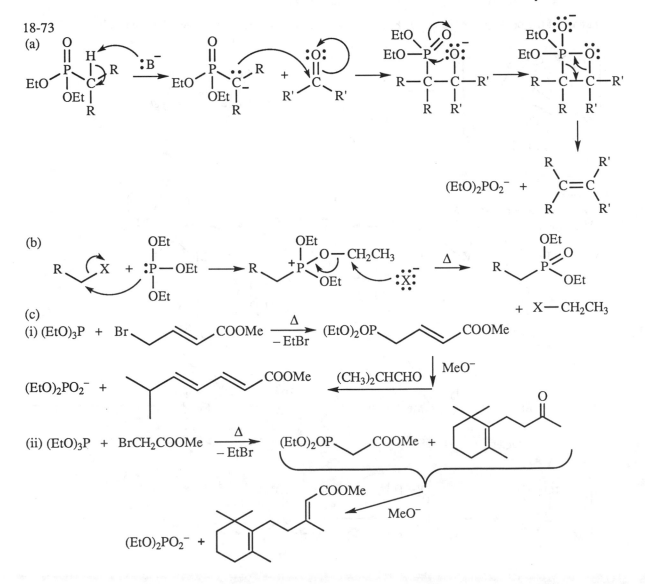

CHAPTER 19—AMINES

19-1 These compounds satisfy the criteria for aromaticity (planar, cyclic π system, and the Huckel number of 4n + 2 π electrons): pyrrole, imidazole, indole, pyridine, 2-methylpyridine, pyrimidine, and purine. The systems with 6 π electrons are: pyrrole, imidazole, pyridine, 2-methylpyridine, and pyrimidine. The systems with 10 π electrons are: indole and purine. The other nitrogen heterocycles shown are not aromatic because they do not have cyclic π systems.

19-2

(a)

(b)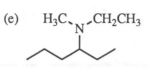

(c)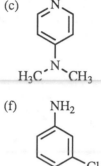

(d)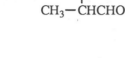

(e) H₃C─N─CH₂CH₃

(f) 3-chloroaniline structure

Note: images 1–4 correspond to structures (a), (b), (c), (d).

19-3

(a) 2-pentanamine
(b) *N*-methyl-2-butanamine
(c) 3-aminophenol (or *meta*-)
(d) 3-methylpyrrole
(e) *trans*-1,2-cyclopentanediamine
(f) *cis*-3-aminocyclohexanecarbaldehyde

19-4
(a) resolvable: there are two chiral carbons; carbon does not invert
(b) not resolvable: the nitrogen is free to invert
(c) not resolvable: it is symmetric
(d) not resolvable: even though the nitrogen is quaternary, one of the groups is a proton which can exchange rapidly, allowing for inversion
(e) resolvable: the nitrogen is quaternary and cannot invert when bonded to carbons

19-5 In order of increasing boiling point (increasing intermolecular hydrogen bonding):
(a) triethylamine and *n*-propyl ether have the same b.p. < di-*n*-propylamine
(b) dimethyl ether < dimethylamine < ethanol
(c) trimethylamine < diethylamine < diisopropylamine

19-6 Listed in order of increasing basicity. (See Appendix 2 for a discussion of acidity and basicity.)

(a) $PhNH_2$ < NH_3 < CH_3NH_2 < NaOH
(b) *p*-nitroaniline < aniline < *p*-methylaniline (*p*-toluidine)
(c) pyrrole < aniline < pyridine
(d) 3-nitropyrrole < pyrrole < imidazole

19-7
(a) secondary amine: one spike in the 3200-3400 cm⁻¹ region, indicating NH
(b) primary amine: two spikes in the 3200-3400 cm⁻¹ region, indicating NH_2
(c) alcohol: strong, broad peak around 3400 cm⁻¹

19-8 A compound with formula $C_4H_{11}N$ has no elements of unsaturation. The NMR doublet at δ 2.5 integrates to 2H; this must be a CH_2 bonded to the nitrogen, and a doublet because of one neighboring H:

$$\underset{\underbrace{\quad\quad}_{\delta\,2.5}}{\overset{\diagdown}{\underset{\diagup}{\text{CH}}}-\text{CH}_2-\overset{\diagup}{\underset{\diagdown}{\text{N}}}} \quad + \;\; 2\,\text{C} \;\; + \;\; 8\,\text{H}$$

The CH appears to be the multiplet at δ 1.5, part of a typical isopropyl pattern with the methyl groups appearing as a 6H doublet at δ 0.9. If any more carbons were bonded to the N, the chemical shift of their protons would also be in the δ 2.5 region, so it is safe to infer that the nitrogen is bonded to two hydrogens (a primary amine), and the remaining two carbons and six hydrogens are two methyl groups on the methine (CH). This would account for the 6H doublet at δ 0.9.

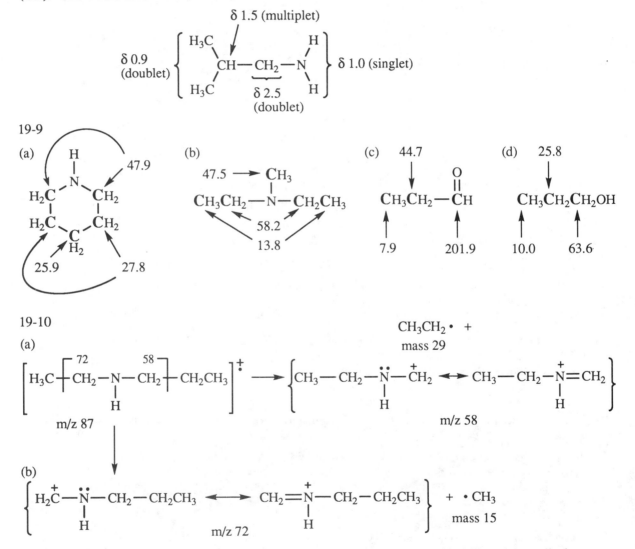

19-9

(a)

(b)

(c)   44.7

(d)   25.8

19-10

(a)

(b)

(c) The fragmentation in (a) occurs more often than the one in (b) because of stability of the radicals produced along with the iminium ions. Ethyl radical is much more stable than methyl radical, so pathway (a) is preferred.

19-11

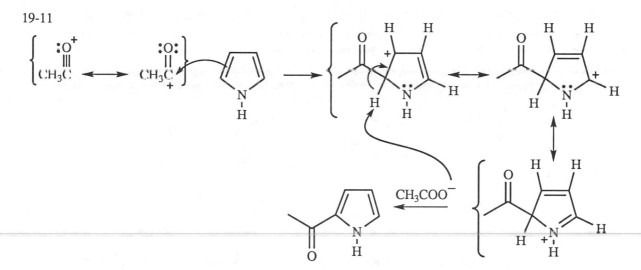

19-12 Pyridine is deactivated in the presence of electrophiles because the pair of electrons on the nitrogen is nucleophilic and reacts with electrophiles before they can attach to the ring. In pyrrole, the pair of electrons that we draw on the nitrogen is actually delocalized in the π system of the aromatic ring. The pyrrole nitrogen is not a good nucleophile, so an electrophile will react with electrons of the ring instead. The preferred site of electrophilic attack (and of protonation by strong acid) is on C-2 of the pyrrole ring.

pyridine—
basic, nucleophilic

pyrrole—
not basic, not
nucleophilic

19-13 Nitration at the 4-position of pyridine is not observed for the same reason that nitration at the 2-position is not observed: the intermediate puts some positive character on an electron-deficient nitrogen, and electronegative nitrogen hates that. (It is important to distinguish this type of positive nitrogen without a complete octet of electrons, from the quaternary nitrogen, also positively charged but with a full octet. It is the number of electrons around atoms that is most important; the charge itself is less important.)

GOOD: —N⁺—   VERY BAD: —N⁺

mechanism

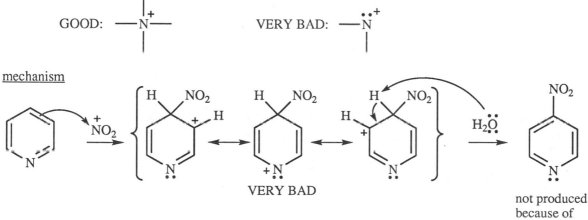

not produced
because of
unfavorable
intermediate

19-14 Any electrophilic attack, including sulfonation, is preferred at the 3-position of pyridine because the intermediate is more stable than the intermediate from attack at either the 2-position or the 4-position. (Resonance forms of the sulfonate group are not shown, but remember that they are important!)

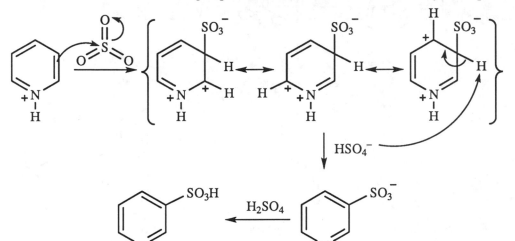

19-15

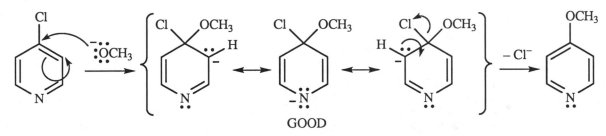

19-16

(a)

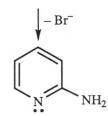

19-16 continued

(b)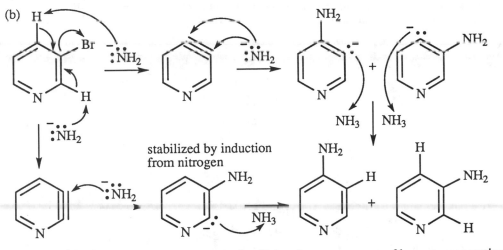

This is a benzyne-type mechanism. (For simplicity above, two steps of benzyne generation are shown as one step: first, a proton is abstracted by amide anion, followed by loss of bromide.) Amide ion is a strong enough base to remove a proton from 3-bromopyridine as it does from a halobenzene. Once a benzyne is generated (two possibilities), the amide ion reacts quickly, forming a mixture of products.

Why does the 3-bromo follow this extreme mechanism while the 2-bromo reacts smoothly by the addition-elimination mechanism? Stability of the intermediate! Negative charge on the electronegative nitrogen makes for a more stable intermediate in the 2-bromo substitution. No such stabilization is possible in the 3-bromo case.

19-17

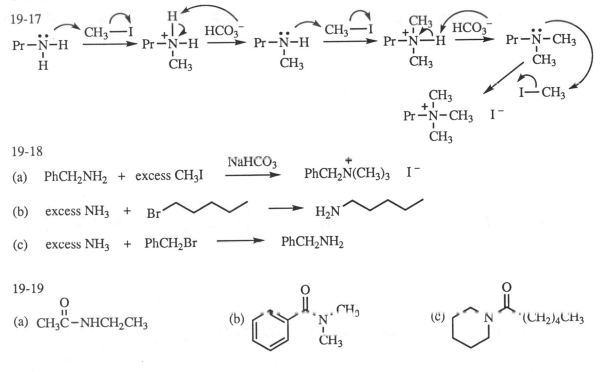

19-18

(a) $PhCH_2NH_2$ + excess CH_3I $\xrightarrow{\text{NaHCO}_3}$ $PhCH_2\overset{+}{N}(CH_3)_3$ I^-

(b) excess NH_3 + Br⟋⟍⟋⟍ ⟶ H_2N⟋⟍⟋⟍

(c) excess NH_3 + $PhCH_2Br$ ⟶ $PhCH_2NH_2$

19-19

(a) $\overset{\displaystyle O}{\underset{\displaystyle }{CH_3\overset{\parallel}{C}-NHCH_2CH_3}}$

(b) [structure: benzoyl group attached to $N(CH_3)_2$ type amide, with C=O, N, CH₃, CH₃]

(c) [structure: piperidine ring N attached to C=O and (CH₂)₄CH₃]

409

19-20 If the amino group were not protected, it would do a nucleophilic substitution on chlorosulfonic acid. Later in the sequence, this group could not be removed without cleaving the other sulfonamide group.

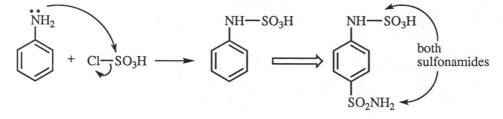

19-21

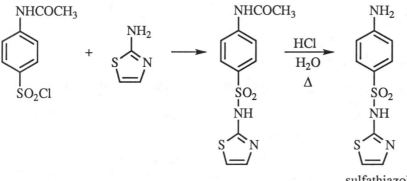

sulfathiazole

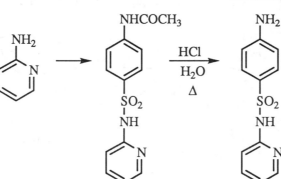

sulfapyridine

19-22

(a)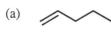

(b) 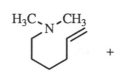 +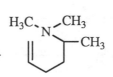

(c)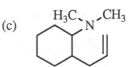

(d)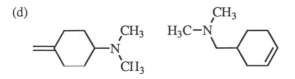

(e)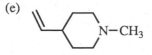

410

19-23 Orientation of the Cope elimination is similar to Hofmann elimination: the *less* substituted alkene is the major product.

(a)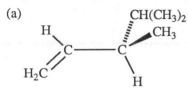

+ (CH₃)₂NOH

(b) CH₂=CH₂ + 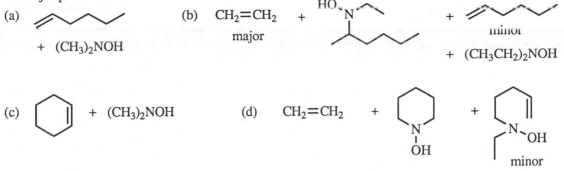 +

major   minor

+ (CH₃CH₂)₂NOH

(c) 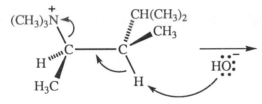 + (CH₃)₂NOH

(d) CH₂=CH₂ + 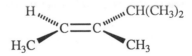 +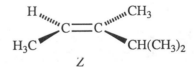

minor

19-24 The key to this problem is to understand that Hofmann elimination occurs via an E2 mechanism requiring *anti* coplanar stereochemistry, whereas Cope elimination requires *syn* coplanar stereochemistry.

(a)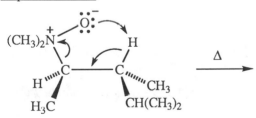

Hofmann orientation loses a hydrogen from the CH₃ and the N(CH₃)₃ group to make the less substituted double bond

(b)

Hofmann elimination

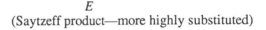

H $\cdots$ C=C $\cdots$ CH(CH₃)₂

H₃C   CH₃

E

(Saytzeff product—more highly substituted)

Cope elimination

H $\cdots$ C=C $\cdots$ CH₃

H₃C   CH(CH₃)₂

Z

(Saytzeff product   more highly substituted)

19-25

(a)

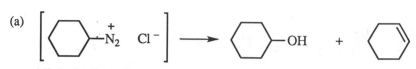

Aliphatic diazonium ions are very unstable, rapidly decomposing to carbocations.

(b)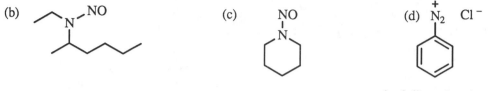

(c)

(d)

Aryl diazonium ions are relatively stable if kept cold.

19-26 The diazonium ion can do aromatic substitution like any other electrophile.

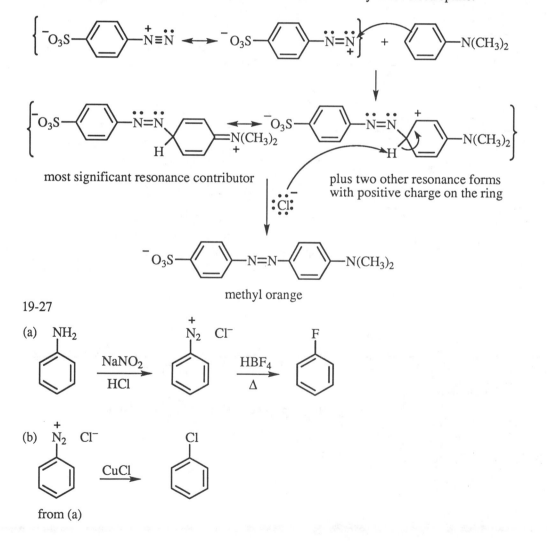

most significant resonance contributor

plus two other resonance forms with positive charge on the ring

methyl orange

19-27

(a)

NaNO₂ / HCl → HBF₄ / Δ →

(b)

CuCl →

from (a)

19-27 continued

(c)

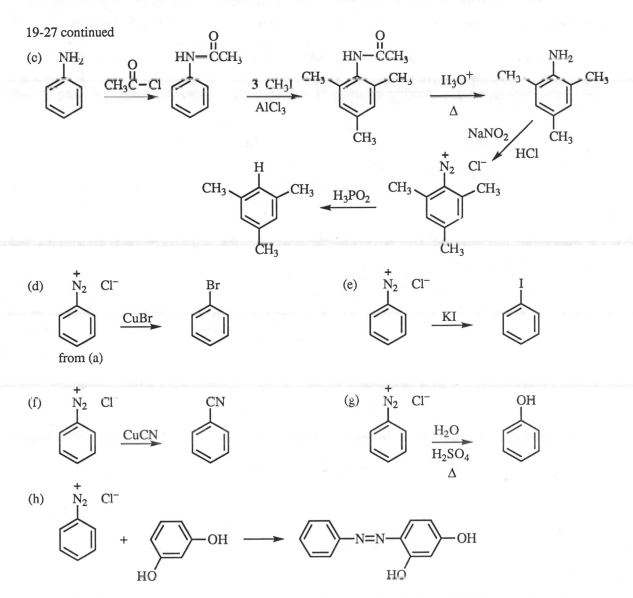

(d)

from (a)

(e)

(f)

(g)

(h)

19-28 General guidelines for choice of reagent for reductive amination: use LiAlH$_4$ when the imine or oxime is isolated. Use NaBH$_3$CN in solution when the imine or iminium ion is not isolated. Alternatively, catalytic hydrogenation works in most cases.

(a)

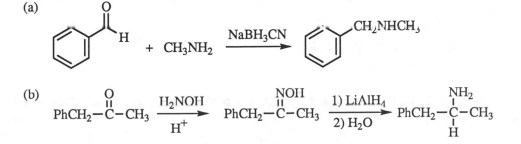

(b)

$$PhCH_2-\overset{O}{\underset{}{C}}-CH_3 \xrightarrow[H^+]{H_2NOH} PhCH_2-\overset{NOH}{\underset{}{C}}-CH_3 \xrightarrow[2)\ H_2O]{1)\ LiAlH_4} PhCH_2-\overset{NH_2}{\underset{H}{C}}-CH_3$$

413

19-28 continued

(c)

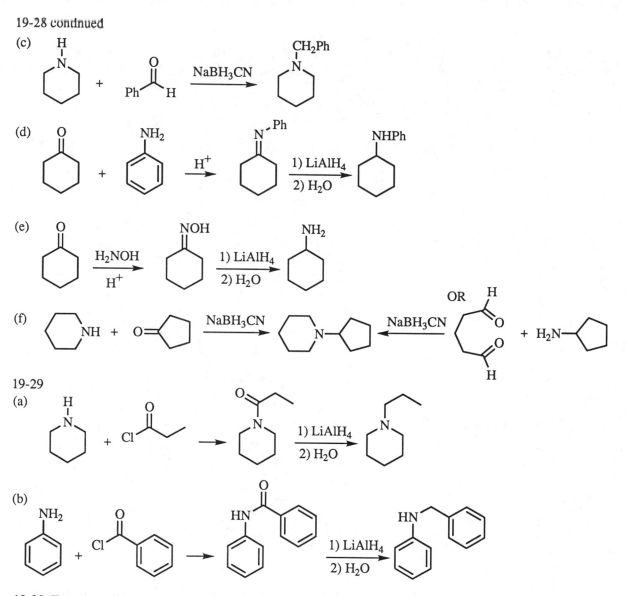

NaBH₃CN

(d)

H⁺

1) LiAlH₄
2) H₂O

(e)

H₂NOH / H⁺

1) LiAlH₄
2) H₂O

(f)

NaBH₃CN

NaBH₃CN

OR

19-29

(a)

1) LiAlH₄
2) H₂O

(b)

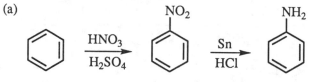

1) LiAlH₄
2) H₂O

19-30 To reduce nitroaromatics, the reducing reagents (H₂ plus a metal catalyst, or a metal plus HCl) can be used virtually interchangeably.

(a)

HNO₃ / H₂SO₄

Sn / HCl

19-30 continued

(b)

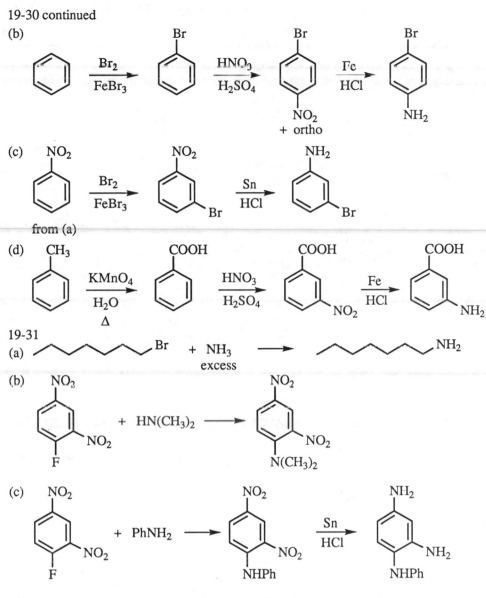

(c)

NO₂ ... NO₂ ... NH₂

from (a)

(d) CH₃ ... COOH ... COOH ... COOH

19-31

(a)

(b)

(c)

19-32 Assume that LiAlH₄ or H₂/catalyst can be used interchangeably.

(a) PhCH₂Br + NaN₃ ⟶ PhCH₂N₃ $\xrightarrow[\text{Pt}]{\text{H}_2}$ PhCH₂NH₂

(b)

415

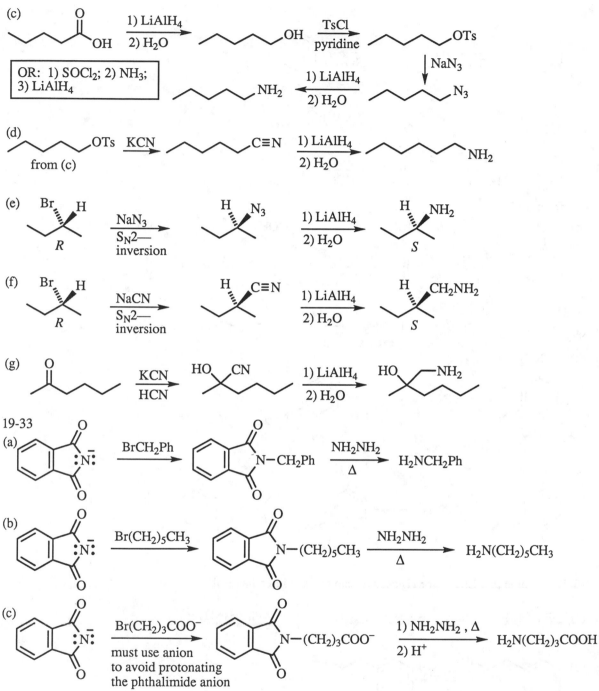

19-34 To make aniline by the Gabriel synthesis, the phthalimide anion would have to do nucleophilic substitution on benzene. Benzene is electron rich and is attacked by electrophiles; nucleophiles cannot attack a benzene without electron-withdrawing substituents.

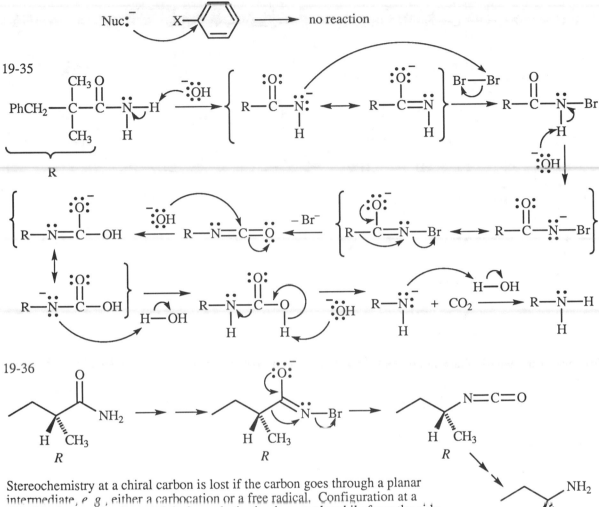

19-35

19-36

Stereochemistry at a chiral carbon is lost if the carbon goes through a planar intermediate, e.g., either a carbocation or a free radical. Configuration at a chiral carbon can be inverted during substitution by a nucleophile from the side opposite the leaving group. However, when the carbon retains all four pairs of electrons, as in this Hofmann rearrangement, it retains its configuration.

19-37

(a) The acyl azide of the Curtius rearrangement is similar to the N-bromo amide of the Hofmann rearrangement in that both have an amide nitrogen with a good leaving group attached. Subsequent alkyl migration to the isocyanate and hydrolysis through the carbamic acid to the amine are identical in both mechanisms.

(b) The leaving group in the Curtius rearrangement is N_2 gas, one of the best leaving groups known "to man or beast", as we used to say.

(c)

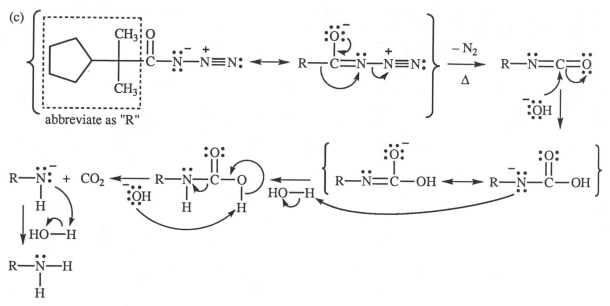

19-38   Please refer to solution 1-19, page 12 of this Solutions Manual.

19-39
(a) primary amine; 2,2-dimethyl-1-propanamine, or neopentylamine
(b) primary amine; 2-propanamine, or isopropylamine
(c) tertiary heterocyclic amine and a nitro group; 3-nitropyridine
(d) quaternary heterocyclic ammonium ion; *N,N*-dimethylpiperidinium iodide
(e) tertiary aromatic amine oxide; *N*-ethyl-*N*-methylaniline oxide
(f) tertiary aromatic amine; *N*-ethyl-*N*-methylaniline
(g) tertiary heterocyclic ammonium ion; pyridinium chloride
(h)   secondary amine; *N*,4-diethyl-3-hexanamine

19-40 Shown in order of increasing basicity. In each set, the aliphatic amine is the strongest base.

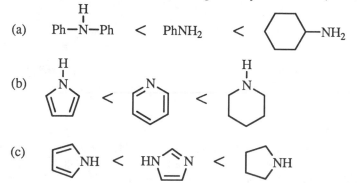

19-41
(a) not resolvable: planar
(b) resolvable: chiral carbon
(c) not resolvable: symmetric
(d) resolvable: nitrogen inversion is very slow
(e) not resolvable: symmetric
(f) resolvable: chiral nitrogen, unable to invert

19-42 The values of pK_b of amines or pK_a of the conjugate acids can be obtained from Table 19-3. The side of the reaction with the weaker acid and base will be favored at equilibrium. (See Appendix 2 for a discussion of acidity and basicity.)

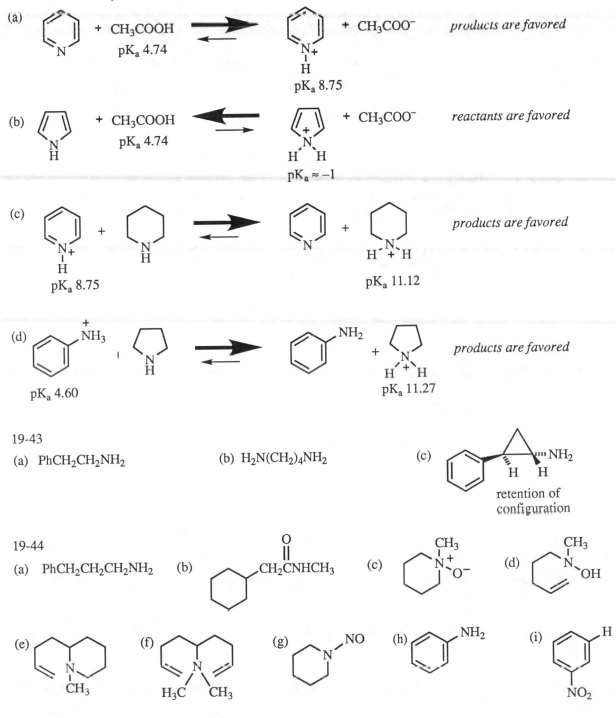

(a)   + CH₃COOH   ⇌   + CH₃COO⁻   *products are favored*

pK_a 4.74 pK_a 8.75

(b)   + CH₃COOH   ⇌   + CH₃COO⁻   *reactants are favored*

pK_a 4.74 $pK_a \approx -1$

(c)   +   ⇌   +   *products are favored*

pK_a 8.75 pK_a 11.12

(d)   +   ⇌   +   *products are favored*

pK_a 4.60 pK_a 11.27

19-43

(a) PhCH₂CH₂NH₂   (b) H₂N(CH₂)₄NH₂   (c)

retention of configuration

19-44

(a) PhCH₂CH₂CH₂NH₂   (b)   (c)   (d)

(e)   (f)   (g)   (h)   (i)

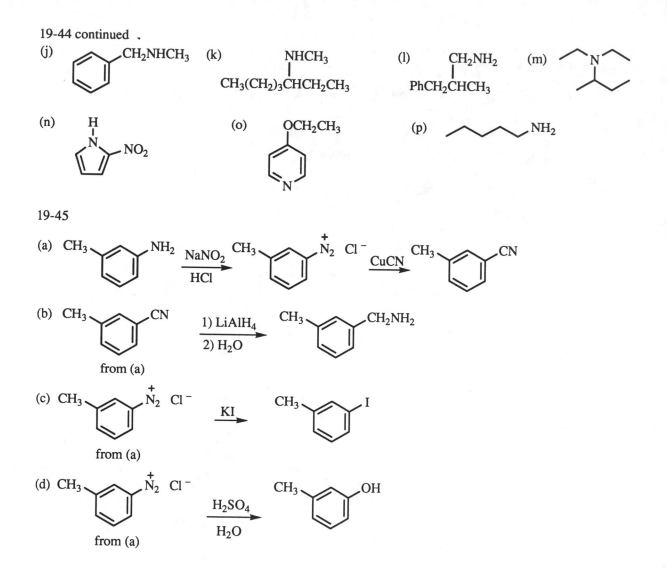

(j) CH₂NHCH₃ (on benzene ring)   (k) NHCH₃ / CH₃(CH₂)₃CHCH₂CH₃   (l) CH₂NH₂ / PhCH₂CHCH₃   (m)

(n) (pyrrole with H on N and NO₂)   (o) OCH₂CH₃ (pyridine)   (p) NH₂

19-45

(a) CH₃ (benzene) NH₂ →(NaNO₂ / HCl)→ CH₃ (benzene) $\overset{+}{N_2}$ Cl⁻ →(CuCN)→ CH₃ (benzene) CN

(b) CH₃ (benzene) CN →(1) LiAlH₄ 2) H₂O)→ CH₃ (benzene) CH₂NH₂
   from (a)

(c) CH₃ (benzene) $\overset{+}{N_2}$ Cl⁻ →(KI)→ CH₃ (benzene) I
   from (a)

(d) CH₃ (benzene) $\overset{+}{N_2}$ Cl⁻ →(H₂SO₄ / H₂O)→ CH₃ (benzene) OH
   from (a)

19-46 This fragmentation is favorable because the iminium ion produced is stabilized by resonance. Also, there are three possible cleavages that give the same ion. Both factors combine to make the cleavage facile, at the expense of the molecular ion.

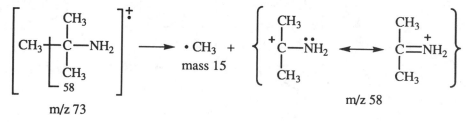

420

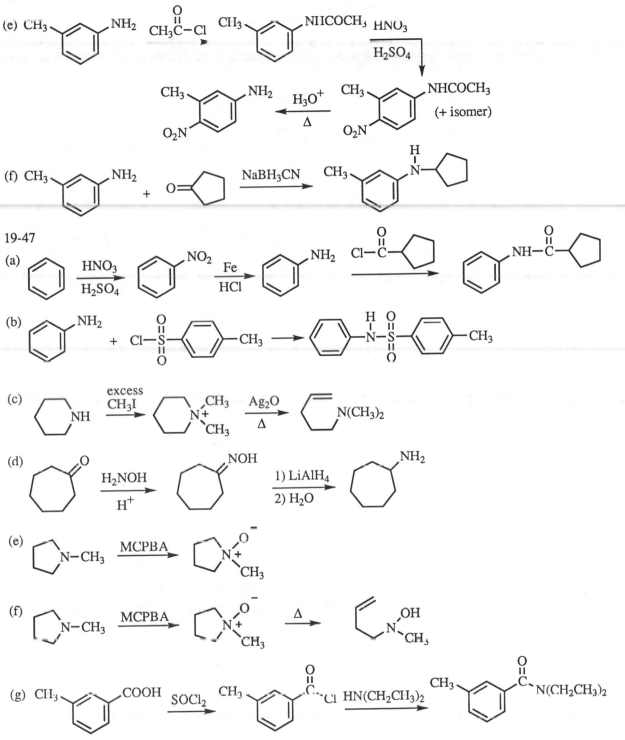

19-48 The problem restricts the starting materials to six carbons or fewer. Always choose starting materials with as many of the necessary functional groups as possible.

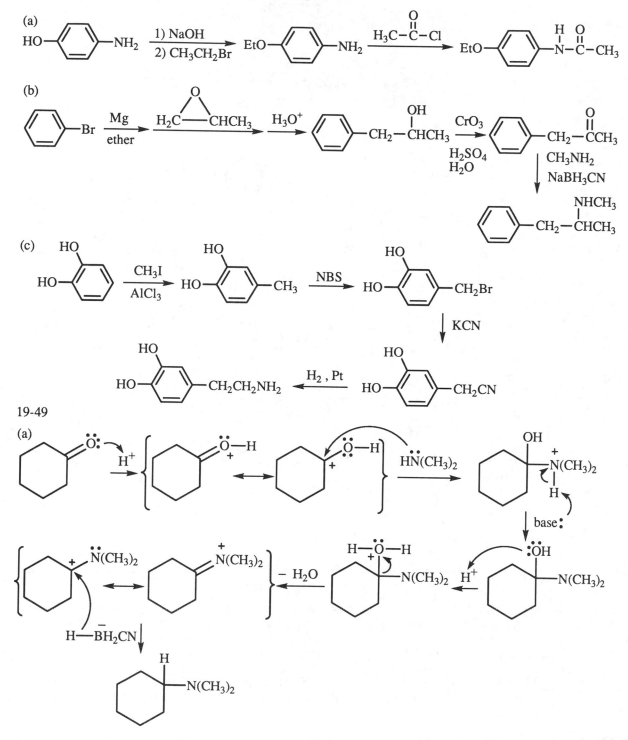

(b)

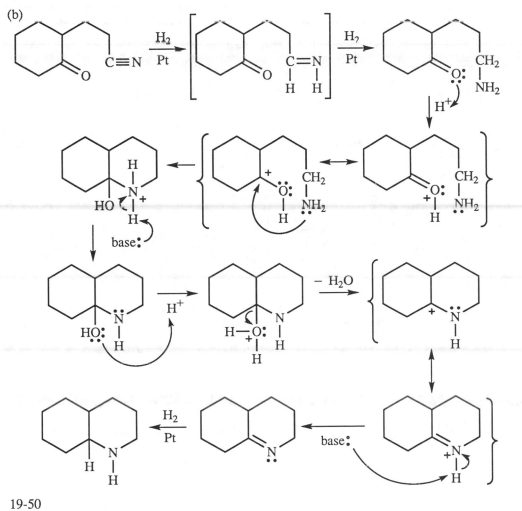

19-50

(a)

(b)

(c)

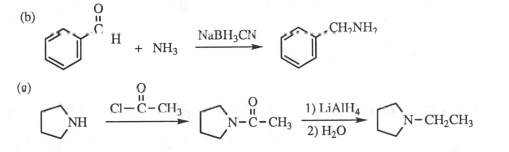

19-50 continued

(d)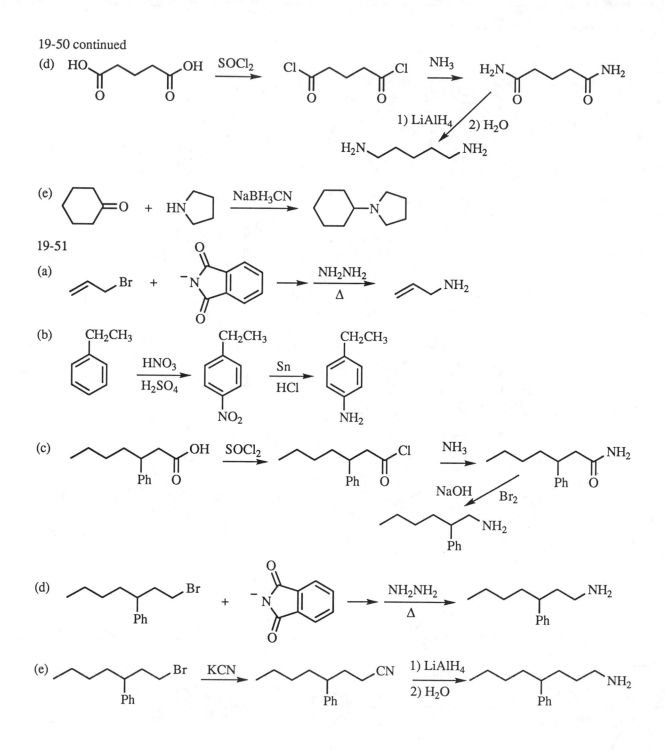

(e)

19-51

(a)

(b)

(c)

(d)

(e)

424

19-52

(a) When guanidine is protonated, the cation is greatly stabilized by resonance, distributing the positive charge over all atoms (except H):

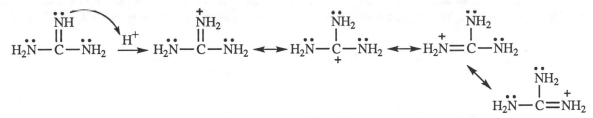

(b) The unprotonated molecule has a resonance form shown below that the protonated molecule cannot have. Therefore, the unprotonated form is stabilized relative to the protonated form. This greater stabilization of the unprotonated form is reflected in weaker basicity.

(c) Anilines are weaker bases than aliphatic amines because the electron pair on the nitrogen is shared with the ring, stabilizing the system. There is a steric requirement, however: the p orbital on the N must be parallel with the p orbitals on the benzene ring in order for the electrons on N to be distributed into the π system of the ring.

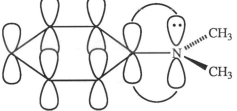

If the orbital on the nitrogen is forced out of this orientation (by substitution on C-2 and C-6, for example), the electrons are no longer shared with the ring. The nitrogen is hybridized sp^3 (no longer any reason to be sp^2), and the electron pair is readily available for bonding $\Rightarrow$ increased basicity.

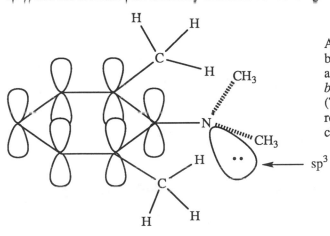

As surprising as it sounds, this aniline is about as basic as a tertiary aliphatic amine, except that the aromatic ring substituent is electron-withdrawing *by induction*, decreasing the basicity slightly. (This phenomenon is called "steric inhibition of resonance". We will see more examples in future chapters.)

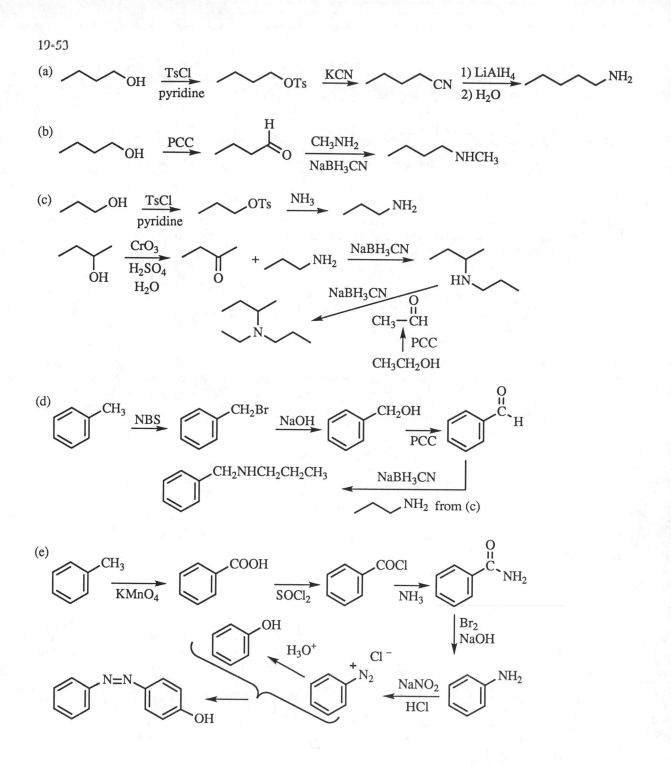

19-54

(a)

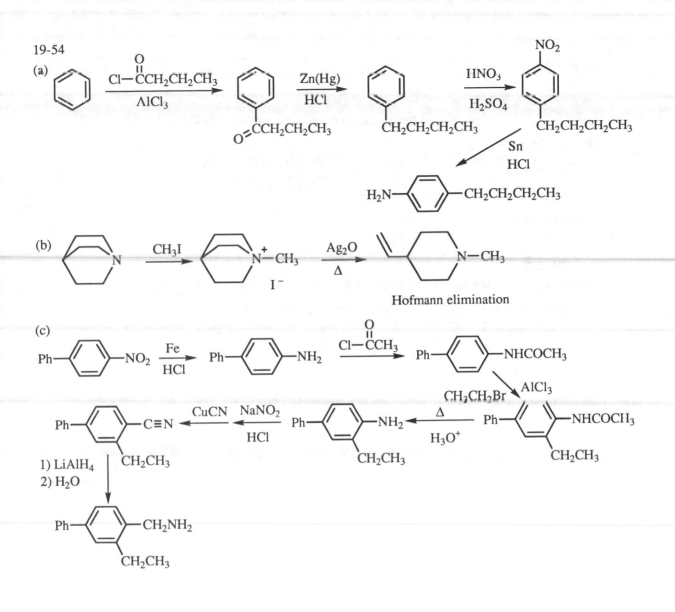

(b)

(c)

1) LiAlH₄
2) H₂O

19-55

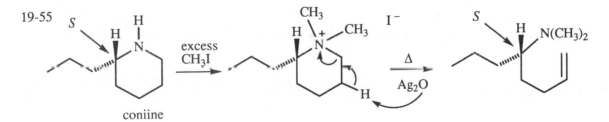

coniine

19-56

(a) <u>unknown **X**</u>

—fishy odor $\Rightarrow$ amine

—molecular weight 101 $\Rightarrow$ odd number of nitrogens

$\Rightarrow$ if one nitrogen and no oxygen, the remainder is C_6H_{15}

<u>Mass spectrum:</u>

—fragment at 86 = M – 15 = loss of methyl $\Rightarrow$ the compound is likely to

have this structural piece: $CH_3 \underset{\underset{\alpha\text{-cleavage}}{\rule{1.2cm}{0.4pt}}}{\rule[0.8ex]{0.8cm}{0.4pt}} C - N$

<u>IR spectrum:</u>

—no OH, no NH $\Rightarrow$ must be a 3° amine

—no C=O or C=C or C≡N or NO_2

<u>NMR spectrum:</u>

—only a triplet and quartet, integration about 3 : 2 $\Rightarrow$ ethyl group(s) only

assemble the evidence: $C_6H_{15}N$, 3° amine, only ethyl in the NMR:

$$CH_3CH_2 - \underset{\underset{CH_2CH_3}{|}}{N} - CH_2CH_3$$

(b) React the triethylamine with HCl. The pure salt is solid and odorless.

$$(CH_3CH_2)_3N \ + \ HCl \ \longrightarrow \ (CH_3CH_2)_3\overset{+}{N}H \quad Cl^-$$
$$\text{salt}$$

(c) Washing her clothing in dilute acid like vinegar (dilute acetic acid) or dilute HCl would form a water-soluble salt as shown in (b). Normal washing will remove the water-soluble salt.

Compound A

Mass spectrum:

—molecular ion at 73 = odd mass = odd number of nitrogens;

if one nitrogen and no oxygen present $\Rightarrow$ molecular formula $C_4H_{11}N$

—base peak at 44 is M - 29 $\Rightarrow$ this fragment must be present:

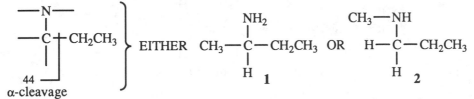

IR spectrum:

—two spikes around 3300 cm^{-1} indicate a 1° amine; no indication of oxygen

NMR spectrum:

—two exchangeable protons suggest NH_2 present

—1H multiplet at δ 2.8 means a CH—NH_2

The structure of **A** must be the same as **1** above:

Compound B
an isomer of **A**, so its molecular formula must also be $C_4H_{11}N$

IR spectrum:

—only one spike at 3300 cm^{-1} $\Rightarrow$ 2° amine

NMR spectrum:

—one exchangeable proton $\Rightarrow$ NH

—two ethyls present
The structure of **B** must be:

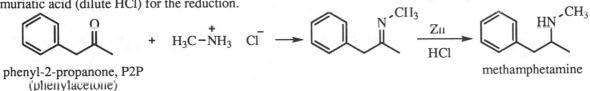

19-58

(a) The acid-catalyzed condensation of P2P (a controlled substance) with methylamine hydrochloride gives an imine which can be reduced to methamphetamine. The suspect was probably planning to use zinc in muriatic acid (dilute HCl) for the reduction.

phenyl-2-propanone, P2P
(phenylacetone)

methamphetamine

(b) The jury acquitted the defendant on the charge of attempted manufacture of methamphetamine. There were legal problems with possible entrapment, plus the fact that he had never opened the bottle of the starting material. The defendant was convicted on several possession charges, however, and was awarded four years of institutional time to study organic chemistry.

19-59

Mass spectrum:

—molecular ion at 87 = odd mass = odd number of nitrogens present

—if one nitrogen and no oxygens ⇒ molecular formula $C_5H_{13}N$

—base peak át m/z 30 ⇒ structure must include this fragment

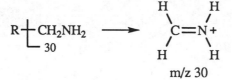

m/z 30

IR spectrum:

—two spikes in the 3300-3400 cm^{-1} region ⇒ 1° amine

NMR spectrum:

—singlet at δ 0.9 for 9H must be a *t*-butyl group

—2H signal at δ 1.0 exchanges with D_2O ⇒ must be protons on N or O

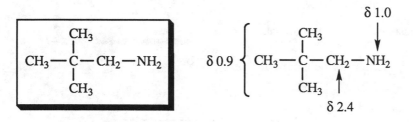

Note that the base peak in the MS arises from cleavage to give these two, relatively stable fragments:

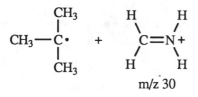

m/z 30

19-60 (a tough problem)

molecular formula $C_{11}H_{16}N_2$ has 5 elements of unsaturation, enough for a benzene ring; no oxygens precludes NO_2 and amide; if C≡N is present, there are not enough elements of unsaturation left for a benzene ring, so benzene and C≡N are mutually exclusive

IR spectrum:
—one spike around 3300 cm^{-1} suggests a 2° amine
—no C≡N
—CH and C=C regions suggest an aromatic ring

Proton NMR spectrum:
—5H multiplet at δ 7.3 indicates a monosubstituted benzene ring (the fact that all the peaks are huddled around 7.3 precludes N being bonded to the ring)
—1H singlet at δ 2.0 is exchangeable ⇒ NH of secondary amine
—2H singlet at δ 3.5 is CH$_2$; the fact that it is so strongly deshielded and unsplit suggests that it is between a nitrogen and the benzene ring

fragments so far:

⬡—CH$_2$—N   + NH + 4 C + 8 H + 1 element of unsaturation

Carbon NMR spectrum:
—four signals around δ 125-138 are the aromatic carbons
—the signal at δ 65 is the CH$_2$ bonded to the benzene
—the other 4 carbons come as two signals at δ 46 and δ 55; each is a triplet, so there are two sets of two equivalent CH$_2$ groups, each bonded to N to shift it downfield

fragments so far:

⬡—CH$_2$—N   + NH + $\genfrac{}{}{0pt}{}{CH_2}{CH_2}$ + $\genfrac{}{}{0pt}{}{CH_2}{CH_2}$ + 1 element of unsaturation

There is no evidence for an alkene in any of the spectra, so the remaining element of unsaturation must be a ring. The simplicity of the NMR spectra indicates a fairly symmetric compound.

Assemble the pieces:

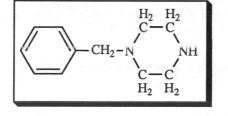

431

20-1

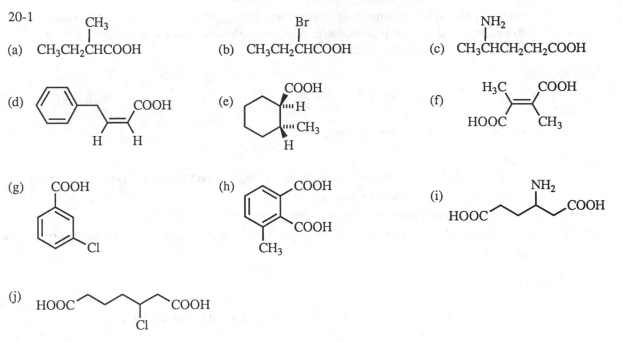

(a) $CH_3CH_2CHCOOH$ with CH_3 substituent

(b) $CH_3CH_2CHCOOH$ with Br substituent

(c) $CH_3CHCH_2CH_2COOH$ with NH_2 substituent

(d) benzyl $CH=CH-COOH$ structure with H, H

(e) cyclohexane with COOH, H, CH₃, H

(f) alkene with H_3C, COOH, HOOC, CH_3

(g) benzoic acid with COOH and Cl

(h) benzene with COOH, COOH, CH_3

(i) HOOC...COOH chain with NH_2

(j) HOOC...COOH chain with Cl

20-2 IUPAC name first; then common name.

(a) 2-iodo-3-methylpentanoic acid; α-iodo-β-methylvaleric acid
(b) (Z)-3,4-dimethyl-3-hexenoic acid
(c) 2,3-dinitrobenzoic acid; no common name
(d) *trans*-1,2-cyclohexanedicarboxylic acid; (*trans*-hexahydrophthalic acid)
(e) 2-chlorobenzene-1,4-dicarboxylic acid; 2-chloroterephthalic acid
(f) 3-methylhexanedioic acid; β-methyladipic acid

20-3 Listed in order of increasing acid strength (weakest acid first). (See Appendix 2 for a review of acidity.)

(a) CH_3CH_2COOH < $CH_3-\underset{Br}{CHCOOH}$ < $CH_3-\overset{Br}{\underset{Br}{C}}COOH$

The greater the number of electron-withdrawing substituents, the greater the stabilization of the carboxylate ion.

(b) $CH_3-\underset{Br}{CHCH_2CH_2COOH}$ < $CH_3CH_2-\underset{Br}{CHCH_2COOH}$ < $CH_3CH_2CH_2-\underset{Br}{CHCOOH}$

The closer the electron-withdrawing group, the greater the stabilization of the carboxylate ion.

(c) CH_3CH_2COOH < $CH_3-\underset{Cl}{CHCOOH}$ < $CH_3-\underset{C\equiv N}{CHCOOH}$ < $CH_3-\underset{NO_2}{CHCOOH}$

The stronger the electron-withdrawing effect of the substituent, the greater the stabilization of the carboxylate ion.

20-4 The principle used to separate a carboxylic acid (a stronger acid) from a phenol (a weaker acid) is to neutralize with a weak base (NaHCO₃)— a base strong enough to ionize the stronger acid but not strong enough to ionize the weaker acid.

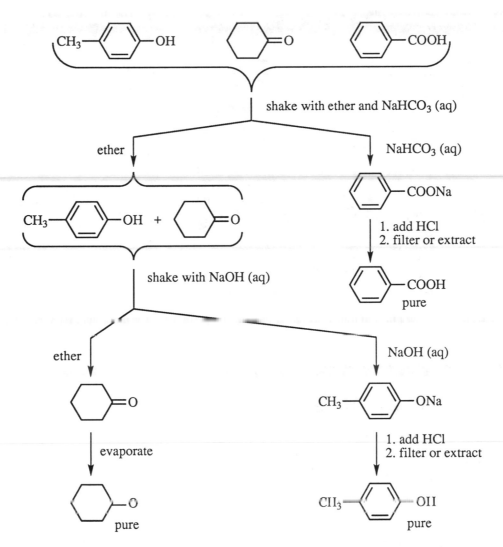

20-5 The reaction mixture includes the initial reactant, reagent, desired product, and the overoxidation product—not unusual for an organic reaction mixture.

(a)

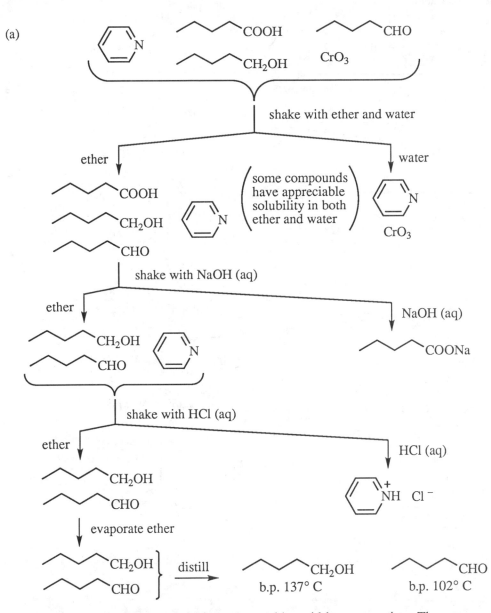

(b) 1-Pentanol cannot be removed from pentanal by acid-base extraction. These two remaining products can be separated by distillation, the alcohol having the higher boiling point because of hydrogen-bonding.

20-6 The COOH has a characteristic IR absorption: a broad peak from 3400-2400 cm^{-1}, with a "shoulder" around 2700 cm^{-1}. The carbonyl stretch at 1695 cm^{-1} is a little lower than the standard 1710 cm^{-1}, suggesting conjugation. The strong alkene absorption at 1650 cm^{-1} also suggests it is conjugated.

20-7

(a) The ethyl pattern is obvious: a 3H triplet at δ 1.0 and a 2H quartet at δ 2.4. The only other peak is the COOH at δ 11.3 (a 5 δ unit offset added to 6.3).

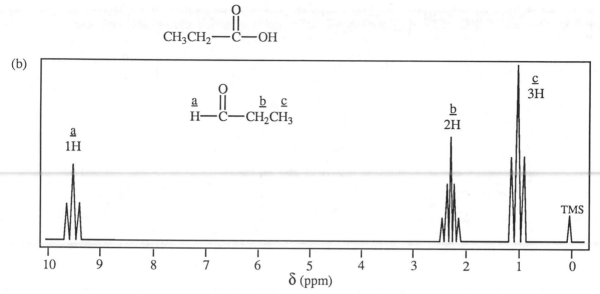

(b)

The multiplet between δ 2 and δ 3 is drawn as a pentet as though it were split equally by the aldehyde proton and the CH$_2$ group. These coupling constants are probably unequal, in which case the actual splitting pattern will be a complex multiplet.

(c) The chemical shift of the aldehyde proton is between δ 9-10, not as far downfield as the carboxylic acid proton. Also, the aldehyde proton is split into a triplet by the CH$_2$, unlike the COOH proton which always appears as a singlet. Finally, the CH$_2$ is split by an extra proton, so it will give a multiplet with complex splitting, instead of the quartet shown in the acid.

20-8

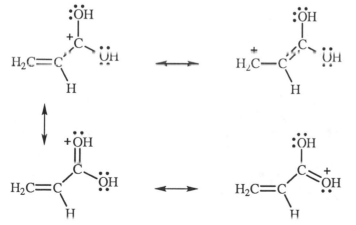

20-9

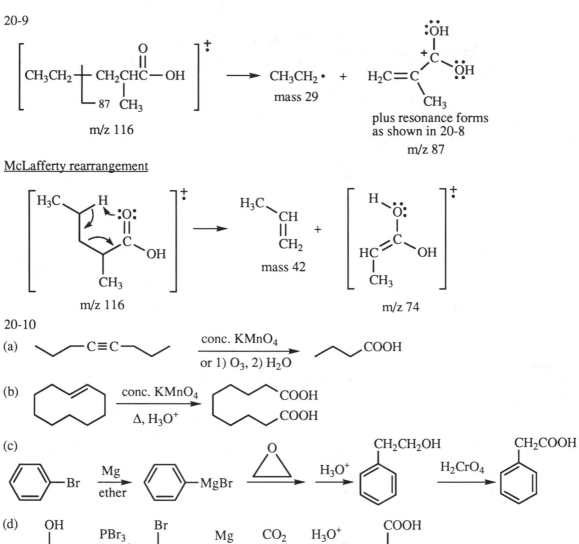

McLafferty rearrangement

20-10

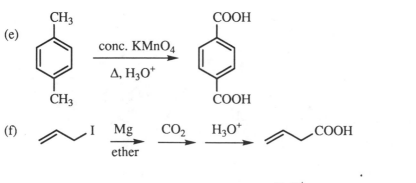

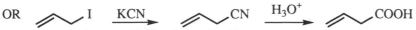

20-11 The mechanism of acid-catalyzed nucleophilic acyl substitution may seem daunting, but it is simply a succession of steps that are already very familiar to you.

Typically, these mechanisms have six steps: four proton transfers (two on, two off), a nucleophilic attack, and a leaving group leaving, with a little resonance stabilization thrown in that makes the whole thing work. The six steps are labeled in the mechanism below:

Step A proton on (resonance stabilization)
Step B nucleophile attacks
Step C proton off
Step D proton on
Step E leaving group leaves (resonance stabilization)
Step F proton off

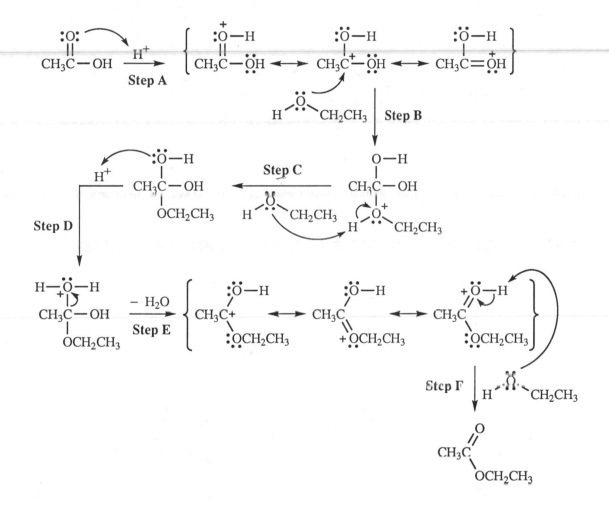

20-12 For the sake of space in this problem, resonance forms will not be drawn, but remember that they are critical!

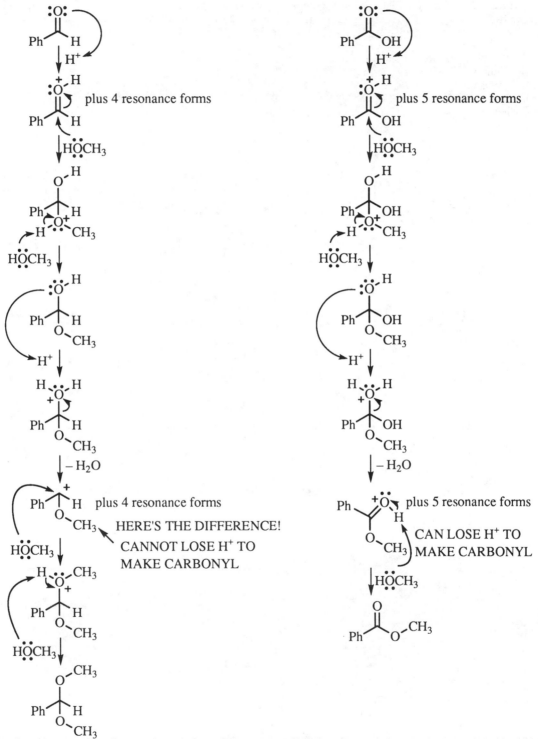

20-13

(a)

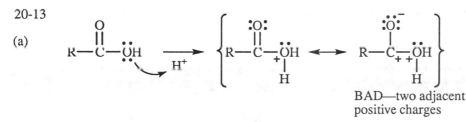

BAD—two adjacent
positive charges

(b) Protonation on the OH gives only two resonance forms, one of which is bad because of adjacent positive charges. Protonation on the C=O is good because of three resonance forms distributing the positive charge over three atoms, with no additional charge separation.

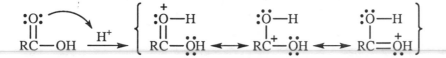

(c) The carbonyl oxygen is more "basic" because, by definition, it reacts with a proton more readily. It does so because the intermediate it produces is more stable than the intermediate from protonation of the OH.

20-14

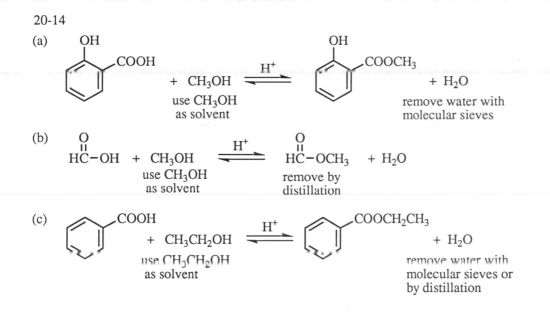

(a)

OH
COOH
+ CH₃OH   →H⁺→   OH
COOCH₃
+ H₂O
use CH₃OH
as solvent
remove water with
molecular sieves

(b)

O
‖
HC–OH + CH₃OH   ⇌H⁺   O
‖
HC–OCH₃ + H₂O
use CH₃OH
as solvent
remove by
distillation

(c)

COOH
+ CH₃CH₂OH   ⇌H⁺   COOCH₂CH₃
+ H₂O
use CH₃CH₂OH
as solvent
remove water with
molecular sieves or
by distillation

(a) and (b)

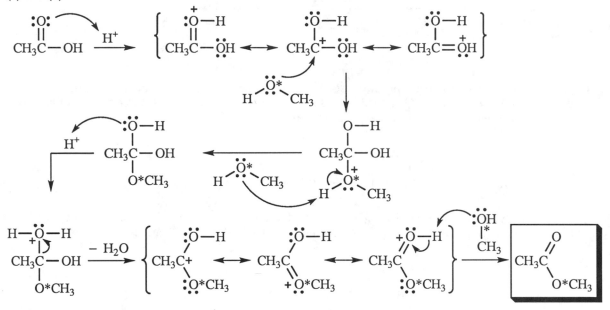

(c) The ^{18}O has two more neutrons, and therefore two more mass units, than ^{16}O. The instrument ideally suited to analyze compounds of different mass is the mass spectrometer.

20-16

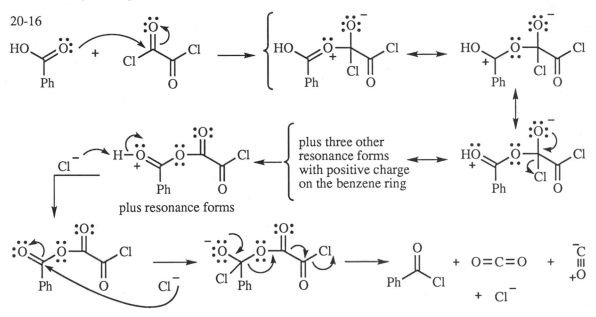

440

20-17

(a)

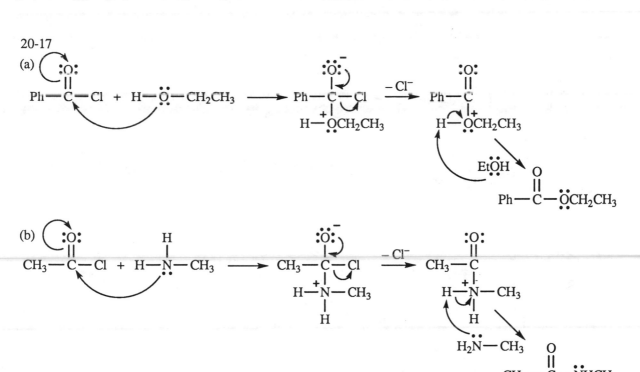

(b)

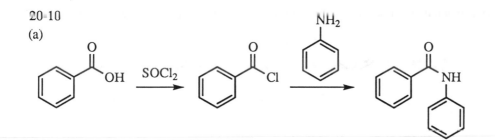

20-10

(a)

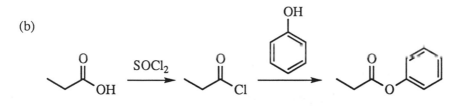

(b)

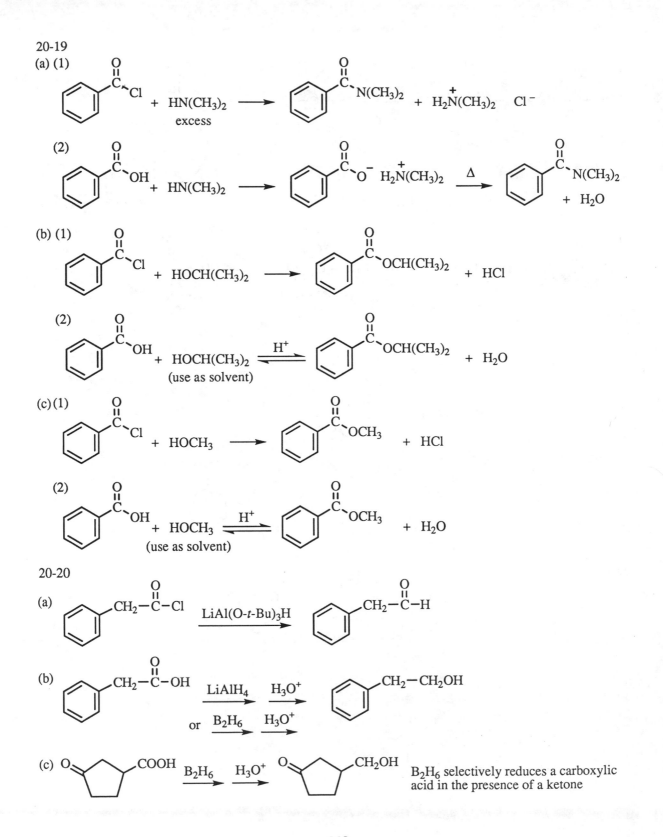

20-19

(a) (1)

(2)

(b) (1)

(2)

(c) (1)

(2)

20-20

(a)

(b)

(c)

B₂H₆ selectively reduces a carboxylic acid in the presence of a ketone

442

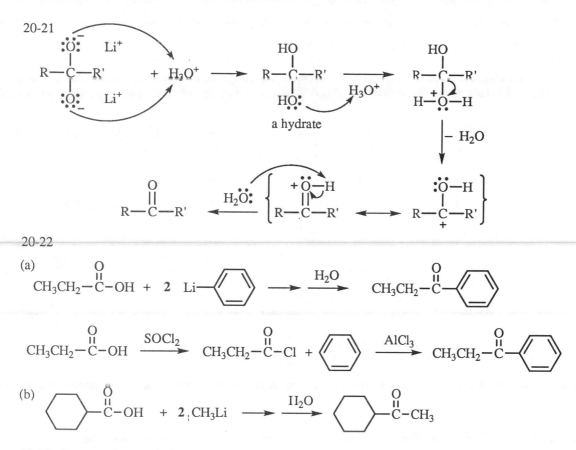

20-21

20-22

(a)

$$CH_3CH_2-\overset{O}{\underset{||}{C}}-OH + 2\ Li-\bigcirc \longrightarrow \overset{H_2O}{\longrightarrow} CH_3CH_2-\overset{O}{\underset{||}{C}}-\bigcirc$$

$$CH_3CH_2-\overset{O}{\underset{||}{C}}-OH \overset{SOCl_2}{\longrightarrow} CH_3CH_2-\overset{O}{\underset{||}{C}}-Cl + \bigcirc \overset{AlCl_3}{\longrightarrow} CH_3CH_2-\overset{O}{\underset{||}{C}}-\bigcirc$$

(b)

$$\bigcirc-\overset{O}{\underset{||}{C}}-OH + 2\ CH_3Li \longrightarrow \overset{II_2O}{\longrightarrow} \bigcirc-\overset{O}{\underset{||}{C}}-CH_3$$

20-23   Please refer to solution 1-19, page 12 of this Solution Manual.

20-24

(a) 3-phenylpropanoic acid
(b) 2-methylbutanoic acid
(c) 2-bromo-3-methylbutanoic acid
(d) 3-methylpentanedioic acid
(e) sodium 2-methylbutanoate
(f) 3-methyl-2-butenoic acid
(g) *trans*-2-methylcyclopentanecarboxylic acid
(h) 2,4,6-trinitrobenzoic acid

20-25

(a) β-phenylpropionic acid
(b) α-methylbutyric acid
(c) α-bromo-β-methylbutyric acid,
   or α-bromoisovaleric acid
(d) β-methylglutaric acid
(e) sodium β-methylbutyrate
(f) β-aminobutyric acid
(g) *o*-bromobenzoic acid
(h) magnesium oxalate

20-26

(a)

$$CH_3-\overset{O}{\underset{||}{C}}-OH$$

(b)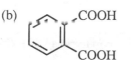

(c)

$$\left(H-\overset{O}{\underset{||}{C}}-O^-\right)_2 Mg^{2+}$$

20-26

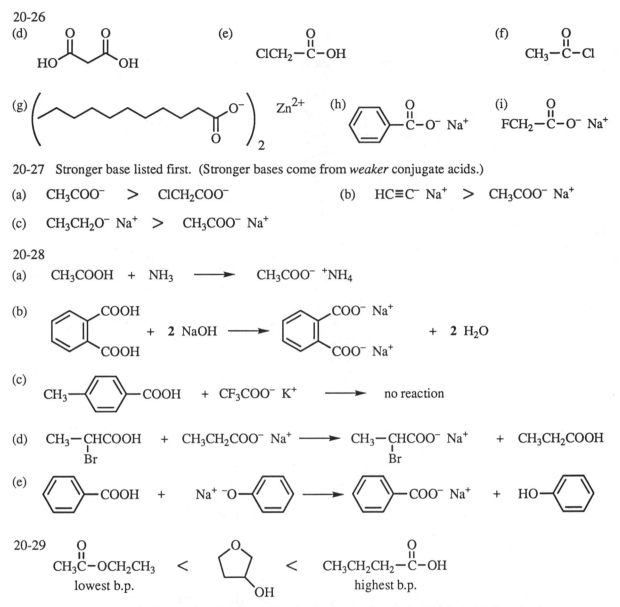

(d)

HO O O OH

(e)

$ClCH_2-\overset{O}{\overset{\|}{C}}-OH$

(f)

$CH_3-\overset{O}{\overset{\|}{C}}-Cl$

(g)

$\left(\right)_2 Zn^{2+}$

(h)

$\overset{O}{\overset{\|}{C}}-O^- \; Na^+$

(i)

$FCH_2-\overset{O}{\overset{\|}{C}}-O^- \; Na^+$

20-27 Stronger base listed first. (Stronger bases come from *weaker* conjugate acids.)

(a) CH_3COO^- > $ClCH_2COO^-$

(b) $HC{\equiv}C^- \; Na^+$ > $CH_3COO^- \; Na^+$

(c) $CH_3CH_2O^- \; Na^+$ > $CH_3COO^- \; Na^+$

20-28

(a) $CH_3COOH \; + \; NH_3 \longrightarrow CH_3COO^- \; {}^+NH_4$

(b)

$\overset{COOH}{\underset{COOH}{}} \; + \; 2 \; NaOH \longrightarrow \overset{COO^- \; Na^+}{\underset{COO^- \; Na^+}{}} \; + \; 2 \; H_2O$

(c)

$CH_3-\overset{}{}-COOH \; + \; CF_3COO^- \; K^+ \longrightarrow$ no reaction

(d) $CH_3-\underset{Br}{CHCOOH} \; + \; CH_3CH_2COO^- \; Na^+ \longrightarrow CH_3-\underset{Br}{CHCOO^-} \; Na^+ \; + \; CH_3CH_2COOH$

(e)

$\overset{}{}-COOH \; + \; Na^+ \; {}^-O-\overset{}{} \longrightarrow \overset{}{}-COO^- \; Na^+ \; + \; HO-\overset{}{}$

20-29

$CH_3\overset{O}{\overset{\|}{C}}-OCH_2CH_3$ <

(ring with OH)

< $CH_3CH_2CH_2-\overset{O}{\overset{\|}{C}}-OH$

lowest b.p.   highest b.p.

The ester cannot form hydrogen bonds and will be the lowest boiling. The alcohol can form hydrogen bonds. The carboxylic acid forms two hydrogen bonds and boils as the dimer, the highest boiling among these three compounds.

20-30 Listed in order of increasing acidity (weakest acid first):

(a) ethanol < phenol < acetic acid
(b) acetic acid < chloroacetic acid < *p*-toluenesulfonic acid
(c) benzoic acid < *m*-nitrobenzoic acid < *o*-nitrobenzoic acid
(d) butyric acid < β-bromobutyric acid < α-bromobutyric acid

20-31 Acetic acid is often used as a test of electronic effects of a series of substituents: they are fairly easily synthesized (or are commercially available), and pK_a values are easily measured by titration.

Substituents on carbon-2 of acetic acid can express only an inductive effect; no resonance effect is possible because the CH_2 is sp^3 hybridized and no pi overlap is possible.

Two conclusions can be drawn from the given pK_a values. First, all four substituents are electron-withdrawing because all four substituted acids are stronger than acetic acid. Second, the magnitude of the electron-withdrawing effect increases in the order: $OH < Cl < CN < NO_2$. (It is always a safe assumption that nitro is the strongest electron-withdrawing group of all the common substituents.)

20-32

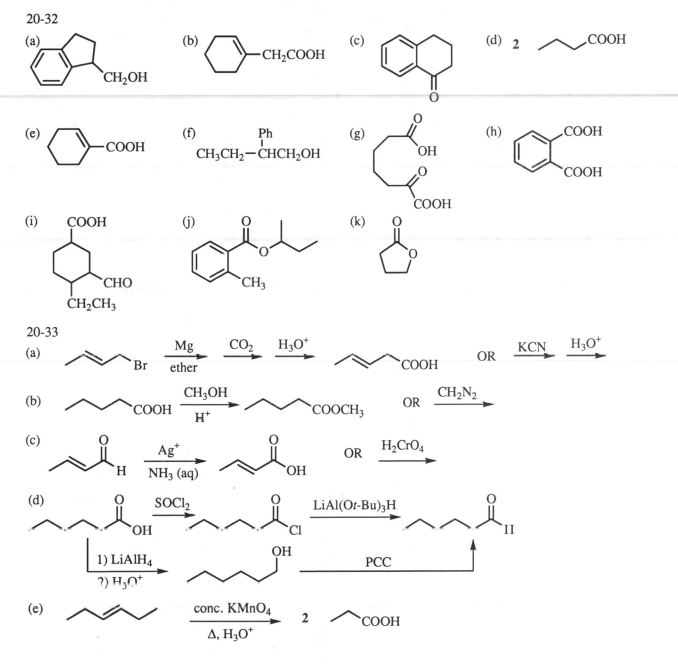

20-33

445

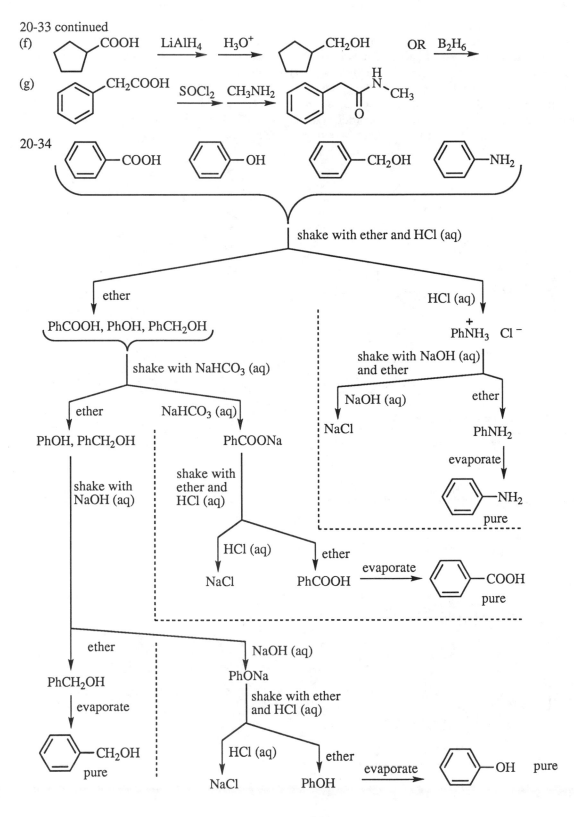

20-33 continued

(f)

(g)

20-34

446

20-35

(a)

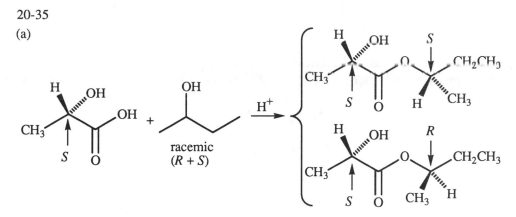

(b) Isomers which are *R,S* and *S,S* are diastereomers.

20-36

(a) PhCH₂CH₂OH $\xrightarrow[\text{pyridine}]{\text{TsCl}}$ $\xrightarrow{\text{KCN}}$ PhCH₂CH₂CN $\xrightarrow[\Delta]{H_3O^+}$ PhCH₂CH₂COOH

$\downarrow$ PBr₃

PhCH₂CH₂Br $\xrightarrow[\text{ether}]{\text{Mg}}$ $\xrightarrow{CO_2}$ $\xrightarrow{H_3O^+}$ PhCH₂CH₂COOH

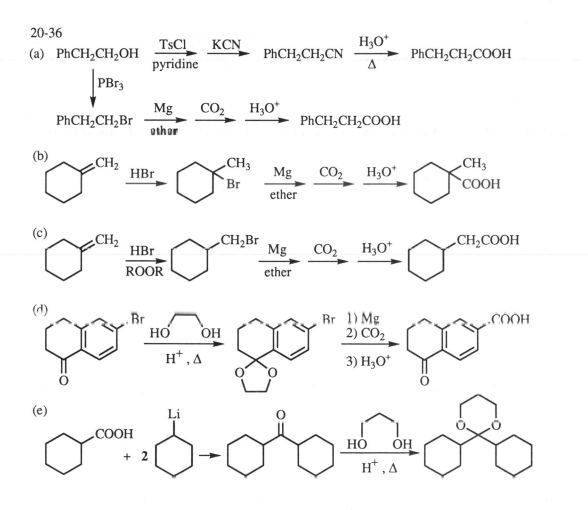

(a) <u>Mass spectrum:</u>

—m/z 152 $\Rightarrow$ molecular ion $\Rightarrow$ molecular weight 152

—m/z 107 $\Rightarrow$ M − 45 $\Rightarrow$ loss of COOH

—m/z 77 $\Rightarrow$ monosubstituted benzene ring,

<u>IR spectrum:</u>

—3400-2400 cm^{-1} , broad $\Rightarrow$ O—H stretch of COOH

—1700 cm^{-1} $\Rightarrow$ C=O

—1240 cm^{-1} $\Rightarrow$ C−O

—1600 cm^{-1} $\Rightarrow$ aromatic C=C

<u>NMR spectrum:</u>

—δ 6.8-7.3, two signals in the ratio of 2H to 3H $\Rightarrow$ monosubstituted benzene ring

—δ 4.6, 2H singlet $\Rightarrow$ CH_2, deshielded

<u>Carbon NMR spectrum:</u>

—δ 170, small peak $\Rightarrow$ carbonyl

—δ 115-157, four peaks $\Rightarrow$ monosubstituted benzene ring; deshielded peak indicates oxygen substitution on the ring

(b) Fragments indicated in the spectra:

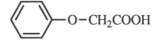

m/z 77

CH$_2$

m/z 14

COOH

m/z 45 and from IR

This appears deceptively simple. The problem is that the mass of these fragments adds to 136, not 152—we are missing 16 mass units $\Rightarrow$ oxygen! Where can the oxygen be? There are only two possibilities:

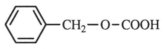

How can we differentiate? Mass spectrometry!

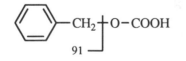

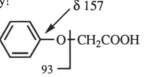

This structure is consistent with the peak at δ 157 in the carbon NMR.

phenoxyacetic acid

The m/z 93 peak in the MS confirms the structure is **phenoxyacetic acid**. The CH_2 is so far downfield in the NMR because it is between two electron-withdrawing groups, the O and the COOH.

(c) The COOH proton is missing from the proton NMR. Either it is beyond 10 and the NMR was not scanned (unlikely), or the peak was broadened beyond detection because of hydrogen bonding with DMSO.

20-38

(a)

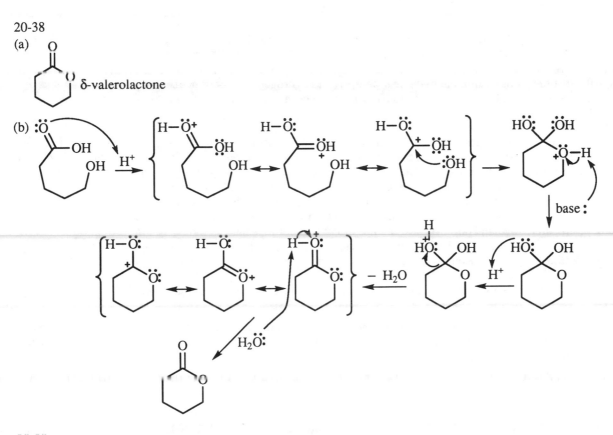

δ-valerolactone

(b)

20-39

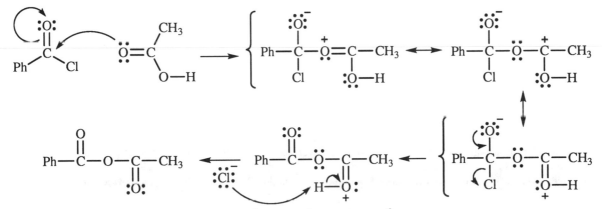

plus two other resonance forms

20-40   A few words about the two types of electronic effects: induction and resonance. Inductive effects are a result of polarized σ bonds, usually because of electronegative atom substituents. Resonance effects work through π systems, requiring overlap of p orbitals to delocalize electrons.

All substituents have an inductive effect compared to hydrogen (the reference). Many groups also have a resonance effect; all that is required to have a resonance effect is that the atom or group have at least one p orbital for overlap.

The most interesting groups have both inductive and resonance effects. In such groups, how can we tell the direction of electron movement, that is, whether a group is electron-donating or electron-withdrawing? And do the resonance and inductive effects reinforce or conflict with each other? We can never "turn off" an inductive effect from a resonance effect; that is, any time a substituent is expressing its resonance effect, it is also expressing its inductive effect. We can minimize a group's inductive effect by moving it farther away; inductive effects decrease with distance. The other side of the coin is more accessible to the experimenter: we can "turn off" a resonance effect in order to isolate an inductive effect. We can do this by interrupting a conjugated π system by inserting an sp^3-hybridized atom, or by making resonance overlap impossible for steric reasons (steric inhibition of resonance).

These three problems are examples of separating inductive effects from resonance effects.

(a) In electrophilic aromatic substitution, the phenyl substituent is an ortho,para-director because it can stabilize the intermediate from electrophilic attack at the ortho and para positions. The phenyl substituent is electron-donating *by resonance*.

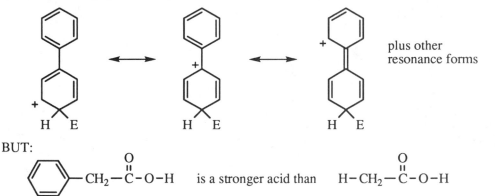

plus other
resonance forms

BUT:

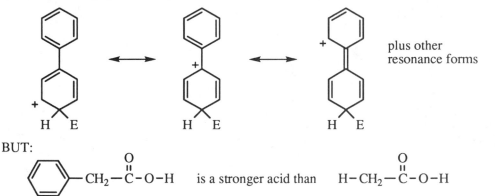 is a stronger acid than

The greater acidity of phenylacetic acid shows that the phenyl substituent is electron-withdrawing, thereby stabilizing the product carboxylate's negative charge. Does this contradict what was said above? Yes and no: What is different is that, since there is no p-orbital overlap between the phenyl group and the carboxyl group because of the CH_2 group in between, the increased acidity must be from a pure *inductive effect*. This structure isolates the inductive effect (which can't be "turned off") from the resonance effect of the phenyl group.

We can conclude three things: (1) phenyl is electron-withdrawing by induction; (2) phenyl is (in this case) electron-donating by resonance; (3) for phenyl, the resonance effect is stronger than the inductive effect (since it is an ortho,para-director).

20-40 continued

(b) The simpler case first—induction only:

$$CH_3O-CH_2-\overset{\overset{\displaystyle O}{\|}}{C}-O-H \quad \text{is a stronger acid than} \quad H-CH_2-\overset{\overset{\displaystyle O}{\|}}{C}-O-H$$

There is no resonance overlap between the methoxy group and the carboxyl group, so this is a pure inductive effect. The methoxy substituent increases the acidity, so methoxy must be electron-withdrawing by induction. This should come as no surprise as oxygen is the second most electronegative element.

The anomaly comes in the decreased acidity of 4-methoxybenzoic acid:

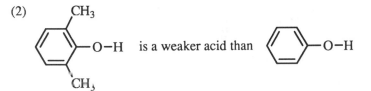

CH_3O-⬡$-\overset{\overset{\displaystyle O}{\|}}{C}-O-H$ is a weaker acid than $H-$⬡$-\overset{\overset{\displaystyle O}{\|}}{C}-O-H$

Through resonance, a pair of electrons from the methoxy oxygen can be donated through the benzene ring to the carboxyl group—a stabilizing effect. However, this electron donation *destabilizes* the carboxylate anion as there is already a negative charge on the carboxyl group; the resonance donation intensifies the negative charge. Since the product of the equilibrium would be destabilized relative to the starting material, the proton donation would be less favorable, which we define as a weaker acid.

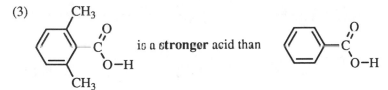

Methoxy is another example of a group which is electron-withdrawing by induction but electron-donating by resonance.

(c) This problem gives three pieces of data to interpret:

(1) $CH_3-CH_2-\overset{\overset{\displaystyle O}{\|}}{C}-O-H$ is a weaker acid than $H-CH_2-\overset{\overset{\displaystyle O}{\|}}{C}-O-H$

Interpretation: the methyl group is electron-donating by induction.

(2)

⬡ with CH₃ groups, —O—H   is a weaker acid than   ⬡—O—H

Interpretation: the methyl group is electron-donating by induction. This interpretation is consistent with (1), as expected, since methyl cannot have any resonance effect.

(3)

⬡ with CH₃ groups, —C(=O)O—H   is a stronger acid than   ⬡—C(=O)O—H

Interpretation: this is the anomaly. Contradictory to the data in (1) and (2), by putting on two methyl groups, the substituent seems to have become electron-withdrawing instead of electron-donating. How?

Quick! Turn the page!

20-40 continued

Steric inhibition of resonance! In benzoic acid, the phenyl ring and the carboxyl group are all in the same plane, and benzene is able to donate electrons by resonance overlap through parallel p orbitals. This stabilizes the starting acid (and destabilizes the carboxylate anion) and makes the acid weaker than it would be without resonance.

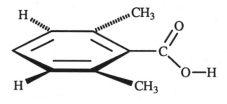 plus other resonance forms

Putting substituents at the 2- and 6-positions prevents the carboxyl or carboxylate from coplanarity with the ring. Resonance is interrupted, and now the carboxyl group sees a phenyl substituent which cannot stabilize the acid through resonance; the stabilization of the acid is lost. At the same time, the *electron-withdrawing inductive effect* of the benzene ring stabilizes the carboxylate anion. These two effects work together to make this acid unusually strong. (Apparently, the slight electron-donating inductive effect of the methyls is overpowered by the stronger electron-withdrawing inductive effect of the benzene ring.)

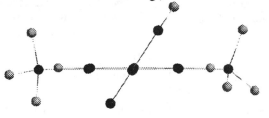

COOH group is perpendicular to the plane of the benzene ring— no resonance interaction.

this three-dimensional view down the C-C bond between the COOH and the benzene ring shows that COOH is twisted out of the benzene plane

20-41
(a)

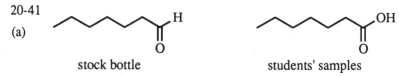

stock bottle   students' samples

(b) The spectrum of the students' samples shows the carboxylic acid present. Contact with oxygen from the air oxidized the sensitive aldehyde group to the acid.

(c) Storing the aldehyde in an inert atmosphere like nitrogen or argon prevents oxidation. Freshly prepared unknowns will avoid the problem.

20-42

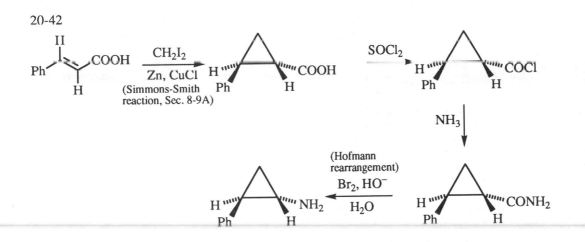

20-43

Spectrum A: $C_9H_{10}O_2$ ⇒ 5 elements of unsaturation

δ 11.8, 1H ⇒ COOH

δ 7.3, 5H ⇒ monosubstituted benzene ring

δ 3.8, 1H quartet ⇒ CHCH₃

δ 1.6, 3H doublet ⇒ CHCH₃

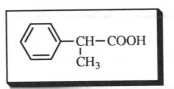

Spectrum B: $C_4H_6O_2$ ⇒ 2 elements of unsaturation

δ 12.1, 1H ⇒ COOH

δ 6.2, 1H singlet ⇒ H−C=C

δ 5.7, 1H singlet ⇒ H−C=C

δ 1.9, 3H singlet ⇒ vinyl CH₃ with no H neighbors CH₃−C=C

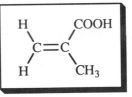

Spectrum C: $C_6H_{10}O_2$ ⇒ 2 elements of unsaturation

δ 12.0, 1H ⇒ COOH

δ 7.0, 1H multiplet ⇒ H−C−C=COOH

δ 5.7, 1H doublet ⇒ C=C−COOH

δ 2.2-0.8 ⇒ CH₂CH₂CH₃

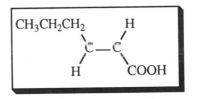

must be *trans* due to large coupling
constant in doublet at δ 5.7

21-1 IUPAC name first; then common name

(a) isobutyl benzoate (both IUPAC and common)
(b) phenyl methanoate; phenyl formate
(c) methyl 2-phenylpropanoate; methyl α-phenylpropionate
(d) N-phenyl-3-methylbutanamide; β-methylbutyranilide
(e) N-benzylethanamide; N-benzylacetamide
(f) 3-hydroxybutanenitrile; β-hydroxybutyronitrile
(g) 3-methylbutanoyl bromide; isovaleryl bromide
(h) dichloroethanoyl chloride; dichloroacetyl chloride
(i) 2-methylpropanoic methanoic anhydride; isobutyric formic anhydride
(j) cyclopentyl cyclobutanecarboxylate (both IUPAC and common)
(k) 5-hydroxyhexanoic acid lactone; δ-caprolactone
(l) N-cyclopentylbenzamide (both IUPAC and common)
(m) propanedioic anhydride; malonic anhydride
(n) 1-hydroxycyclopentanecarbonitrile; cyclopentanone cyanohydrin
(o) cis-4-cyanocyclohexanecarboxylic acid; no common name
(p) 3-bromobenzoyl chloride; m-bromobenzoyl chloride
(q) N-methyl-5-aminoheptanoic acid lactam; no common name
(r) N-ethanoylpiperidine; N-acetylpiperidine

21-2 An aldehyde has a C—H absorption (usually 2 peaks) at 2700-2800 cm^{-1}. A carboxylic acid has a strong, broad absorption between 2400-3400 cm^{-1}. The spectrum of methyl benzoate has no peaks in this region.

21-3 The C—O single bond stretch in ethyl octanoate appears at 1170 cm^{-1}, while methyl benzoate shows this absorption at 1120 and 1280 cm^{-1}.

21-4

(a) acid chloride: single C=O peak at 1800 cm^{-1}; no other carbonyl comes so high

(b) primary amide: C=O at 1650 cm^{-1} and two N—H peaks between 3200-3400 cm^{-1}

(c) anhydride: two C=O absorptions at 1750 and 1820 cm^{-1}

21-5

(a) The formula C$_3$H$_5$NO has two elements of unsaturation. The IR spectrum shows two peaks between 3200-3400 cm^{-1}, an NH$_2$ group. The strong peak at 1670 cm^{-1} is a C=O, and the peak at 1610 cm^{-1} is a C=C. This accounts for all of the atoms.

$$H_2C = CH - \overset{\overset{\displaystyle O}{\|}}{C} - NH_2$$

The NMR corroborates the assignment. The 1H multiplet at δ 5.8 is the vinyl H next to the carbonyl. The 2H multiplet at δ 6.3 is the vinyl hydrogen pair on carbon-3. The 2H singlet at δ 4.8 is the amide hydrogens.

21-5 continued

(b) The formula $C_5H_8O_2$ has two elements of unsaturation. The IR spectrum shows no significant OH, so this compound is neither an alcohol nor a carboxylic acid. The strong peak at 1730 cm^{-1} is likely an ester carbonyl. The C—O appears between 1050-1250 cm^{-1}. The IR shows no C=C absorption, so the other element of unsaturation is likely a ring. The carbon NMR spectrum shows the carbonyl carbon at δ 171, the C—O carbon at δ 69, and three more carbons in the aliphatic region, but no carbons in the vinyl region between δ 100-150, so there can be no C=C. The proton NMR shows multiplets of 2H at δ 4.3 and 2.5, most likely CH_2 groups next to oxygen and carbonyl respectively.

The only structure with an ester, four CH_2 groups, and a ring, is δ-valerolactone:

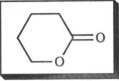

21-6

(a)

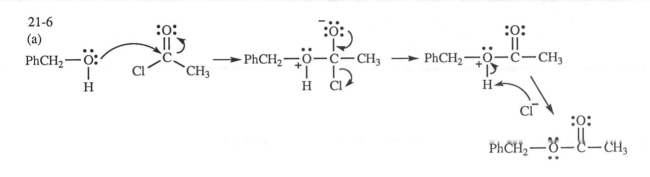

(b)

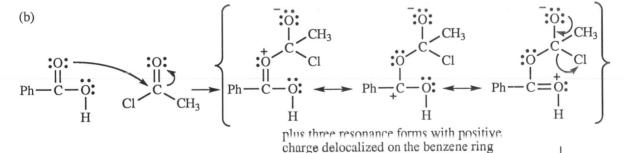

the carbonyl oxygen is more nucleophilic than the single-bonded oxygen because the product is resonance stabilized

plus three resonance forms with positive charge delocalized on the benzene ring

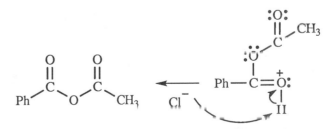

plus all the resonance forms as above

21-6 continued

(c)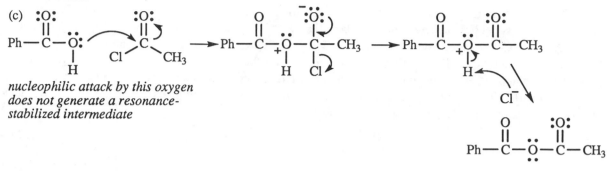

nucleophilic attack by this oxygen
does not generate a resonance-
stabilized intermediate

21-7 **Figure 21-9 is critical!** Reactions which go from a more reactive functional group to a less reactive functional group ("downhill reactions") will occur readily.
(a) amide to acid chloride will NOT occur—it is an "uphill" transformation
(b) acid chloride to amide will occur rapidly
(c) amide to ester will NOT occur—another "uphill" transformation
(d) acid chloride to anhydride will occur rapidly
(e) anhydride to amide will occur rapidly

21-8

(a)

$$CH_3CH_2-\overset{O}{\overset{\|}{C}}-Cl \ + \ HOCH_2CH_3 \longrightarrow CH_3CH_2-\overset{O}{\overset{\|}{C}}-OCH_2CH_3 \ + \ HCl$$

(b)

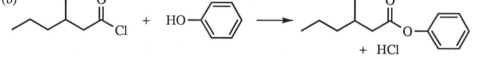

+ HCl

(c)

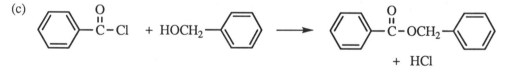

+ HCl

(d)

21-9

(a)

$$H_3C-\overset{O}{\overset{\|}{C}}-Cl \ + \ HN(CH_3)_2 \longrightarrow H_3C-\overset{O}{\overset{\|}{C}}-N(CH_3)_2 \ + \ HCl$$

(b)

$$H_3C-\overset{O}{\overset{\|}{C}}-Cl \ + \ H_2N\!-\!\!\bigcirc \longrightarrow H_3C-\overset{O}{\overset{\|}{C}}-NH\!-\!\!\bigcirc \ + \ HCl$$

456

21-9 continued

(c)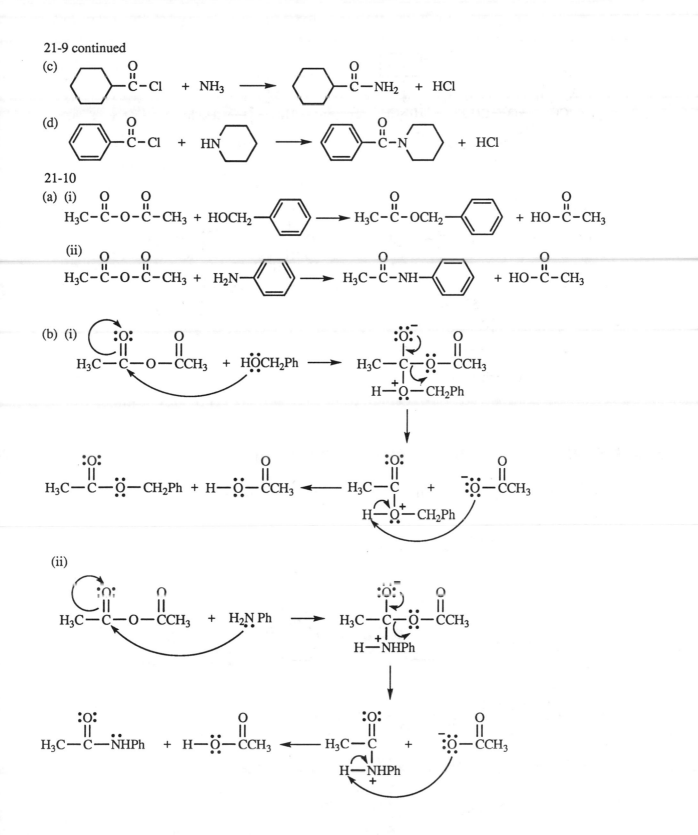

(d)

21-10

(a) (i)

(ii)

(b) (i)

(ii)

457

21-11

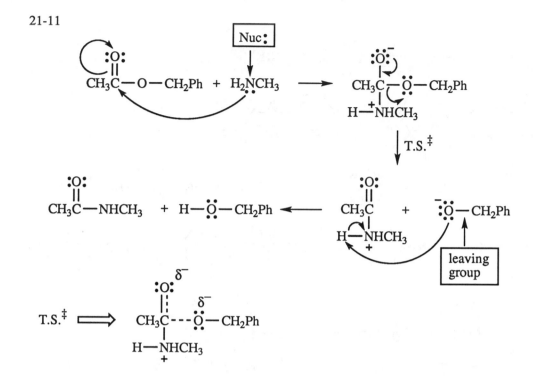

21-12

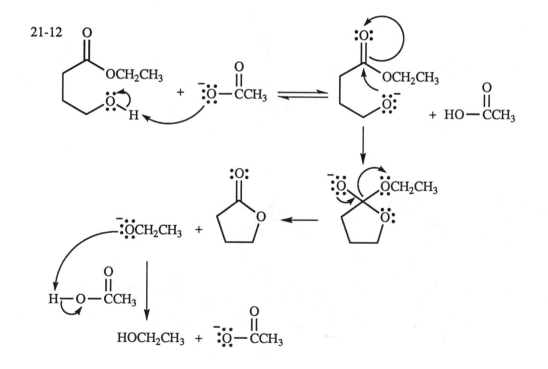

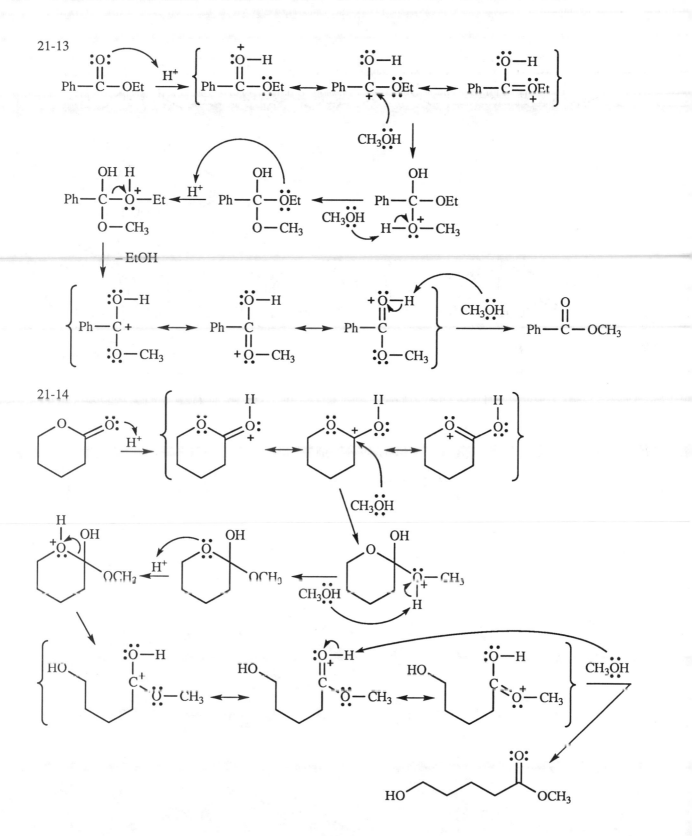

21-15

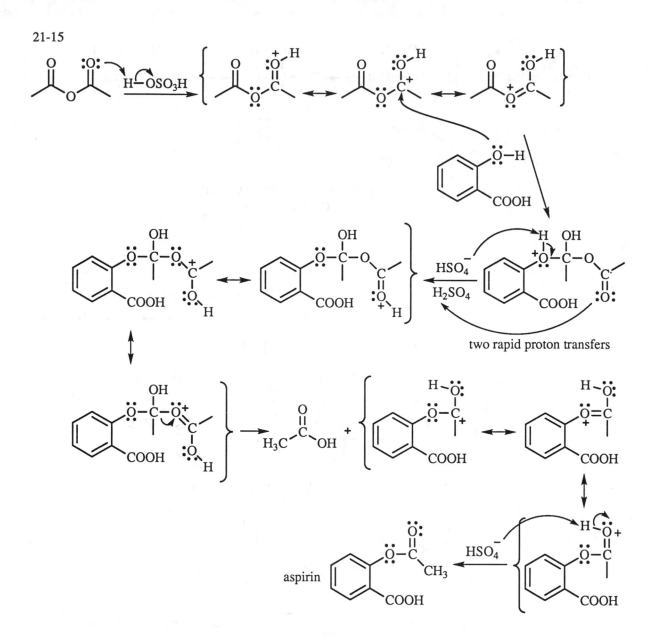

two rapid proton transfers

aspirin

460

21-16 The asterisk (*) will denote ^{18}O.

(a)

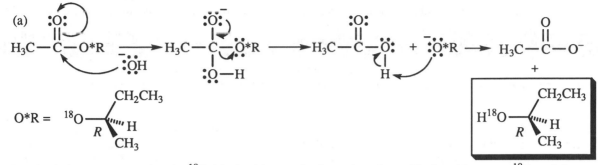

$$O*R = \quad {}^{18}O\overset{CH_2CH_3}{\underset{CH_3}{\overset{|}{\underset{|}{C}}}}\!\!\!\!\!\!\!\!{}_{\textstyle R}{}^{\textstyle,,,,H}$$

The alcohol product contains the ^{18}O label, with none in the carboxylate. The bond between ^{18}O and the tetrahedral carbon with *(R)* configuration did not break, so the configuration is retained.

(b) The products are identical regardless of mechanism.

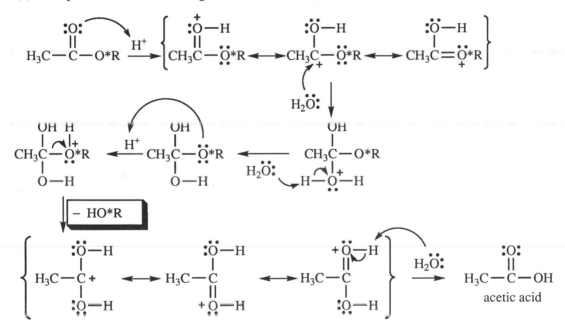

(c) The ^{18}O has 2 more neutrons in its nucleus than ^{16}O. Mass spectra of these products would show the molecular ion of acetic acid at its standard value of m/z 60, whereas the molecular ion of 2-butanol would appear at m/z 76 instead of m/z 74, proving that the heavy isotope of oxygen went with the alcohol. This demonstrates that the bond between oxygen and the carbonyl carbon is broken, not the bond between the oxygen and the alkyl carbon.

To show if the alcohol was chiral or racemic, measuring its optical activity in a polarimeter and comparing to known values would prove its configuration. (The heavy oxygen isotope has a negligible effect on optical rotation.)

21-17

(a) A catalyst is defined as a chemical species that speeds a reaction but is not consumed in the reaction. In the acidic hydrolysis, acid is used in the first and fourth steps of the mechanism but is regenerated in the third and last steps. Acid is not consumed; the final concentration of acid is the same as the initial concentration. In the basic hydrolysis, however, the hydroxide that initially attacks the carbonyl is never regenerated. An alkoxide leaves from the carbonyl, but it quickly neutralizes the carboxylic acid. For every molecule of ester, one molecule of hydroxide is consumed; the base *promotes* the reaction but does not *catalyze* the reaction.

(b) Basic hydrolysis is not reversible. Once an ester molecule is hydrolyzed in base, the carboxylate cannot form an ester. Acid catalysis, however, is an equilibrium: the mixture will always contain some ester, and the yield will never be as high as in basic hydrolysis. Second, long chain fatty acids are not soluble in water until they are ionized; they are soluble only as their sodium salts (soap). Basic hydrolysis is preferred for higher yield and greater solubility of the product.

21-18

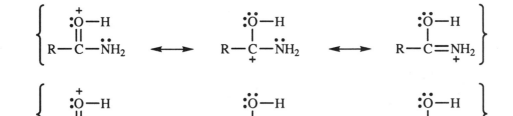

21-19

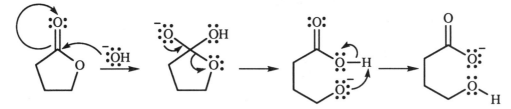

21-20

(a)

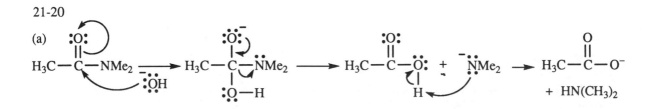

21-20

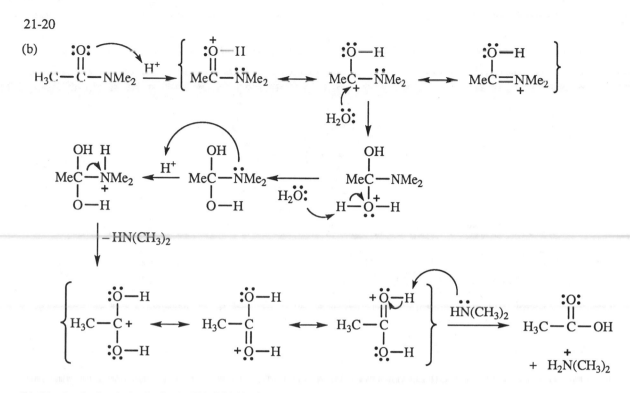

21-21 In the basic hydrolysis (21-20(a)), the step that drives the reaction to completion is the final step, the deprotonation of the carboxylic acid by the amide anion. In the acidic hydrolysis (21-20(b)), protonation of the amine by acid is exothermic and it prevents the reverse reaction by tying up the pair of electrons on the nitrogen so that the amine is no longer nucleophilic.

21-22

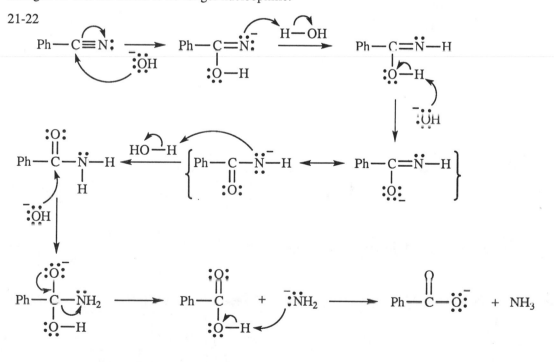

21-23

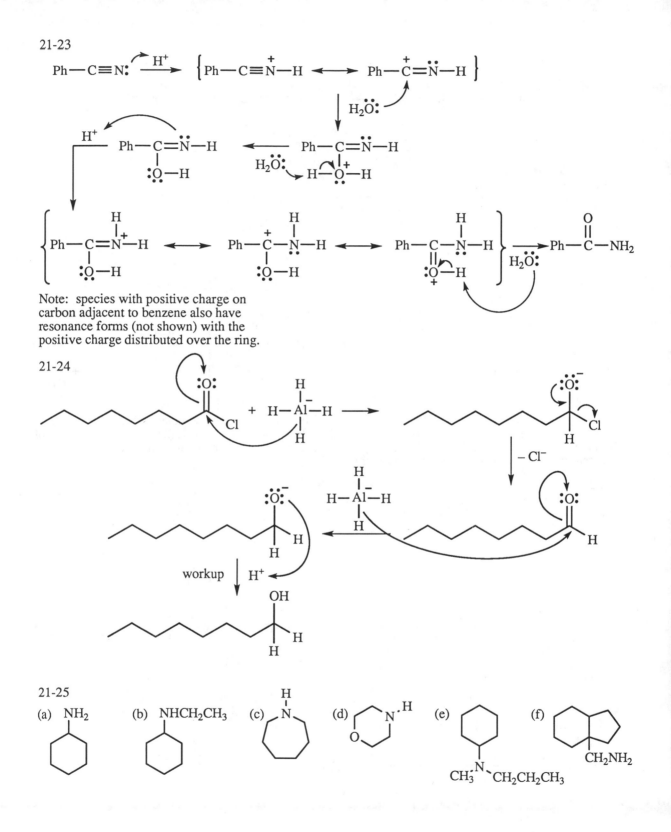

Note: species with positive charge on carbon adjacent to benzene also have resonance forms (not shown) with the positive charge distributed over the ring.

21-24

workup

21-25

(a) NH₂

(b) NHCH₂CH₃

(c)

(d)

(e)

(f) CH₂NH₂

21-26

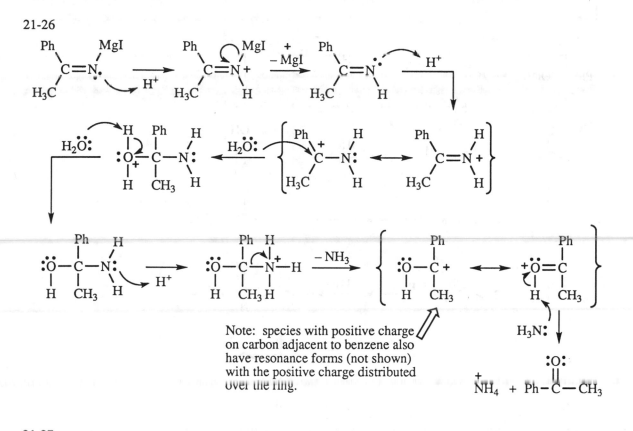

Note: species with positive charge on carbon adjacent to benzene also have resonance forms (not shown) with the positive charge distributed over the ring.

21-27

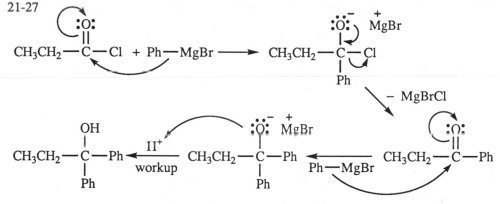

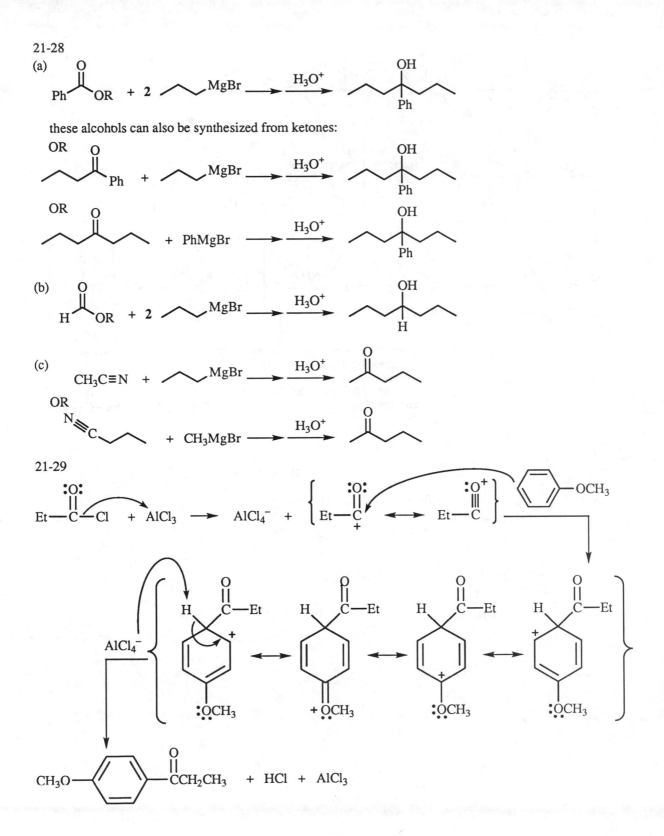

21-30

(a)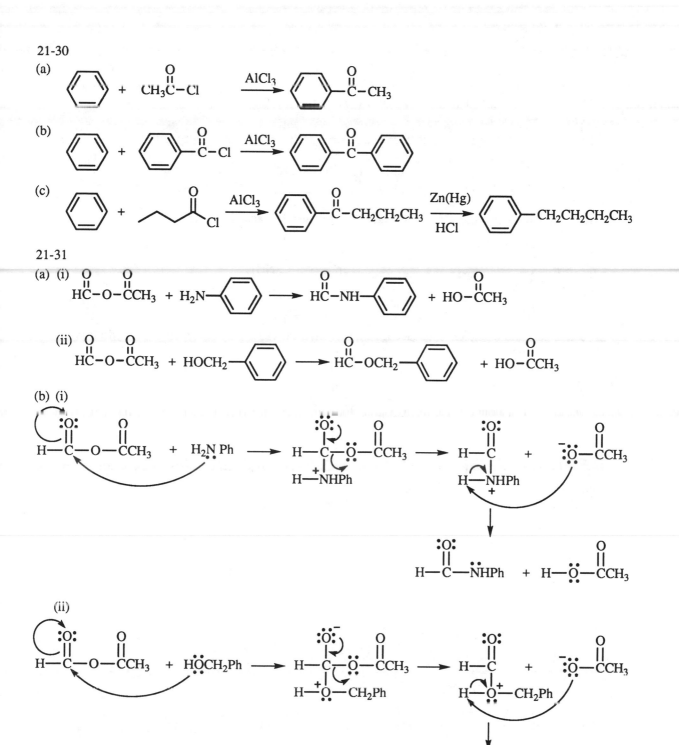

(b)

(c)

21-31

(a) (i)

(ii)

(b) (i)

(ii)

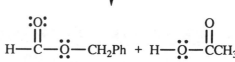

21-32

(a)

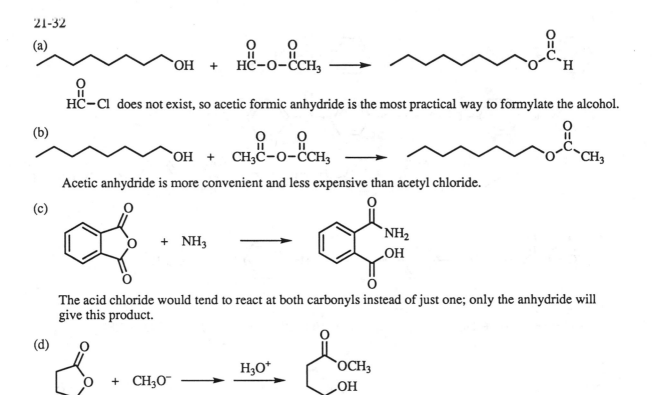

$$\text{HC-Cl}$$ does not exist, so acetic formic anhydride is the most practical way to formylate the alcohol.

(b)

Acetic anhydride is more convenient and less expensive than acetyl chloride.

(c)

+ NH₃ →

The acid chloride would tend to react at both carbonyls instead of just one; only the anhydride will give this product.

(d)

+ CH₃O⁻ → H₃O⁺ →

The acid chloride would tend to react at both carbonyls instead of just one; only the anhydride will give this product.

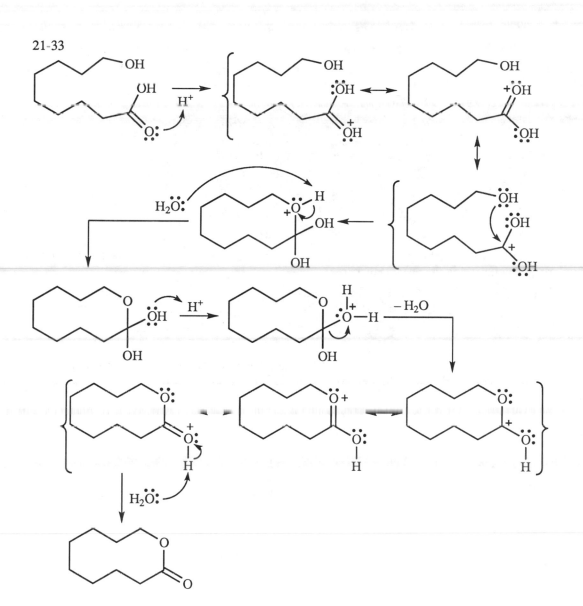

21-33

21-34

(a)

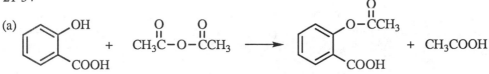

Generally, acetic anhydride is the optimum reagent for the preparation of acetate esters. Acetyl chloride would also react with the carboxylic acid to form a mixed anhydride.

(b)

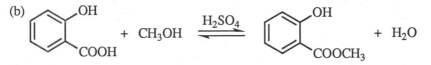

Fischer esterification works well to prepare simple carboxylic esters. The diazomethane method would also react with the phenol, making the phenyl ether.

(c)

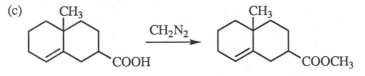

Fischer esterification would make the ester, but in the process, the acidic conditions would risk migrating the double bond into conjugation with the carbonyl group. The diazomethane reaction is run under neutral conditions where double bond migration will not occur.

21-35 Syntheses may have more than one correct approach.

(a)

$$Ph-\overset{O}{\overset{\|}{C}}-OCH_3 + 2\ PhMgBr \xrightarrow{\text{ether}} \xrightarrow{H_3O^+} Ph-\overset{OH}{\overset{|}{\underset{|}{C}}}-Ph$$
$$\underset{Ph}{}$$

(b)

$$H-\overset{O}{\overset{\|}{C}}-OCH_2CH_3 + 2\ PhCH_2MgBr \xrightarrow{\text{ether}} \xrightarrow{H_3O^+} H-\overset{OH}{\overset{|}{\underset{CH_2Ph}{C}}}-CH_2Ph$$

(c)

$$Ph-\overset{O}{\overset{\|}{C}}-OCH_3 + H_2NCH_2CH_3 \longrightarrow Ph-\overset{O}{\overset{\|}{C}}-NHCH_2CH_3$$

(d)

$$H-\overset{O}{\overset{\|}{C}}-OCH_2CH_3 + 2\ PhMgBr \xrightarrow{\text{ether}} \xrightarrow{H_3O^+} H-\overset{OH}{\overset{|}{\underset{Ph}{C}}}-Ph$$

(e)

$$Ph-\overset{O}{\overset{\|}{C}}-OCH_3 + LiAlH_4 \xrightarrow{\text{ether}} \xrightarrow{H_3O^+} PhCH_2OH$$

(f)

$$Ph-\overset{O}{\overset{\|}{C}}-OCH_3 \xrightarrow{H_3O^+} Ph-\overset{O}{\overset{\|}{C}}-OH + CH_3OH$$

21-35 continued

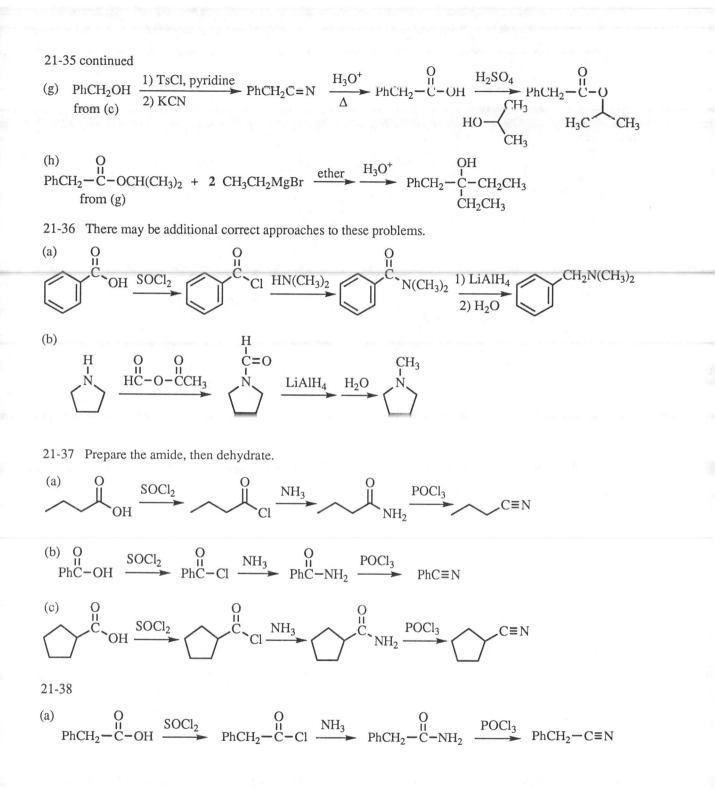

(g) PhCH₂OH $\xrightarrow[\text{2) KCN}]{\text{1) TsCl, pyridine}}$ PhCH₂C≡N $\xrightarrow[\Delta]{\text{H}_3\text{O}^+}$ PhCH₂—C—OH $\xrightarrow{\text{H}_2\text{SO}_4}$ PhCH₂—C—O
from (c)

(h) PhCH₂—C—OCH(CH₃)₂ + **2** CH₃CH₂MgBr $\xrightarrow{\text{ether}}$ $\xrightarrow{\text{H}_3\text{O}^+}$ PhCH₂—C—CH₂CH₃
from (g)

21-36 There may be additional correct approaches to these problems.

(a)

(b)

21-37 Prepare the amide, then dehydrate.

(a)

(b) PhC—OH $\xrightarrow{\text{SOCl}_2}$ PhC—Cl $\xrightarrow{\text{NH}_3}$ PhC—NH₂ $\xrightarrow{\text{POCl}_3}$ PhC≡N

(c)

21-38

(a) PhCH₂—C—OH $\xrightarrow{\text{SOCl}_2}$ PhCH₂—C—Cl $\xrightarrow{\text{NH}_3}$ PhCH₂—C—NH₂ $\xrightarrow{\text{POCl}_3}$ PhCH₂—C≡N

471

21-38 continued

(b)

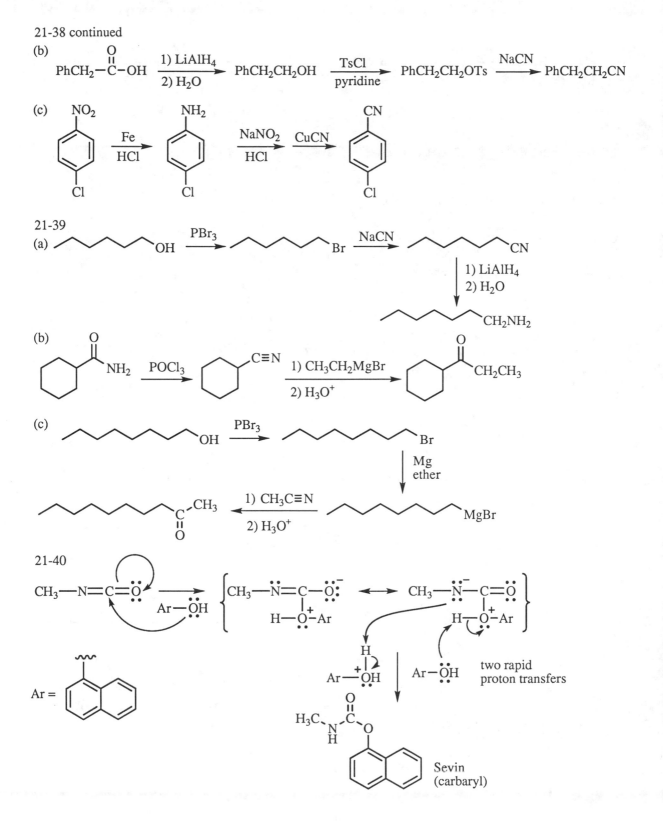

21-39

(a)

(b)

(c)

21-40

Ar =

21-41

(a)

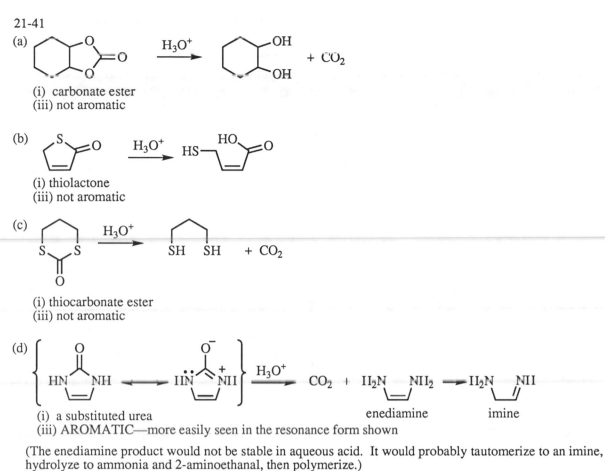

(i) carbonate ester
(iii) not aromatic

(b)

(i) thiolactone
(iii) not aromatic

(c)

(i) thiocarbonate ester
(iii) not aromatic

(d)

(i) a substituted urea
(iii) AROMATIC—more easily seen in the resonance form shown

(The enediamine product would not be stable in aqueous acid. It would probably tautomerize to an imine, hydrolyze to ammonia and 2-aminoethanal, then polymerize.)

(e) At first glance, this AROMATIC compound does not appear to be an acid derivative. Like any enol, however, its tautomer must be considered.

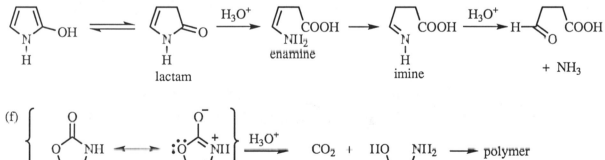

(f)

(i) a carbamate or urethane
(iii) AROMATIC—more easily seen in the resonance form

21-42

(a) Carbamoyl phosphate is a mixed anhydride between carbamic acid and phosphoric acid. It would react easily with an amine to form an amide bond (technically, a urea), with phosphate as the leaving group.
(b) N-Carbamoylaspartate has a carbonyl with two nitrogens on either side; this group is a urea derivative.
(c) The NH_2 group on one end replaces the OH of a carboxylic acid on the other end; this reaction is a nucleophilic acyl substitution.
(d) Orotate is aromatic as can be seen readily in the tautomer. It is called a "pyrimidine base" because of its structural similarity to the pyrimidine ring.

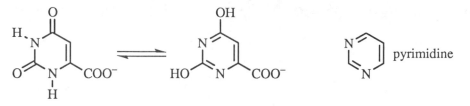

21-43   Please refer to solution 1-19, page 12 of this Solutions Manual.

21-44

(a) 3-methylpentanoyl chloride
(b) benzoic formic anhydride
(c) acetanilide; N-phenylethanamide
(d) N-methylbenzamide
(e) phenyl acetate; phenyl ethanoate
(f) methyl benzoate

(g) benzonitrile
(h) 4-phenylbutane nitrile; γ-phenylbutyronitrile
(i) dimethyl isophthalate, or dimethyl benzene-1,3-dicarboxylate
(j) N,N-diethyl-3-methylbenzamide
(k) 4-hydroxypentanoic acid lactone; γ-valerolactone
(l) 3-aminopentanoic acid lactam; β-valerolactam

21-45

(a) O
      ‖
   PhC−OCH₂CH₃

(b) O   O
      ‖   ‖
   PhC−O−CCH₃

(c) O
      ‖
   PhC−N
         |
         H

(d) [structure with H_3CO and Ph]

(e) OH
      |
   Ph−C−Ph
      |
      Ph

21-46 When a carboxylic acid is treated with a basic reagent, the base removes the acidic proton rather than attacking at the carbonyl (proton transfers are much faster than formation or cleavage of other types of bonds). Once the carboxylate anion is formed, the carbonyl is no longer susceptible to nucleophilic attack: nucleophiles do not attack sites of negative charge. By contrast, in acidic conditions, the protonated carbonyl has a positive charge and is activated to nucleophilic attack.

basic conditions

$$R−\overset{O}{\overset{\|}{C}}−OH \quad + \quad {}^-OR' \longrightarrow R−\overset{O}{\overset{\|}{C}}−O^- \quad + \quad HOR'$$

anion—not susceptible
to nucleophilic attack

acidic conditions

$$R−\overset{O}{\overset{\|}{C}}−OH \quad + \quad H^+ \longrightarrow R−\overset{OH}{\overset{|}{\underset{+}{C}}}−OH$$

rapidly attacked
by R'OH nucleophile

474

21-47

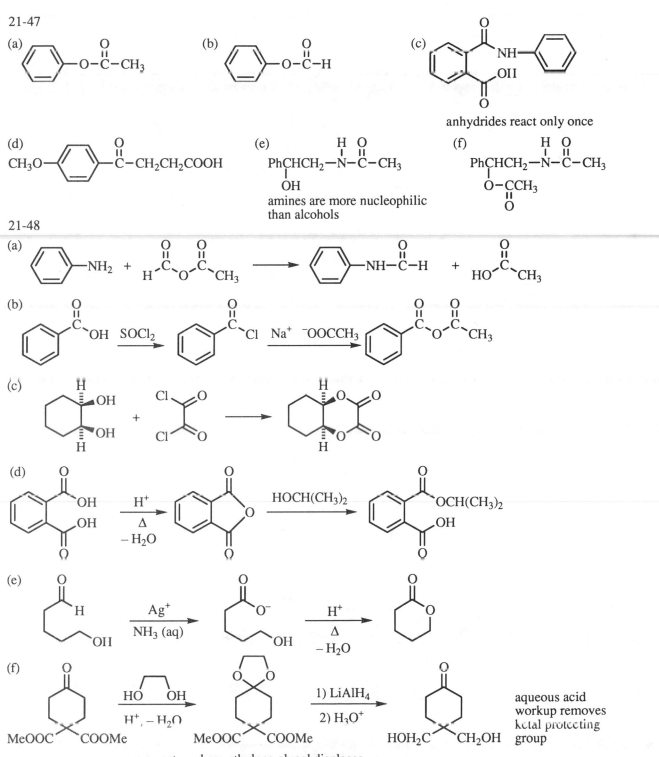

(a) [phenyl] —O—C(=O)—CH₃

(b) [phenyl] —O—C(=O)—H

(c) [benzene ring with C(=O)—NH—phenyl and C(=O)OH]
anhydrides react only once

(d) CH₃O—[benzene]—C(=O)—CH₂CH₂COOH

(e) PhCHCH₂—N(H)—C(=O)—CH₃
      |
      OH
amines are more nucleophilic than alcohols

(f) PhCHCH₂—N(H)—C(=O)—CH₃
      |
      O—CCH₃
          ‖
          O

21-48

(a) [aniline]—NH₂ + H—C(=O)—O—C(=O)—CH₃ ⟶ [phenyl]—NH—C(=O)—H + HO—C(=O)—CH₃

(b) [benzoic acid]—C(=O)OH →SOCl₂→ [benzoyl chloride]—C(=O)Cl →Na⁺ ⁻OOCCH₃→ [benzoic acetic anhydride]—C(=O)—O—C(=O)—CH₃

(c) [cyclohexane-diol cis] + Cl—C(=O)—C(=O)—Cl ⟶ [fused bicyclic dioxodioxine]

(d) [phthalic acid] →H⁺, Δ, −H₂O→ [phthalic anhydride] →HOCH(CH₃)₂→ [monoisopropyl phthalate]—C(=O)—OCH(CH₃)₂ and C(=O)—OH

(e) [5-hydroxypentanal] →Ag⁺ / NH₃ (aq)→ [carboxylate + OH] →H⁺, Δ, −H₂O→ [δ-valerolactone]

(f) [4-oxocyclohexane-1,1-dicarboxylate MeOOC COOMe] →HO—OH / H⁺, −H₂O→ [ketal MeOOC COOMe] →1) LiAlH₄ 2) H₃O⁺→ [4-oxocyclohexane with HOH₂C CH₂OH]
aqueous acid workup removes ketal protecting group

any ester where ethylene glycol displaces methanol will be reduced with LiAlH₄

475

21-49

(a)

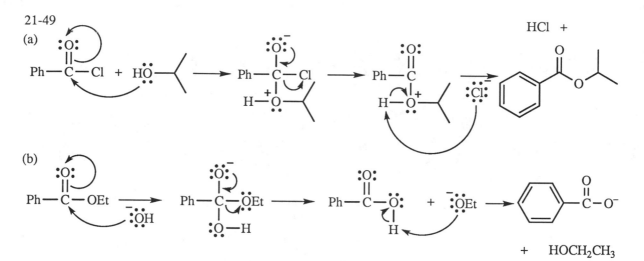

HCl +

(b)

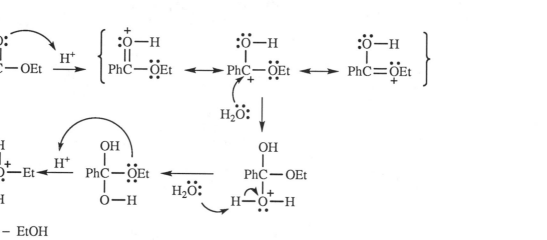

+ HOCH₂CH₃

(c)

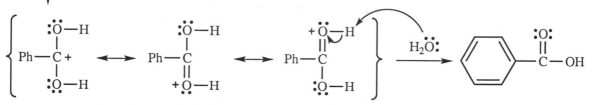

Note: species with positive charge
on carbon adjacent to benzene also
have resonance forms (not shown)
with the positive charge distributed
over the ring.

21-49 continued

(d)

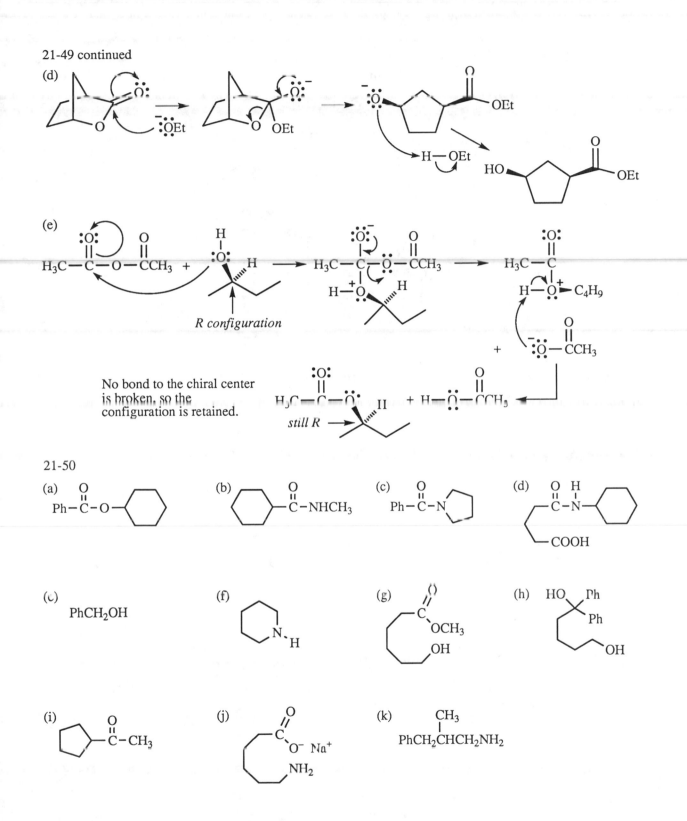

(e)

No bond to the chiral center is broken, so the configuration is retained.

still R →

21-50

(a) Ph—C(=O)—O—(cyclohexyl)

(b) (cyclohexyl)—C(=O)—NHCH₃

(c) Ph—C(=O)—N(pyrrolidine)

(d) C(=O)—N—H (cyclohexyl), with —COOH

(e) PhCH₂OH

(f) (piperidine) N—H

(g) C(=O)—OCH₃ ring with OH

(h) HO, Ph, Ph with OH

(i) (cyclopentyl)—C(=O)—CH₃

(j) C(=O)—O⁻ Na⁺ with NH₂

(k) PhCH₂CHCH₂NH₂ with CH₃

477

21-51 Products after adding dilute acid in the workup:

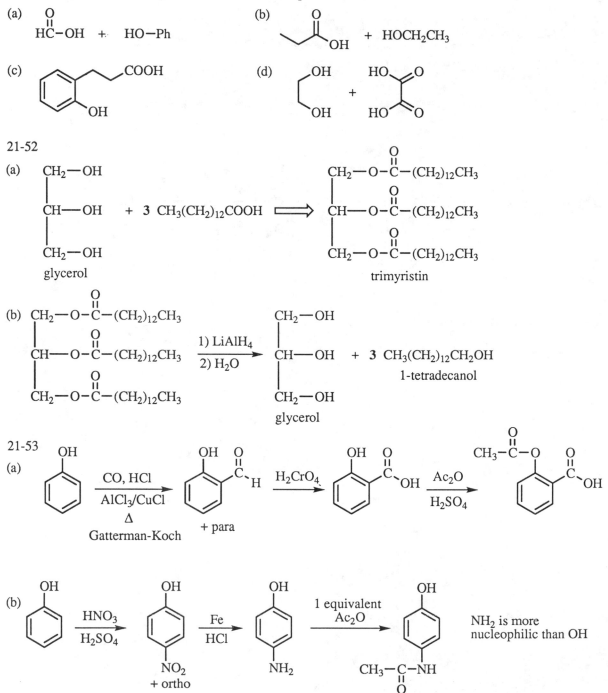

(a)

HC–OH   +   HO–Ph

(b)

+ HOCH₂CH₃

(c)

(d)

21-52

(a)   glycerol   + 3 CH₃(CH₂)₁₂COOH ⟹ trimyristin

(b)   1) LiAlH₄   2) H₂O   →   glycerol + 3 CH₃(CH₂)₁₂CH₂OH   1-tetradecanol

21-53

(a)   CO, HCl / AlCl₃/CuCl / Δ   Gatterman-Koch   + para   →   H₂CrO₄   →   Ac₂O / H₂SO₄

(b)   HNO₃ / H₂SO₄   + ortho   →   Fe / HCl   →   1 equivalent Ac₂O   →   NH₂ is more nucleophilic than OH

478

21-54

(a)

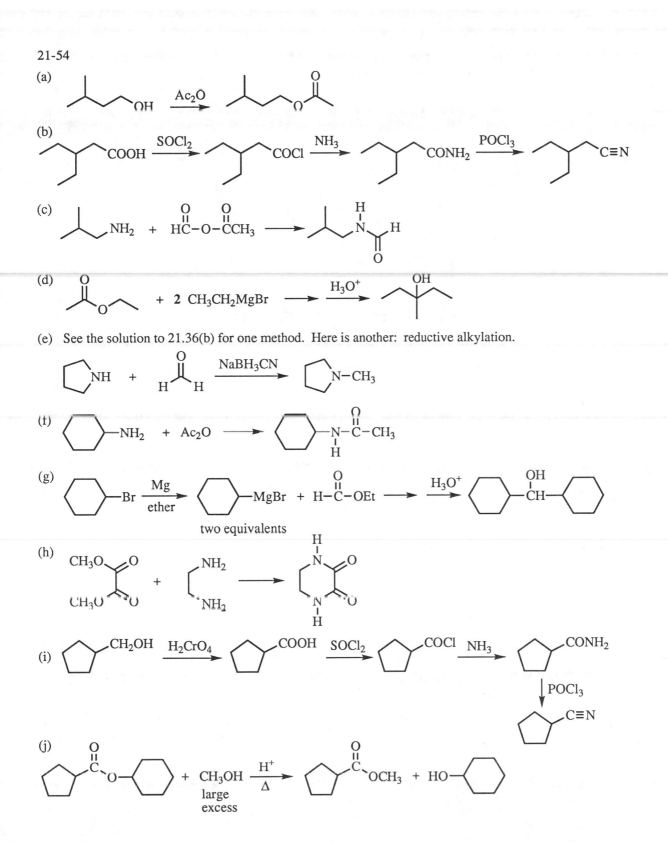

(b)

(c)

(d)

(e) See the solution to 21.36(b) for one method. Here is another: reductive alkylation.

(f)

(g)

two equivalents

(h)

(i)

(j)

large
excess

479

21-55 Diethyl carbonate has *two* leaving groups on the carbonyl. It can undergo *two* nucleophilic acyl substitutions, followed by one nucleophilic addition.

(a)

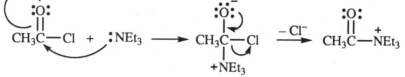

(b) $CH_3CH_2Br \xrightarrow[\text{ether}]{Mg} CH_3CH_2MgBr$

$$3\ CH_3CH_2MgBr\ +\ EtO-\overset{\overset{\displaystyle O}{\|}}{C}-OEt\ \xrightarrow{\quad\quad} \xrightarrow{H_3O^+}\ CH_3CH_2-\overset{\overset{\displaystyle OH}{|}}{\underset{\underset{\displaystyle CH_2CH_3}{|}}{C}}-CH_2CH_3$$

21-56 Triethylamine is nucleophilic, but it has no H on nitrogen to lose, so it forms a salt instead of a stable amide.

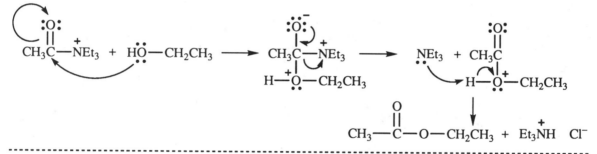

When ethanol is added, it attacks the carbonyl of the salt, with triethylamine as the leaving group.

$$CH_3-\overset{\overset{\displaystyle O}{\|}}{C}-O-CH_2CH_3\ +\ Et_3\overset{+}{N}H\quad Cl^-$$

An alternate mechanism explains the same products, and is more likely with hindered bases:

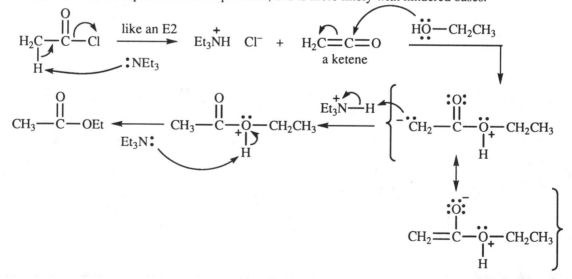

480

21-57

(a)

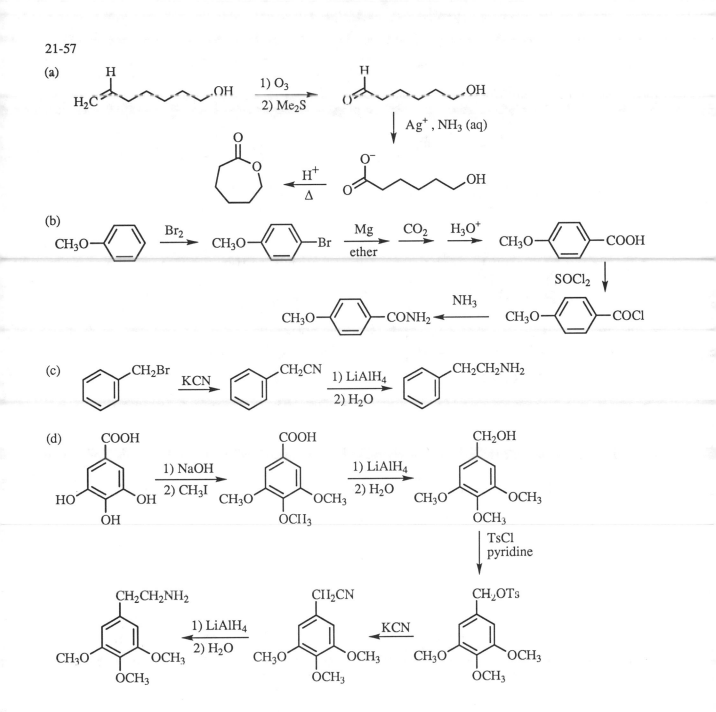

(b)

(c)

(d)

21-58

(a)

CH₃CH₂O—C—OCH₂CH₃

(b)

CH₃NH—C—NHCH₃

(c)

CH₃O—C—N—phenyl

(d)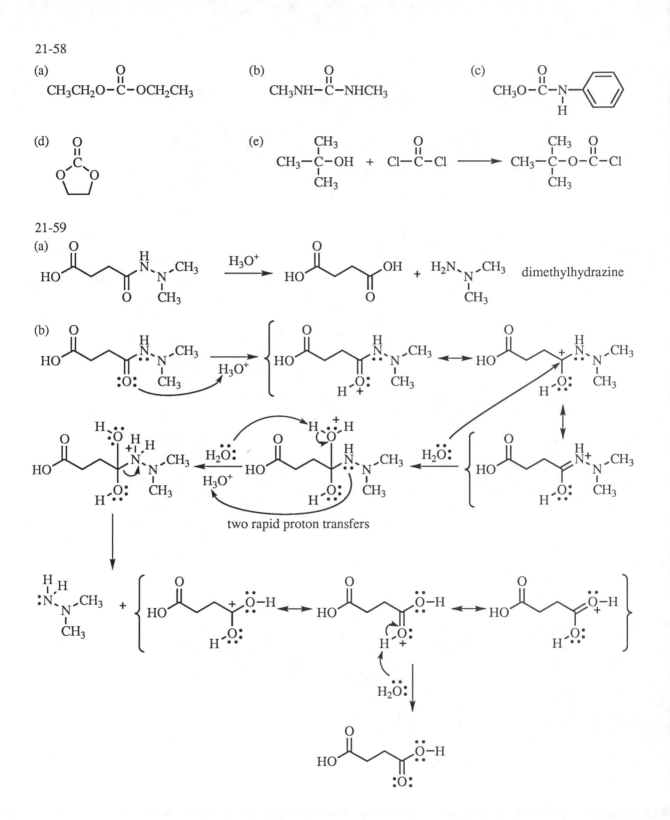

(e)

CH₃—C(CH₃)₂—OH + Cl—C—Cl ⟶ CH₃—C(CH₃)₂—O—C—Cl

21-59

(a)

(b)

two rapid proton transfers

21-60

(a)

$$CH_3—N=C=O \ + \ H_2O \longrightarrow H_3C—\underset{H}{\overset{O}{\underset{|}{N}}}—\overset{\|}{C}—OH \longrightarrow CH_3NH_2 \ (g) \ + \ CO_2 \ (g)$$

methyl isocyanate   a carbamic acid—unstable

Both of these reactions are exothermic. In a closed vessel like an industrial reactor, the production of gaseous products causes a large pressure increase, risking an explosion.

(b)

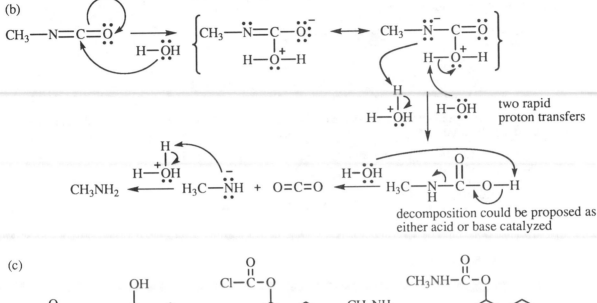

two rapid
proton transfers

decomposition could be proposed as
either acid or base catalyzed

(c)

phosgene

21-61

(a) (i) The repeating functional group is an ester, so the polymer is a polyester (named Kodel®).

   (ii) hydrolysis products:

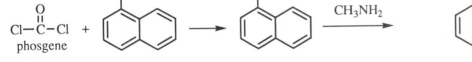

   (iii) The monomers could be the same as the hydrolysis products, or else some reactive derivative of the dicarboxylic acid, like an acid chloride or an ester derivative.

(b) (i) The repeating functional group is an amide, so the polymer is a polyamide (named Nylon 6).

   (ii) hydrolysis product:

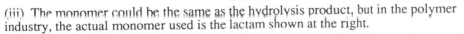

   (iii) The monomer could be the same as the hydrolysis product, but in the polymer industry, the actual monomer used is the lactam shown at the right.

483

21-61 continued

(c) (i) The repeating functional group is a carbonate, so the polymer is a polycarbonate (named Lexan®).
(ii) hydrolysis products:

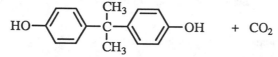

(iii) The phenol monomer would be the same as the hydrolysis product; phosgene or a carbonate ester would be the other monomer.

(d) (i) The repeating functional group is an amide, so the polymer is a polyamide.
(ii) hydrolysis product:

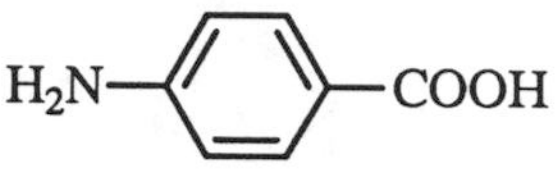

H₂N—⟨benzene ring⟩—COOH   p-aminobenzoic acid, PABA, used in sunscreens

(iii) The monomer could be the same as the hydrolysis product; a reactive derivative of the acid like an· ester could also be used.

(e) (i) The repeating functional group is a urethane, so the polymer is a polyurethane.
(ii) hydrolysis products:

HOCH₂CH₂OH + CO₂ +

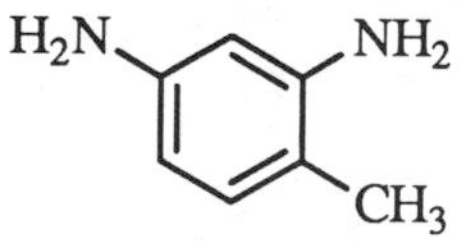

(iii) monomers:

HOCH₂CH₂OH +

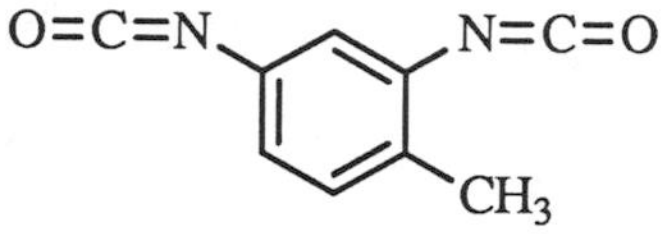

21-62

(a) Both structures are β-lactam antibiotics, a penicillin and a cephalosporin.
(b) "Cephalosporin N" has a 5-membered, sulfur-containing ring. This belongs in the penicillin class of antibiotics.

484

21-63 The rate of a reaction depends on its activation energy, that is, the difference in energy between starting material and the transition state. The transition state in saponification is similar in structure, and therefore in energy, to the tetrahedral intermediate:

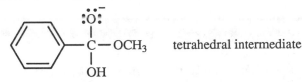

The tetrahedral carbon has no resonance overlap with the benzene ring, so any resonance effect of a substituent on the ring will have very little influence on the energy of the transition state.

What will have a big influence on the activation energy is whether a substituent stabilizes or destabilizes the starting material. Anything that stabilizes the starting material will therefore increase the activation energy, slowing the reaction; anything that destabilizes the starting material will decrease the activation energy, speeding the reaction.

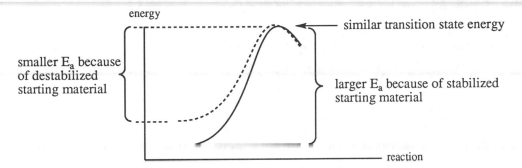

(a) One of the resonance forms of methyl *p*-nitrobenzoate has a positive charge on the benzene carbon adjacent to the positive carbonyl carbon. This resonance form destabilizes the starting material, lowering the activation energy, speeding the reaction.

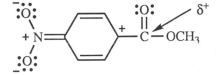

poor resonance contributor, destabilizing the starting material; no effect in the transition state

(b) One of the resonance forms of methyl *p*-methoxybenzoate has all atoms with full octets, and negative charge on the most electronegative atom. This resonance form stabilizes the starting material, increasing the activation energy, slowing the reaction.

good resonance contributor, stabilizing the starting material; no effect in the transition state

21-64

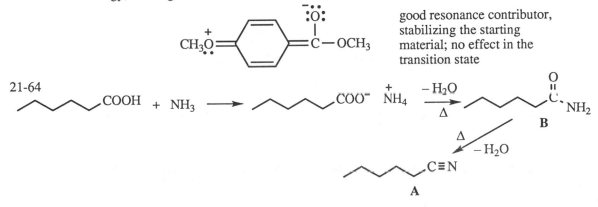

21-65 A singlet at δ 2.15 is H on carbon next to carbonyl, the only type of proton in the compound. The IR spectrum shows no OH, and shows two carbonyl absorptions at high frequency, characteristic of an anhydride. Mass of the molecular ion at 102 proves that the anhydride must be acetic anhydride, a reagent commonly used in aspirin synthesis.

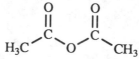

Acetic anhydride can be disposed of by hydrolyzing (carefully! exothermic!) and neutralizing in aqueous base.

21-66

IR spectrum:

—sharp spike at 2250 cm^{-1} ⇒ C≡N

—1750 cm^{-1} ⇒ C=O ⎫
—1200 cm^{-1} ⇒ C—O ⎬ maybe an ester
 ⎭

NMR spectrum:

—triplet and quartet ⇒ CH_3CH_2

—this quartet at δ 4.3 ⇒ CH_3CH_2O

—2H singlet at δ 3.5 ⇒ isolated CH_2

$$CH_3CH_2O \ + \ \overset{O}{\underset{\|}{C}} \ + \ CH_2 \ + \ C≡N$$

sum of the masses is 113, consistent with the MS

The fragments can be combined in only two possible ways:

$$CH_3CH_2O-\overset{O}{\underset{\|}{C}}-CH_2\,C≡N \qquad\qquad CH_3CH_2OCH_2-\overset{O}{\underset{\|}{C}}-C≡N$$
$$\textbf{A} \qquad\qquad\qquad\qquad\qquad\qquad \textbf{B}$$

The NMR proves the structure to be **A**. If the structure were **B**, the CH_2 between oxygen and the carbonyl would come farther downfield than the CH_2 of the ethyl (deshielded by oxygen and carbonyl instead of by oxygen alone). As this is not the case, the structure cannot be **B**.

The peak in the mass spectrum at m/z 68 is due to α-cleavage of the ester:

$$\left[CH_3CH_2O\!\!-\!\!\overset{O}{\underset{\underset{\underset{m/z\ 113}{68}}{\|}}{C}}\!\!-\!\!CH_2\,C≡N \right]^{+\cdot} \longrightarrow \left\{ \overset{\ddot{:}\ddot{O}\ddot{:}}{\underset{m/z\ 68}{\|}} + CCH_2CN \longleftrightarrow \overset{:O^+}{\underset{}{\equiv}} CCH_2CN \right\}$$

$$+ \ CH_3CH_2O\cdot \quad mass\ 45$$

21-67

The formula C_5H_9NO has 2 elements of unsaturation.

IR spectrum: The strongest peak at 1670 cm^{-1} comes low in the carbonyl region; in the absence of conjugation (no alkene peak observed), a carbonyl this low is almost certainly an amide. There is one broad peak in the NH/OH region, hinting at the likelihood of a secondary amide.

NMR spectrum: The broad peak at δ 7.55 is exchangeable with D_2O; this is an amide proton. A broad, 2H peak at δ 3.3 is a CH_2 next to nitrogen. A broad, 2H peak at δ 2.4 is a CH_2 next to carbonyl. The 4H peak at δ 1.8 is probably two more CH_2 groups. There appears to be coupling among these protons but it is not resolved enough to be useful for interpretation. This is often the case when the compound is cyclic, with restricted rotation around carbon-carbon bonds, giving *non-equivalent* (axial and equatorial) hydrogens on the same carbon.

$$CH_2-\overset{\overset{\displaystyle O}{\|}}{C}-\overset{\overset{\displaystyle H}{|}}{N}-CH_2 \quad + \quad CH_2 \quad + \quad CH_2$$

One element of unsaturation is the π bond in the carbonyl. There is no alkene so the other element of unsaturation must be a ring. The most consistent structure:

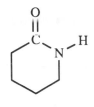

δ-valerolactam

IR spectrum: A strong carbonyl peak at 1720 cm^{-1}, in conjunction with the C—O peak at 1200 cm^{-1}, suggests the presence of an ester. An alkene peak appears at 1660 cm^{-1}.

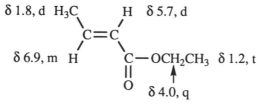

NMR spectrum: The typical ethyl pattern stands out: 3H triplet at δ 1.2 and 2H quartet at δ 4. The chemical shift of the CH$_2$ suggests it is bonded to an oxygen. The other groups are: a 3H doublet at δ 1.8, likely to be a CH$_3$ next to one H; a 1H doublet at δ 5.7, a vinyl hydrogen with one neighboring H; and a 1H multiplet at δ 6.9, another vinyl H with many neighbors. The large coupling constant in the doublet at δ 5.7 shows that the two vinyl hydrogens are *trans*.

There is only one possible way to assemble these pieces:

Mass spectrum: This structure has mass 114, consistent with the molecular ion.
Major fragmentations:

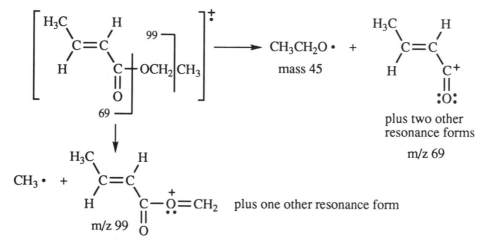

21-69 If you got this problem, put a gold star on your forehead.

The formula $C_6H_8O_3$ indicates 3 elements of unsaturation.

<u>IR spectrum</u>: The absence of strong OH peaks shows that the compound is neither an alcohol nor a carboxylic acid. There are two carbonyl absorptions: the one about 1770 cm^{-1} is likely a strained cyclic ester (reinforced with the C—O peak around 1150 cm^{-1}), while the one at 1720 cm^{-1} is probably a ketone. (An anhydride also has two peaks, but they are of higher frequency than the ones in this spectrum.)

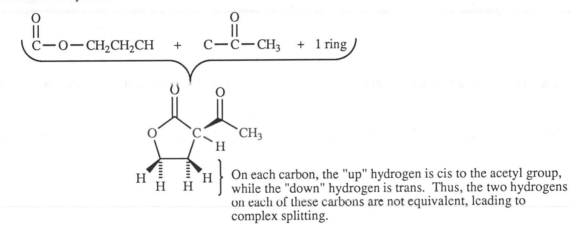

<u>Proton NMR spectrum</u>: The NMR shows four types of protons. The 2H triplet at δ 4.3 is a CH_2 group next to an oxygen on one side, with a CH_2 on the other. The 1H multiplet at δ 3.7 is also strongly deshielded (probably by two carbonyls), a CH next to a CH_2. The 3H singlet at δ 2.45 is a CH_3 on one of the carbonyls. The remaining two hydrogens are highly coupled, a CH_2 where the two hydrogens are not equivalent. There are no vinyl hydrogens (and no alkene carbon in the carbon NMR), so the remaining element of unsaturation must be a ring.

Assemble the pieces:

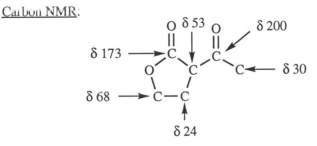

On each carbon, the "up" hydrogen is cis to the acetyl group, while the "down" hydrogen is trans. Thus, the two hydrogens on each of these carbons are not equivalent, leading to complex splitting.

<u>Carbon NMR</u>:

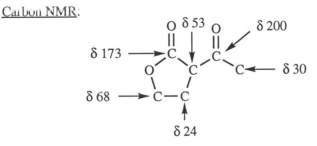

22-1

(a)

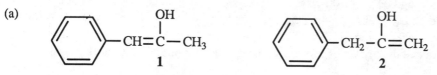

(b) Enol **1** will predominate at equilibrium as its double bond is conjugated with the benzene ring, making it more stable than **2**.

(c)

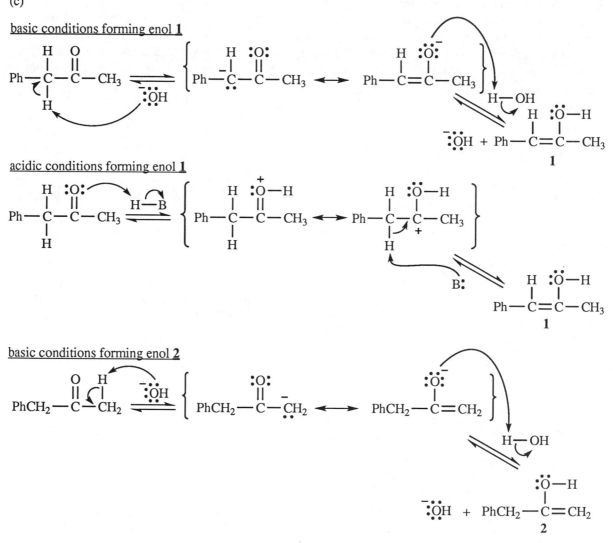

22-1 (c) continued

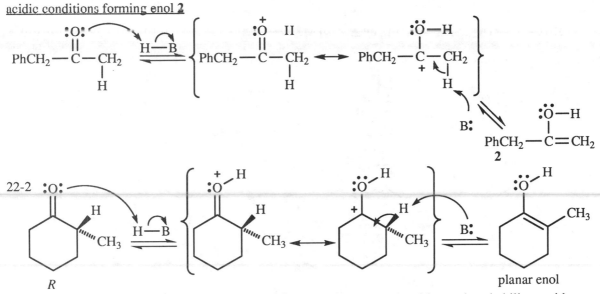

22-2

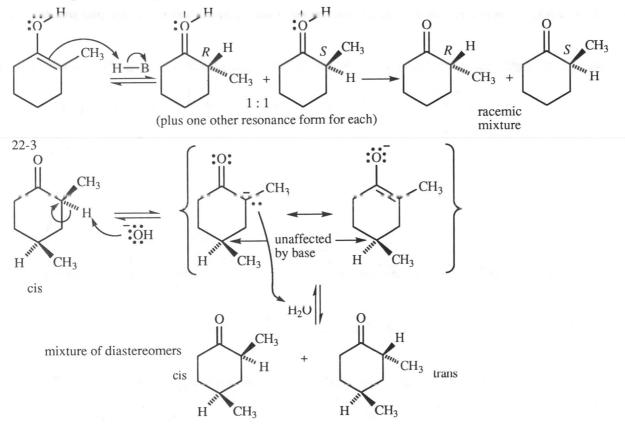

This planar enol intermediate has lost all chirality. Protonation can occur with equal probability at either face of the pi bond leading to racemic product.

491

22-4

(a)

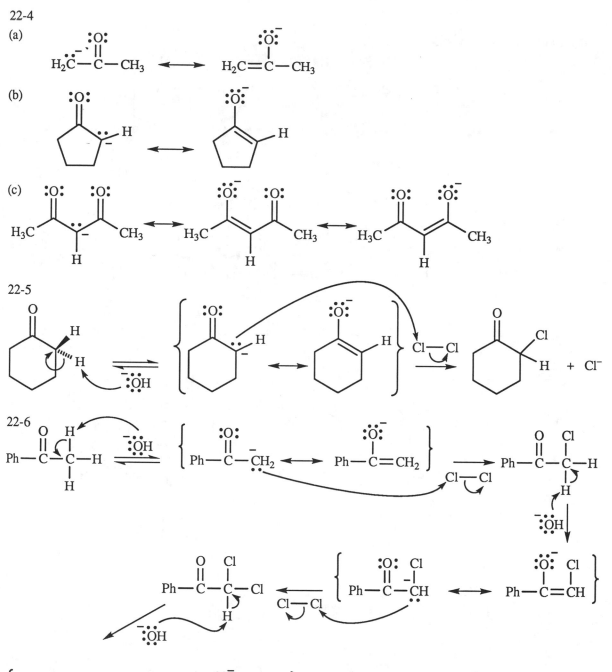

(b)

(c)

22-5

22-6

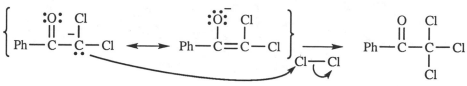

22-7 For this problem, the cyclohexyl group is abbreviated "Cy".

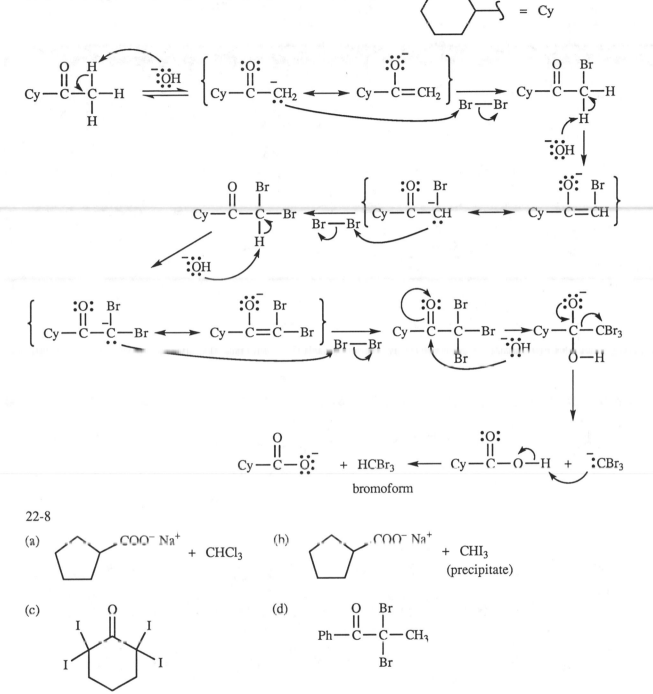

bromoform

22-8

(a) + CHCl₃

(h) + CHI₃ (precipitate)

(c)

(d) Ph—C—C—CH₃

22-9 Methyl ketones, and alcohols which are oxidized to methyl ketones, will give a positive iodoform test. All of the compounds in this problem except 3-pentanone (part (d)) will give a positive iodoform test.

22-10

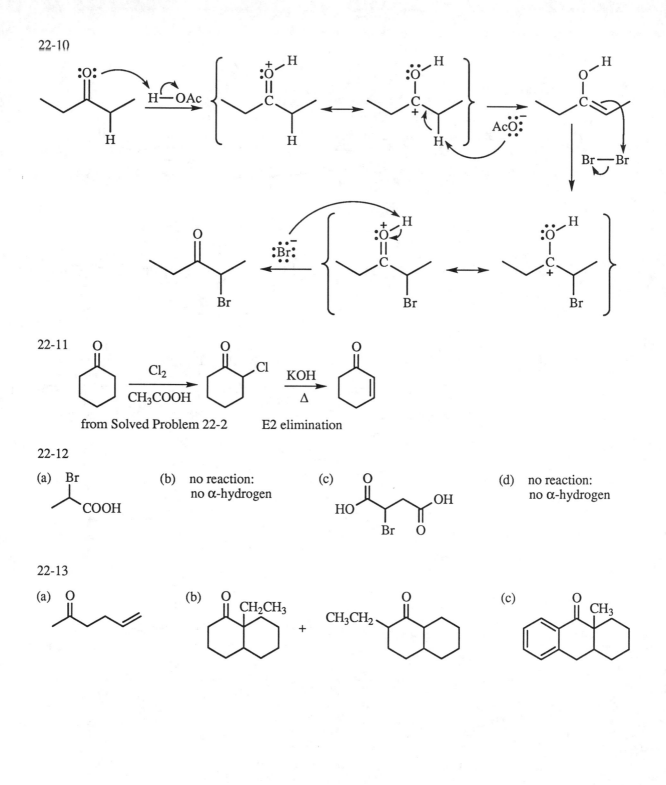

22-11

from Solved Problem 22-2   E2 elimination

22-12

(a)

(b) no reaction:
     no α-hydrogen

(c)

(d) no reaction:
     no α-hydrogen

22-13

(a)

(b)

+

(c)

494

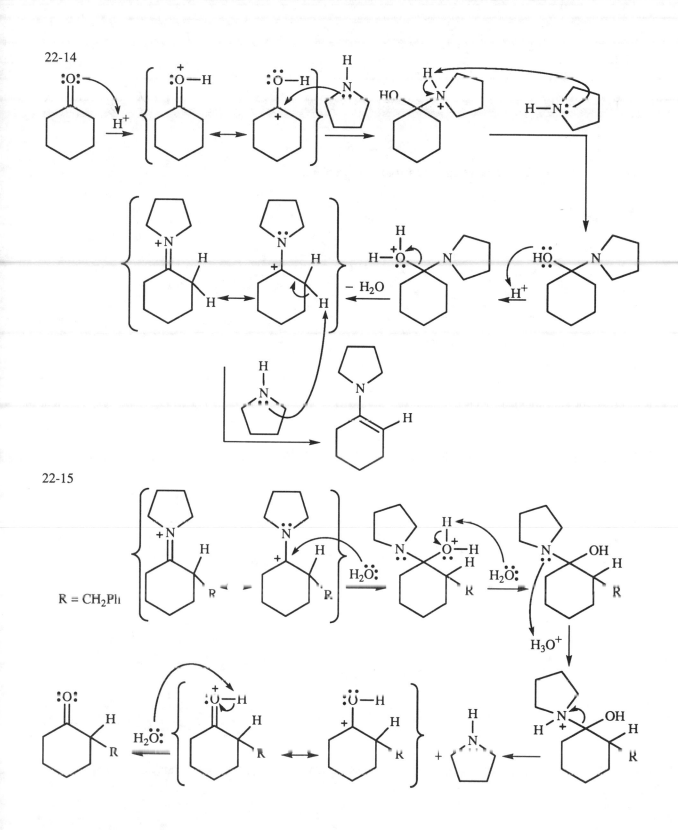

22-14

22-15

R = CH₂Ph

22-16

(a)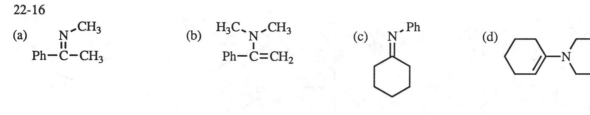

22-17 Any 2° aliphatic amines can be used for this problem.

(a)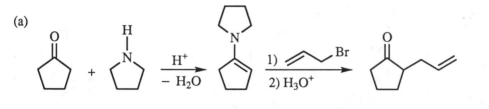

(b)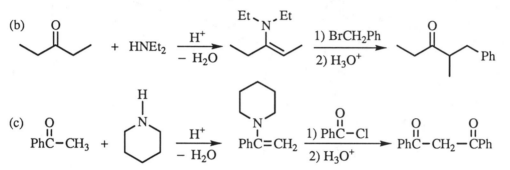

(c)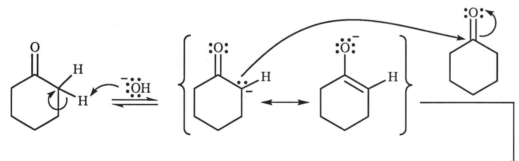

22-18

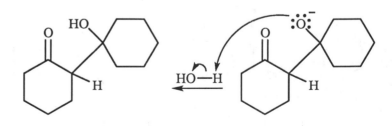

22-19

(a)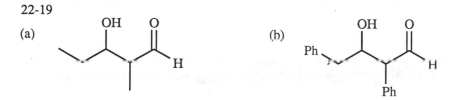

(h)

22-20 All the steps in the aldol condensation are reversible. Adding base to diacetone alcohol promoted the reverse aldol reaction. The equilibrium greatly favors acetone.

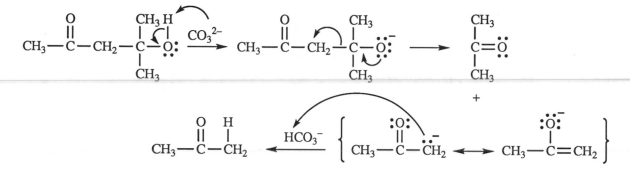

22-21

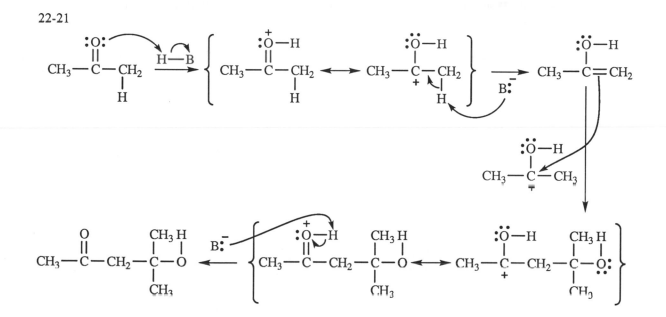

22-22

(a) <u>acidic conditions</u>

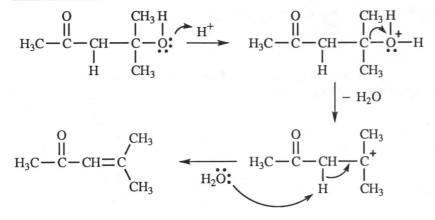

(b) <u>basic conditions</u>

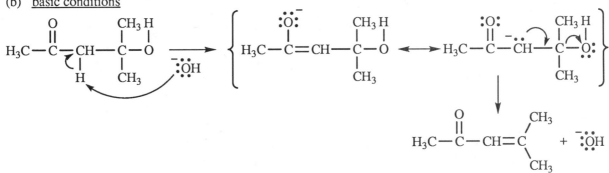

22-23

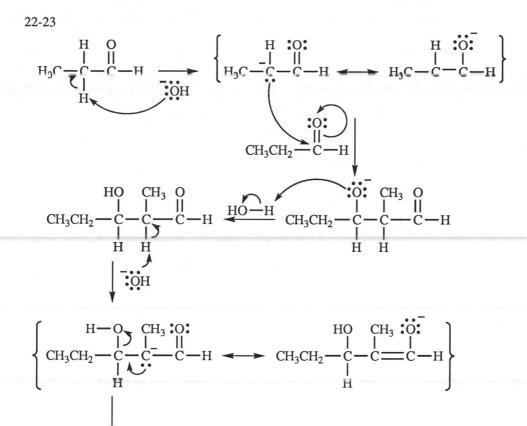

22-24

(a)

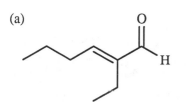

(b)

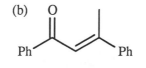

(c)

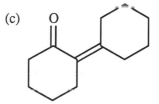

22-25

(a)

Step 1: carbon skeletons

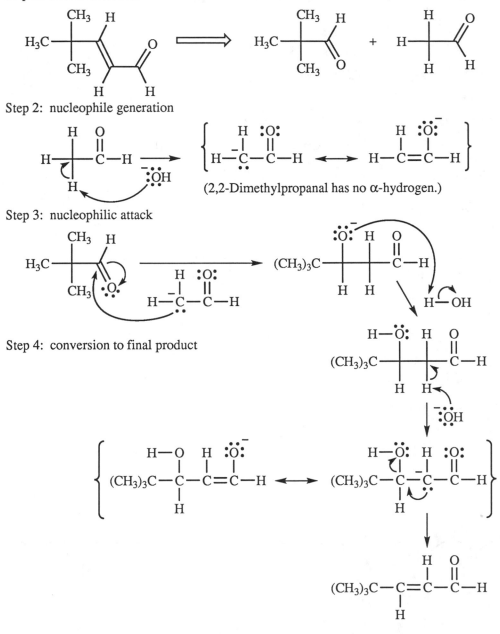

Step 2: nucleophile generation

(2,2-Dimethylpropanal has no α-hydrogen.)

Step 3: nucleophilic attack

Step 4: conversion to final product

Step 5: combine Steps 2, 3, and 4 to complete the mechanism

22-25 continued

(b)

Step 1: carbon skeletons

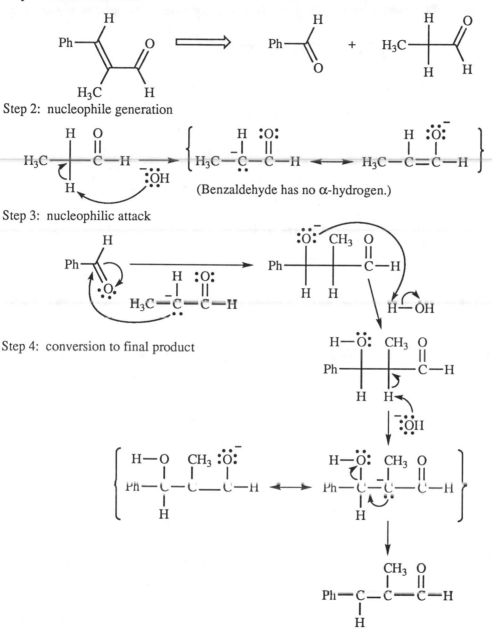

Step 2: nucleophile generation

(Benzaldehyde has no α-hydrogen.)

Step 3: nucleophilic attack

Step 4: conversion to final product

Step 5: combine Steps 2, 3, and 4 to complete the mechanism

22-26 This solution presents the sequence of reactions leading to the product, following the format of the Problem-Solving feature. This is not a complete mechanism.

Step 2: generation of the nucleophile

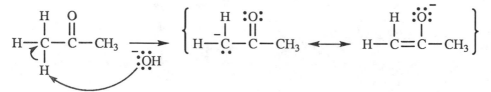

Step 3: nucleophilic attack

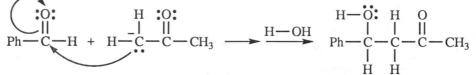

Step 4: dehydration

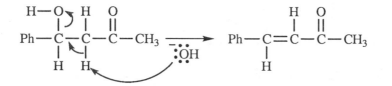

The same sequence of steps occurs on the other side.

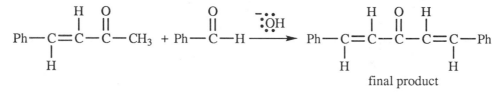

final product

22-27

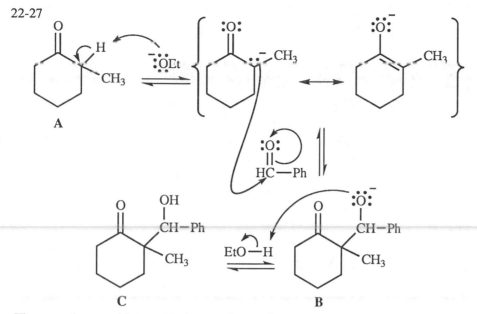

There are three problems with the reaction as shown:

1. Hydrogen on a 3° carbon (structure **A**) is less acidic than hydrogen on a 2° carbon. The 3° hydrogen will be removed at a slower rate than the 2° hydrogen.

2. Nucleophilic attack by the 3° carbon will be more hindered, and therefore slower, than attack by the 2° carbon. Structure **B** is quite hindered.

3. Once a normal aldol product is formed, dehydration gives a conjugated system which has great stability. The aldol product **C** cannot dehydrate because no α-hydrogen remains. Some **C** will form, but eventually the reverse-aldol process will return **C** to starting materials which, in turn, will react at the other α-carbon to produce the conjugated system. (This reason is the Kiss of Death for **C**.)

22-28
(a)   (b)

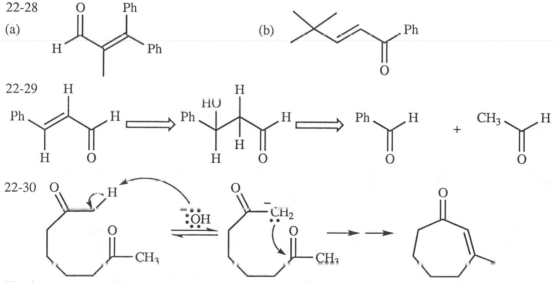

The formation of a seven-membered ring is unfavorable for entropy reasons: the farther apart the nucleophile and the electrophile, the harder time they will have finding each other. If the molecule has a possibility of forming a 5- or a 7-membered ring, it will almost always prefer to form the 5-membered ring.

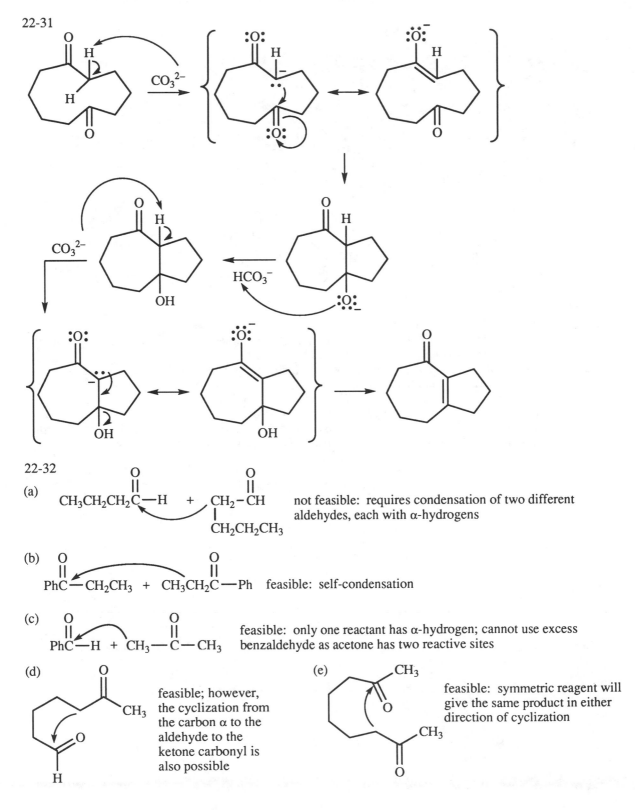

22-31

22-32

(a) $CH_3CH_2CH_2\overset{\overset{\displaystyle O}{\|}}{C}$—H   + CH_2—$\overset{\overset{\displaystyle O}{\|}}{C}H$ not feasible: requires condensation of two different
  aldehydes, each with α-hydrogens

$CH_2CH_2CH_3$

(b) $Ph\overset{\overset{\displaystyle O}{\|}}{C}$—$CH_2CH_3$ + $CH_3CH_2\overset{\overset{\displaystyle O}{\|}}{C}$—Ph   feasible: self-condensation

(c) $Ph\overset{\overset{\displaystyle O}{\|}}{C}$—H + CH_3—$\overset{\overset{\displaystyle O}{\|}}{C}$—$CH_3$ feasible: only one reactant has α-hydrogen; cannot use excess
  benzaldehyde as acetone has two reactive sites

(d) feasible; however, the cyclization from the carbon α to the aldehyde to the ketone carbonyl is also possible

(e) feasible: symmetric reagent will give the same product in either direction of cyclization

22-33 (a) and (b)

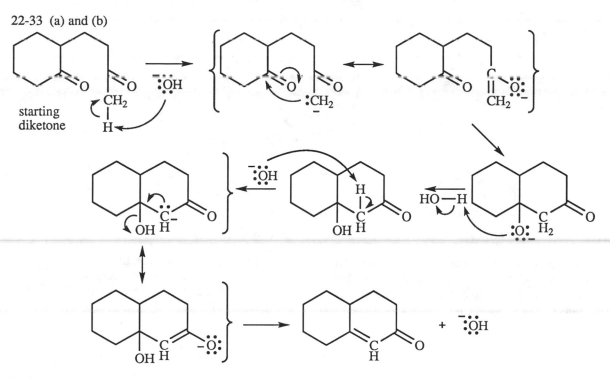

starting diketone

22-34

(a) The side reaction with sodium methoxide is transesterification. The starting material, and therefore the product, would be a mixture of methyl and ethyl esters.

$$H_3C-\overset{\overset{\displaystyle O}{\|}}{C}-OCH_2CH_3 \quad + \quad NaOCH_3 \quad \rightleftharpoons \quad H_3C-\overset{\overset{\displaystyle O}{\|}}{C}-OCH_3 \quad + \quad NaOCH_2CH_3$$

(b) Sodium hydroxide would irreversibly saponify the ester, completely stopping the Claisen condensation as the carbonyl no longer has a leaving group attached to it.

$$H_3C-\overset{\overset{\displaystyle O}{\|}}{C}-OCH_2CH_3 \quad + \quad NaOH \quad \longrightarrow \quad H_3C-\overset{\overset{\displaystyle O}{\|}}{C}-O^-\ Na^+ \quad + \quad HOCH_2CH_3$$

22-35

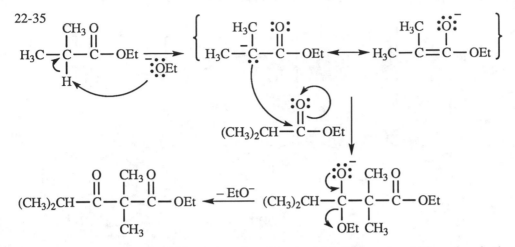

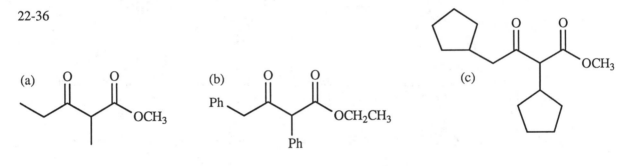

There are two reasons why this reaction gives a poor yield. The nucleophilic carbon in the enolate is 3° and attack is hindered. More important, the final product has no hydrogen on the α-carbon, so the deprotonation by base which is the driving force in other Claisen condensations cannot occur here. What is produced is an *equilibrium mixture* of product and starting materials; the conversion to product is low.

22-36

(a)

(b)

(c)

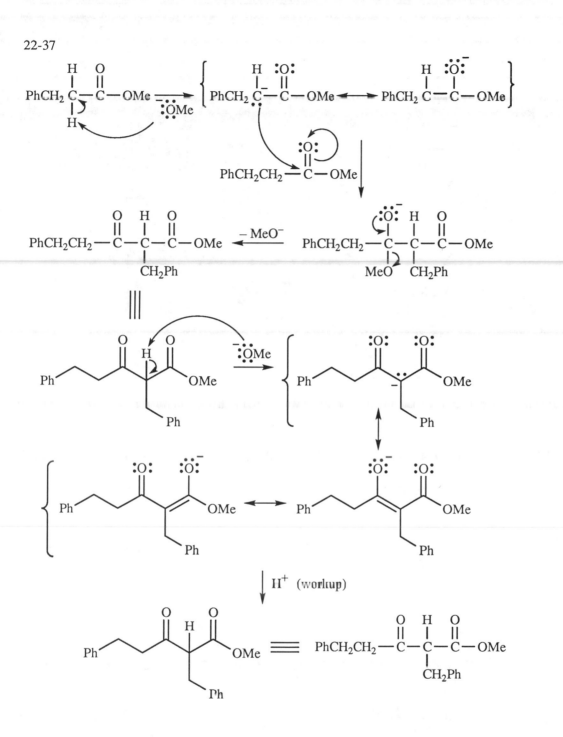

507

22-38

(a)

$$CH_3CH_2CH_2\overset{\displaystyle O}{\overset{\displaystyle \|}{C}}-OCH_2CH_3$$

(b)

$$PhCH_2\overset{\displaystyle O}{\overset{\displaystyle \|}{C}}-OCH_3$$

(c)

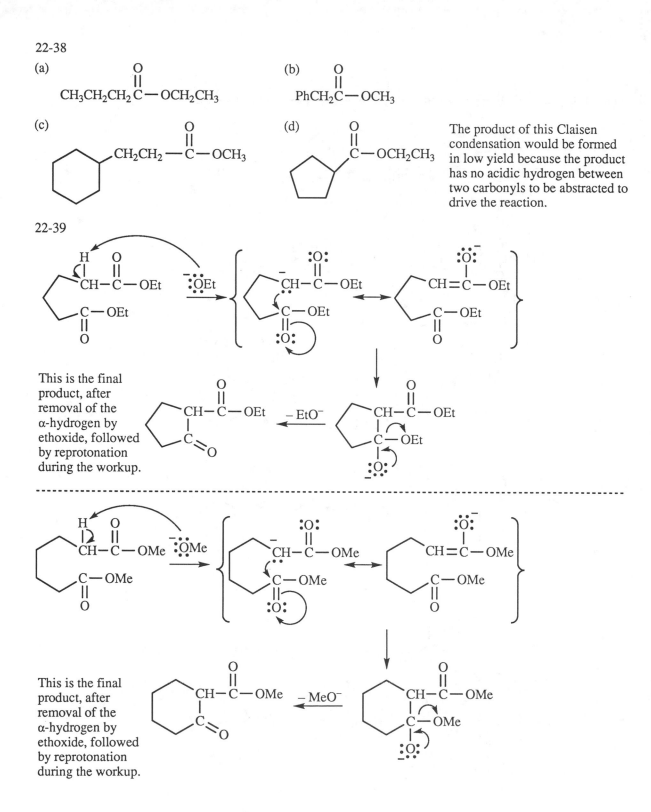

(d)

The product of this Claisen condensation would be formed in low yield because the product has no acidic hydrogen between two carbonyls to be abstracted to drive the reaction.

22-39

This is the final product, after removal of the α-hydrogen by ethoxide, followed by reprotonation during the workup.

This is the final product, after removal of the α-hydrogen by ethoxide, followed by reprotonation during the workup.

508

22-40

(a) not possible by Dieckmann—not a β-keto ester

(b)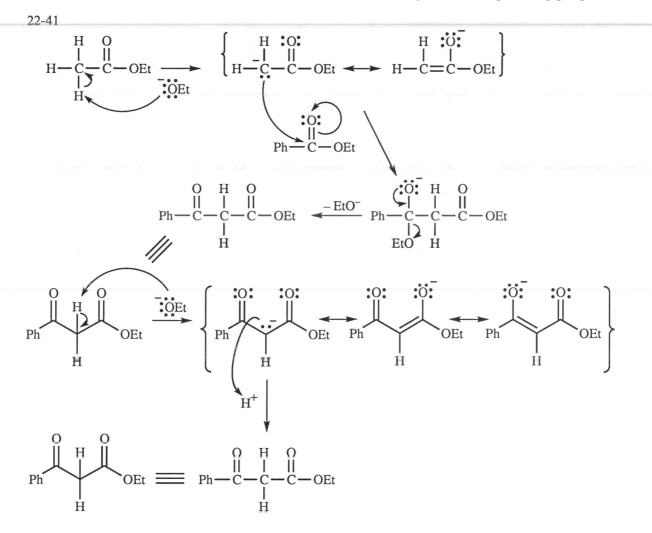

+ NaOCH₃ (mixture of products results)

(c) 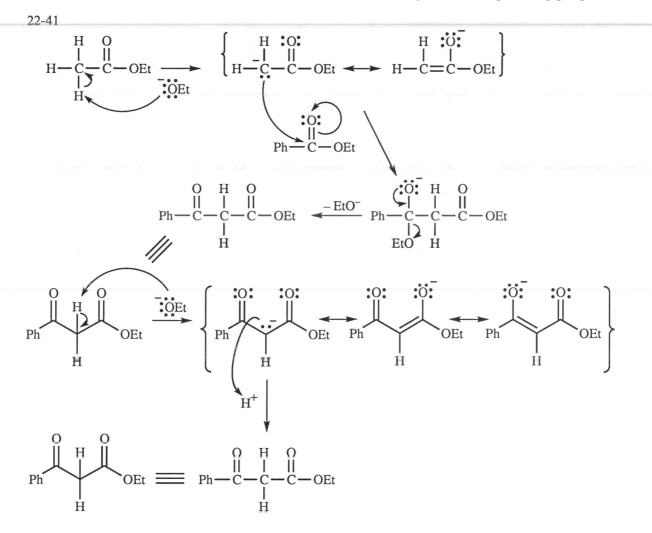 + NaOCH₃

(d) + NaOCH₂CH₃

The protecting group is necessary to prevent aldol condensation. Aqueous acid workup removes the protecting group.

22-41

22-42

(a)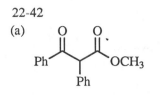

(b)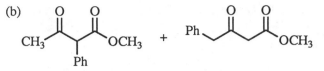

plus 2 self-condensation products—a poor choice
because both esters have α-hydrogens

(c)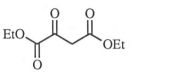

(d)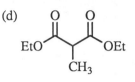

22-43

(a)

$$Ph-\overset{O}{\underset{\|}{C}}-OEt \;+\; CH_3CH_2-\overset{O}{\underset{\|}{C}}-OEt$$

(b)

$$PhCH_2-\overset{O}{\underset{\|}{C}}-OMe \;+\; MeO-\overset{O}{\underset{\|}{C}}-\overset{O}{\underset{\|}{C}}-OMe$$

(c)

$$EtO-\overset{O}{\underset{\|}{C}}-CH_2Ph \;+\; EtO-\overset{O}{\underset{\|}{C}}-OEt$$

(d)

$$(CH_3)_3C-\overset{O}{\underset{\|}{C}}-OMe \;+\; CH_3CH_2CH_2CH_2-\overset{O}{\underset{\|}{C}}-OMe$$

22-44

(a)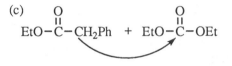

actually present in the enol form:

(b)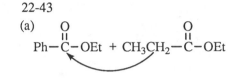

(c)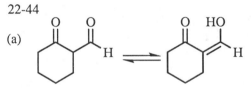

22-45

(a) two ways: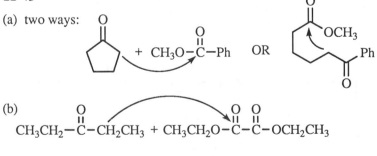

(b)

$$CH_3CH_2-\overset{O}{\underset{\|}{C}}-CH_2CH_3 \;+\; CH_3CH_2O-\overset{O}{\underset{\|}{C}}-\overset{O}{\underset{\|}{C}}-OCH_2CH_3$$

(c) (d) two ways:

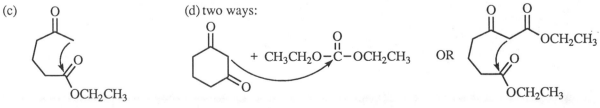

510

22-46

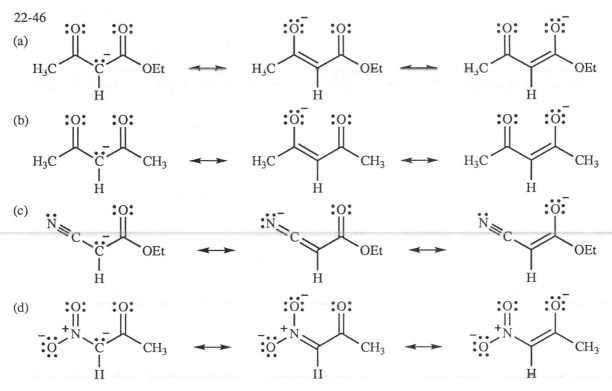

(other resonance forms of the nitro group are not shown)

22-47 In the products, the wavy lines indicate the bonds that must be made by alkylation, before hydrolysis and decarboxylation produce the substituted acetic acid.

(a)

EtO—C(O)—CH₂—C(O)—OEt $\xrightarrow[\text{2) PhCH}_2\text{Br}]{\text{1) NaOEt}}$ EtO—C(O)—CH(CH₂Ph)—C(O)—OEt $\xrightarrow[\Delta]{\text{H}_3\text{O}^+}$ CH₂Ph—CH₂—C(O)—OH + CO₂ + 2 EtOH

(b)

EtO—C(O)—CH₂—C(O)—OEt $\xrightarrow[\text{2) CH}_3\text{I}]{\text{1) NaOEt}}$ $\xrightarrow[\text{2) CH}_3\text{I}]{\text{1) NaOEt}}$ EtO—C(O)—C(CH₃)₂—C(O)—OEt $\xrightarrow[\Delta]{\text{H}_3\text{O}^+}$ —C(O)—OH + CO₂ + 2 EtOH

(c)

EtO—C(O)—CH₂—C(O)—OEt $\xrightarrow[\text{2) Ph(CH}_2)_2\text{Br}]{\text{1) NaOEt}}$ EtO—C(O)—CH(CH₂CH₂Ph)—C(O)—OEt $\xrightarrow[\Delta]{\text{H}_3\text{O}^+}$ Ph—CH₂—CH₂—CH₂—C(O)—OH + CO₂ + 2 EtOH

(d)

EtO—C(O)—CH₂—C(O)—OEt $\xrightarrow[\text{2) Br(CH}_2)_4\text{Br}]{\text{1) 2 NaOEt}}$ EtO—C(O)—C(cyclopentane)—C(O)—OEt $\xrightarrow[\Delta]{\text{H}_3\text{O}^+}$ cyclopentane—COOH + CO₂ + 2 EtOH

22-48

(a) PhCH₂CH₂$-$C$-$CH₃ (with C=O above)
+ CO₂ + EtOH

(b) 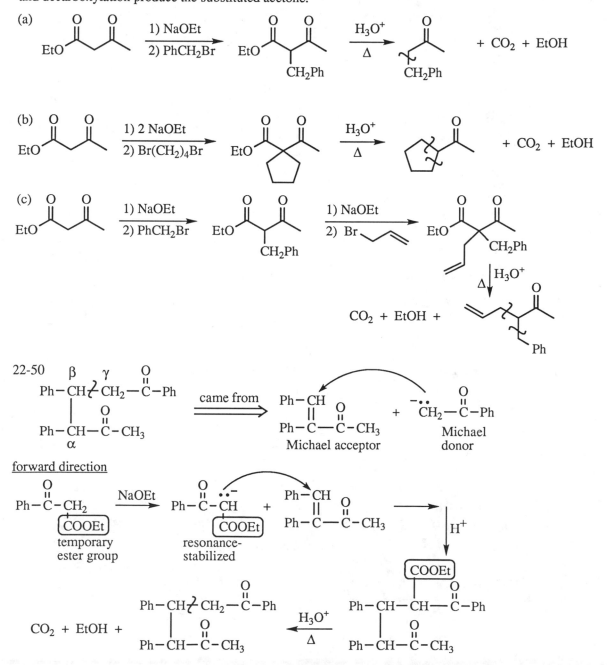 + CO₂ + EtOH

(c) + CO₂ + EtOH

22-49 In the products, the wavy lines indicate the bonds that must be made by alkylation, before hydrolysis and decarboxylation produce the substituted acetone.

(a)

EtO ...
1) NaOEt
2) PhCH₂Br
→ EtO ... CH₂Ph
H₃O⁺ / Δ
→ ... CH₂Ph + CO₂ + EtOH

(b)

EtO ...
1) 2 NaOEt
2) Br(CH₂)₄Br
→ EtO ...
H₃O⁺ / Δ
→ ... + CO₂ + EtOH

(c)

EtO ...
1) NaOEt
2) PhCH₂Br
→ EtO ... CH₂Ph
1) NaOEt
2) Br⌁
→ EtO ... CH₂Ph
Δ H₃O⁺
CO₂ + EtOH + ... Ph

22-50

β   γ   O
Ph$-$CH⌁CH₂$-$C$-$Ph
  |
  O
  ‖
Ph$-$CH$-$C$-$CH₃
  α

came from ⟹

Ph$-$CH
    ‖   O
Ph$-$C$-$C$-$CH₃
Michael acceptor

+ :CH₂$-$C$-$Ph (with O above)
Michael donor

forward direction

O
‖
Ph$-$C$-$CH₂
[COOEt]
temporary ester group

NaOEt →

O
‖
Ph$-$C$-$CH :⁻
[COOEt]
resonance-stabilized

+

Ph$-$CH
    ‖   O
Ph$-$C$-$C$-$CH₃

→ H⁺ ↓

[COOEt]   O
    | ‖
Ph$-$CH$-$CH$-$C$-$Ph
    |
    O
    ‖
Ph$-$CH$-$C$-$CH₃

CO₂ + EtOH +

O
‖
Ph$-$CH⌁CH₂$-$C$-$Ph
    |
    O
    ‖
Ph$-$CH$-$C$-$CH₃

H₃O⁺ / Δ ←

512

22-51 First, you might wonder why this sequence does not make the desired product:

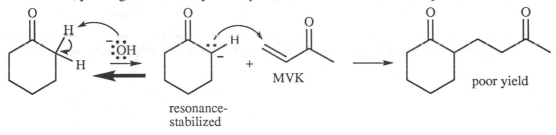

resonance-
stabilized

The poor yield in this conjugate addition is due primarily to the numerous competing reactions: the ketone enolate can self-condense (aldol), can condense with the ketone of MVK (aldol), or can deprotonate the methyl of MVK to generate a new nucleophile. The complex mixture of products makes this route practically useless.

What permits enamines (or other stabilized enolates) to work are: a) the certainty of which atom is the nucleophile, and b) the lack of self-condensation. Enamines can also do conjugate addition:

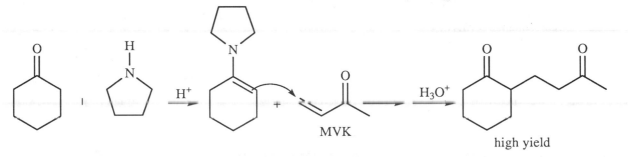

22-52 The enolate of acetoacetic ester can be used in a Michael addition to an α,β-unsaturated ketone like MVK.

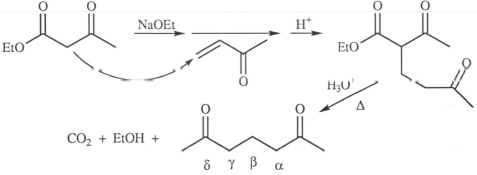

513

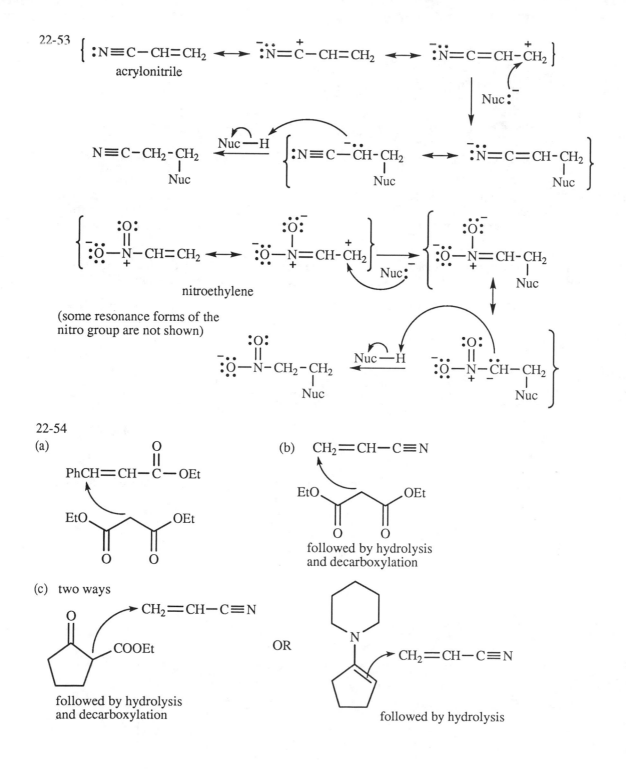

22-53

acrylonitrile

nitroethylene

(some resonance forms of the
nitro group are not shown)

22-54

(a)

(b) followed by hydrolysis
and decarboxylation

(c) two ways

followed by hydrolysis
and decarboxylation

OR

followed by hydrolysis

22-54 continued

(d)

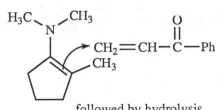
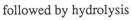

followed by hydrolysis

(f)

$$PhCH=CH-\overset{\overset{\displaystyle O}{\|}}{C}-OEt$$

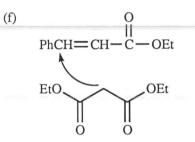

followed by hydrolysis
and decarboxylation

(e)

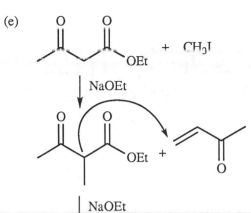

NaOEt

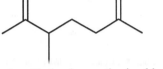

NaOEt

Δ | H_3O^+

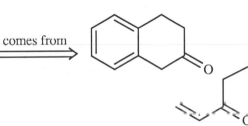

(could also be synthesized by the
Stork enamine reactions)

22-55

Step 1: carbon skeleton

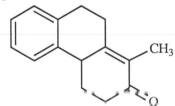 comes from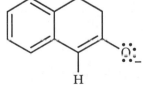

Step 2: nucleophile generation

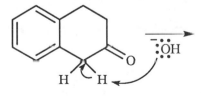 $\xrightarrow{:OH}$ 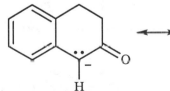 ↔

plus resonance forms
with negative charge
on the benzene ring

continued on next page

515

22-55 continued

Step 3: nucleophilic attack (Michael addition)

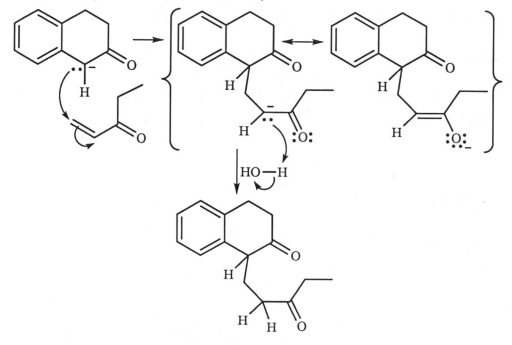

Step 4: conversion to final product

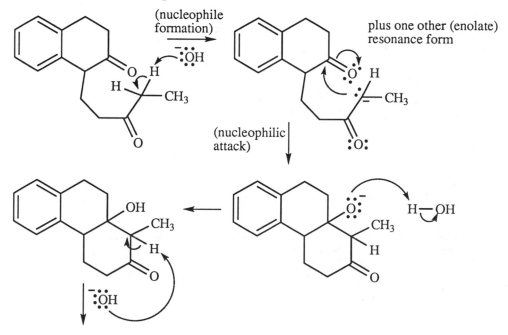

continued on next page

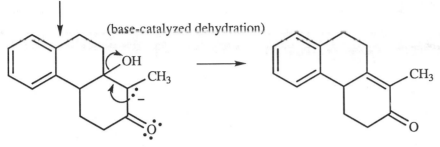

(base-catalyzed dehydration)

plus one other (enolate)
resonance form

Step 5: The complete mechanism is the combination of Steps 2, 3, and 4. Notice that this mechanism is simply described by:
1) Enolate formation, followed by Michael addition;
2) Aldol condensation, followed by dehydration.

22-56

Step 1: carbon skeleton

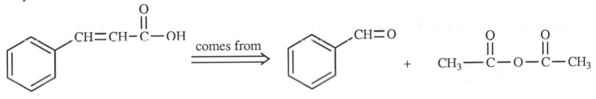

comes from

Step 2: nucleophile generation

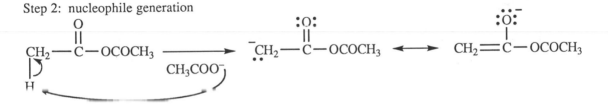

Step 3: nucleophilic attack

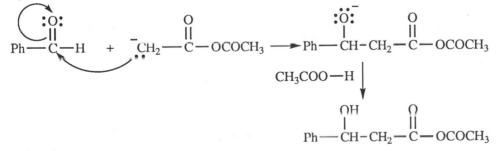

continued on next page

22-56 continued

Step 4: conversion to final product

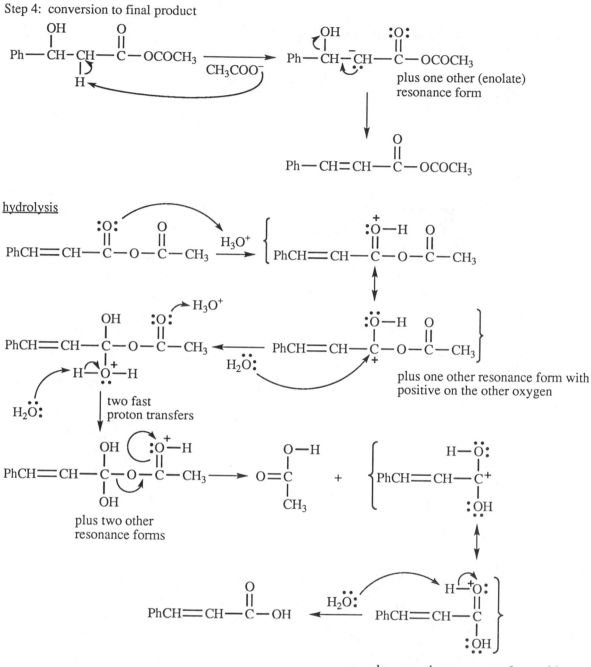

plus one other (enolate)
resonance form

hydrolysis

plus one other resonance form with
positive on the other oxygen

two fast
proton transfers

plus two other
resonance forms

plus one other resonance form with
positive on the other oxygen

Step 5: The complete mechanism is the combination of Steps 2, 3, and 4.

22-57 The Robinson annulation consists of a Michael addition followed by aldol cyclization with dehydration. In the retrosynthetic direction, disconnect the alkene formed in the aldol/dehydration, then disconnect the Michael addition to discover the reactants.

(a) aldol and dehydration forms the α,β double bond:

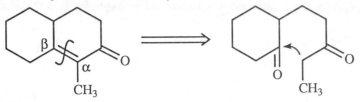

Michael addition forms a bond to the β' carbon:

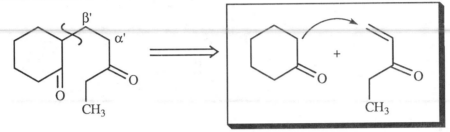

(b) aldol and dehydration forms the α,β double bond:

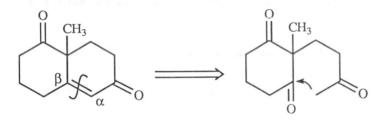

Michael addition forms a bond to the β' carbon:

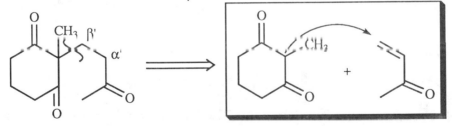

22-58 Please refer to solution 1-19, page 12 of this Solutions Manual.

22-59 The most acidic hydrogens are shown in boldface.   (See Appendix 2 for a review of acidity.)

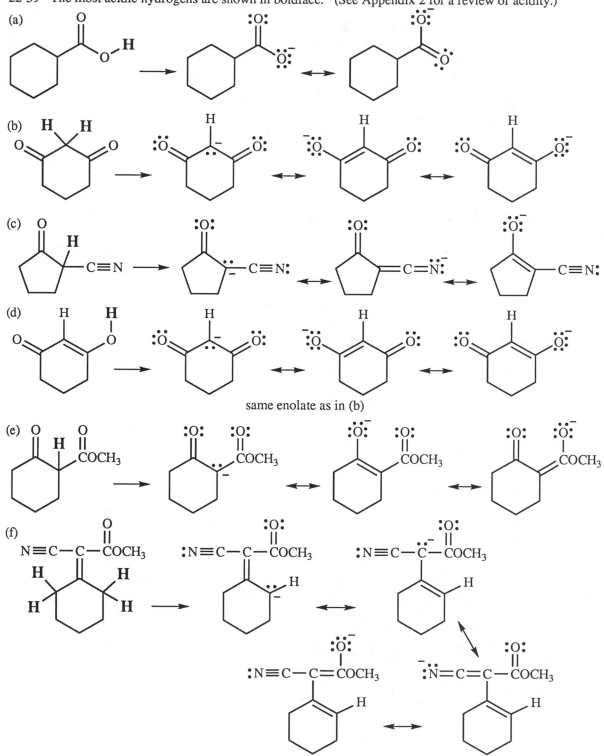

same enolate as in (b)

22-59 continued

(g)

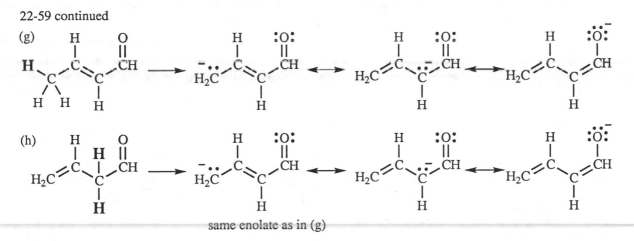

(h)

same enolate as in (g)

22-60 In order of increasing acidity. The most acidic protons are shown in boldface. (The approximate pK_a values are shown for comparison.) See Appendix 2 for a review of acidity.

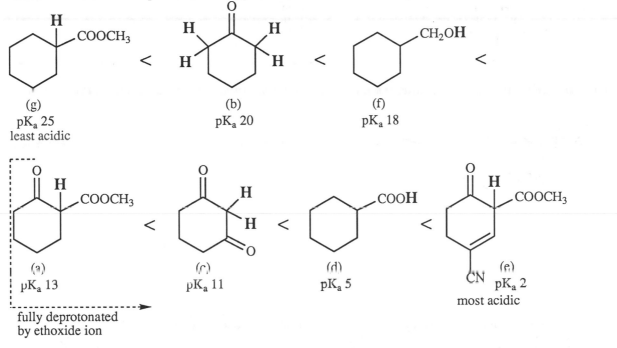

(g)
pK_a 25
least acidic

<
(b)
pK_a 20

<
(f)
pK_a 18

<

(a)
pK_a 13

fully deprotonated
by ethoxide ion

<
(c)
pK_a 11

<
(d)
pK_a 5

<
(e)
pK_a 2
most acidic

22-61

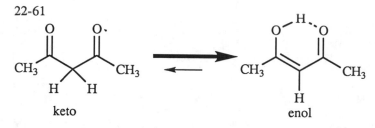

keto   enol

The enol form is stable because of the conjugation and because of intramolecular hydrogen-bonding in a six-membered ring.

In dicarbonyl compounds in general, the weaker the electron-donating ability of the group G, the more it will exist in the enol form: aldehydes (G = H) are almost completely enolized, then ketones, esters, and finally amides which have virtually no enol content.

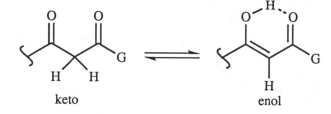

keto   enol

22-62 The wavy line lies across the bond formed in the aldol condensation.

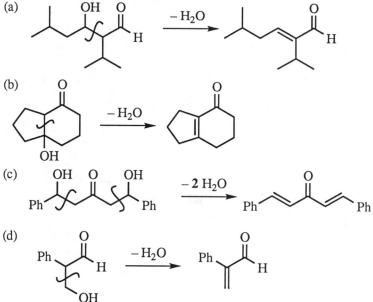

22-63 The wavy line lies across the bond formed in the Claisen condensation.

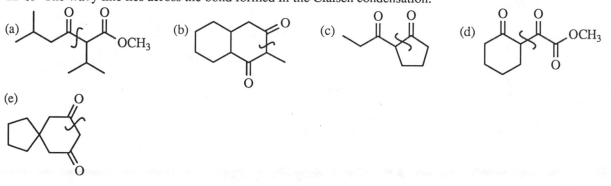

522

22-64

(a) mechanism of aldol condensation in problem 22-62(a)

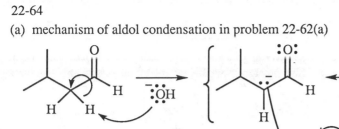

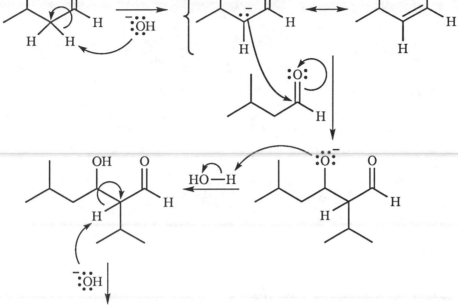

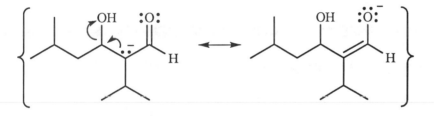

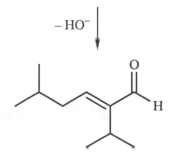

22-64 continued

(b) mechanism of aldol condensation in problem 22-62(b)

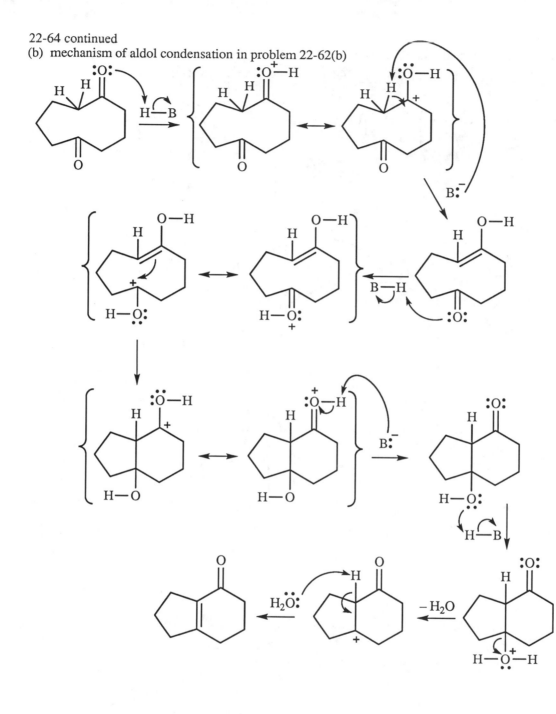

22-64 continued

(c) mechanism of Claisen condensation in problem 22-63(a)

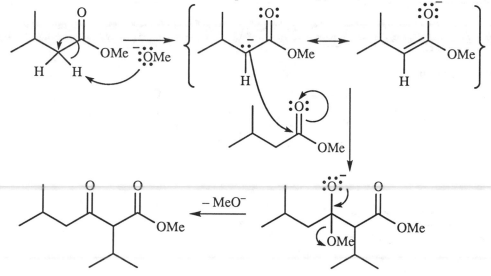

(this product will be deprotonated by methoxide,
but regenerated upon acidic workup)

(d) mechanism of Claisen condensation in problem 22-63(b)

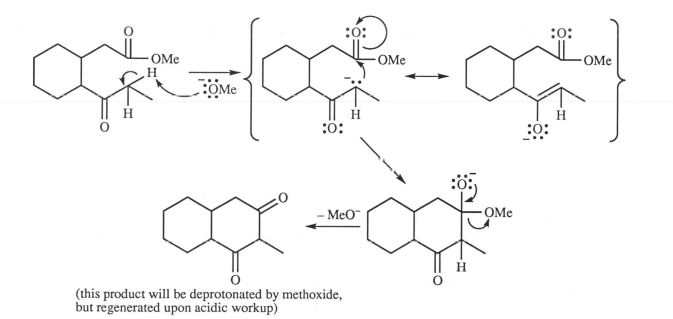

(this product will be deprotonated by methoxide,
but regenerated upon acidic workup)

22-65 All products shown are after acidic workup.

(a) aldol self-condensation

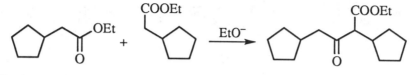

(b) Claisen self-condensation

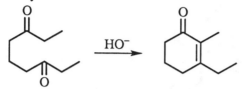

(c) aldol cyclization

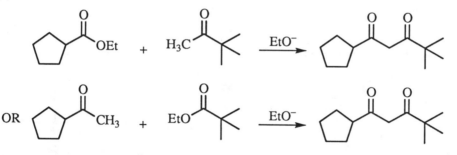

(d) mixed Claisen

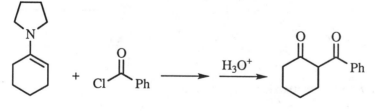

(e) mixed aldol

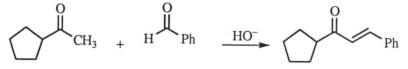

(f) enamine acylation—attempt at aldol would give self-condensation

22-66

(a) $CH_3CH_2-\overset{O}{\underset{\|}{C}}-CH_2CH_3$
  $+ CO_2 + CH_3OH$

(b) 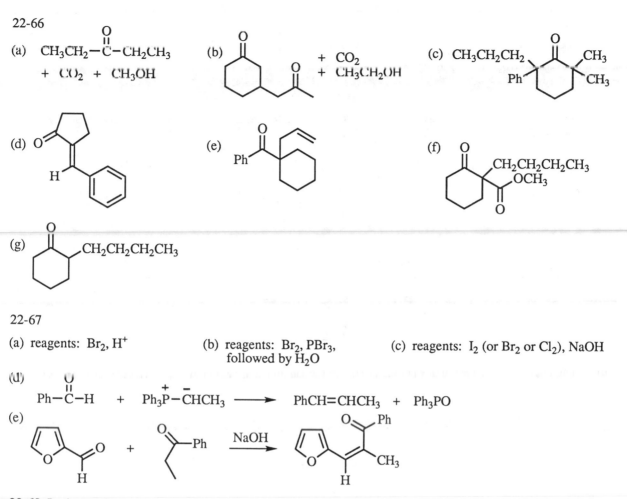 $+ CO_2$
  $+ CH_3CH_2OH$

(c) $CH_3CH_2CH_2$ ... CH_3
  Ph ... CH_3

(d)

(e) Ph ...

(f) $CH_2CH_2CH_2CH_3$
  OCH_3

(g) $CH_2CH_2CH_2CH_3$

22-67

(a) reagents: Br_2, H^+

(b) reagents: Br_2, PBr_3, followed by H_2O

(c) reagents: I_2 (or Br_2 or Cl_2), NaOH

(d)

$Ph-\overset{O}{\underset{\|}{C}}-H \quad + \quad Ph_3\overset{+}{P}-\overset{-}{C}HCH_3 \longrightarrow PhCH=CHCH_3 \quad + \quad Ph_3PO$

(e)

22-68 In the products, the wavy lines indicate the bonds that must be made by alkylation, before hydrolysis and decarboxylation produce the substituted acetic acid.

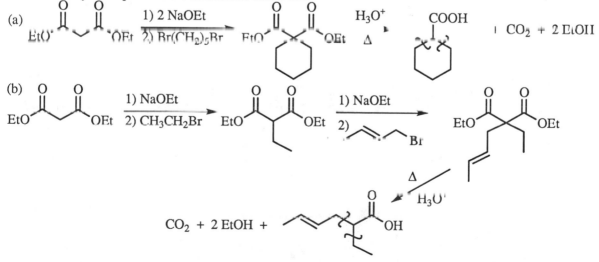

527

22-68 continued

(c)

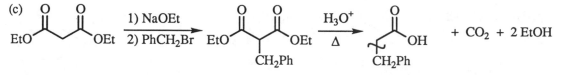

22-69 In the products, the wavy lines indicate the bonds that must be made by alkylation, before hydrolysis and decarboxylation produce the substituted acetone.

(a)

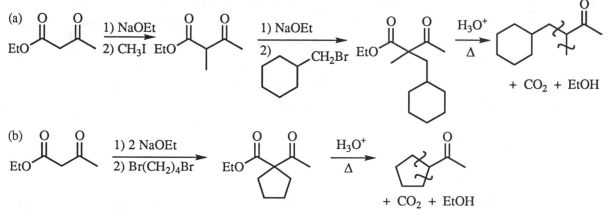

(b)

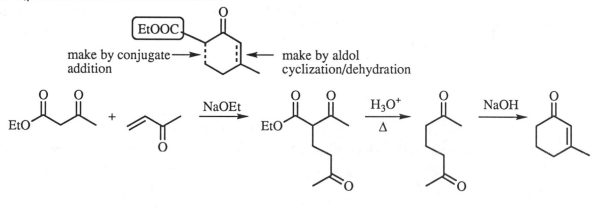

(c) The acetoacetic ester synthesis makes substituted acetone, so where is the acetone in this product?

substituted acetone

The single bond to this substituted acetone can be made by the acetoacetic ester synthesis. How can we make the α,β double bond? Aldol condensation!

make by conjugate → addition
make by aldol ← cyclization/dehydration

22-70 These compounds are made by aldol condensations followed by other reactions. The key is to find the skeleton make by the aldol.

(a) Where is the possible α,β-unsaturated carbonyl in this skeleton?

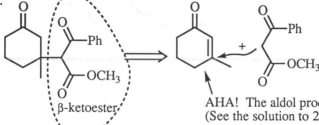

forward synthesis

(b) The aldol skeleton is not immediately apparent in this formidable product. What can we see from it? Most obvious is the β-dicarbonyl (β-ketoester) which we know to be a good nucleophile, capable of substitution or Michael addition. In this case, Michael addition is most likely as the site of attack is β to another carbonyl.

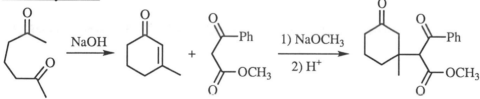

AHA! The aldol product reveals itself.
(See the solution to 22-69 (c).)

forward synthesis

(c) The key in this product is the α-nitroketone, the equivalent of a β-dicarbonyl system, capable of doing Michael addition to the β-carbon of the other carbonyl.

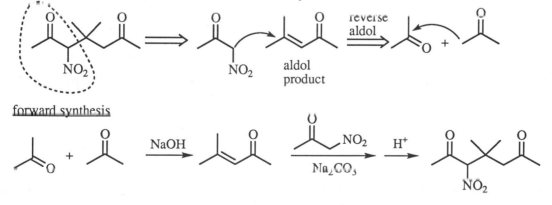

forward synthesis

529

22-71

(a)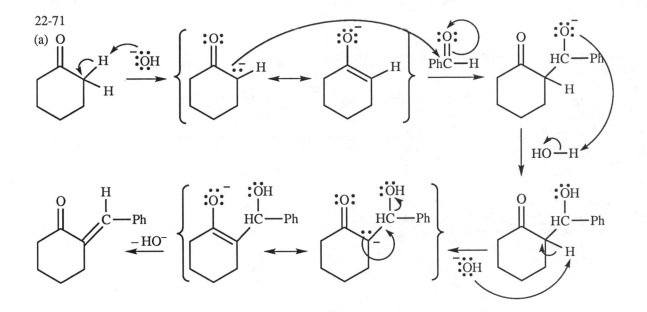

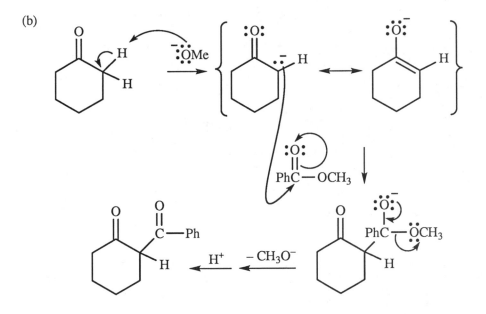

(b)

22-71 continued

(c) Robinson annulations are explained most easily by remembering that the first step is a Michael addition, followed by aldol cyclization with dehydration.

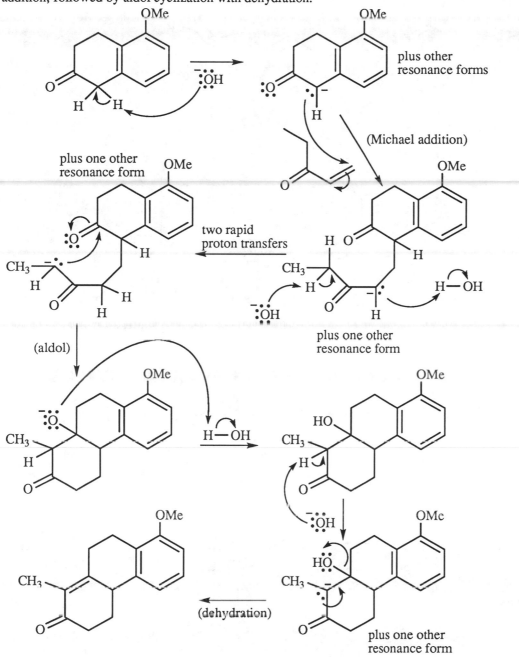

(e)

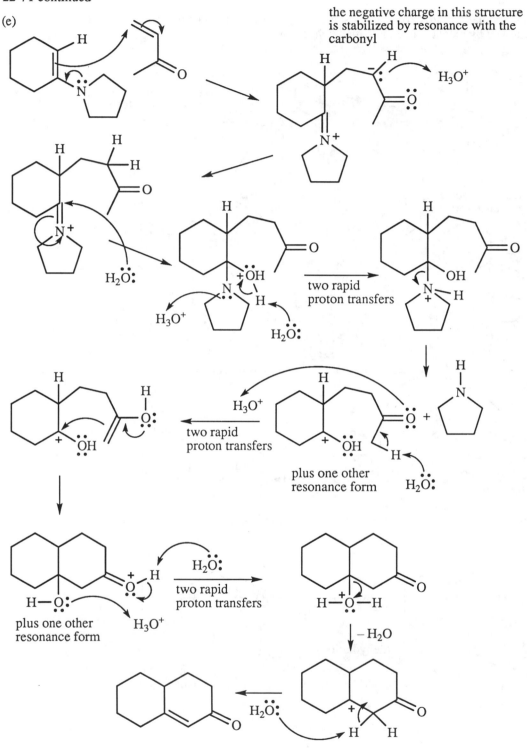

the negative charge in this structure is stabilized by resonance with the carbonyl

two rapid proton transfers

two rapid proton transfers

plus one other resonance form

plus one other resonance form

two rapid proton transfers

$-H_2O$

22-72

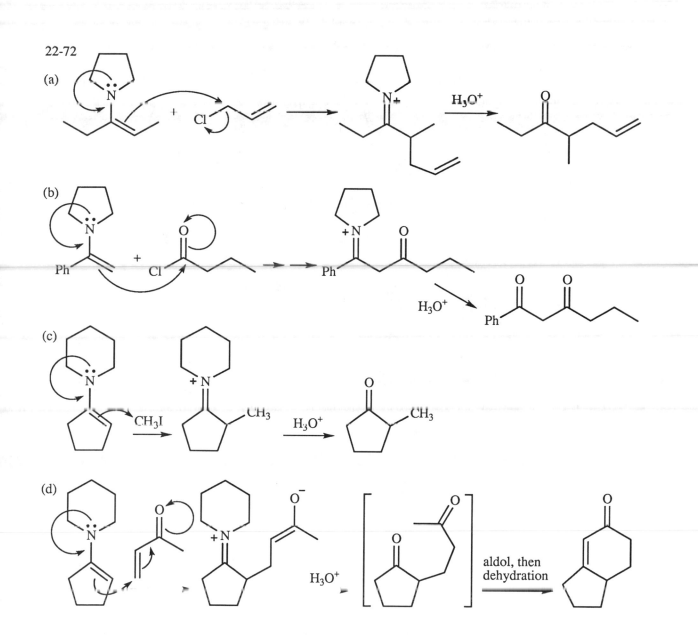

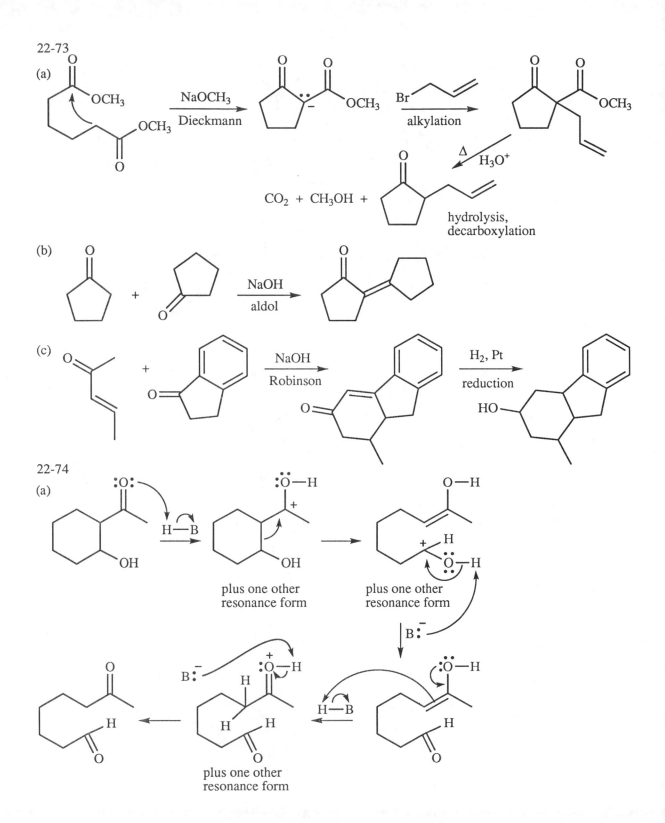

22-73

(a)

(b)

(c)

22-74

(a)

plus one other
resonance form

plus one other
resonance form

plus one other
resonance form

534

(b)

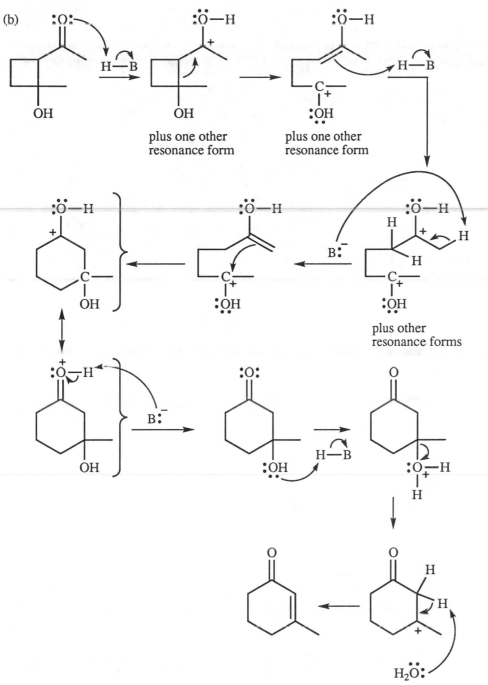

plus one other
resonance form

plus one other
resonance form

plus other
resonance forms

22-74 continued

(c)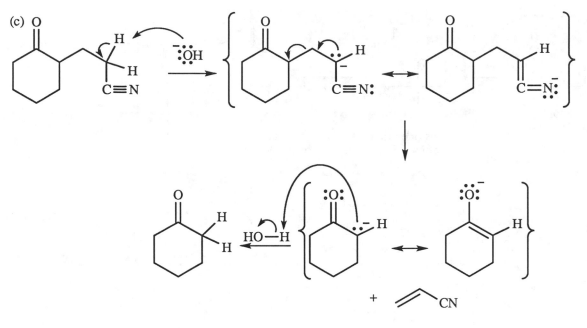

22-75 All of these Robinson annulations are catalyzed by NaOH.

(a)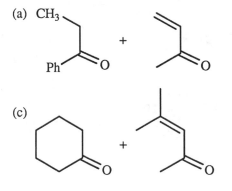

(b)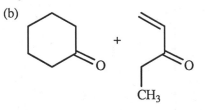

(c)

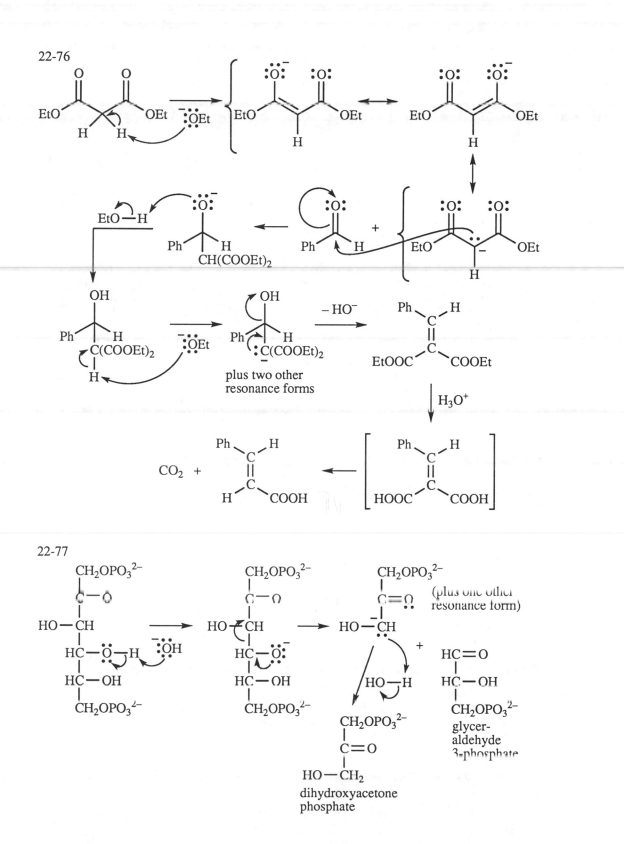

22-76

22-77

22-78 This is an aldol condensation. **P** stands for a protein chain in this problem.

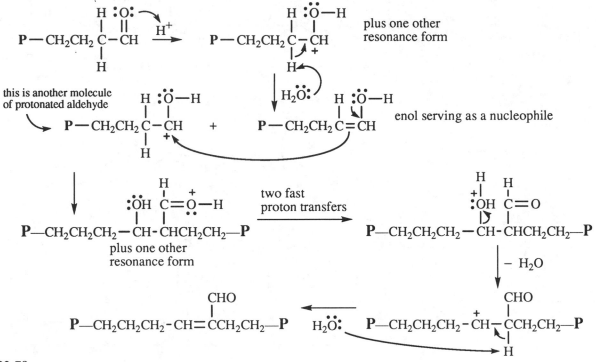

22-79

(a) aldol followed by Michael

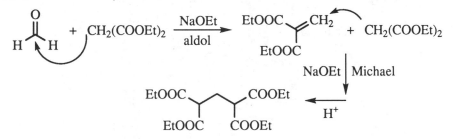

(b) Michael followed by Claisen condensation; hydrolysis and decarboxylation

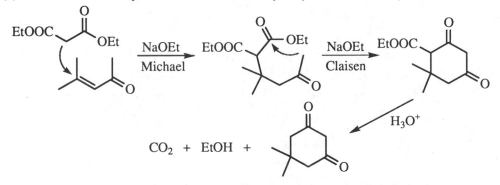

22-79 continued

(c) aldol, Michael, aldol cyclization, decarboxylation

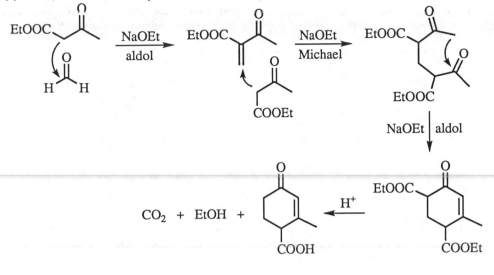

CAUTION! This chapter of the Solutions Manual contains a significant deviation from standard organic notation. Historically, chemists working with carbohydrates have used bonds with no atoms on the end to indicate a bond to *hydrogen*, not to methyl as we have seen until now. For consistency with tradition in structures of carbohydrates, and for simplicity of these sometimes cluttered pictures, the structures in this chapter and the structures *of carbohydrates* in subsequent chapters will follow this symbolism. Examples:

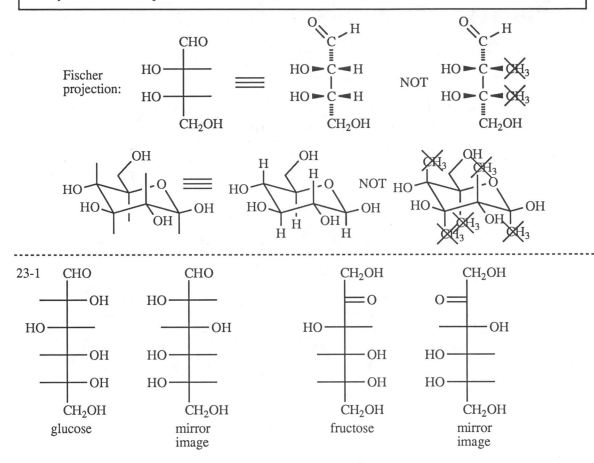

23-1

glucose

mirror image

fructose

mirror image

All four of these compounds are chiral and optically active.

23-2

(a)

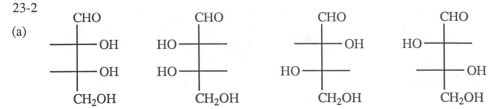

two chiral centers ⇒ four stereoisomers (two pairs of enantiomers) if none are meso

23-2 continued

(b)

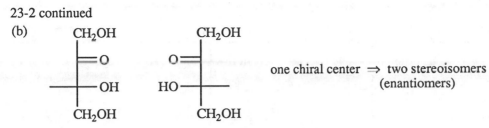

one chiral center $\Rightarrow$ two stereoisomers
(enantiomers)

(c) An aldohexose has four chiral carbons and sixteen stereoisomers. A ketohexose has three chiral carbons and eight stereoisomers.

23-3

(a)

```
   CH₂OH
     |
     C=O
     |
   CH₂OH
```

(b)

```
  CHO              CHO
   |                |
   ─OH         HO─
   |                |
  CH₂OH            CH₂OH
```

23-4

```
   CHO
    |
HO─
    |
   ─OH
    |
HO─
    |
HO─
    |
  CH₂OH

L-(−)-glucose
```

```
   CHO
    |
   ─OH
    |
HO─
    |
HO─
    |
  CH₂OH

L-(+)-arabinose
```

```
   CHO
    |
HO─
    |
HO─
    |
  CH₂OH

L-(+)-erythrose
```

```
   CHO
    |
HO─
    |
  CH₂OH

L-(−)-glyceraldehyde
```

23-5

```
        2
       CHO
        ⋮
 4 H ► C ◄ OH  1
        ⋮
      CH₂OH
        3

     D = R
```

```
        2
       CHO
        ⋮
1 HO ► C ◄ H  4
        ⋮
      CH₂OH
        3

     L = S
```

23-6

```
   CH₂CH₂CH₃
     |
     ─OH
     |
HO─
     |
   CH₂OH
```

```
   CH₂CH₂CH₃
     |
HO─
     |
     ─OH
     |
   CH₂OH
```

23-7

(a)
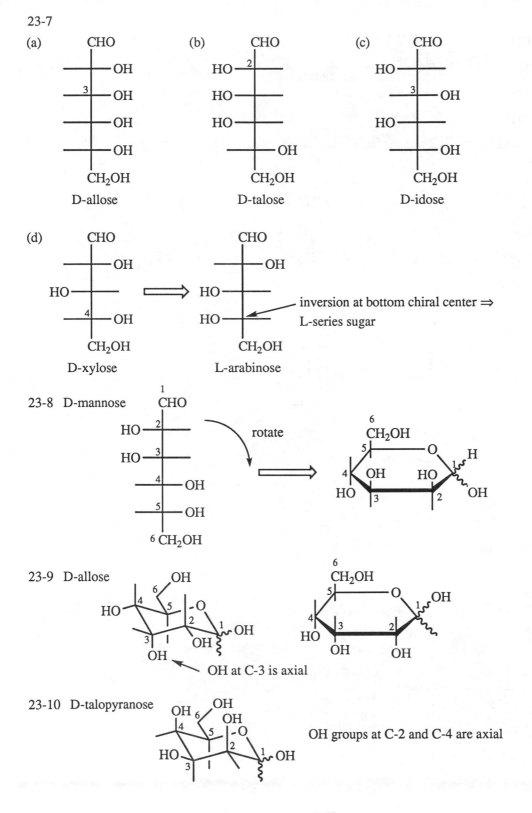
D-allose

(b)
CHO
HO — 2
HO
HO
OH
CH₂OH
D-talose

(c)
CHO
HO
HO — 3 — OH
HO
OH
CH₂OH
D-idose

(d)
CHO
OH
HO
4 — OH
CH₂OH
D-xylose

⟹

CHO
OH
HO
HO
CH₂OH
L-arabinose

inversion at bottom chiral center ⟹
L-series sugar

23-8 D-mannose

1 CHO
HO — 2
HO — 3
4 — OH
5 — OH
6 CH₂OH

rotate ⟹

OH groups drawing

23-9 D-allose

OH at C-3 is axial

23-10 D-talopyranose

OH groups at C-2 and C-4 are axial

542

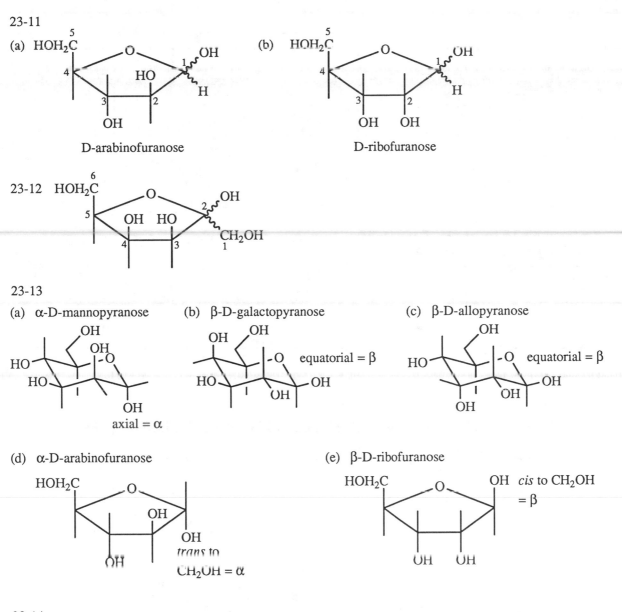

23-11

(a) D-arabinofuranose

(b) D-ribofuranose

23-12

23-13

(a) α-D-mannopyranose

axial = α

(b) β-D-galactopyranose

equatorial = β

(c) β-D-allopyranose

equatorial = β

(d) α-D-arabinofuranose

OH
trans to
CH$_2$OH = α

(e) β-D-ribofuranose

OH *cis* to CH$_2$OH
= β

23-14

a = fraction of galactose as the α anomer; b = fraction of galactose as the β anomer

a (+ 150.7°) + b (+ 52.8°) = + 80.2°

a + b = 1; b = 1 − a

a (+ 150.7°) + (1 - a) (+ 52.8°) = + 80.2° $\xrightarrow{\text{solve for "a"}}$ a = 0.28 ; b = 0.72

The equilibrium mixture contains 28% of the α anomer and 72% of the β anomer.

543

23-15

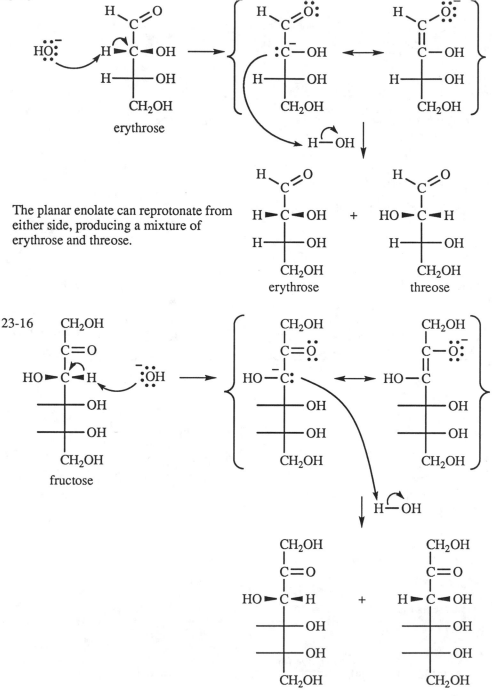

The planar enolate can reprotonate from either side, producing a mixture of erythrose and threose.

23-16

544

23-17

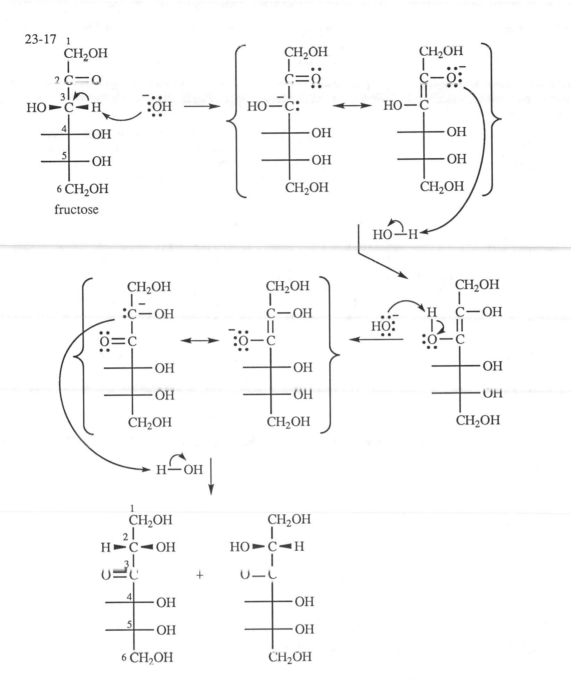

fructose

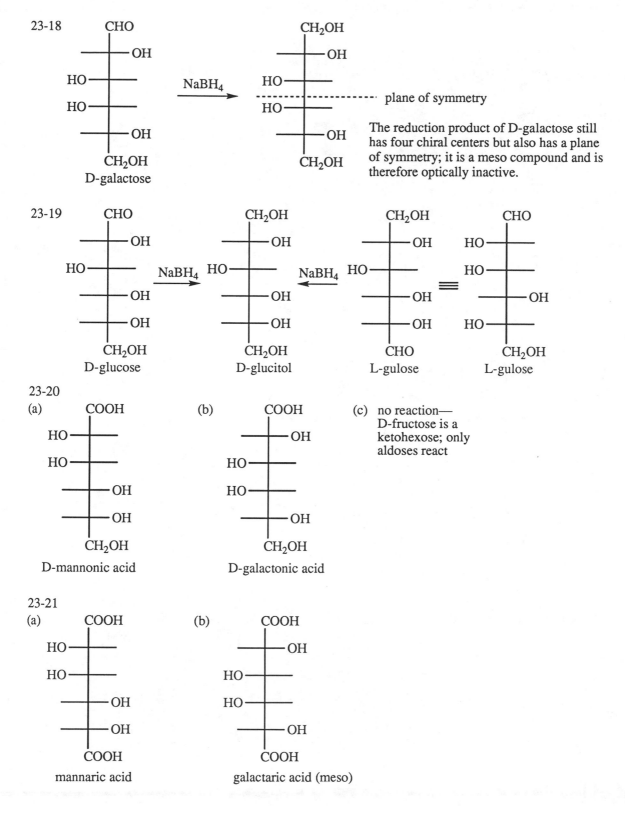

23-18

CHO
—OH
HO—
HO—
—OH
CH₂OH
D-galactose

NaBH₄ →

CH₂OH
—OH
HO—
HO— -------- plane of symmetry
—OH
CH₂OH

The reduction product of D-galactose still has four chiral centers but also has a plane of symmetry; it is a meso compound and is therefore optically inactive.

23-19

CHO
—OH
HO—
—OH
—OH
CH₂OH
D-glucose

NaBH₄ →

CH₂OH
—OH
HO—
—OH
—OH
CH₂OH
D-glucitol

← NaBH₄

CH₂OH
—OH
HO—
—OH
—OH
CHO
L-gulose

≡

CHO
HO—
HO—
—OH
HO—
CH₂OH
L-gulose

23-20

(a)
COOH
HO—
HO—
—OH
—OH
CH₂OH
D-mannonic acid

(b)
COOH
—OH
HO—
HO—
—OH
CH₂OH
D-galactonic acid

(c) no reaction—
D-fructose is a
ketohexose; only
aldoses react

23-21

(a)
COOH
HO—
HO—
—OH
—OH
COOH
mannaric acid

(b)
COOH
—OH
HO—
HO—
—OH
COOH
galactaric acid (meso)

23-22

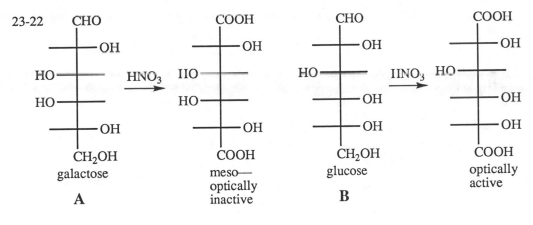

galactose

A

meso—
optically
inactive

glucose

B

optically
active

23-23

(a) not reducing: an acetal ending in "oside"
(b) reducing: a hemiacetal ending in "ose"
(c) reducing: a hemiacetal ending in "ose"
(d) not reducing: an acetal ending in "oside"
(e) reducing: one of the rings has a hemiacetal
(f) not reducing: all anomeric carbons are in acetal form; suc<u>rose</u>, a common name, is misleading as it suggests a hemiacetal form

23-24

(a)

axial = α

(c)

axial = α

(d)

cis to
CH_2OH = β

23-25

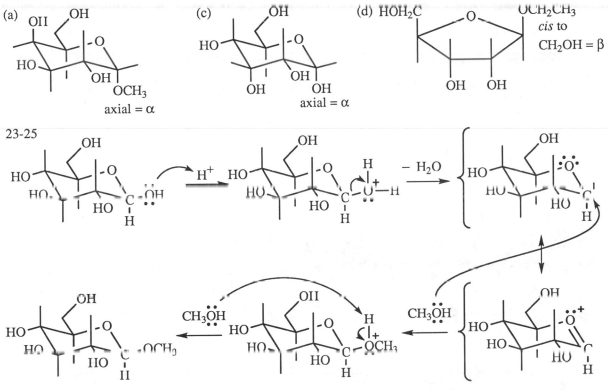

547

23-26

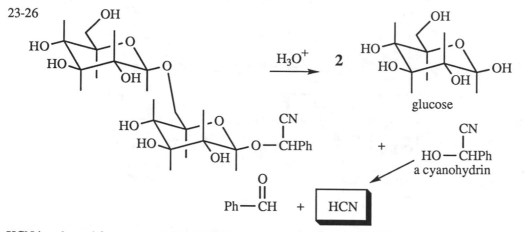

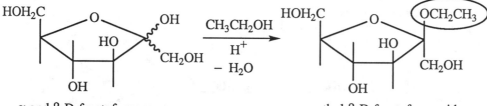

HCN is released from amygdalin. HCN is a potent cytotoxic (cell-killing) agent, particularly toxic to nerve cells.

23-27

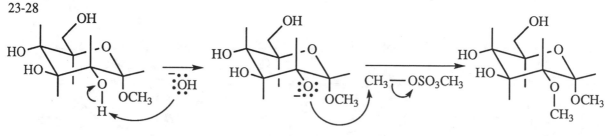

α and β-D-fructofuranose

ethyl β-D-fructofuranoside

The aglycone in each product is circled.

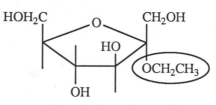

ethyl α-D-fructofuranoside

23-28

HOH₂C ... (reaction scheme)

23-29

(a)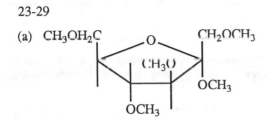

(h)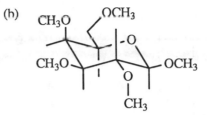

23-30

(a)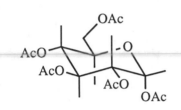

(b)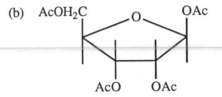

23-31

(a)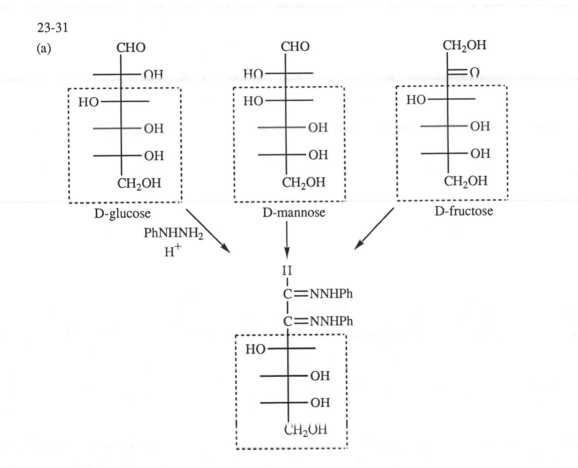

23-31 continued

(b)

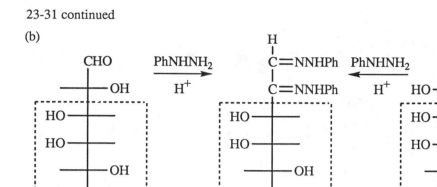

D-Talose must be the C-2 epimer of D-galactose.

23-32   Reagents for the Ruff degradation are: 1. Br_2, H_2O; 2. H_2O_2, $Fe_2(SO_4)_3$.

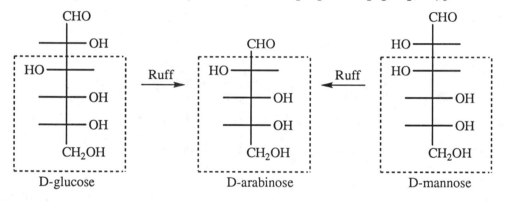

23-33   Reagents for the Ruff degradation are: 1. Br_2, H_2O; 2. H_2O_2, $Fe_2(SO_4)_3$.

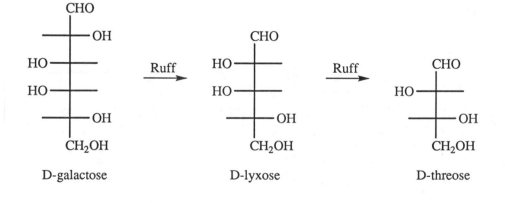

23-34 Reagents for the Ruff degradation are: 1. Br_2, H_2O; 2. H_2O_2, $Fe_2(SO_4)_3$.

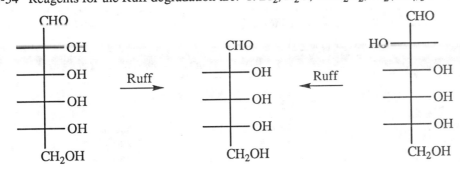

D-allose   D-altrose

23-35

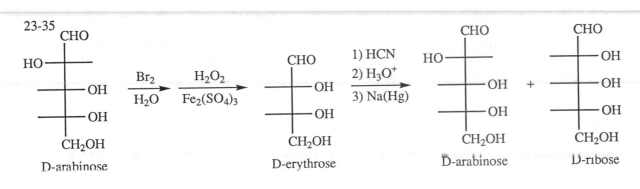

D-arabinose   D-erythrose   D-arabinose   D-ribose

23-36

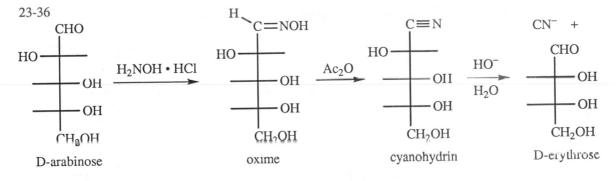

D-arabinose   oxime   cyanohydrin   D-erythrose

23-37 Solve this problem by working backward from (+)-glyceraldehyde.

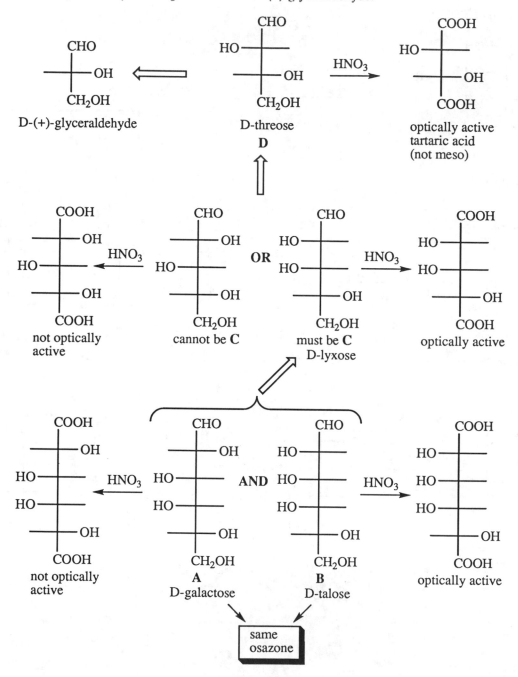

23-38 Solve this problem by working backward from (+)-glyceraldehyde.

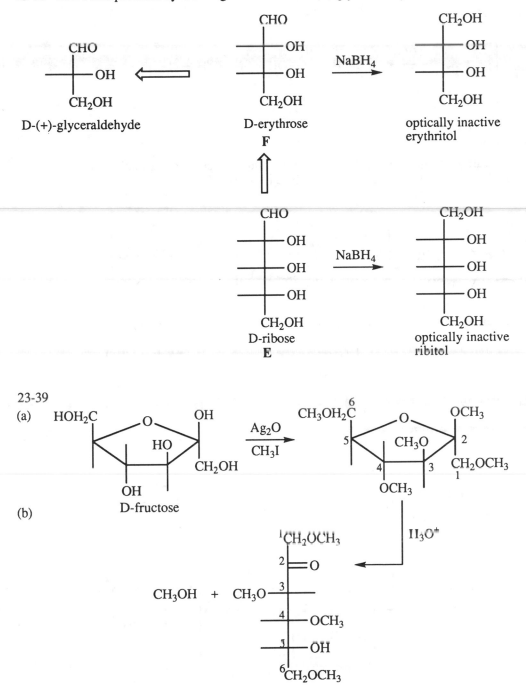

23-39

(a)

(b)

(c) Determining that the open chain form of fructose is a ketone at C-2 with a free OH at C-5 shows that fructose exists as a furanose hemiacetal.

23-40

(a)

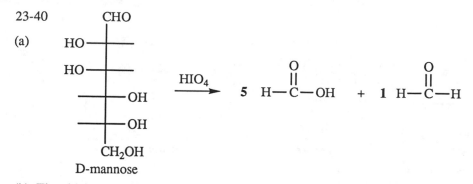

D-mannose

(b) The aldehyde carbonyl will give a molecule of formic acid upon periodate oxidation. The ketone carbonyl will produce a molecule of CO_2 that will bubble out of solution.

23-41

(a)

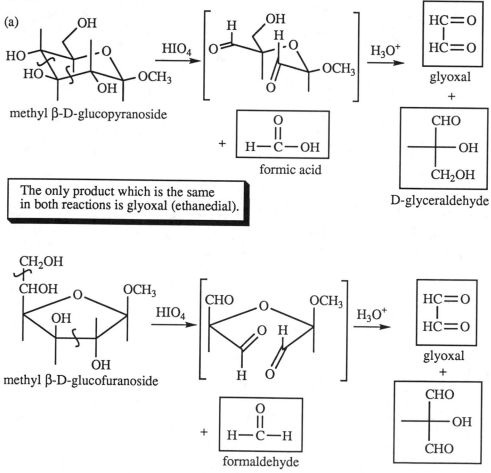

methyl β-D-glucopyranoside

formic acid

glyoxal

D-glyceraldehyde

The only product which is the same in both reactions is glyoxal (ethanedial).

methyl β-D-glucofuranoside

formaldehyde

glyoxal

23-41 continued

(b)

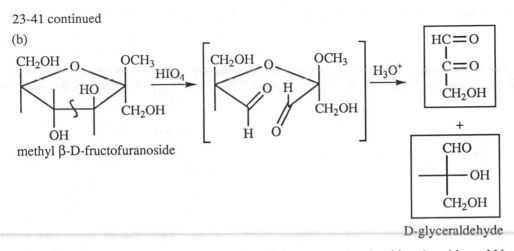

methyl β-D-fructofuranoside

D-glyceraldehyde

Had the fructose acetal been in a six-membered ring, one molecule of formic acid would have been produced, as in part (a). Isolation of two three-carbon molecules proves the presence of a furanoside.

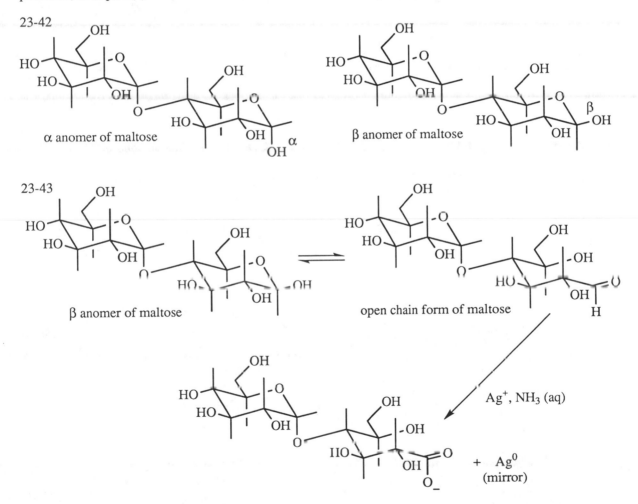

23-42

α anomer of maltose

β anomer of maltose

23-43

β anomer of maltose

open chain form of maltose

Ag^+, NH_3 (aq)

+ Ag^0
(mirror)

23-44 Lactose is a hemiacetal. Therefore, it can mutarotate and is a reducing sugar.

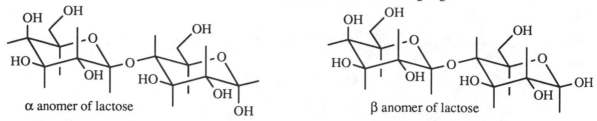

α anomer of lactose   β anomer of lactose

23-45 Gentiobiose is a hemiacetal; in water, the hemiacetal is in equilibrium with the open-chain form and can react as an aldehyde. Gentiobiose can mutarotate and is a reducing sugar.

23-46 Trehalose must be two glucose molecules connected by an α-1,1'-glycoside.

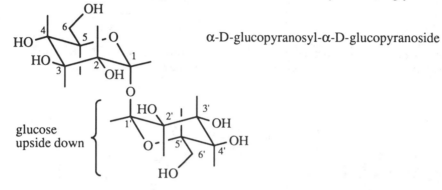

α-D-glucopyranosyl-α-D-glucopyranoside

glucose upside down

23-47

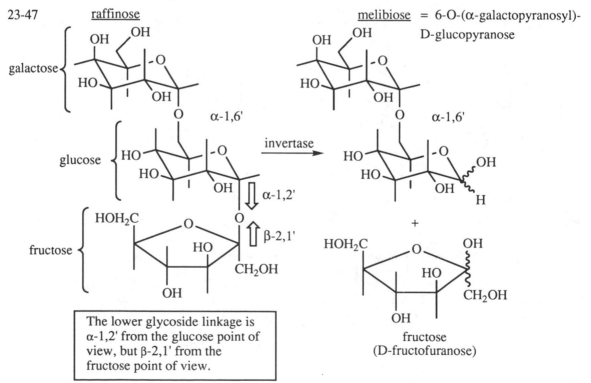

raffinose   melibiose = 6-O-(α-galactopyranosyl)-D-glucopyranose

galactose

glucose

α-1,6'

invertase

α-1,2'

β-2,1'

fructose

The lower glycoside linkage is α-1,2' from the glucose point of view, but β-2,1' from the fructose point of view.

+

fructose (D-fructofuranose)

23-48 cellulose acetate

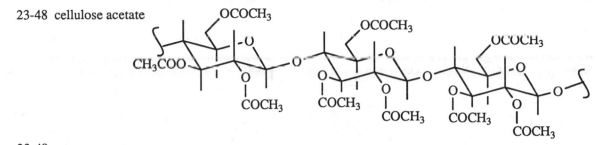

23-49

cytosine:

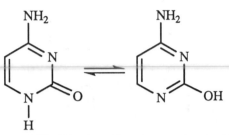

uracil:

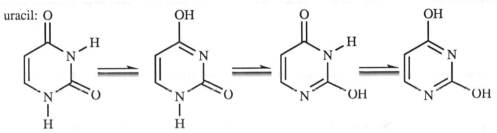

guanine:

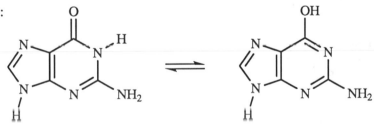

557

23-50 Aminoglycosides, including nucleosides, are similar to acetals: stable to base, cleaved by acid.

(a)

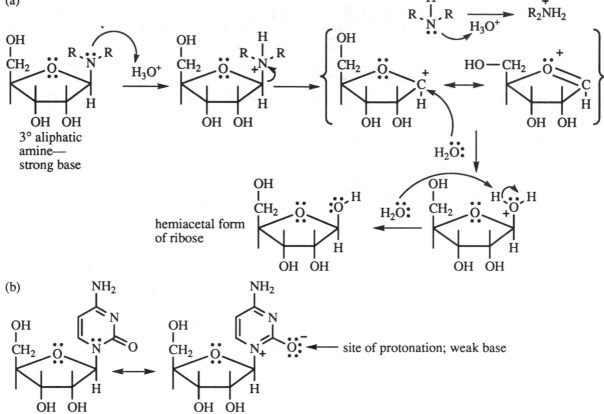

(b)

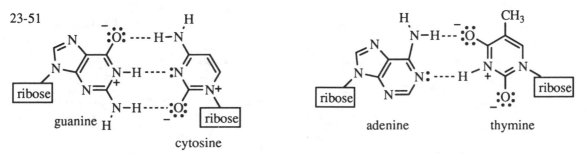

Nucleosides are less rapidly hydrolyzed in aqueous acid because the site of protonation (in cytidine, the oxygen shown with the negative charge in the second resonance form) is much less basic than the aliphatic amine in an aminoglycoside. Nucleosides require stronger acid, or longer time and higher temperature, to be hydrolyzed.

This is important in living systems as it would cause genetic damage or even death of an organism if its DNA or RNA were too easily decomposed. Organisms go to great length and expend considerable energy to maintain the structural integrity of their DNA.

23-51

The polar resonance forms show how the hydrogen bonds are particularly strong. Each oxygen has significant negative charge, and in each pair, one H—N is polarized more strongly because the N has positive charge.

23-52 Please refer to solution 1-19, page 12 of this Solutions Manual.

23-53

(a)

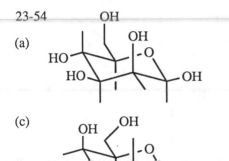

(b)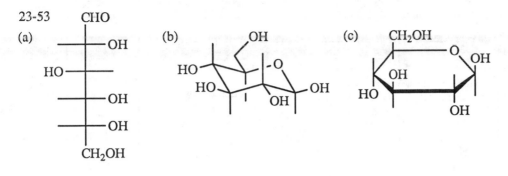

(c)

23-54

(a)

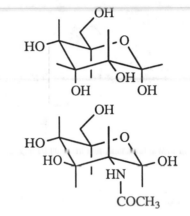

(b)

(c)

(d)

23-55

(a) D-aldohexose
(b) D-aldopentose
(c) L-ketohexose
(d) L-aldohexose
(e) D-ketopentose
(f) L-aldotetrose

23-56

(a)

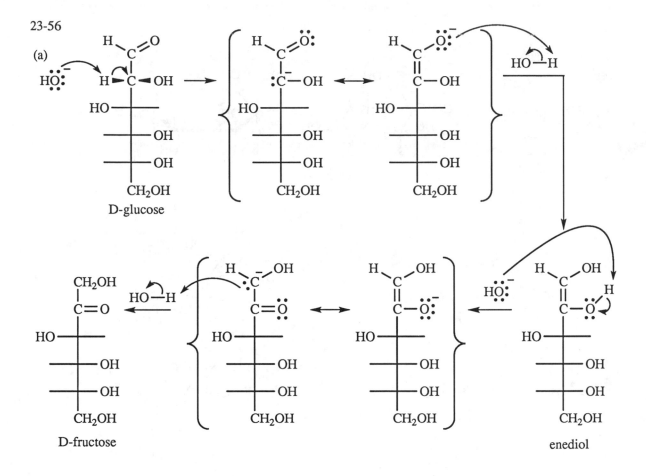

D-glucose

enediol

D-fructose

(b) The α isomer has the anomeric OH *trans* to the CH₂OH off of C-5. The β-anomer has these groups *cis*.

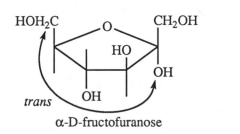

α-D-fructofuranose

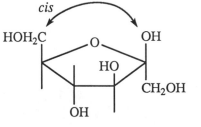

β-D-fructofuranose

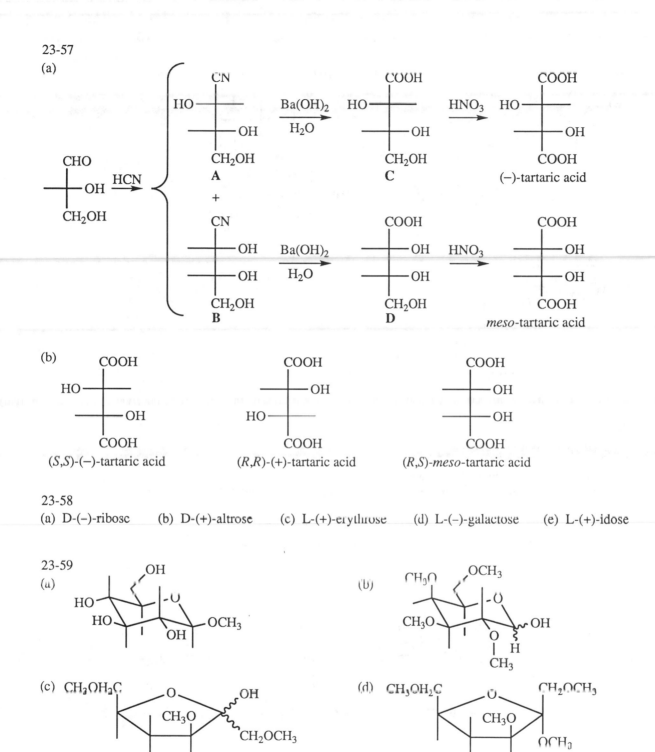

23-57

(a)

A + B

$\xrightarrow{\text{Ba(OH)}_2 / \text{H}_2\text{O}}$ C (→ (−)-tartaric acid via HNO₃) and D (→ *meso*-tartaric acid via HNO₃)

(b)

(S,S)-(−)-tartaric acid   (R,R)-(+)-tartaric acid   (R,S)-*meso*-tartaric acid

23-58

(a) D-(−)-ribose   (b) D-(+)-altrose   (c) L-(+)-erythrose   (d) L-(−)-galactose   (e) L-(+)-idose

23-59

(a)

(b)

(c)

(d)

23-60

(a)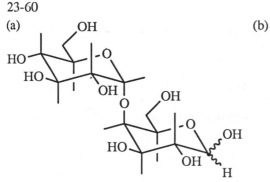

(b)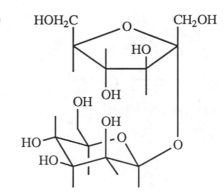

(c)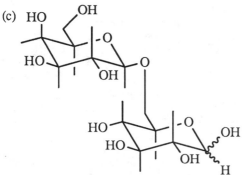

23-61

(a) methyl β-D-fructofuranoside
(b) 3,6-di-O-methyl-β-D-mannopyranose
(c) 4-O-(α-D-fructofuranosyl)-β-D-galactopyranose
(d) β-D-N-acetylgalactopyranosamine, or 2-acetamido-2-deoxy-β-D-galactopyranose

23-62 These are reducing sugars and would undergo mutarotation:

—in problem 23-59: (b) and (c);
—in problem 23-60: (a) and (c);
—in problem 23-61: (b), (c), and (d)

23-63

(a)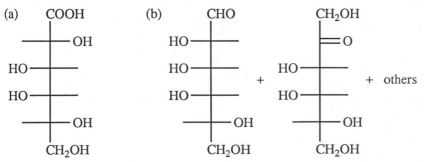

(b)

(c)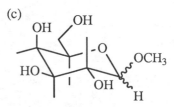

23-63 continued

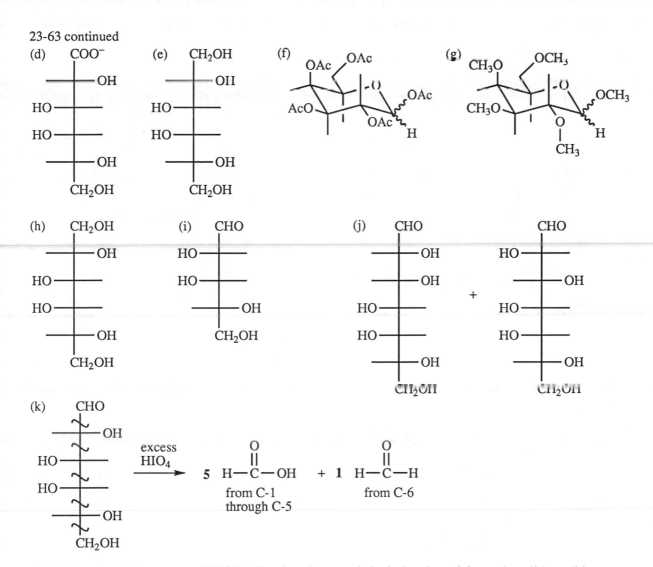

(d)
COO⁻
— OH
HO —
HO —
— OH
CH₂OH

(e)
CH₂OH
— OH
HO —
HO —
— OH
CH₂OH

(f) OAc OAc
AcO—
OAc
OAc
H

(g) CH₃O OCH₃
CH₃O—
OCH₃
O
CH₃
H

(h)
CH₂OH
— OH
HO —
HO —
— OH
CH₂OH

(i)
CHO
HO —
HO —
— OH
CH₂OH

(j)
CHO
— OH
— OH
HO —
HO —
— OH
CH₂OH

+

CHO
HO —
— OH
HO —
HO —
— OH
CH₂OH

(k)
CHO
— OH
HO —
HO —
— OH
CH₂OH

excess HIO₄ →

$$5 \; H-\overset{O}{\underset{||}{C}}-OH \;+\; 1 \; H-\overset{O}{\underset{||}{C}}-H$$

from C-1 through C-5   from C-6

23-64 Use the milder reagent, CH₃I/Ag₂O, when the sugar is in the hemiacetal form; the mild conditions prevent isomerization. When the carbohydrate is present as an acetal (a glycoside), use the more basic reagent, NaOH/(CH₃)₂SO₄; an acetal is stable to basic conditions.

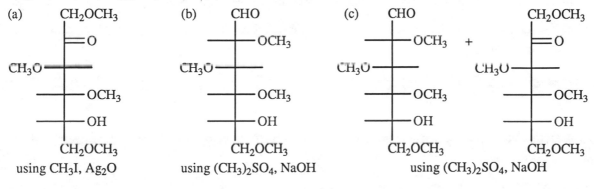

(a)
CH₂OCH₃
= O
CH₃O —
— OCH₃
— OH
CH₂OCH₃
using CH₃I, Ag₂O

(b)
CHO
— OCH₃
CH₃O —
— OCH₃
— OH
CH₂OCH₃
using (CH₃)₂SO₄, NaOH

(c)
CHO
— OCH₃
CH₃O —
— OCH₃
— OH
CH₂OCH₃

+

CH₂OCH₃
= O
CH₃O —
— OCH₃
— OH
CH₂OCH₃
using (CH₃)₂SO₄, NaOH

563

23-64 continued

(d)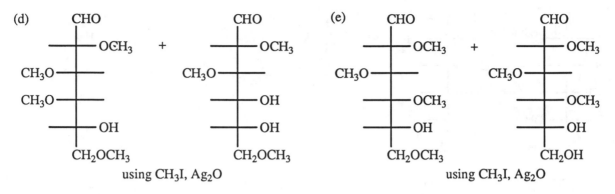

using CH₃I, Ag₂O

(e)

using CH₃I, Ag₂O

(f)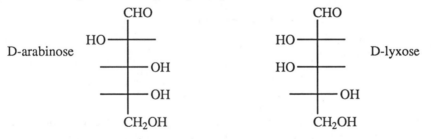

using $(CH_3)_2SO_4$, NaOH

23-65

(a) These D-aldopentoses will give optically active aldaric acids.

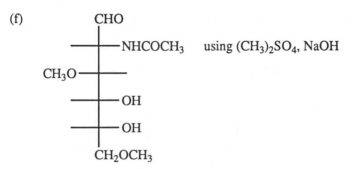

D-arabinose

D-lyxose

(b) Only D-threose (of the aldotetroses) will give an optically active aldaric acid.

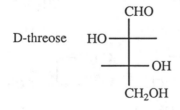

D-threose

(c) **X** is D-galactose.

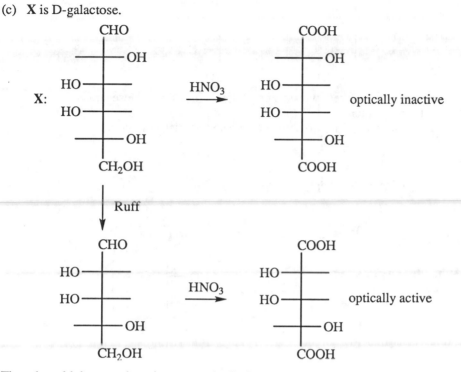

The other aldohexose that gives an optically inactive aldaric acid is D-allose, with all OH groups on the right side of the Fischer projection. Ruff degradation followed by nitric acid gives an optically *inactive* aldaric acid, however, so **X** cannot be D-allose.

(d) The optically active, five-carbon aldaric acid comes from the optically active pentose, not from the optically inactive, six-carbon aldaric acid. The principle is not violated.

(e)

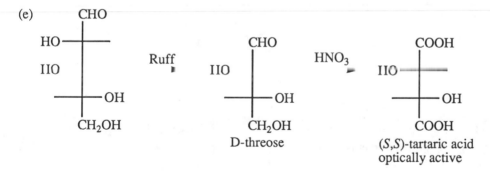

23-66
(a)

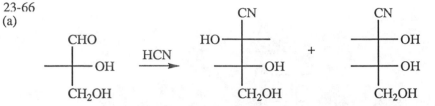

(b) The products are diastereomers with different physical properties. They could be separated by crystallization, distillation, or chromatography.
(c) Both products are optically active. Each has two chiral centers and no plane of symmetry.

23-67

(a) The Tollens reaction is run in aqueous base which isomerizes carbohydrates.

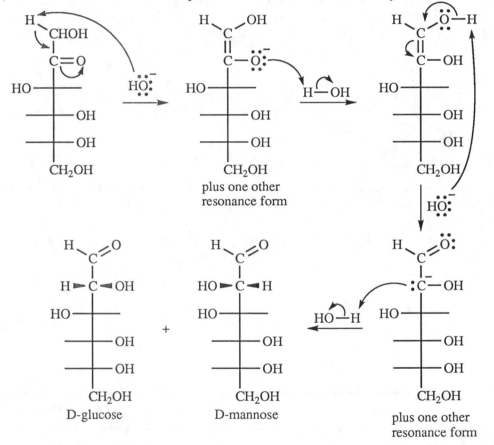

plus one other
resonance form

D-glucose   D-mannose

plus one other
resonance form

(b) Bromine water is acidic, not basic like the Tollens reagent. Carbohydrates isomerize quickly in base, but only very slowly in acid, so bromine water can oxidize without isomerization.

23-68 Tagatose is a monosaccharide, a ketohexose, that is found in the pyranose form.

(a)

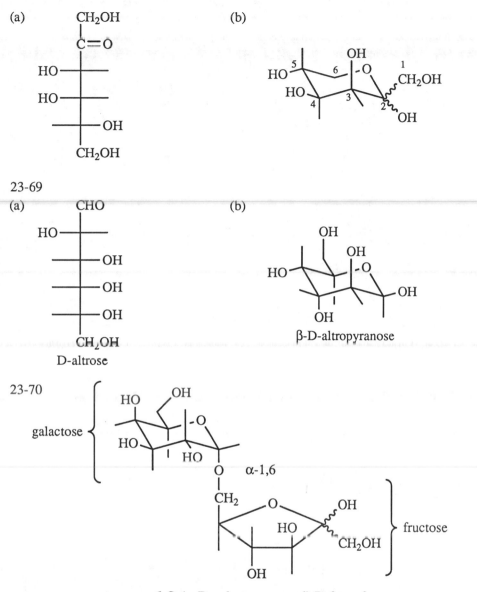

(b)

23-69

(a)

D-altrose

(b)

β-D-altropyranose

23-70

galactose

α-1,6

fructose

6-O-(α-D-galactopyranosyl)-D-fructofuranose

23-71

(a)

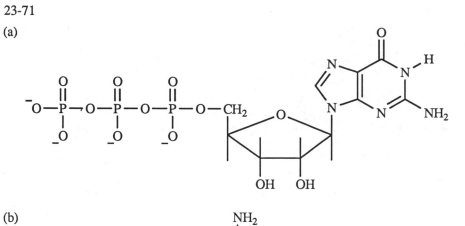

(b)

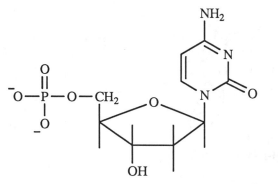

(c)

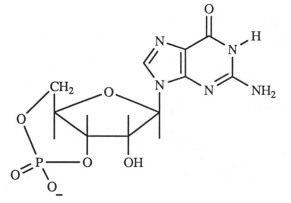

23-72 Bonds to H are omitted for simplicity.

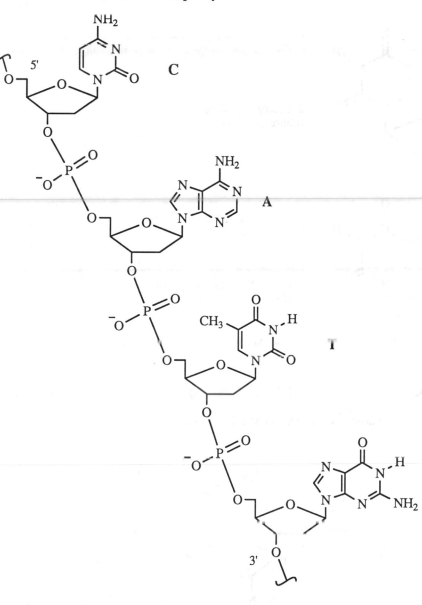

23-73

(a) No, there is no relation between the amount of G and A.

(b) Yes, this must be true mathematically.

(c) Chargaff's rule must apply only to double-stranded DNA. For each G in one strand, there is a complementary C in the opposing strand, but there is no correlation between G and C *in the same strand.*

23-74

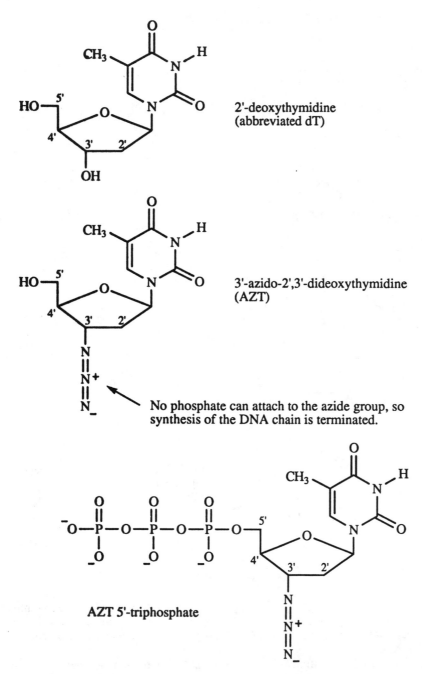

2'-deoxythymidine
(abbreviated dT)

3'-azido-2',3'-dideoxythymidine
(AZT)

No phosphate can attach to the azide group, so
synthesis of the DNA chain is terminated.

AZT 5'-triphosphate

570

23-75 Recall from section 19-17 that nitrous acid is unstable, generating nitrosonium ion.

(a)

$$H-O-N=O + H^+ \longrightarrow H_2O + \overset{+}{N}=O$$

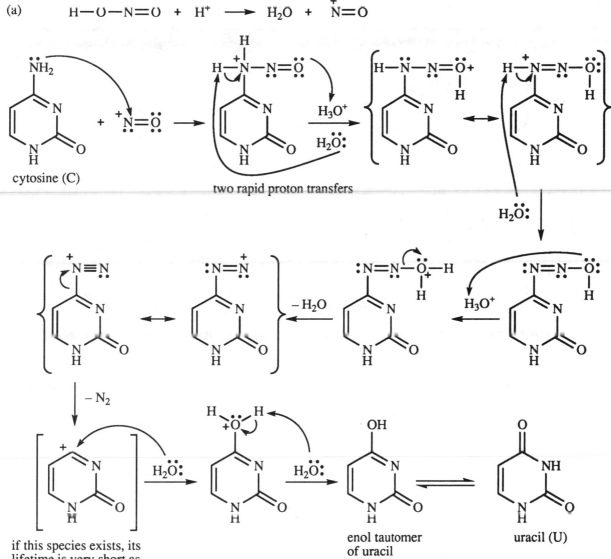

if this species exists, its
lifetime is very short as
water will attack quickly

enol tautomer
of uracil

uracil (U)

two rapid proton transfers

(b) In base pairing, cytosine pairs with guanine. If cytosine is converted to uracil, however, each replication will not carry the complement of cytosine (guanine) but instead will carry the complement of uracil (adenine). This is the definition of a mutation, where the wrong base is inserted in a nucleic acid chain.

(c) In RNA, the transformation of cytosine (C) to uracil (U) is not repaired because U is a base normally found in RNA. In DNA, however, thymine (with an extra methyl group) is used instead of uracil. If cytosine is diazotized to uracil, the DNA repair enzymes detect it as a mutation and correct it.

24-1

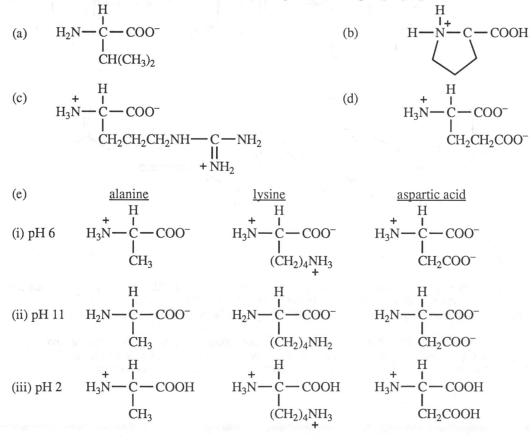

(a)

(b)

(c)

(d)

24-2 In their evolution, plants have needed to be more resourceful than animals in developing biochemical mechanisms for survival. Thus, plants make more of their own required compounds than animals do. The amino acid phenylalanine is produced by plants but required in the diet of mammals. To interfere with a plant's production of phenylalanine is fatal to the plant, but since humans do not produce phenylalanine, glyphosate is virtually non-toxic to us.

24-3 Here is a simple way of determining if a group will be protonated: *at solution pH below the group's* pK_a *value, the group will be protonated; at pH higher than the group's* pK_a *value, it will not be protonated.*

(a)

$$H_2N - \overset{\overset{\displaystyle H}{|}}{\underset{\underset{\displaystyle CH(CH_3)_2}{|}}{C}} - COO^-$$

(b)

$$H - \overset{\overset{\displaystyle H}{|} +}{N} - C - COOH$$

(c)

$$\overset{+}{H_3N} - \overset{\overset{\displaystyle H}{|}}{\underset{\underset{\displaystyle CH_2CH_2CH_2NH - \underset{\underset{\displaystyle + NH_2}{||}}{C} - NH_2}{|}}{C}} - COO^-$$

(d)

$$\overset{+}{H_3N} - \overset{\overset{\displaystyle H}{|}}{\underset{\underset{\displaystyle CH_2CH_2COO^-}{|}}{C}} - COO^-$$

(e)

	alanine	lysine	aspartic acid
(i) pH 6	$H_3\overset{+}{N} - \overset{H}{\underset{CH_3}{C}} - COO^-$	$H_3\overset{+}{N} - \overset{H}{\underset{(CH_2)_4\overset{+}{NH_3}}{C}} - COO^-$	$H_3\overset{+}{N} - \overset{H}{\underset{CH_2COO^-}{C}} - COO^-$
(ii) pH 11	$H_2N - \overset{H}{\underset{CH_3}{C}} - COO^-$	$H_2N - \overset{H}{\underset{(CH_2)_4NH_2}{C}} - COO^-$	$H_2N - \overset{H}{\underset{CH_2COO^-}{C}} - COO^-$
(iii) pH 2	$H_3\overset{+}{N} - \overset{H}{\underset{CH_3}{C}} - COOH$	$H_3\overset{+}{N} - \overset{H}{\underset{(CH_2)_4\overset{+}{NH_3}}{C}} - COOH$	$H_3\overset{+}{N} - \overset{H}{\underset{CH_2COOH}{C}} - COOH$

24-4

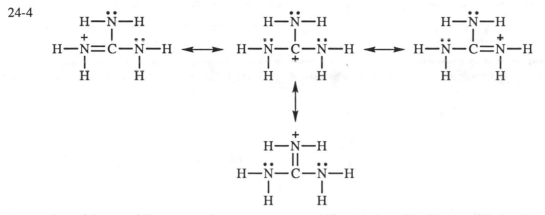

Protonation of the guanidino group gives a resonance-stabilized cation with all octets filled and the positive charge delocalized over three nitrogen atoms. Arginine's strongly basic isoelectric point reflects the unusual basicity of the guanidino group due to this resonance stabilization in the protonated form. (See Problems 1-38 and 19-52(a).)

24-5

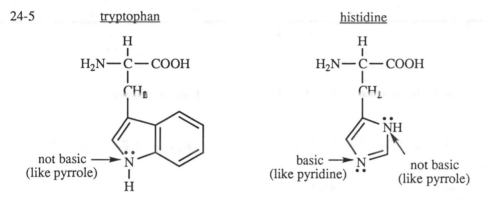

The basicity of any nitrogen depends on its electron pair's availability for bonding with a proton. In tryptophan, the nitrogen's electron pair is part of the aromatic π system; without this electron pair in the π system, the molecule would not be aromatic. Using this electron pair for bonding to a proton would therefore destroy the aromaticity—not a favorable process.

In the imidazole ring of histidine, the electron pair of one nitrogen is also part of the aromatic π system and is unavailable for bonding; this nitrogen is not basic. The electron pair on the other nitrogen, however, is in an sp^2 orbital available for bonding, and is about as basic as pyridine.

24-6 At pH 9.7, alanine (isoelectric point (IEP) 6.0) has a charge of –1 and will migrate to the anode. Lysine (IEP 9.7) is at its isoelectric point and will not move. Aspartic acid (IEP 2.8) has a charge of –2 and will also migrate to the anode, faster than alanine.

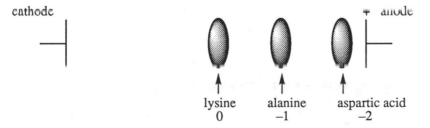

24-7 At pH 6.0, tryptophan (IEP 5.9) has a charge of zero and will not migrate. Cysteine (IEP 5.0) has a partial negative charge and will move toward the anode. Histidine (IEP 7.6) has a partial positive charge and will move toward the cathode.

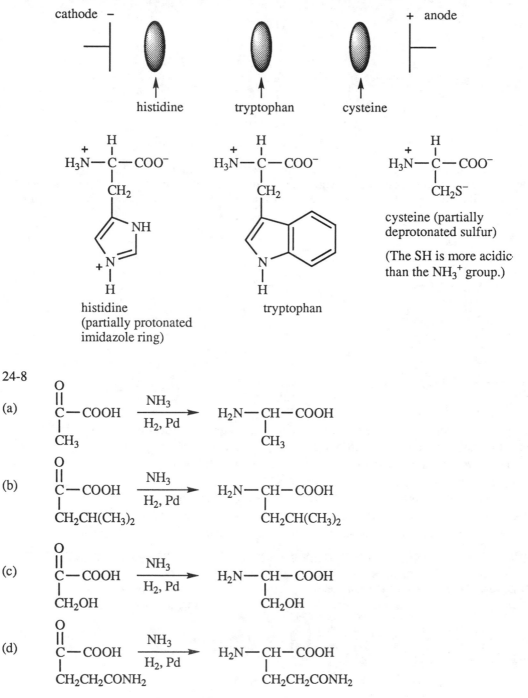

24-8

(a)
$$CH_3-C(=O)-COOH \xrightarrow[\text{H}_2, \text{Pd}]{\text{NH}_3} H_2N-CH(CH_3)-COOH$$

(b)
$$(CH_3)_2CHCH_2-C(=O)-COOH \xrightarrow[\text{H}_2, \text{Pd}]{\text{NH}_3} H_2N-CH(CH_2CH(CH_3)_2)-COOH$$

(c)
$$HOCH_2-C(=O)-COOH \xrightarrow[\text{H}_2, \text{Pd}]{\text{NH}_3} H_2N-CH(CH_2OH)-COOH$$

(d)
$$H_2NOCCH_2CH_2-C(=O)-COOH \xrightarrow[\text{H}_2, \text{Pd}]{\text{NH}_3} H_2N-CH(CH_2CH_2CONH_2)-COOH$$

24-9 All of these reactions use: first arrow: Br_2/PBr_3, followed by H_2O workup; second arrow: excess NH_3, followed by neutralizing workup.

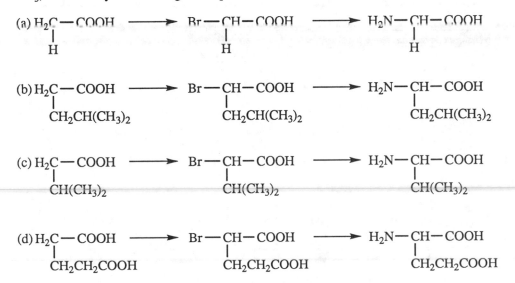

(a) $H_2C-COOH$ ⟶ $Br-CH-COOH$ ⟶ $H_2N-CH-COOH$
 with H substituent

(b) $H_2C-COOH$ ⟶ $Br-CH-COOH$ ⟶ $H_2N-CH-COOH$
 with $CH_2CH(CH_3)_2$ substituent

(c) $H_2C-COOH$ ⟶ $Br-CH-COOH$ ⟶ $H_2N-CH-COOH$
 with $CH(CH_3)_2$ substituent

(d) $H_2C-COOH$ ⟶ $Br-CH-COOH$ ⟶ $H_2N-CH-COOH$
 with CH_2CH_2COOH substituent

In part (d), care must be taken to avoid reaction α to the other COOH. In practice, this would be accomplished by using less than one-half mole of bromine per mole of the diacid.

24-10

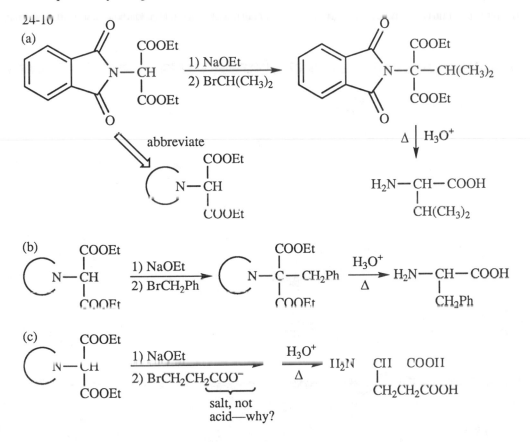

(a)
$$\text{phthalimide}-N-CH\begin{cases}COOEt\\COOEt\end{cases} \xrightarrow[\text{2) BrCH(CH}_3)_2]{\text{1) NaOEt}} \text{phthalimide}-N-C(CH(CH_3)_2)\begin{cases}COOEt\\COOEt\end{cases}$$

abbreviate

$$N-CH\begin{cases}COOEt\\COOEt\end{cases}$$

$$\Delta \Big| H_3O^+$$

$$H_2N-CH-COOH$$
$$\quad\quad\; CH(CH_3)_2$$

(b)
$$N-CH\begin{cases}COOEt\\COOEt\end{cases} \xrightarrow[\text{2) BrCH}_2Ph]{\text{1) NaOEt}} N-C(CH_2Ph)\begin{cases}COOEt\\COOEt\end{cases} \xrightarrow[\Delta]{H_3O^+} H_2N-CH-COOH$$
$$\quad\quad\quad\quad\quad\quad\quad\quad\quad\quad\quad\quad\quad\quad\quad\quad\quad\quad\quad\quad\quad\quad\quad\quad\quad CH_2Ph$$

(c)
$$N-CH\begin{cases}COOEt\\COOEt\end{cases} \xrightarrow[\text{2) BrCH}_2CH_2COO^-]{\text{1) NaOEt}} \xrightarrow[\Delta]{H_3O^+} H_2N-CH-COOH$$
$$\quad\quad\quad\quad\quad\quad\quad\quad\quad\quad\quad\quad\quad\quad\quad\quad\quad\quad\quad\quad\quad CH_2CH_2COOH$$

salt, not
acid—why?

24-10 continued

(d)

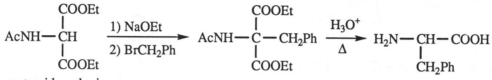

24-11

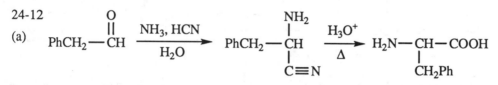

acetamidomalonic
ester

24-12

(a)

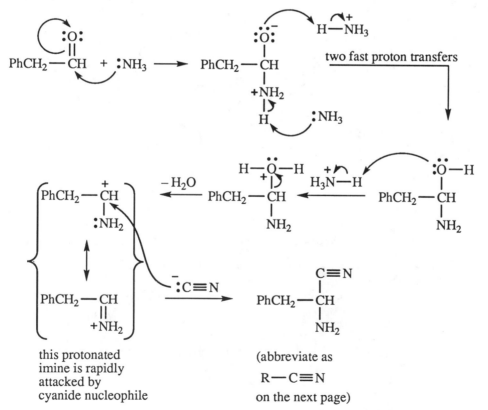

(b) While the solvent for the Strecker synthesis is water, the proton acceptor is ammonia and the proton donor is ammonium ion.

this protonated
imine is rapidly
attacked by
cyanide nucleophile

(abbreviate as

R—C≡N

on the next page)

24-12 (b) continued

mechanism of acid hydrolysis of the nitrile

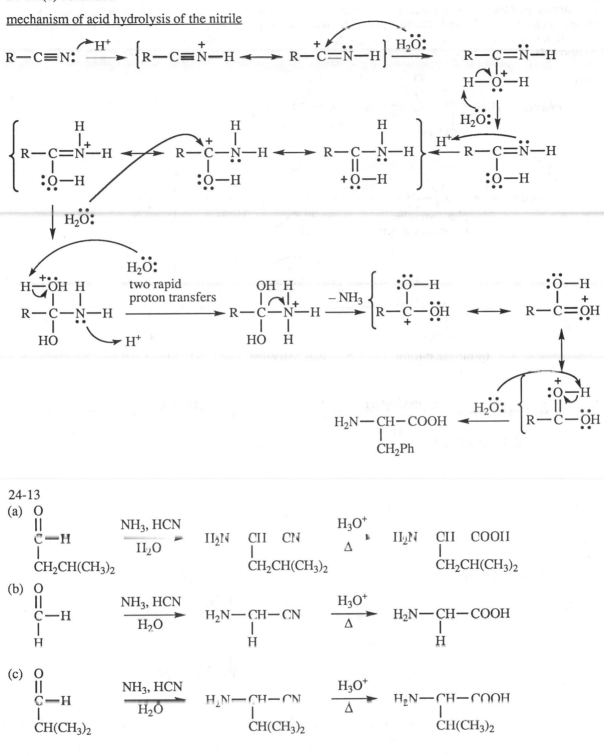

24-13

(a)

$$\underset{\underset{CH_2CH(CH_3)_2}{|}}{\overset{\overset{O}{\|}}{C}}-H \quad \xrightarrow[\text{H}_2\text{O}]{\text{NH}_3,\ \text{HCN}} \quad H_2N-\underset{\underset{CH_2CH(CH_3)_2}{|}}{\overset{|}{CH}}-CN \quad \xrightarrow[\Delta]{\text{H}_3\text{O}^+} \quad H_2N-\underset{\underset{CH_2CH(CH_3)_2}{|}}{\overset{|}{CH}}-COOH$$

(b)

$$\underset{\underset{H}{|}}{\overset{\overset{O}{\|}}{C}}-H \quad \xrightarrow[\text{H}_2\text{O}]{\text{NH}_3,\ \text{HCN}} \quad H_2N-\underset{\underset{H}{|}}{\overset{|}{CH}}-CN \quad \xrightarrow[\Delta]{\text{H}_3\text{O}^+} \quad H_2N-\underset{\underset{H}{|}}{\overset{|}{CH}}-COOH$$

(c)

$$\underset{\underset{CH(CH_3)_2}{|}}{\overset{\overset{O}{\|}}{C}}-H \quad \xrightarrow[\text{H}_2\text{O}]{\text{NH}_3,\ \text{HCN}} \quad H_2N-\underset{\underset{CH(CH_3)_2}{|}}{\overset{|}{CH}}-CN \quad \xrightarrow[\Delta]{\text{H}_3\text{O}^+} \quad H_2N-\underset{\underset{CH(CH_3)_2}{|}}{\overset{|}{CH}}-COOH$$

24-14 In acid solution, the free amino acid will be protonated, with a positive charge, and probably soluble in water as are other organic ions. The acylated amino acid, however, is not basic since the nitrogen is present as an amide. In acid solution, the acylated amino acid is neutral and not soluble in water. Water extraction or ion-exchange chromatography (Figure 24-11) would be practical techniques to separate these compounds.

24-15

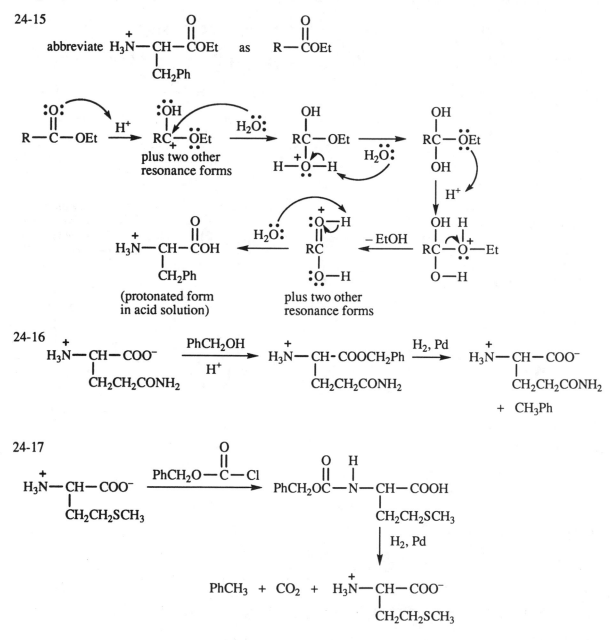

24-16

$$H_3\overset{+}{N}-CH-COO^- \xrightarrow[H^+]{PhCH_2OH} H_3\overset{+}{N}-CH\text{-}COOCH_2Ph \xrightarrow{H_2, Pd} H_3\overset{+}{N}-CH-COO^-$$
with CH_2CH_2CONH_2 substituents, and $+ CH_3Ph$

24-17

$$H_3\overset{+}{N}-CH-COO^- \xrightarrow{PhCH_2O-\overset{O}{\overset{||}{C}}-Cl} PhCH_2O\overset{O}{\overset{||}{C}}-\overset{H}{\overset{|}{N}}-CH-COOH$$
with CH_2CH_2SCH_3 substituents

$$\downarrow H_2, Pd$$

$$PhCH_3 + CO_2 + H_3\overset{+}{N}-CH-COO^-$$
with CH_2CH_2SCH_3

24-18

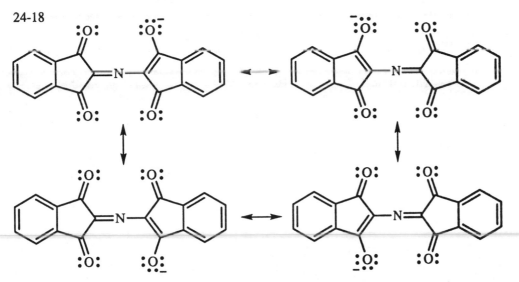

These are the most significant resonance contributors in which the electronegative oxygens carry the negative charge. There are also two other forms in which the negative charge is on the carbons bonded to the nitrogen, plus the usual resonance forms involving the alternate Kekulé structures of the benzene rings.

24-19

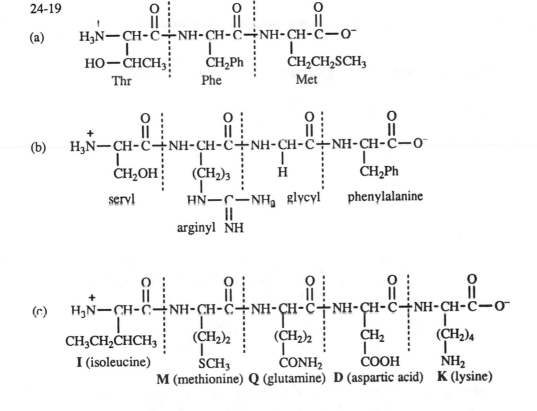

24-20

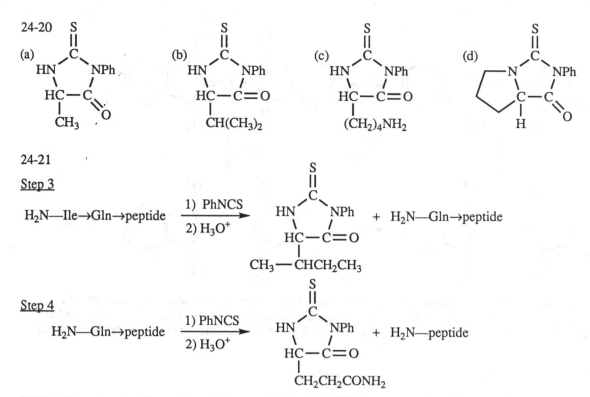

24-21

Step 3

$H_2N—Ile \rightarrow Gln \rightarrow peptide$ $\xrightarrow{\text{1) PhNCS}}{\text{2) }H_3O^+}$ [structure] $+ H_2N—Gln \rightarrow peptide$

Step 4

$H_2N—Gln \rightarrow peptide$ $\xrightarrow{\text{1) PhNCS}}{\text{2) }H_3O^+}$ [structure] $+ H_2N—peptide$

24-22 Abbreviate the N-terminus of the peptide chain as NH_2R .

(a) This is a nucleophilic aromatic substitution by the addition-elimination mechanism. The presence of two nitro groups makes this reaction feasible under mild conditions.

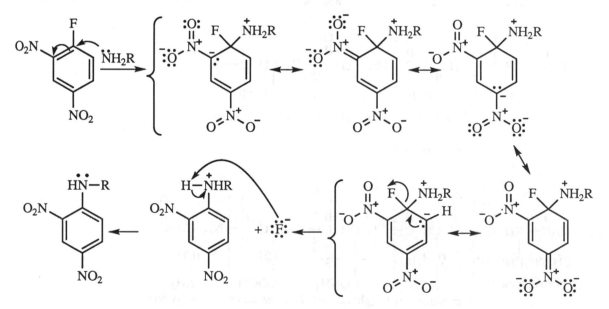

(b) The main drawback of the Sanger method is that only one amino acid is analyzed per sample of protein. The Edman degradation can usually analyze more than 20 amino acids per sample of protein.

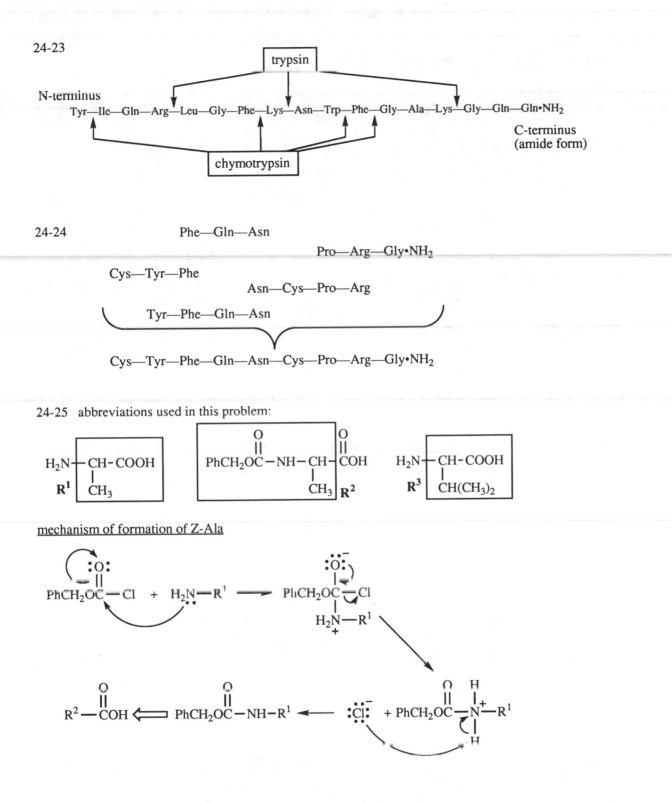

24-23

trypsin

N-terminus

Tyr—Ile—Gln—Arg—Leu—Gly—Phe—Lys—Asn—Trp—Phe—Gly—Ala—Lys—Gly—Gln—Gln•NH₂

C-terminus
(amide form)

chymotrypsin

24-24   Phe—Gln—Asn

Pro—Arg—Gly•NH₂

Cys—Tyr—Phe

Asn—Cys—Pro—Arg

Tyr—Phe—Gln—Asn

Cys—Tyr—Phe—Gln—Asn—Cys—Pro—Arg—Gly•NH₂

24-25   abbreviations used in this problem:

mechanism of formation of Z-Ala

581

24-25 continued

two possible mechanisms of ethyl chloroformate activation

mechanism 1

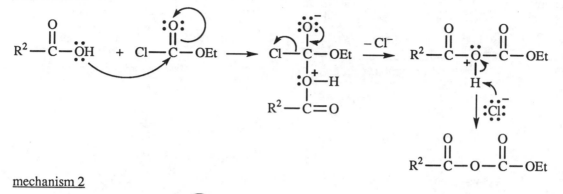

mechanism 2

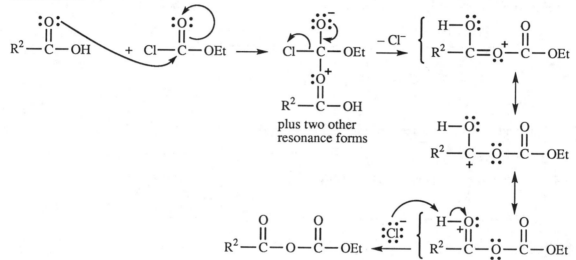

plus two other
resonance forms

mechanism of the coupling with valine

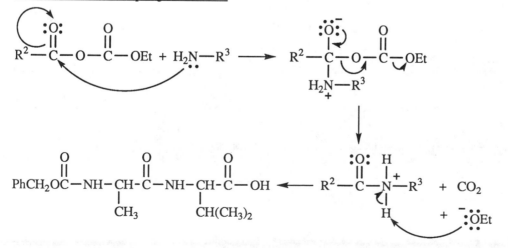

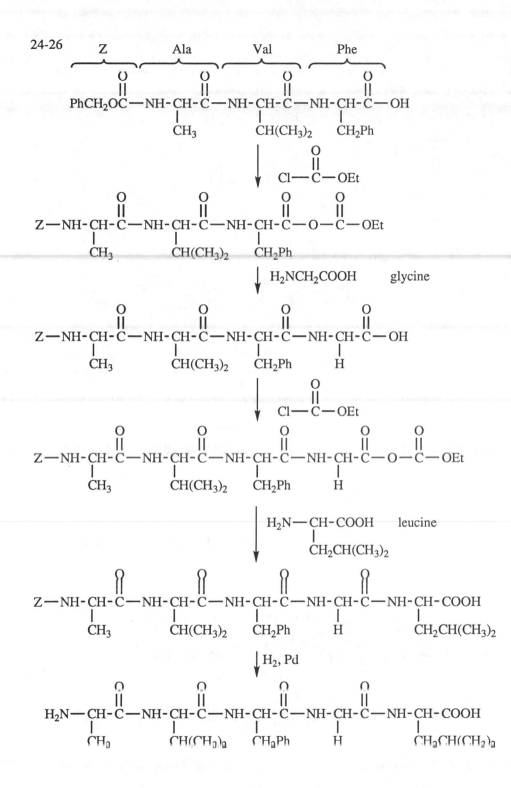

583

24-27

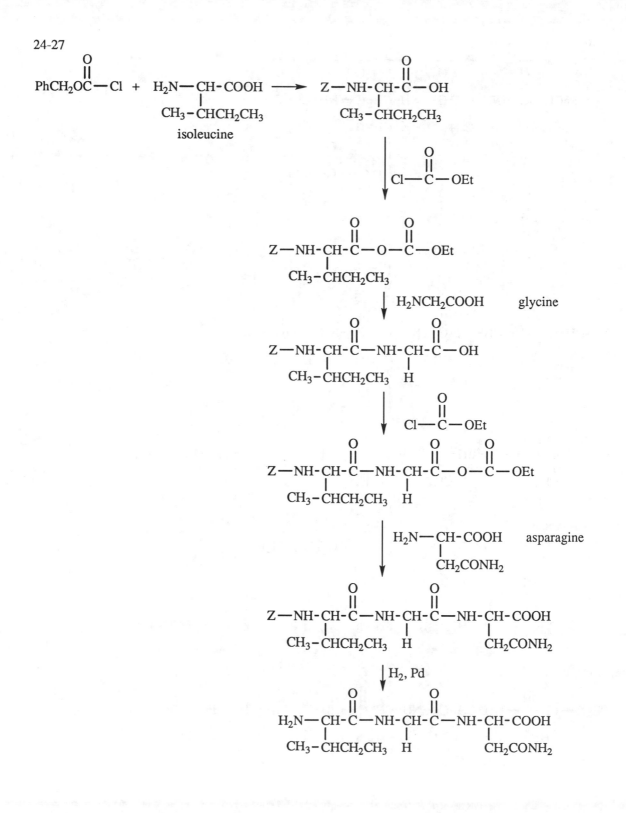

24-28 In this problem, "Cy" stands for "cyclohexyl".

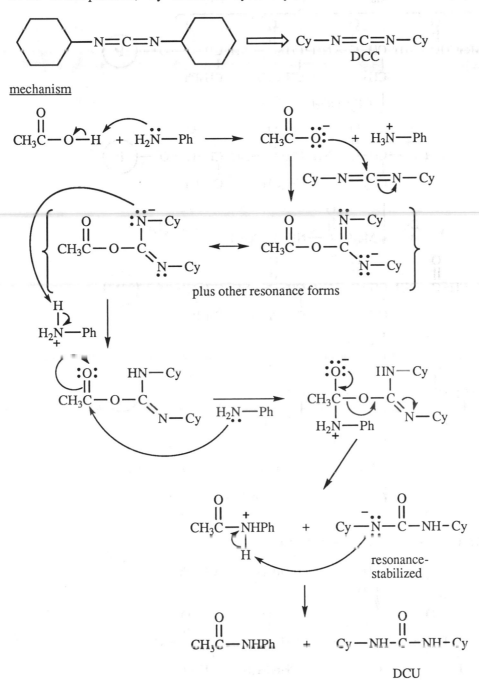

mechanism

24-29

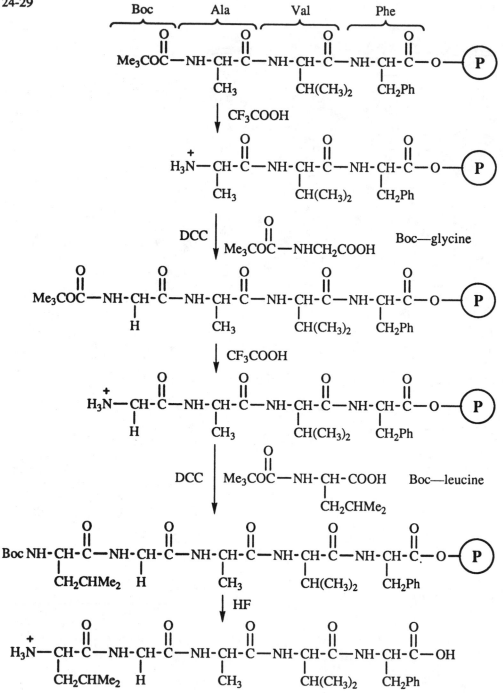

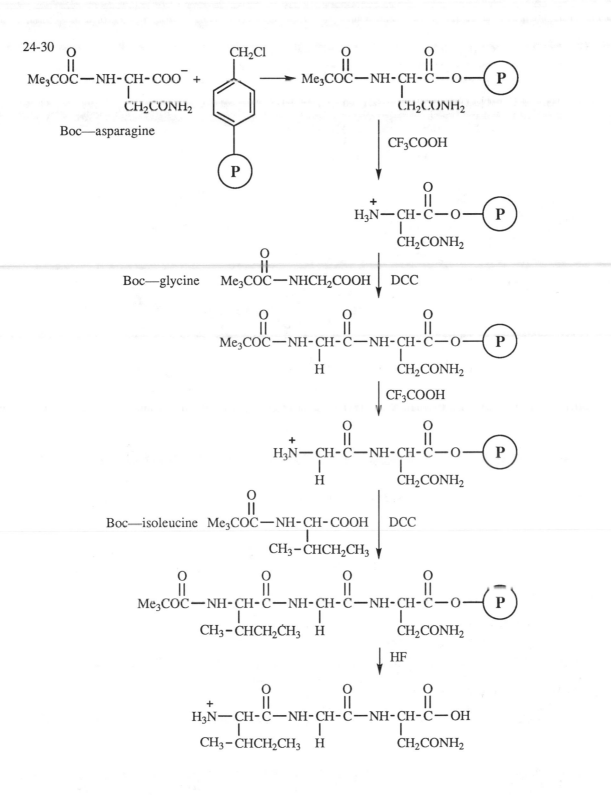

24-31 Please refer to solution 1-19, page 12 of this Solutions Manual.

24-32

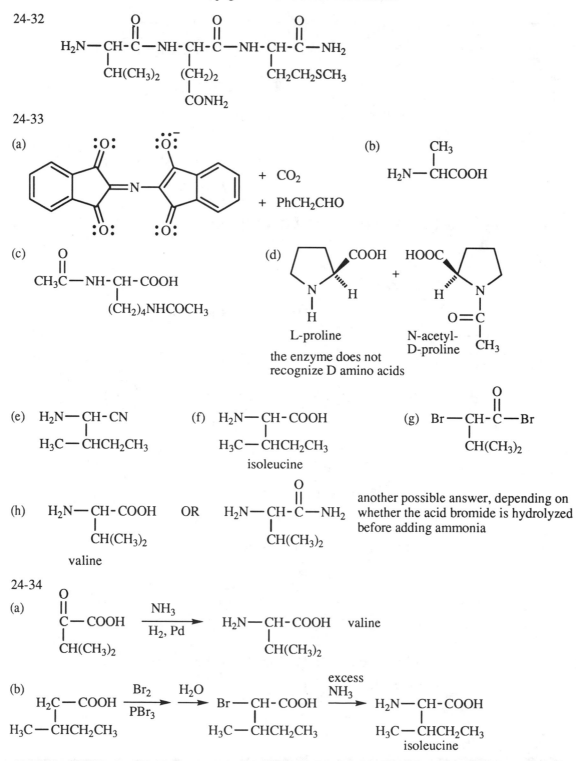

24-33

(a) + CO_2

+ $PhCH_2CHO$

(b)
$$H_2N-\underset{\underset{CH_3}{|}}{CH}COOH$$

(c)
$$CH_3\overset{\overset{O}{\parallel}}{C}-NH\text{-}\underset{\underset{(CH_2)_4NHCOCH_3}{|}}{CH}\text{-}COOH$$

(d) L-proline + N-acetyl-D-proline

the enzyme does not recognize D amino acids

(e)
$$H_2N-\underset{\underset{H_3C-CHCH_2CH_3}{|}}{CH}\text{-}CN$$

(f)
$$H_2N-\underset{\underset{H_3C-CHCH_2CH_3}{|}}{CH}\text{-}COOH$$
isoleucine

(g)
$$Br-\underset{\underset{CH(CH_3)_2}{|}}{CH}\text{-}\overset{\overset{O}{\parallel}}{C}-Br$$

(h)
$$H_2N-\underset{\underset{CH(CH_3)_2}{|}}{CH}\text{-}COOH \quad OR \quad H_2N-\underset{\underset{CH(CH_3)_2}{|}}{CH}\text{-}\overset{\overset{O}{\parallel}}{C}-NH_2$$
valine

another possible answer, depending on whether the acid bromide is hydrolyzed before adding ammonia

24-34

(a)
$$\underset{\underset{CH(CH_3)_2}{|}}{\overset{\overset{O}{\parallel}}{C}}-COOH \xrightarrow[\text{H}_2,\ \text{Pd}]{\text{NH}_3} H_2N-\underset{\underset{CH(CH_3)_2}{|}}{CH}\text{-}COOH \quad \text{valine}$$

(b)
$$\underset{\underset{H_3C-CHCH_2CH_3}{|}}{H_2C}-COOH \xrightarrow[\text{PBr}_3]{\text{Br}_2} \xrightarrow{\text{H}_2\text{O}} Br-\underset{\underset{H_3C-CHCH_2CH_3}{|}}{CH}\text{-}COOH \xrightarrow[\text{NH}_3]{\text{excess}} H_2N-\underset{\underset{H_3C-CHCH_2CH_3}{|}}{CH}\text{-}COOH$$
isoleucine

24-34 continued

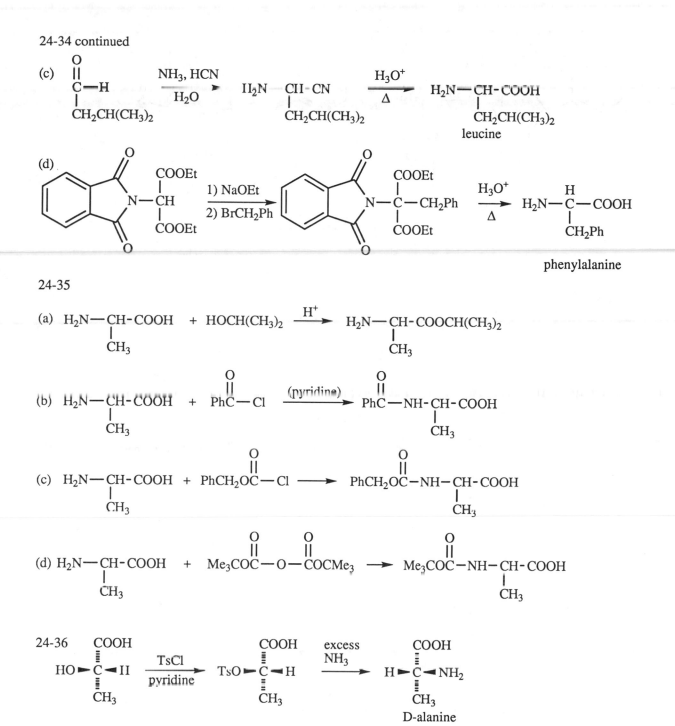

(c)

$$\underset{CH_2CH(CH_3)_2}{\overset{\overset{O}{\|}}{C}-H} \xrightarrow[H_2O]{NH_3, HCN} \underset{CH_2CH(CH_3)_2}{H_2N-CH-CN} \xrightarrow[\Delta]{H_3O^+} \underset{\underset{leucine}{CH_2CH(CH_3)_2}}{H_2N-CH-COOH}$$

(d)

24-35

(a) $\underset{CH_3}{H_2N-CH-COOH} + HOCH(CH_3)_2 \xrightarrow{H^+} \underset{CH_3}{H_2N-CH-COOCH(CH_3)_2}$

(b) $\underset{CH_3}{H_2N-CH-COOH} + \overset{O}{\overset{\|}{PhC}}-Cl \xrightarrow{(pyridine)} \underset{CH_3}{\overset{O}{\overset{\|}{PhC}}-NH-CH-COOH}$

(c) $\underset{CH_3}{H_2N-CH-COOH} + \overset{O}{\overset{\|}{PhCH_2OC}}-Cl \longrightarrow \underset{CH_3}{\overset{O}{\overset{\|}{PhCH_2OC}}-NH-CH-COOH}$

(d) $\underset{CH_3}{H_2N-CH-COOH} + \overset{O}{\overset{\|}{Me_3COC}}-O-\overset{O}{\overset{\|}{COCMe_3}} \longrightarrow \underset{CH_3}{\overset{O}{\overset{\|}{Me_3COC}}-NH-CH-COOH}$

24-36

$$\underset{CH_3}{\overset{COOH}{HO \blacktriangleright C \blacktriangleleft H}} \xrightarrow[pyridine]{TsCl} \underset{CH_3}{\overset{COOH}{TsO \blacktriangleright C \blacktriangleleft H}} \xrightarrow[NH_3]{excess} \underset{\underset{D\text{-alanine}}{CH_3}}{\overset{COOH}{H \blacktriangleright C \blacktriangleleft NH_2}}$$

589

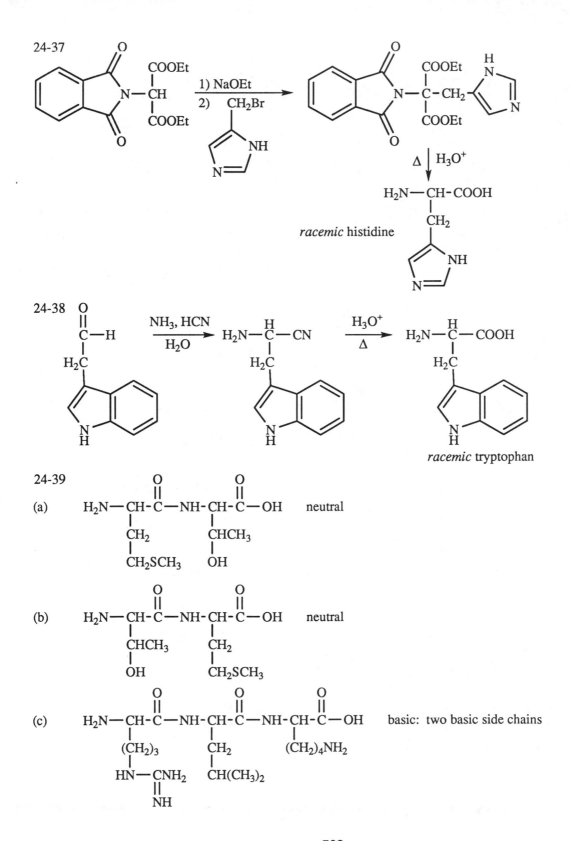

24-37

1) NaOEt
2) CH₂Br

Δ | H₃O⁺

H₂N—CH-COOH
CH₂

racemic histidine

24-38

NH₃, HCN / H₂O

H₃O⁺ / Δ

H₂N—C—CN

H₂N—C—COOH

racemic tryptophan

24-39

(a)   H₂N—CH-C—NH-CH-C—OH   neutral
              CH₂   CHCH₃
              CH₂SCH₃   OH

(b)   H₂N—CH-C—NH-CH-C—OH   neutral
              CHCH₃   CH₂
              OH   CH₂SCH₃

(c)   H₂N—CH-C—NH-CH-C—NH-CH-C—OH   basic: two basic side chains
              (CH₂)₃   CH₂   (CH₂)₄NH₂
              HN—CNH₂   CH(CH₃)₂
              ∥
              NH

590

24-39 continued

(d) 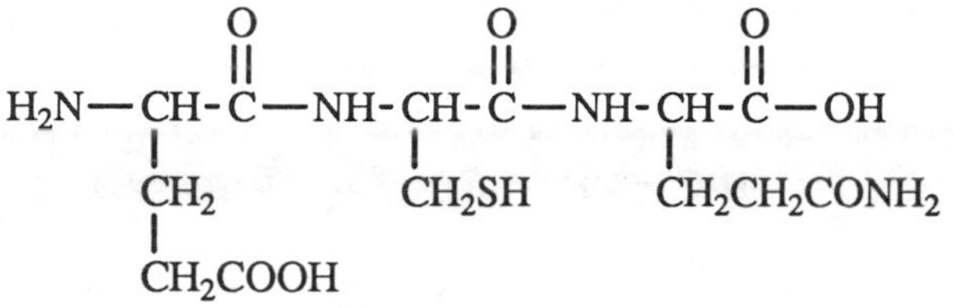

acidic: carboxylic acid side chain, and the SH is weakly acidic

24-40

(a), (b), (c)

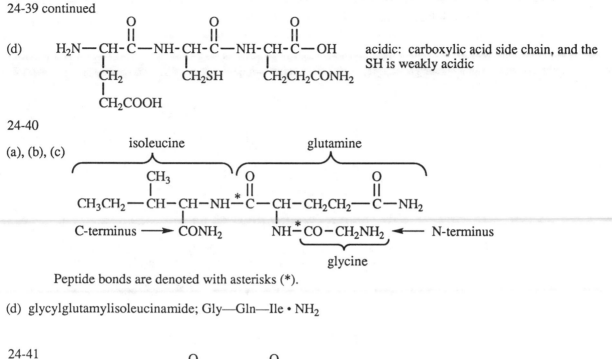

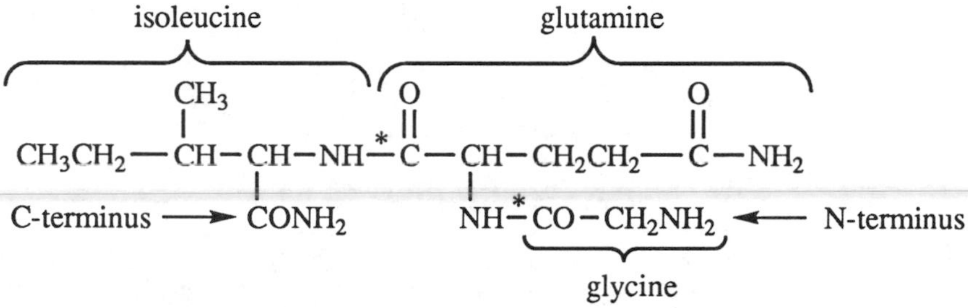
isoleucine
glutamine
C-terminus
N-terminus
glycine

Peptide bonds are denoted with asterisks (*).

(d) glycylglutamylisoleucinamide; Gly—Gln—Ile • NH_2

24-41

Aspartame: 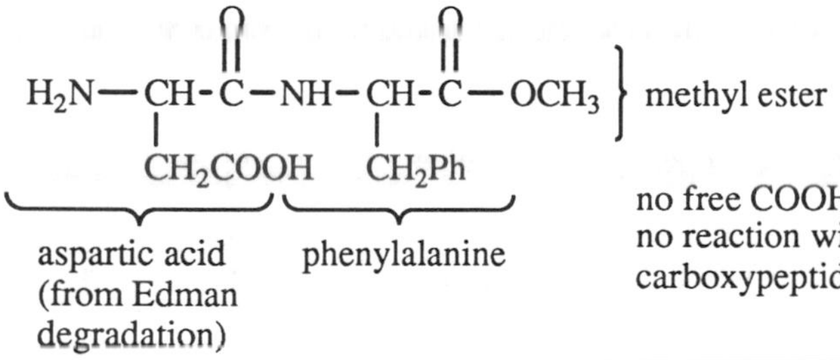
methyl ester
aspartic acid
(from Edman
degradation)
phenylalanine

no free COOH ⇒
no reaction with
carboxypeptidase

Aspartame is aspartylphenylalanine methyl ester.

24-42

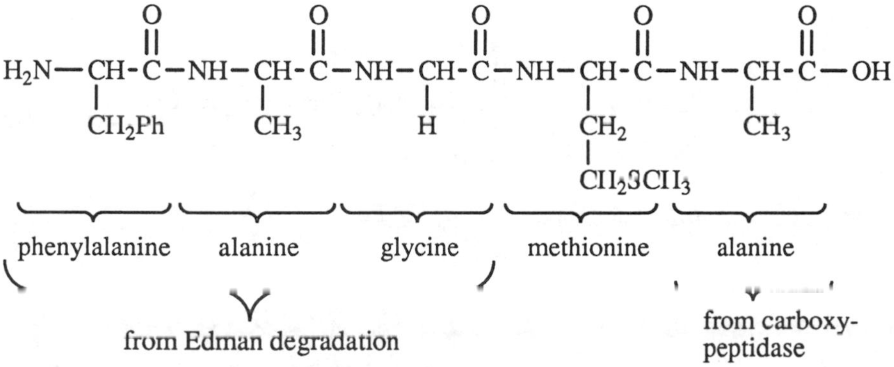
phenylalanine
alanine
glycine
methionine
alanine
from Edman degradation
from carboxy-
peptidase

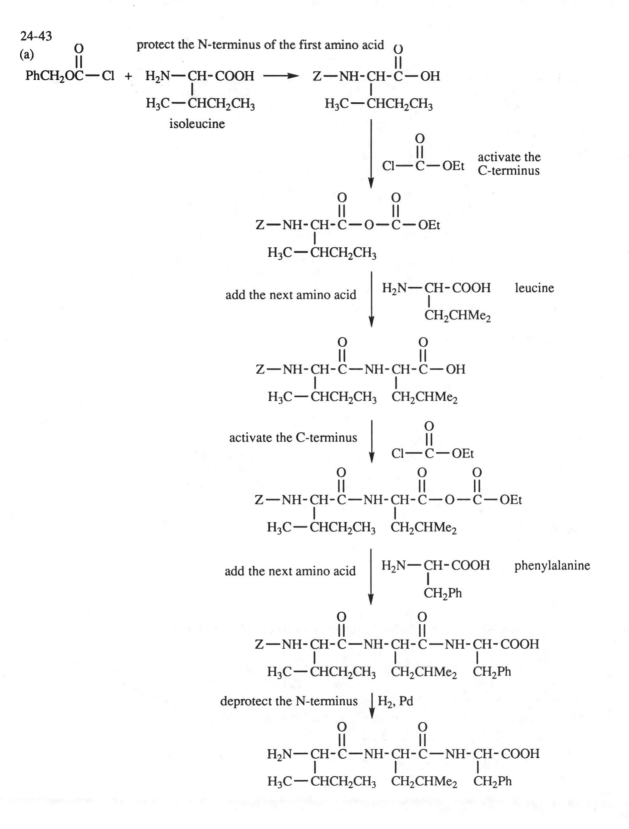

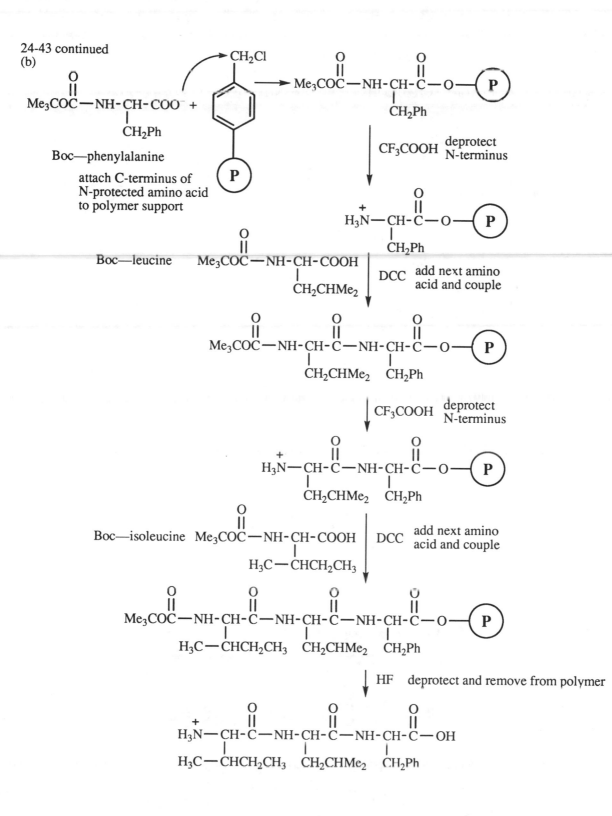

24-43 continued
(b)

24-44

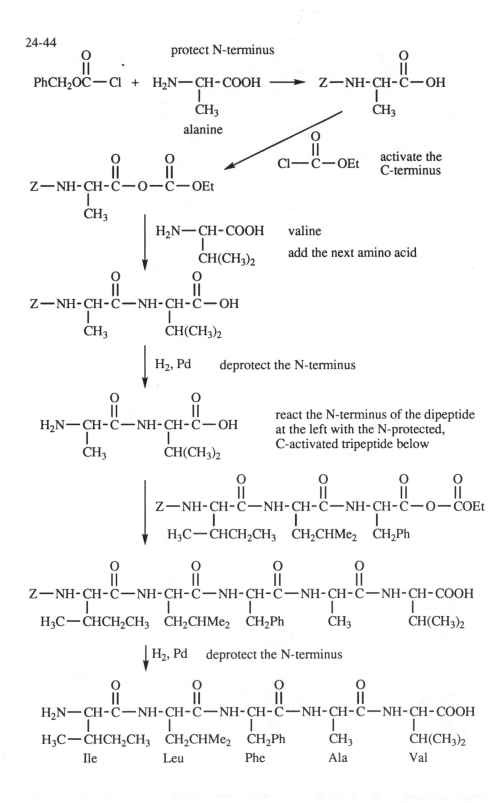

24-45

(a) There are two possible sources of ammonia in the hydrolysate. The C-terminus could have been present as the amide instead of the carboxyl, or the glutamic acid could have been present as its amide, glutamine.

(b) The C-terminus is present as the amide. The N-terminus is present as the lactam (cyclic amide) combining the amino group with the carboxyl group of the glutamic acid side chain.

(c) The fact that hydrolysis does not release ammonia implies that the C-terminus is not an amide. Yet, carboxypeptidase treatment gives no reaction, showing that the C-terminus is not a free carboxyl group. Also, treatment with phenyl isothiocyanate gives no reaction, suggesting no free amine at the N-terminus. The most plausible explanation is that the N-terminus has reacted with the C-terminus to produce a cyclic amide, a lactam. (These large rings, called macrocycles, are often found in nature as hormones or antibiotics.)

24-46

(a) Lipoic acid is a mild oxidizing agent. In the process of oxidizing another reactant, lipoic acid is reduced.

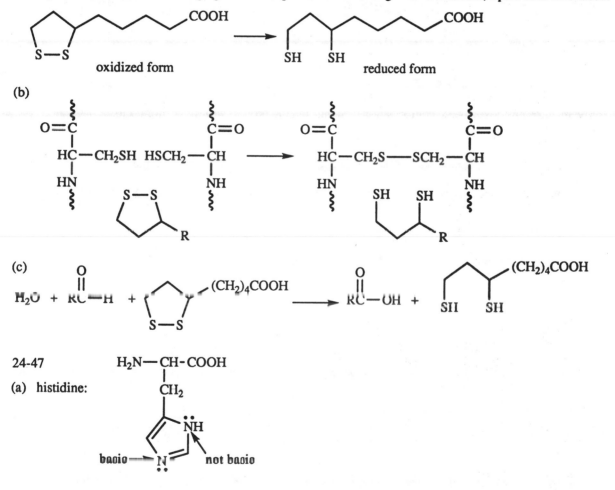

24-47

(a) histidine:

595

24-47 continued

(b)

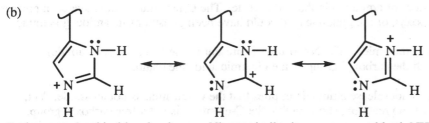

In the protonated imidazole, the two N's are similar in structure, and both NH groups are acidic.

(c)

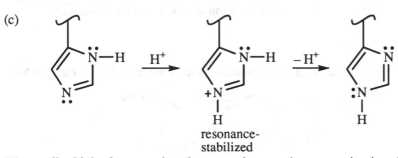

resonance-
stabilized

We usually think of protonation-deprotonation reactions occurring in solution where protons can move with solvent molecules. In an enzyme active-site, there is no "solvent", so there must be another mechanism for movement of protons. Often, conformational changes in the protein will move atoms closer or farther. Histidine serves the function of moving a proton toward or away from a particular site by using its different nitrogens in concert as a proton acceptor and a proton donor.

24-48 The high isoelectric point suggests a strongly basic side chain as in lysine. The N—CH$_2$ bond in the side chain of arginine is likely to have remained intact during the metabolism. (Can you propose a likely mechanism for this reaction?)

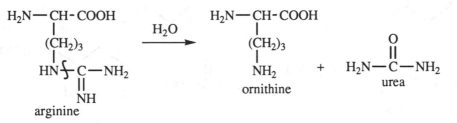

596

24-49

(a) glutathione:

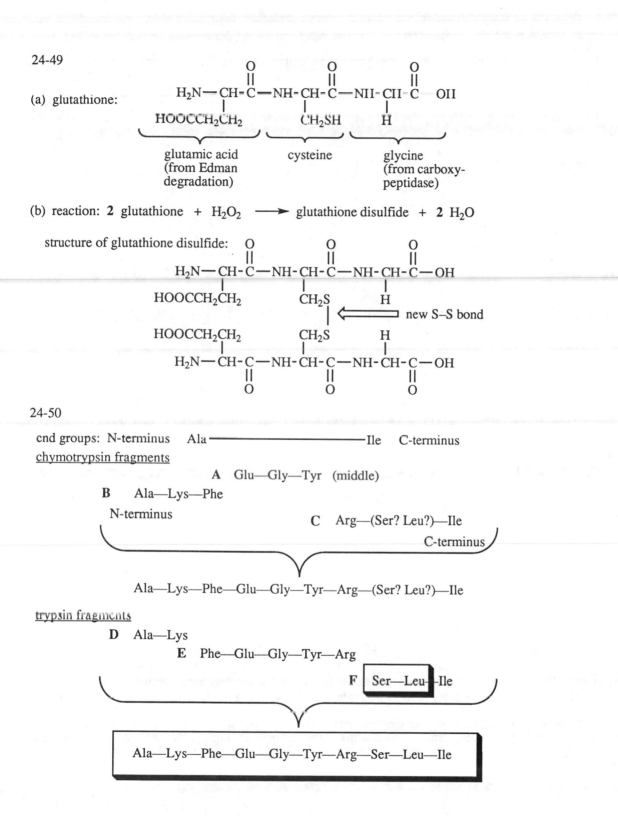

$$\underset{\substack{\text{glutamic acid}\\\text{(from Edman}\\\text{degradation)}}}{\underbrace{\underset{HOOCCH_2CH_2}{H_2N-CH-\overset{\overset{\displaystyle O}{\|}}{C}}}}-\underset{\substack{\text{cysteine}}}{\underbrace{\underset{CH_2SH}{NH-CH-\overset{\overset{\displaystyle O}{\|}}{C}}}}-\underset{\substack{\text{glycine}\\\text{(from carboxy-}\\\text{peptidase)}}}{\underbrace{\underset{H}{NII-CII-\overset{\overset{\displaystyle O}{\|}}{C}}}}\;OII$$

(b) reaction: **2** glutathione + H_2O_2 $\longrightarrow$ glutathione disulfide + **2** H_2O

structure of glutathione disulfide:

$$\underset{HOOCCH_2CH_2}{H_2N-CH-\overset{\overset{\displaystyle O}{\|}}{C}}-\underset{CH_2S}{NH-CH-\overset{\overset{\displaystyle O}{\|}}{C}}-\underset{H}{NH-CH-\overset{\overset{\displaystyle O}{\|}}{C}}-OH$$

$\longleftarrow$ new S–S bond

$$\underset{HOOCCH_2CH_2}{H_2N-CH-\underset{\underset{\displaystyle O}{\|}}{C}}-\underset{CH_2S}{NH-CH-\underset{\underset{\displaystyle O}{\|}}{C}}-\underset{H}{NH-CH-\underset{\underset{\displaystyle O}{\|}}{C}}-OH$$

24-50

end groups: N-terminus   Ala ——————————— Ile   C-terminus

chymotrypsin fragments

  A Glu—Gly—Tyr (middle)

B   Ala—Lys—Phe
      N-terminus

            C Arg—(Ser? Leu?)—Ile
                        C-terminus

Ala—Lys—Phe—Glu—Gly—Tyr—Arg—(Ser? Leu?)—Ile

trypsin fragments

  D Ala—Lys

      E Phe—Glu—Gly—Tyr—Arg

            F Ser—Leu—Ile

Ala—Lys—Phe—Glu—Gly—Tyr—Arg—Ser—Leu—Ile

25-1

trimyristin

$$CH_2-O-\underset{\underset{O}{\|}}{C}-(CH_2)_{12}CH_3$$
$$CH-O-\underset{\underset{O}{\|}}{C}-(CH_2)_{12}CH_3$$
$$CH_2-O-\underset{O}{\overset{O}{\|}}{C}-(CH_2)_{12}CH_3$$

25-2

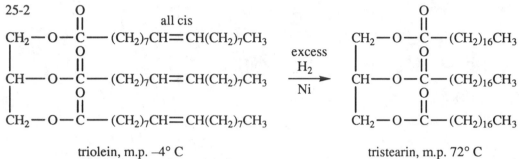

triolein, m.p. −4° C
(liquid at room temperature)

tristearin, m.p. 72° C
(solid at room temperature)

25-3

(a)

$$2 \ CH_3(CH_2)_{16}-\underset{O}{\overset{O}{\|}}{C}-O^-\ Na^+ \ + \ Ca^{2+} \longrightarrow \left[CH_3(CH_2)_{16}-\underset{O}{\overset{O}{\|}}{C}-O \right]_2 Ca \ + \ 2 \ Na^+$$

(b)

$$2 \ CH_3(CH_2)_{16}-\underset{O}{\overset{O}{\|}}{C}-O^-\ Na^+ \ + \ Mg^{2+} \longrightarrow \left[CH_3(CH_2)_{16}-\underset{O}{\overset{O}{\|}}{C}-O \right]_2 Mg \ + \ 2 \ Na^+$$

(c)

$$3 \ CH_3(CH_2)_{16}-\underset{O}{\overset{O}{\|}}{C}-O^-\ Na^+ \ + \ Fe^{3+} \longrightarrow \left[CH_3(CH_2)_{16}-\underset{O}{\overset{O}{\|}}{C}-O \right]_3 Fe \ + \ 3 \ Na^+$$

25-4

(a) Both sodium carbonate (its old name is "washing soda") and sodium phosphate will increase the pH above 6, so that the carboxyl group of the soap molecule will remain ionized, thus preventing precipitation.

(b) In the presence of calcium, magnesium, and ferric ions, the carboxylate group of soap will form precipitates called "hard-water scum", or as scientists label it, "bathtub ring". Both carbonate and phosphate ions will form complexes or precipitates with these cations, thereby preventing precipitation of the soap from solution.

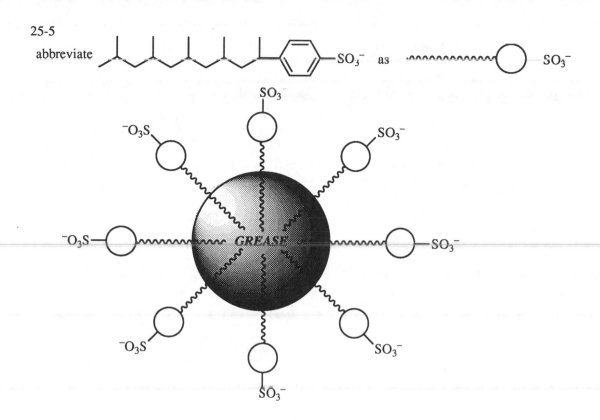

25-5

abbreviate

as

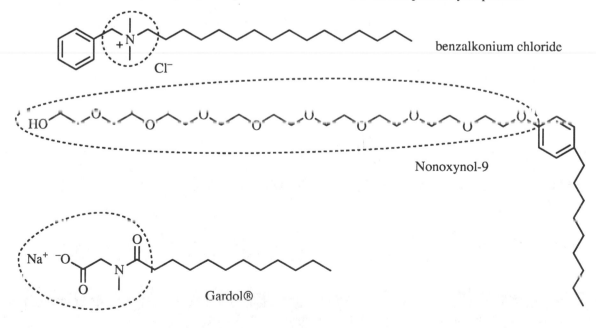

. 25-6 In each structure, the hydrophilic portion is circled. The uncircled part is hydrophobic.

benzalkonium chloride

Nonoxynol-9

Gardol®

25-7

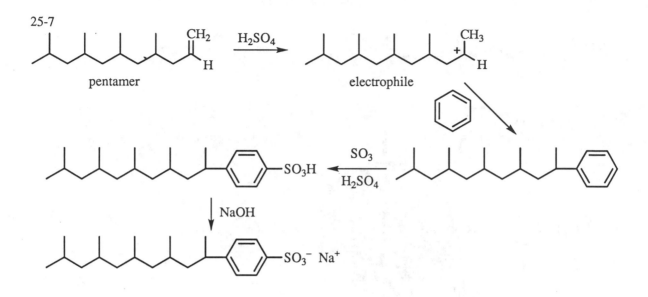

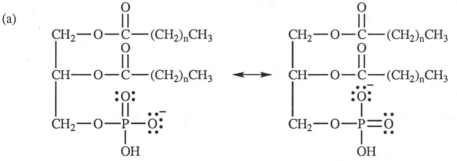

25-8

(a)

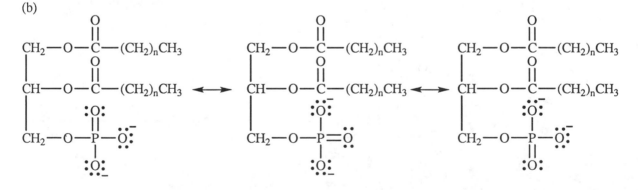

(b)

25-9 Estradiol is a phenol and can be ionized with aqueous NaOH. Testosterone does not have any hydrogens acidic enough to react with NaOH. Treatment of a solution of estradiol and testosterone in organic solvent with aqueous base will extract the phenoxide form of estradiol into the aqueous layer, leaving testosterone in the organic layer. Acidification of the aqueous base will precipitate estradiol which can be filtered. Evaporation of the organic solvent will leave testosterone.

25-10 Models may help. Abbreviations: "ax" = axial; "eq" = equatorial. Note that substituents at *cis*-fused ring junctures are axial to one ring and equatorial to another.

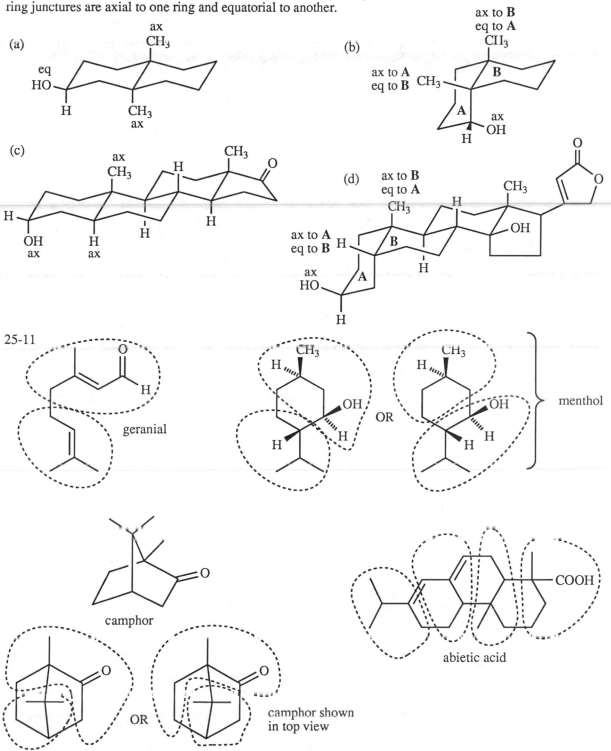

25-11

25-12　β-carotene

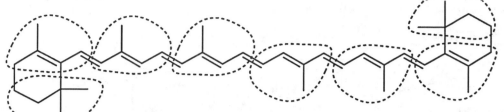

25-13

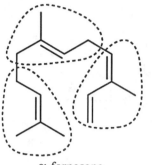

α-farnesene
sesquiterpene

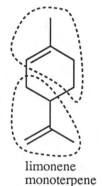

limonene
monoterpene
(Limonene can also be circled
in the other direction around
the ring—see menthol in 25-11.)

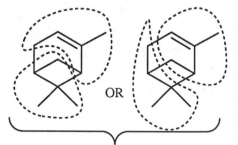

OR

α-pinene
monoterpene

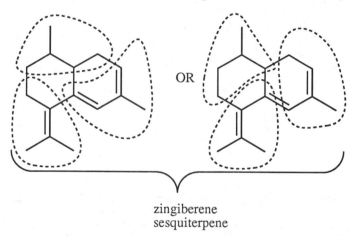

OR

zingiberene
sesquiterpene

25-14　Please refer to problem 1-19, page 12 of this Solutions Manual.

25-15

(a) triglyceride　　(b) synthetic detergent　　(c) wax　　(d) sesquiterpene　　(e) steroid

25-16

(a) **3** $CH_3(CH_2)_7CH=CH(CH_2)_7COO^- Na^+$ + $HOCH(CH_2OH)_2$
　　　　　　　　soap　　　　　　　　　　　　glycerol

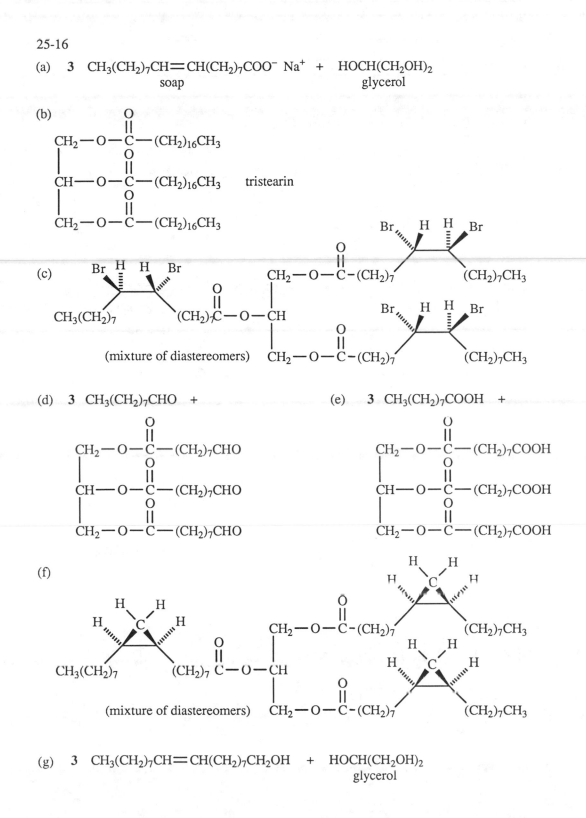

(b)

$CH_2-O-\overset{\overset{O}{\|}}{C}-(CH_2)_{16}CH_3$

$CH-O-\overset{\overset{O}{\|}}{C}-(CH_2)_{16}CH_3$　　tristearin

$CH_2-O-\overset{\overset{O}{\|}}{C}-(CH_2)_{16}CH_3$

(c) (mixture of diastereomers)

(d) **3** $CH_3(CH_2)_7CHO$ +

$CH_2-O-\overset{\overset{O}{\|}}{C}-(CH_2)_7CHO$

$CH-O-\overset{\overset{O}{\|}}{C}-(CH_2)_7CHO$

$CH_2-O-\overset{\overset{O}{\|}}{C}-(CH_2)_7CHO$

(e) **3** $CH_3(CH_2)_7COOH$ +

$CH_2-O-\overset{\overset{O}{\|}}{C}-(CH_2)_7COOH$

$CH-O-\overset{\overset{O}{\|}}{C}-(CH_2)_7COOH$

$CH_2-O-\overset{\overset{O}{\|}}{C}-(CH_2)_7COOH$

(f) (mixture of diastereomers)

(g) **3** $CH_3(CH_2)_7CH=CH(CH_2)_7CH_2OH$ + $HOCH(CH_2OH)_2$
　　　　　　　　　　　　　　　　　　　　　　　glycerol

25-17

(a) $CH_3(CH_2)_7CH=CH(CH_2)_7COOH$ $\xrightarrow{\text{H}_2}{\text{Ni}}$ $\xrightarrow[\text{2) H}_3\text{O}^+]{\text{1) LiAlH}_4}$ $CH_3(CH_2)_{16}CH_2OH$

(b) $CH_3(CH_2)_7CH=CH(CH_2)_7COOH$ $\xrightarrow{\text{H}_2}{\text{Ni}}$ $CH_3(CH_2)_{16}COOH$

(c) $CH_3(CH_2)_{16}COOH$ + $HOCH_2(CH_2)_{16}CH_3$ $\xrightarrow[\Delta]{\text{H}^+}$ $CH_3(CH_2)_{16}COOCH_2(CH_2)_{16}CH_3$
from (b)   from (a)

(d) $CH_3(CH_2)_7CH=CH(CH_2)_7COOH$ $\xrightarrow[\text{2) Me}_2\text{S}]{\text{1) O}_3}$ $CH_3(CH_2)_7CH=O$ + $O=CH(CH_2)_7COOH$
           nonanal

(e) $CH_3(CH_2)_7CH=CH(CH_2)_7COOH$ $\xrightarrow[\text{H}_2\text{O, }\Delta]{\text{KMnO}_4}$ $CH_3(CH_2)_7COOH$ + $HOOC(CH_2)_7COOH$
           nonanedioic acid

(f) $CH_3(CH_2)_7CH=CH(CH_2)_7COOH$ $\xrightarrow[\text{PBr}_3]{\substack{\text{excess}\\ \text{Br}_2}}$ $\xrightarrow{\text{H}_2\text{O}}$

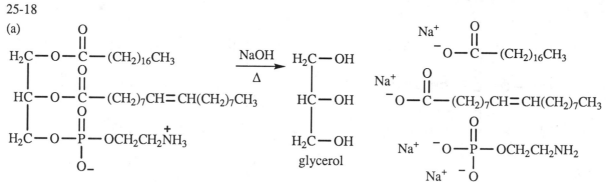

25-18

(a)

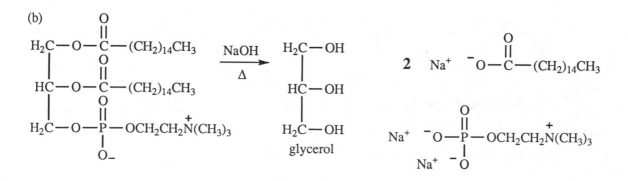

(b)

H₂C—O—C—(CH₂)₁₄CH₃ ... (structure)

604

25-19

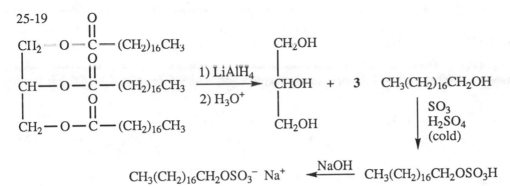

25-20 Reagents in parts (a), (b), and (d) would react with alkenes. If both samples contained alkenes, these reagents could not distinguish the samples. Saponification (part (c)), however, is a reaction of an ester, so only the vegetable oil would react, not the hydrocarbon oil mixture.

25-21 (a) Add an aqueous solution of calcium ion or magnesium ion. Sodium stearate will produce a precipitate, while the sulfonate will not precipitate.
(b) Beeswax, an ester, can be saponified with NaOH. Paraffin wax is a solid mixture of alkanes and will not react.
(c) Myristic acid will dissolve (or be emulsified) in dilute aqueous base. Trimyristin will remain unaffected.
(d) Triolein (an unsaturated oil) will decolorize bromine in CCl_4, but trimyristin (a saturated fat) will not.

25-22
(a)

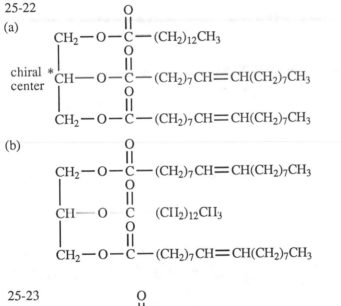

(b)

25-23

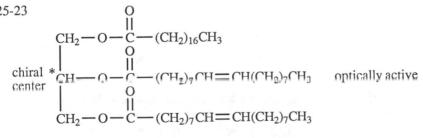

chiral center optically active

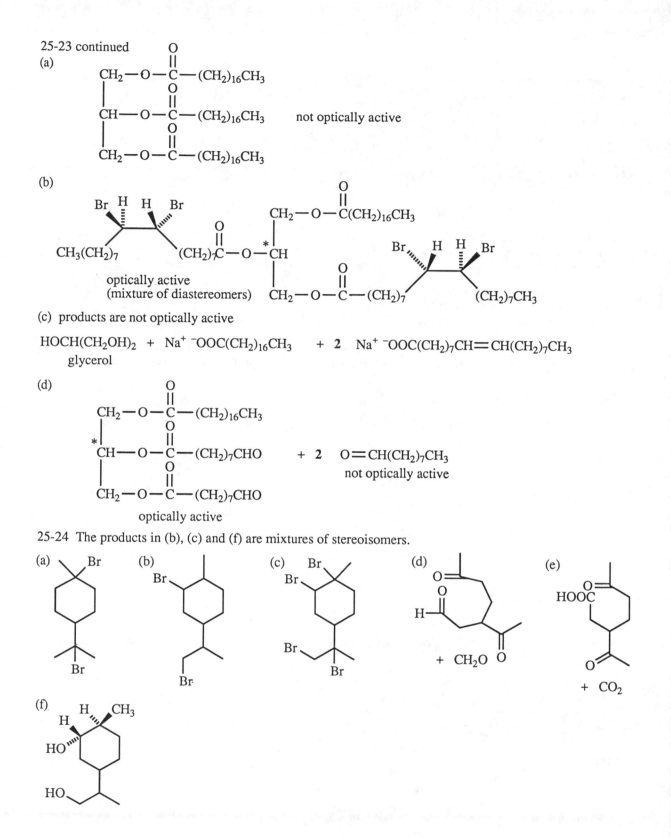

25-23 continued

(a)

CH₂—O—C—(CH₂)₁₆CH₃
CH—O—C—(CH₂)₁₆CH₃   not optically active
CH₂—O—C—(CH₂)₁₆CH₃

(b)

optically active
(mixture of diastereomers)

(c) products are not optically active

HOCH(CH₂OH)₂ + Na⁺ ⁻OOC(CH₂)₁₆CH₃ + 2 Na⁺ ⁻OOC(CH₂)₇CH=CH(CH₂)₇CH₃
  glycerol

(d)

CH₂—O—C—(CH₂)₁₆CH₃
CH—O—C—(CH₂)₇CHO   + 2 O=CH(CH₂)₇CH₃
CH₂—O—C—(CH₂)₇CHO   not optically active

optically active

25-24 The products in (b), (c) and (f) are mixtures of stereoisomers.

(a) (b) (c) (d) + CH₂O (e) + CO₂

(f)

606

25-25 Is it any wonder that Olestra cannot be digested!

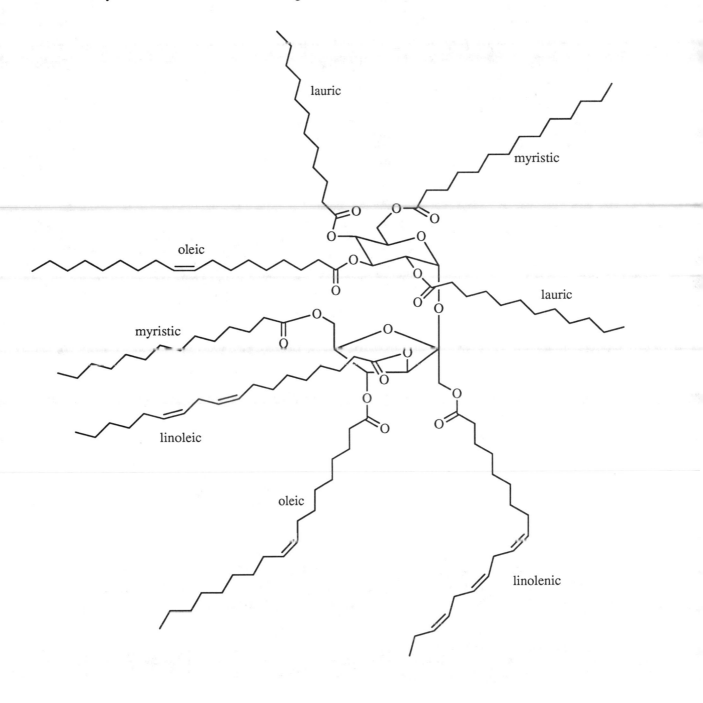

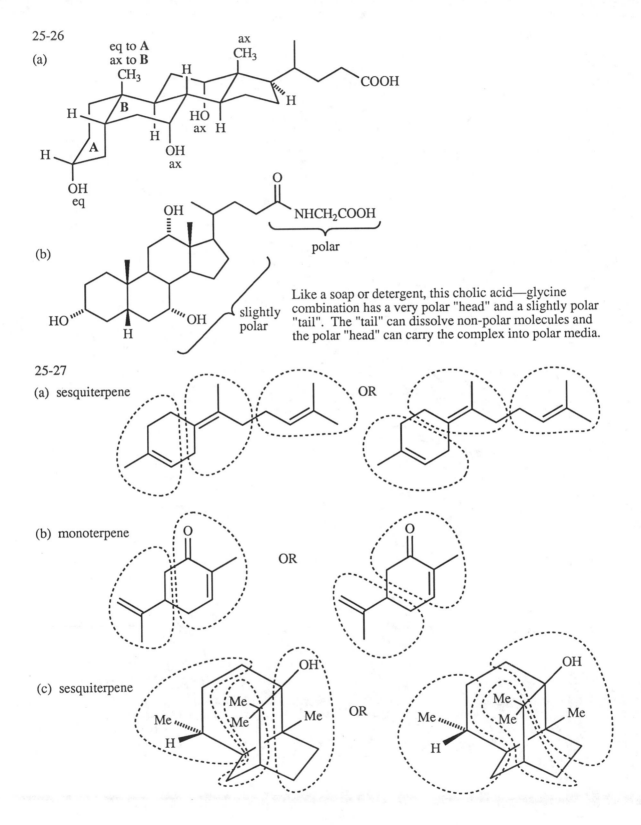

25-26

(a)

(b)

Like a soap or detergent, this cholic acid—glycine combination has a very polar "head" and a slightly polar "tail". The "tail" can dissolve non-polar molecules and the polar "head" can carry the complex into polar media.

25-27

(a) sesquiterpene   OR

(b) monoterpene   OR

(c) sesquiterpene   OR

(d) sesquiterpene

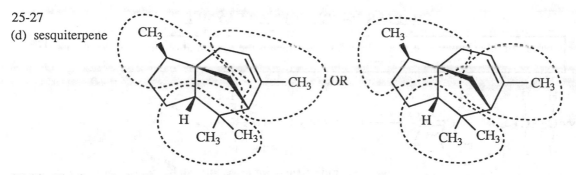

25-28 The formula $C_{18}H_{34}O_2$ has two elements of unsaturation; one is the carbonyl, so the other must be an alkene or a ring. Catalytic hydrogenation gives stearic acid, so the carbon cannot include a ring; it must contain an alkene. The products from $KMnO_4$ oxidation determine the location of the alkene:

$$HOOC(CH_2)_4COOH \; + \; HOOC(CH_2)_{10}CH_3 \Longrightarrow HOOC(CH_2)_4CH{=}CH(CH_2)_{10}CH_3$$

If the alkene were trans, the coupling constant for the vinyl protons would be about 15 Hz; a 10 Hz coupling constant indicates a cis alkene. The 7 Hz coupling is from the vinyl H's to the neighboring CH_2's.

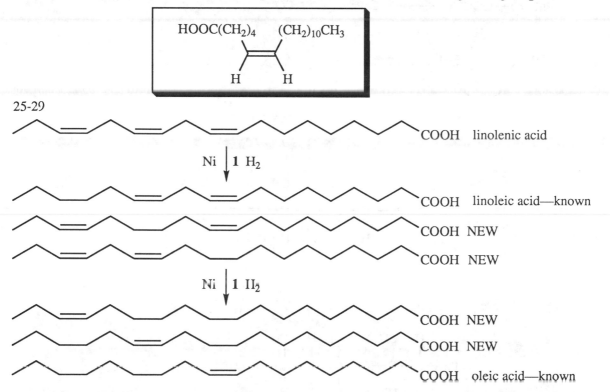

25-29

25-30 The cetyl glycoside would be a good emulsifying agent. It has a polar end (glucose) and a non-polar end (the hydrocarbon chain), so it can dissolve non-polar molecules, then carry them through polar media in micelles. This is found in Nature where non-polar molecules such as steroid hormones, antibiotics, and other physiologically-active compounds are carried through the bloodstream (aqueous) by attaching saccharides (usually mono, di, or tri), making the non-polar group water soluble.

Note: In this chapter, the "wavy bond" symbol means the continuation of a polymer chain.

26-1

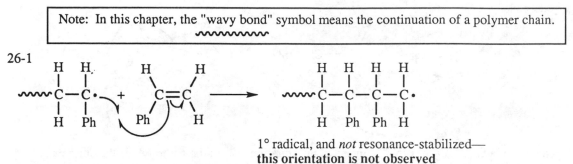

1° radical, and *not* resonance-stabilized—
this orientation is not observed

Orientation of addition always generates the more stable intermediate; the energy difference between a 1° radical (shown above) and a benzylic radical is huge. The phenyl substituents must necessarily be on alternating carbons because the orientation of attack is always the same—not a random process.

26-2

$$PhCOO-OOCPh \longrightarrow 2\ Ph\cdot\ +\ 2\ CO_2$$

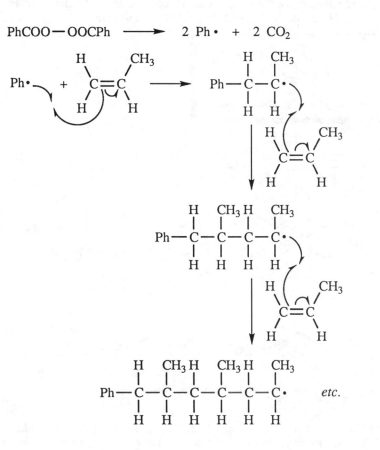

26-3 The benzylic hydrogen will be abstracted in preference to a 2° hydrogen because the benzylic radical is both 3° and resonance-stabilized, and the 2° radical is neither.

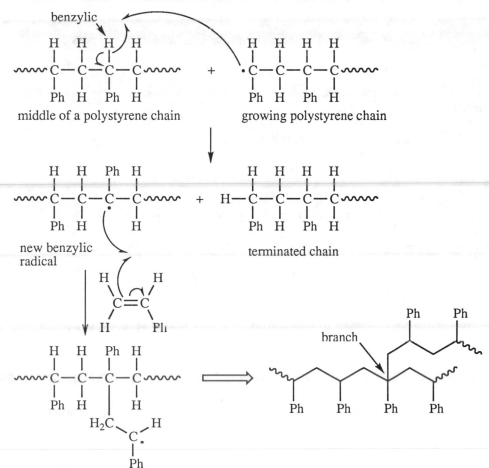

26-4 Addition occurs with the orientation giving the more stable intermediate. In the case of isobutylene, the growing chain will bond at the less substituted carbon to generate the more highly substituted carbocation

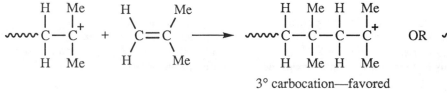

3° carbocation—favored

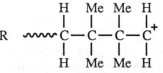

1° carbocation—disfavored
(also more steric hindrance)

26-5

(a)   chlorine can stabilize a carbocation intermediate by resonance

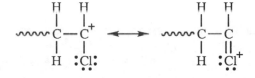

(b) CH_3 can stabilize the carbocation intermediate by induction

$$\text{wwww}\overset{\overset{\displaystyle H}{|}}{\underset{\underset{\displaystyle H}{|}}{C}}-\overset{\overset{\displaystyle H}{|}}{\underset{\underset{\displaystyle CH_3}{|}}{\overset{+}{C}}} \qquad 2° \text{ (not the best case imaginable, but still possible)}$$

(c) terrible for cationic polymerization: both substituents are electron-withdrawing and would *destabilize* the carbocation intermediate

$$\text{wwww}\overset{\overset{\displaystyle H}{|}}{\underset{\underset{\displaystyle H}{|}}{C}}-\overset{\overset{\displaystyle COOCH_3}{|}}{\underset{\underset{\displaystyle C\equiv N}{|}}{\overset{+}{C}}} \qquad \textit{destabilized} \text{ carbocation}$$

26-6 benzylic

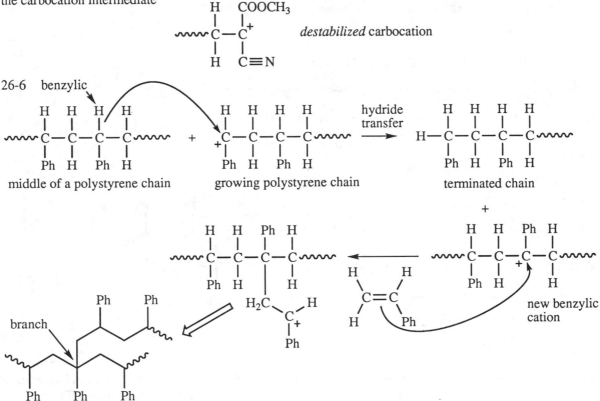

Polystyrene is particularly susceptible to branching because the 3° benzylic cation produced by a hydride transfer is so stable. In poly(isobutylene), there is no hydrogen on the carbon with the stabilizing substituents; any hydride transfer would generate a 2° carbocation at the expense of a 3° carbocation at the end of a growing chain—this is an increase in energy and therefore unfavorable.

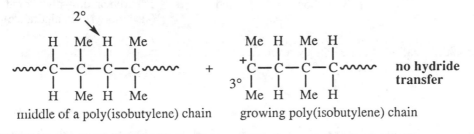

middle of a poly(isobutylene) chain   growing poly(isobutylene) chain

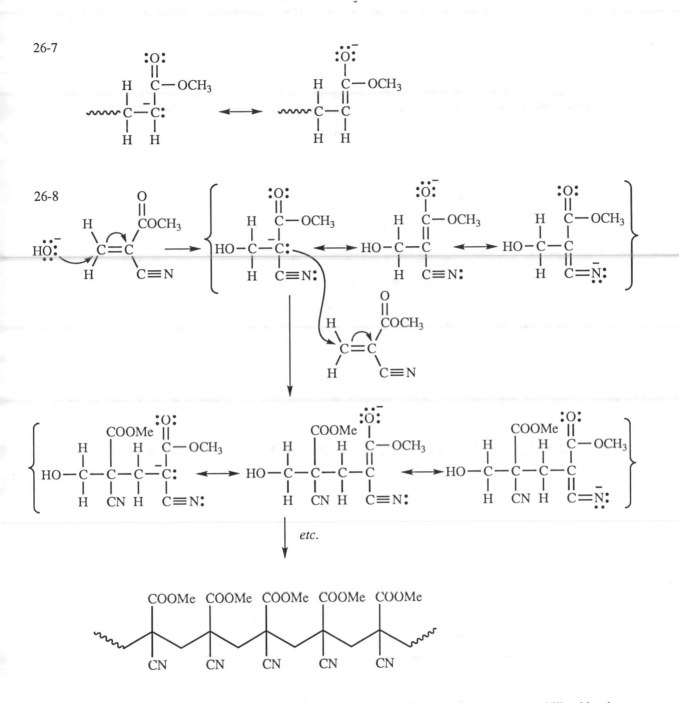

This polymerization goes so quickly because the anionic intermediate is highly resonance stabilized by the carbonyl and the cyano groups. A stable intermediate suggests a low activation energy which translates to a fast reaction.

26-9

(a)

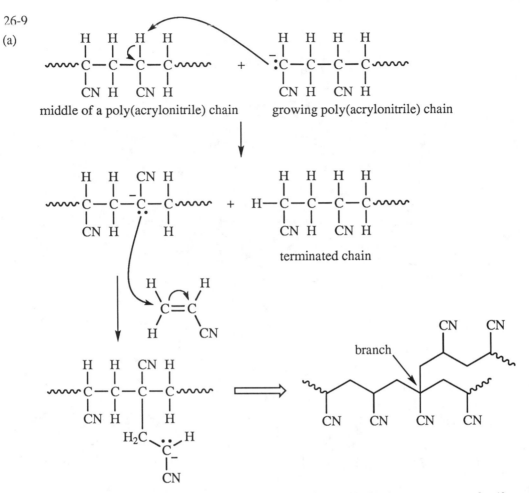

middle of a poly(acrylonitrile) chain   growing poly(acrylonitrile) chain

terminated chain

branch

(b) The chain-branching hydride transfer (from a cationic mechanism) or proton transfer (from an anionic mechanism) ends a less-highly-substituted end of a chain and generates an intermediate on a more-highly-substituted middle of a chain (a 3° carbon in these mechanisms). This stabilizes a carbocation, but greater substitution *destabilizes* a carbanion. Branching can and does happen in anionic mechanisms, but it is less likely than in cationic mechanisms.

26-10   isotactic poly(acrylonitrile)

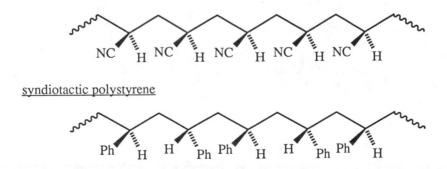

syndiotactic polystyrene

26-11

(a)

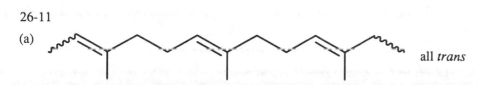

all *trans*

(b) The *trans* double bonds in gutta-percha allow for more ordered packing of the chains, that is, a higher degree of crystallinity. (Recall how *cis* double bonds in fats and oils lower the melting points because the *cis* orientation disrupts the ordering of the packing of the chains.) The more crystalline a polymer is, the less elastic it is.

26-12   Whether the alkene is cis or trans is not specified.

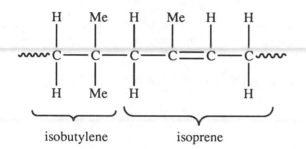

26-13   The repeating unit in each polymer is boxed.

(a) Nomex®

(b) Kevlar®

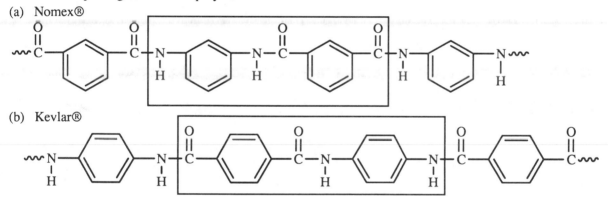

26-14   Kodel® polyester (only one repeating unit shown)

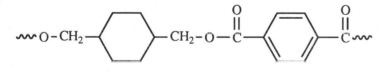

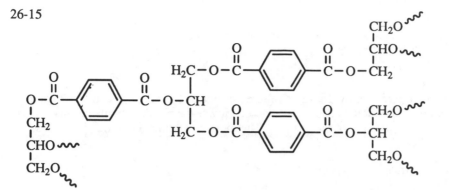

Glycerol is a trifunctional molecule, so not only does it grow in two directions to make a chain, it grows in three directions. All of its chains are cross-linked, forming a three-dimensional lattice with very little motion possible. The more cross-linked the polymer is, the more rigid it is.

26-16 For simplicity in this problem, bisphenol A will be abbreviated as a substituted phenol.

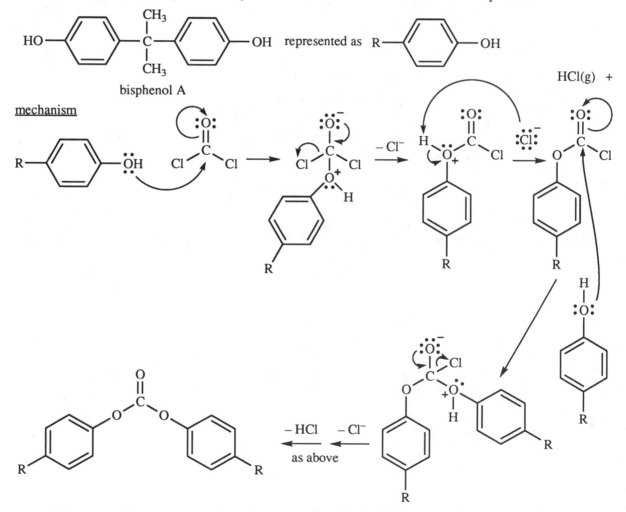

26-17 Bisphenol A is made by condensing two molecules of phenol with one molecule of acetone, with loss of a molecule of water. This is an electrophilic aromatic subsitution (more specifically, a Friedel-Crafts alkylation), and would require an acid catalyst to generate the carbocation. While a Lewis acid could be used, the mechanism below shows a protic acid.

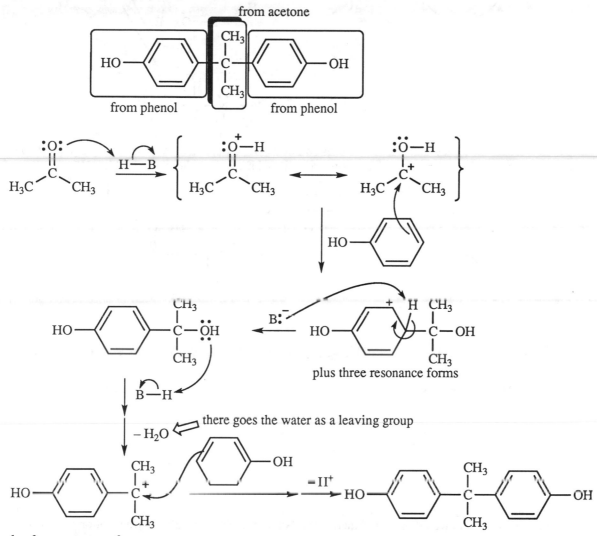

plus four resonance forms

26-18

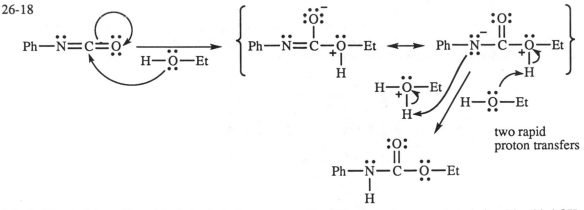

26-19 Glycerol is a trifunctional alcohol. It uses two of its OH groups in a growing chain. The third OH group cross-links with another chain. The more cross-linked a polymer, the more rigid it is.

26-20

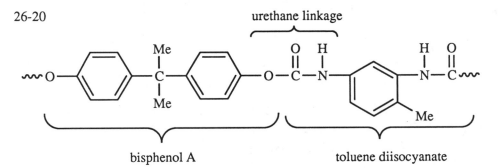

bisphenol A   toluene diisocyanate

26-21   Please refer to solution 1-19, page 12 of this Solutions Manual.

26-22

(a)

~~~C—C—C—C—C—C~~~

with H, Me, H, Me, H, Me substituents top and bottom

(b) Polyisobutylene is an addition polymer. No small molecule is lost, so this cannot be a condensation polymer.

(c) Either cationic polymerization or free-radical polymerization would be appropriate. The carbocation or free-radical intermediate would be 3° and therefore relatively stable. Anionic polymerization would be inappropriate as there is no electron-withdrawing group to stabilize the anion.

26-23

(a) It is a polyurethane.

(b) As with all polyurethanes, it is a condensation polymer.

(c)

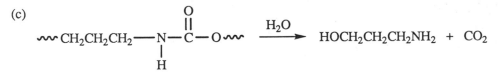

26-24

(a) It is a polyester.

(b) As with all polyesters, it is a condensation polymer.

(c)

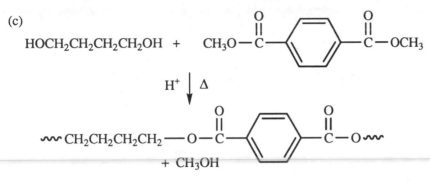

Using the dicarboxylic acid instead of the ester would produce water as the small neutral molecule lost in this condensation.

26-25

(a) Urylon® is a polyurea.

(b) A polyurea is a condensation polymer.

(c)

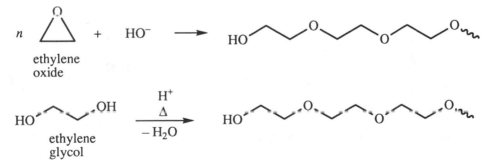

26-26

(a) Polyethylene glycol, abbreviated PEG, is a polyether.

(b) PEG is usually made from ethylene oxide (first reaction shown). In theory, PEG could also be made by intermolecular dehydration of ethylene glycol (second reaction shown), but the yields are low and the chains are short.

(c) Basic catalysts are most likely as they open the epoxide to generate a new nucleophile. Acid catalysts are possible but they risk dehydration and ether cleavage.

26-26 continued

(d) Mechanism of ethylene oxide polymerization (showing hydroxide as the base):

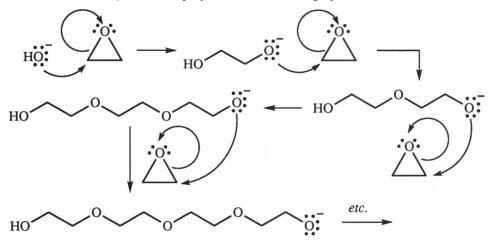

26-27

(a) Polychloroprene (Neoprene®) is an addition polymer.
(b) Polychloroprene comes from the diene, chloroprene, just as natural rubber comes from isoprene:

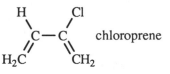

chloroprene

26-28

(a)

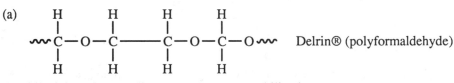

Delrin® (polyformaldehyde)

(b) All of these intermediates are resonance-stabilized.

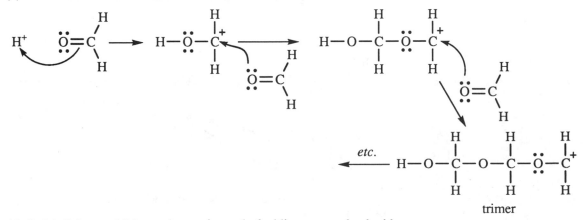

trimer

(c) Delrin® is an addition polymer; instead of adding across the double bond of an alkene, addition occurs across the double bond of a carbonyl group.

620

26-29

(a) *cis*

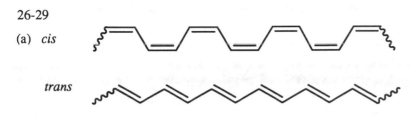

trans

(b) Each structure has a fully conjugated chain. It is reasonable to expect electrons to be able to be transferred through the π system, just as resonance effects can work over long distances through conjugated systems.

(c) It is not surprising that the conductivity is directional. Electrons must flow along the π system of the chain, so if the chains were aligned, conductivity would be greater in the direction parallel to the polymer chains. (It is possible, though less likely, that electrons could pass from the π system of one chain to the π system of another, that is, perpendicular to the direction of the chain; we would expect reduced conductivity in that direction.)

26-30

(a) A Nylon is a polyamide. Amides can be hydrolyzed in aqueous acid, cleaving the polymer chain in the process.

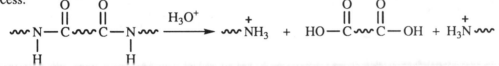

(b) A polyester can be saponified in aqueous base, cleaving the polymer chain in the process.

26-31

(a)

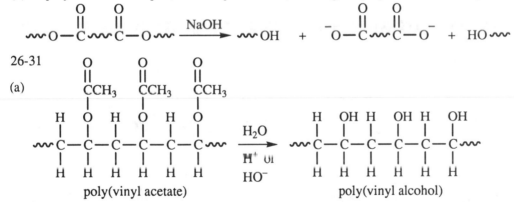

poly(vinyl acetate)   poly(vinyl alcohol)

(b) A polyester is a condensation polymer in which monomer units are linked through ester groups as part of the polymer chain. Poly(vinyl acetate) is really a substituted polyethylene, an **addition** polymer, with only carbons in the chain; the ester groups are in the side chains, not in the polymer backbone.

(c) Hydrolysis of the esters in poly(vinyl acetate) does not affect the chain because the ester groups do not occur in the chain as they do in Dacron®.

(d) Vinyl alcohol cannot be polymerized because it is unstable, tautomerizing to acetaldehyde.

26-32

(a)

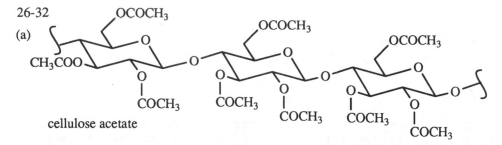

cellulose acetate

(b) Cellulose has three OH groups per glucose monomer, which form hydrogen bonds with other polar groups. Transforming these OH groups into acetates makes the polymer much less polar and therefore more soluble in organic solvents.

(c) The acetone dissolved the cellulose acetate in the fibers. As the acetone evaporated, the cellulose acetate remained but no longer had the fibrous, woven structure of cloth. It recrystallized as white fluff.

(d) Any article of clothing made from synthetic fibers is susceptible to the ravages of organic solvents. Solvent splashes leave dimples or blotches on Corfam shoes. (Yet, Corfam shoes could still provide protection for the toenail polish!)

26-33

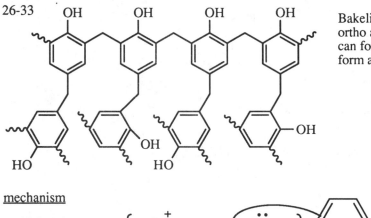

Bakelite is highly cross-linked through the ortho and para positions of phenol; each phenol can form a chain at two ring positions, then form a branch at the third position.

mechanism

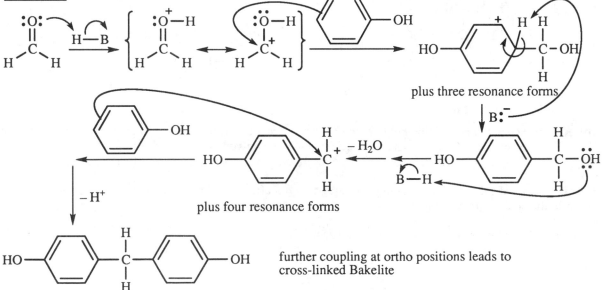

plus three resonance forms

plus four resonance forms

further coupling at ortho positions leads to cross-linked Bakelite

26-34

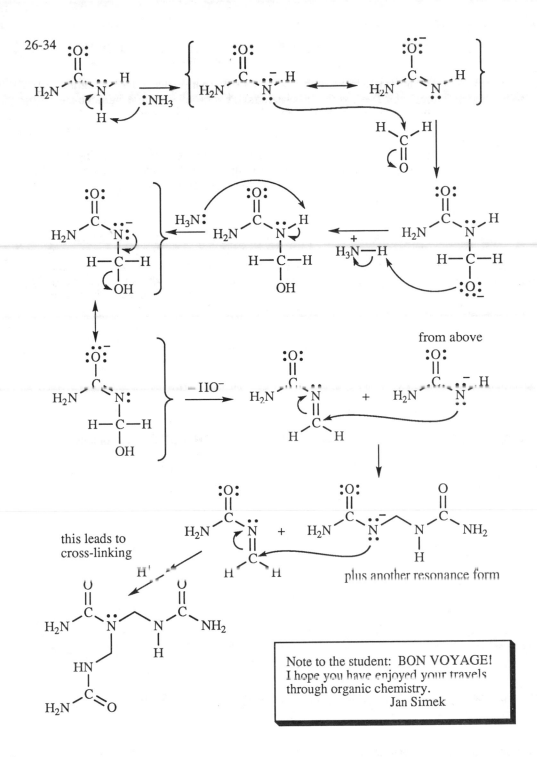

this leads to
cross-linking

plus another resonance form

from above

Note to the student: BON VOYAGE!
I hope you have enjoyed your travels
through organic chemistry.
Jan Simek

Appendix 1: Summary of IUPAC Nomenclature of Organic Compounds

Introduction

The purpose of the IUPAC system of nomenclature is to establish an international standard of naming compounds to facilitate communication. The goal of the system is to give each structure a unique and unambiguous name, and to correlate each name with a unique and unambiguous structure.

I. Fundamental Principle

IUPAC nomenclature is based on naming a molecule's longest chain of carbons connected by single bonds, whether in a continuous chain or in a ring. All deviations, either multiple bonds or atoms other than carbon and hydrogen, are indicated by prefixes or suffixes according to a specific set of priorities.

II. Alkanes and Cycloalkanes

Alkanes are the family of saturated hydrocarbons, that is, molecules containing carbon and hydrogen connected by single bonds only. These molecules can be in continuous chains (called linear or acyclic), or in rings (called cyclic or alicyclic). The names of alkanes and cycloalkanes are the root names of organic compounds. Beginning with the five-carbon alkane, the number of carbons in the chain is indicated by the Greek or Latin prefix. Rings are designated by the prefix "cyclo". (In the geometrical symbols for rings, each apex represents a carbon with the number of hydrogens required to fill its valence.)

C_1	CH_4	methane		C_{12}	$CH_3(CH_2)_{10}CH_3$	dodecane
C_2	CH_3CH_3	ethane		C_{13}	$CH_3(CH_2)_{11}CH_3$	tridecane
C_3	$CH_3CH_2CH_3$	propane		C_{14}	$CH_3(CH_2)_{12}CH_3$	tetradecane
C_4	$CH_3(CH_2)_2CH_3$	butane		C_{20}	$CH_3(CH_2)_{18}CH_3$	icosane
C_5	$CH_3(CH_2)_3CH_3$	pentane		C_{21}	$CH_3(CH_2)_{19}CH_3$	henicosane
C_6	$CH_3(CH_2)_4CH_3$	hexane		C_{22}	$CH_3(CH_2)_{20}CH_3$	docosane
C_7	$CH_3(CH_2)_5CH_3$	heptane		C_{23}	$CH_3(CH_2)_{21}CH_3$	tricosane
C_8	$CH_3(CH_2)_6CH_3$	octane		C_{30}	$CH_3(CH_2)_{28}CH_3$	triacontane
C_9	$CH_3(CH_2)_7CH_3$	nonane		C_{31}	$CH_3(CH_2)_{29}CH_3$	hentriacontane
C_{10}	$CH_3(CH_2)_8CH_3$	decane		C_{40}	$CH_3(CH_2)_{38}CH_3$	tetracontane
C_{11}	$CH_3(CH_2)_9CH_3$	undecane		C_{50}	$CH_3(CH_2)_{48}CH_3$	pentacontane

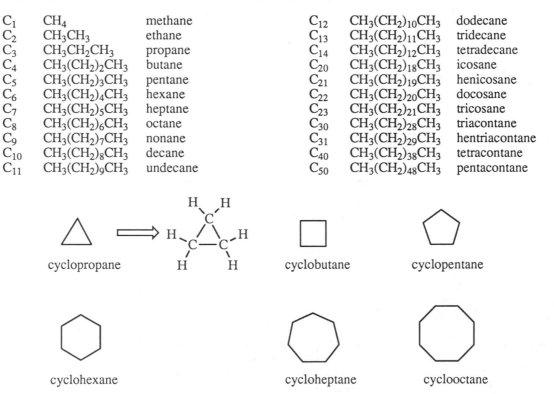

cyclopropane   cyclobutane   cyclopentane

cyclohexane   cycloheptane   cyclooctane

III. Nomenclature of Molecules Containing Substituents and Functional Groups

A. Priorities of Substituents and Functional Groups
LISTED HERE FROM HIGHEST TO LOWEST PRIORITY, except that the substituents within Group C have equivalent priority.

Group A—Functional Groups Named By Prefix Or Suffix

Functional Group	Structure	Prefix	Suffix
Carboxylic Acid	$R-\overset{\overset{O}{\|\|}}{C}-OH$	carboxy-	-oic acid (-carboxylic acid)
Aldehyde	$R-\overset{\overset{O}{\|\|}}{C}-H$	oxo- (formyl)	-al (carbaldehyde)
Ketone	$R-\overset{\overset{O}{\|\|}}{C}-R$	oxo-	-one
Alcohol	$R-O-H$	hydroxy-	-ol
Amine	$R-N\diagup$	amino-	-amine

Group B—Functional Groups Named By Suffix Only

Functional Group	Structure	Prefix	Suffix
Alkene	$\diagdown C = C \diagup$	--------	-ene
Alkyne	$-C \equiv C-$	--------	-yne

Group C—Substituent Groups Named By Prefix Only

Substituent	Structure	Prefix	Suffix
Alkyl (see next page)	$R-$	alkyl-	--------
Alkoxy	$R-O-$	alkoxy-	--------
Halogen	$F-$	fluoro-	--------
	$Cl-$	chloro-	--------
	$Br-$	bromo	
	$I-$	iodo-	--------

Miscellaneous substituents and their prefixes

$-NO_2$	$-CH=CH_2$	$-CH_2CH=CH_2$	
nitro	vinyl	allyl	phenyl

Appendix 1, Summary of IUPAC Nomenclature, continued

Common alkyl groups—replace "ane" ending of alkane name with "yl". Alternate names for complex substituents are given in brackets.

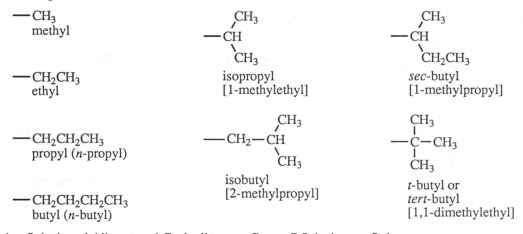

—CH₃
methyl

—CH₂CH₃
ethyl

—CH₂CH₂CH₃
propyl (*n*-propyl)

—CH₂CH₂CH₂CH₃
butyl (*n*-butyl)

isopropyl
[1-methylethyl]

sec-butyl
[1-methylpropyl]

isobutyl
[2-methylpropyl]

t-butyl or
tert-butyl
[1,1-dimethylethyl]

B. Naming Substituted Alkanes and Cycloalkanes—Group C Substituents Only

Organic compounds containing substituents from Group C are named following this sequence of steps, as indicated on the examples below:

•Step 1. Find the longest continuous carbon chain. Determine the root name for this parent chain. In cyclic compounds, the ring is usually considered the parent chain, unless it is attached to a longer chain of carbons; indicate a ring with the prefix "cyclo" before the root name. (When there are two longest chains of equal length, use the chain with the greater number of substituents.)
•Step 2. Number the chain in the direction such that the position number of the first substituent is the smaller number. If the first substituents have the same number, then number so that the second substituent has the smaller number, *etc*.
•Step 3. Determine the name and position number of each substituent. (A substituent on a nitrogen is designated with an "*N*" instead of a number; see Section **III.D.1.** below.)
•Step 4. Indicate the number of identical groups by the prefixes di, tri, tetra, *etc*.
•Step 5. Place the position numbers and names of the substituent groups, in alphabetical order, before the root name. In alphabetizing, ignore prefixes like *sec*-, *tert*-, di, tri, *etc*., but include iso and cyclo. Always include a position number for each substituent, regardless of redundancies.

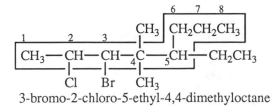

3-bromo-2-chloro-5-ethyl-4,4-dimethyloctane

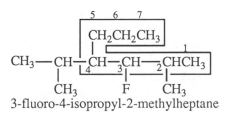

3-fluoro-4-isopropyl-2-methylheptane

H₃C–CHCH₂CH₃

1-*sec*-butyl-3-nitrocyclohexane
(numbering determined by the alphabetical order of substituents)

626

Appendix 1, Summary of IUPAC Nomenclature, continued

C. Naming Molecules Containing Functional Groups from Group B—Suffix Only

1. Alkenes—Follow the same steps as for alkanes, except:
a. Number the chain of carbons *that includes the alkene* so that the C=C has the lower position number, since it has a higher priority than any substituents;
b. Change "ane" to "ene" and assign a position number to the first carbon of the alkene;
c. Designate geometrical isomers with a *cis,trans* or *E,Z* prefix.

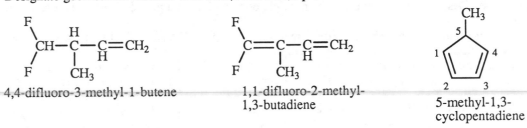

4,4-difluoro-3-methyl-1-butene

1,1-difluoro-2-methyl-
1,3-butadiene

5-methyl-1,3-
cyclopentadiene

Special case: When the chain cannot include an alkene, a substituent name is used. See Section V.A.2.a.

3-vinyl-1-cyclohexene

2. Alkynes—Follow the same steps as for alkanes, except:
a. Number the chain of carbons *that includes the alkyne* so that the alkyne has the lower position number;
b. Change "ane" to "yne" and assign a position number to the first carbon of the alkyne.
Note: The Group B functional groups (alkene and alkyne) are considered to have equal priority: in a molecule with both an ene and an yne, whichever is closer to the end of the chain determines the direction of numbering. In the case where each would have the same position number, the alkene takes the lower number. In the name, "ene" comes before "yne" because of alphabetization.

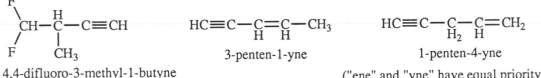

4,4-difluoro-3-methyl-1-butyne

3-penten-1-yne

1-penten-4-yne

("ene" and "yne" have equal priority unless they have the same position number, when "ene" takes the lower number)

(Notes: 1. An "e" is dropped if the letter following it is a vowel: "3-penten-1-yne", not "3-pentene-1-yne".
2. An "a" is added if inclusion of di, tri, *etc.*, would put two consonants together: "1,3-butadiene", not "1,3-butdiene".)

D. Naming Molecules Containing Functional Groups from Group A—Prefix or Suffix

In naming molecules containing one or more of the functional groups in Group A, the group of highest priority is indicated by suffix; the others are indicated by prefix, with priority equivalent to any other substituents. The table in Section III.A. defines the priorities; they are discussed on the following pages in order of increasing priority.

Appendix 1, Summary of IUPAC Nomenclature, continued

Now that the functional groups and substituents from Groups A, B, and C have been described, a modified set of steps for naming organic compounds can be applied to all simple structures:

•Step 1. Find the highest priority functional group. Determine and name the longest continuous carbon chain that includes this group.
•Step 2. Number the chain so that the highest priority functional group is assigned the lower number.
•Step 3. If the carbon chain includes multiple bonds (Group B), replace "ane" with "ene" for an alkene or "yne" for an alkyne. Designate the position of the multiple bond with the number of the first carbon of the multiple bond.
•Step 4. If the molecule includes Group A functional groups, replace the last "e" with the suffix of the highest priority functional group, and include its position number.
•Step 5. Indicate all Group C substituents, and Group A functional groups of lower priority, with a prefix. Place the prefixes, with appropriate position numbers, in alphabetical order before the root name.

1. Amines: prefix: amino-; suffix: -amine—substituents on nitrogen denoted by "*N*"

$CH_3CH_2CH_2$—NH_2

1-propanamine

3-methoxy-1-cyclohexanamine
("1" is optional in this case)

N,N-diethyl-3-buten-2-amine

2. Alcohols: prefix: hydroxy-; suffix: -ol

CH_3CH_2—OH

ethanol

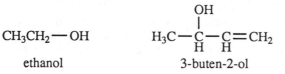

3-buten-2-ol

2-amino-1-cyclobutanol
("1" is optional in this case)

3. Ketones: prefix: oxo-; suffix: -one (pronounced "own")

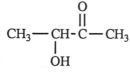

3-hydroxy-2-butanone

3-cyclohexen-1-one
("1" is optional in this case)

4-amino-*N,N*-dimethyl-4-penten-2-one

4. Aldehydes: prefix: oxo-, or formyl- (O=CH-); suffix: -al (abbreviation: —CHO)

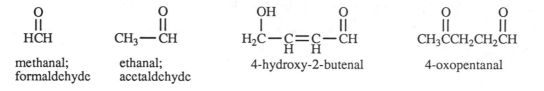

methanal;
formaldehyde

ethanal;
acetaldehyde

4-hydroxy-2-butenal

4-oxopentanal

Appendix 1, Summary of IUPAC Nomenclature, continued

<u>Special case:</u> When the chain cannot include the carbon of the aldehyde, the suffix "carbaldehyde" is used:

 cyclohexanecarbaldehyde

5. Carboxylic Acids: prefix: carboxy-; suffix: -oic acid (abbreviation: —COOH)
(Note: Chemists traditionally use, and IUPAC accepts, the names "formic acid" and "acetic acid" in place of "methanoic acid" and "ethanoic acid".)

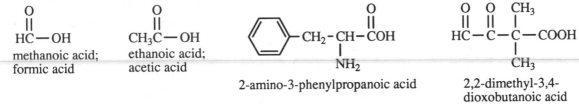

methanoic acid;   ethanoic acid;
formic acid   acetic acid   2-amino-3-phenylpropanoic acid   2,2-dimethyl-3,4-dioxobutanoic acid

<u>Special case:</u> When the chain numbering cannot include the carbon of the carboxylic acid, the suffix "carboxylic acid" is used:

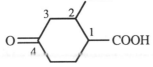

2-formyl-4-oxocyclohexanecarboxylic acid
("formyl" is used to indicate an aldehyde as a substituent when its carbon cannot be in the chain numbering)

<u>E. Naming Carboxylic Acid Derivatives</u>
The six common groups derived from carboxylic acids are, in decreasing priority after carboxylic acids: salts, anhydrides, esters, acyl halides, amides, and nitriles.

1. Salts of Carboxylic Acids
Salts are named with cation first, followed by the anion name of the carboxylic acid, where **"ic acid"** is replaced by **"ate"** :

acet**ic acid**	becomes	acet**ate**
butano**ic acid**	becomes	butano**ate**
cyclohexanebarboxy**lic acid**	becomes	cyclohexanecarboxyl**ate**

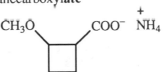

lithium 2-aminopropanoate   sodium chloroacetate   ammonium 2-methoxy-cyclobutanecarboxylate

2. Anhydrides: "oic acid" is replaced by "oic anhydride"

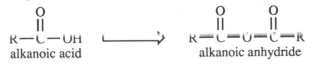

alkanoic acid   alkanoic anhydride

Appendix 1, Summary of IUPAC Nomenclature, continued

3. Esters

Esters are named as "organic salts" that is, the alkyl name comes first, followed by the name of the carboxylate anion. (common abbreviation: —COOR)

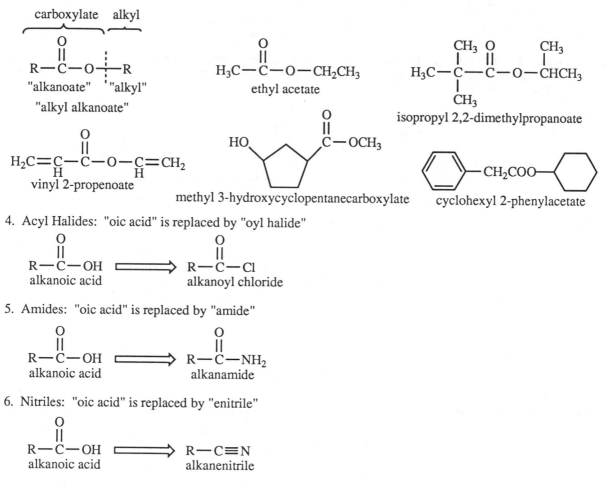

carboxylate   alkyl

"alkanoate"   "alkyl"

"alkyl alkanoate"

ethyl acetate

isopropyl 2,2-dimethylpropanoate

vinyl 2-propenoate

methyl 3-hydroxycyclopentanecarboxylate

cyclohexyl 2-phenylacetate

4. Acyl Halides: "oic acid" is replaced by "oyl halide"

$$R-\overset{\overset{\displaystyle O}{\|}}{C}-OH \implies R-\overset{\overset{\displaystyle O}{\|}}{C}-Cl$$

alkanoic acid   alkanoyl chloride

5. Amides: "oic acid" is replaced by "amide"

$$R-\overset{\overset{\displaystyle O}{\|}}{C}-OH \implies R-\overset{\overset{\displaystyle O}{\|}}{C}-NH_2$$

alkanoic acid   alkanamide

6. Nitriles: "oic acid" is replaced by "enitrile"

$$R-\overset{\overset{\displaystyle O}{\|}}{C}-OH \implies R-C\equiv N$$

alkanoic acid   alkanenitrile

IV. Nomenclature of Aromatic Compounds

"Aromatic" compounds are those derived from benzene and similar ring systems. As with aliphatic nomenclature described above, the process is: determining the root name of the parent ring; determining priority, name, and position number of substituents; and assembling the name in alphabetical order. *Functional group priorities are the same in aliphatic and aromatic nomenclature.*

A. Common Parent Ring Systems

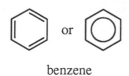

benzene

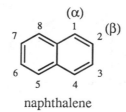

naphthalene

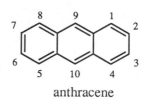

anthracene

Appendix 1, Summary of IUPAC Nomenclature, continued

B. Monosubstituted Benzenes
1. Most substituents keep their designation, followed by the word "benzene":

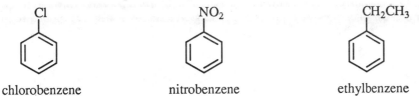

chlorobenzene   nitrobenzene   ethylbenzene

2. Some common substituents change the root name of the ring. IUPAC accepts these as root names, listed here in decreasing priority:

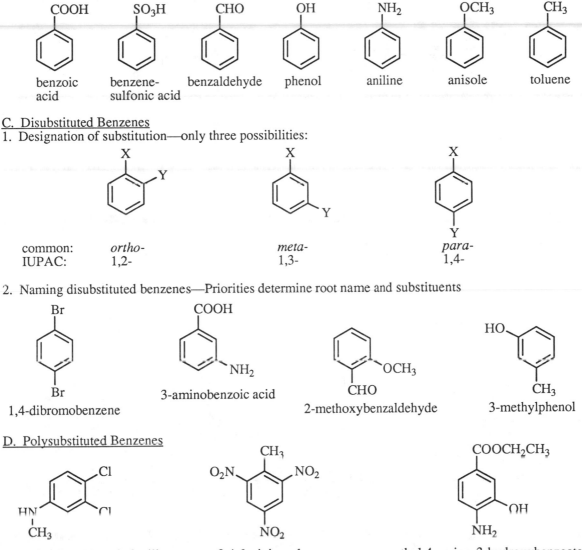

benzoic   benzene-   benzaldehyde   phenol   aniline   anisole   toluene
acid   sulfonic acid

C. Disubstituted Benzenes
1. Designation of substitution—only three possibilities:

common: *ortho-*   *meta-*   *para-*
IUPAC: 1,2-   1,3-   1,4-

2. Naming disubstituted benzenes—Priorities determine root name and substituents

1,4-dibromobenzene   3-aminobenzoic acid   2-methoxybenzaldehyde   3-methylphenol

D. Polysubstituted Benzenes

3,4-dichloro-*N*-methylaniline   2,4,6-trinitrotoluene   ethyl 4-amino-3-hydroxybenzoate
                                  (TNT)

Appendix 1, Summary of IUPAC Nomenclature, continued

E. Aromatic Ketones

A special group of aromatic compounds are ketones where the carbonyl is attached to at least one benzene ring. Such compounds are named as "phenones", the prefix depending on the size and nature of the group on the other side of the carbonyl. These are the common examples:

acetophenone   propiophenone

butyrophenone   benzophenone

Nomenclature Problem Set (answers on p. 634)

1.

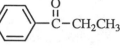

2.

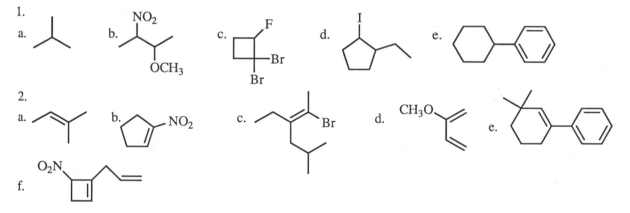

3. Alkyne carbons are shown for clarity.

4.

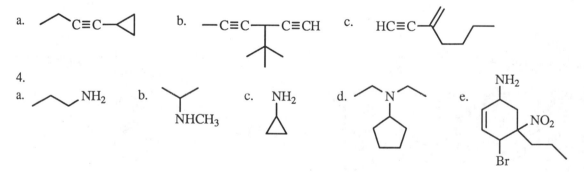

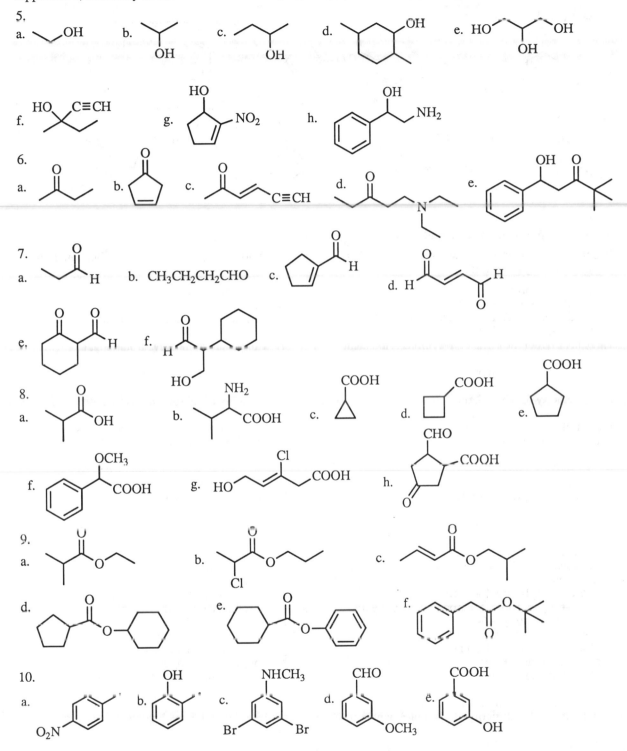

5.
a. [structure: CH₂CH₂OH]   b. [structure: (CH₃)₂CHOH]   c. [structure]   d. [cyclohexane with OH and two methyls]   e. HO~~~OH with OH

f. HO C≡CH [structure]   g. HO [cyclopentene with NO₂]   h. OH [benzene with CH(OH)CH₂NH₂], NH₂

6.
a. [structure with O]   b. [cyclopentenone]   c. [structure with O and C≡CH]   d. [structure with O and N(ethyl)₂]   e. OH O [structure]

7.
a. [structure with O and H]   b. CH₃CH₂CH₂CHO   c. [cyclopentene-CHO], H   d. H [structure O and H]

e. [cyclohexanone with CHO], H   f. H [structure with O, HO, cyclohexane]

8.
a. [structure O, OH]   b. NH₂ [structure COOH]   c. [cyclopropane COOH]   d. [cyclobutane COOH]   e. COOH [cyclopentane]

f. OCH₃ [benzene CH COOH]   g. Cl HO~~~COOH   h. CHO [cyclopentanone ring with COOH], O

9.
a. [structure O, O, ethyl]   b. Cl [structure O, O, propyl]   c. [structure O, O, isobutyl]

d. [cyclopentane C(O)O cyclohexane]   e. [cyclohexane C(O)O phenyl]   f. [benzene CH₂ C(O)O tert-butyl]

10.
a. O₂N [benzene]   b. OH [benzene]   c. NHCH₃ [benzene with Br, Br]   d. CHO [benzene OCH₃]   e. COOH [benzene OH]

633

Appendix 1, Summary of IUPAC Nomenclature, continued

10. continued

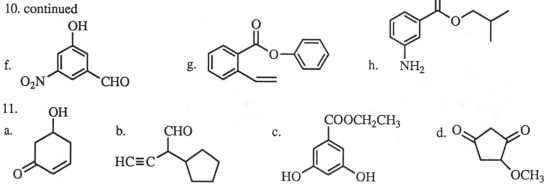

f.

g.

h.

11.

a.

b.

c.

d.

Answers to Nomenclature Problems

In some answers, an unnecessary "1" is included for clarity. It is not wrong to include a "1", but sometimes is can be wrong to omit it.

1. a. 2-methylpropane
   b. 2-methoxy-3-nitrobutane
   c. 1,1-dibromo-2-fluorocyclobutane
   d. 1-ethyl-2-iodocyclopentane
   e. 1-phenylcyclohexane or 1-cyclohexylbenzene
2. a. 2-methyl-2-butene
   b. 1-nitro-1-cyclopentene
   c. Z-2-bromo-3-ethyl-5-methyl-2-hexene
   d. 2-methoxy-1,3-butadiene
   e. 3,3-dimethyl-1-phenyl-1-cyclohexene
   f. 1-allyl-4-nitro-1-cyclobutene
3. a. 1-cyclopropyl-1-butyne
   b. 3-t-butyl-1,4-hexadiyne
   c. 2-n-butyl-1-buten-3-yne
4. a. 1-propanamine
   b. N-methyl-2-propanamine
   c. 1-cyclopropanamine
   d. N,N-diethyl-1-cyclopentanamine
   e. 4-bromo-5-nitro-5-n-propyl-2-cyclohexen-1-amine
5. a. ethanol
   b. 2-propanol
   c. 2-butanol
   d. 2,5-dimethyl-1-cyclohexanol
   e. 1,2,3-propanetriol cyclopentanecarboxylate
   f. 3-methyl-1-pentyn-3-ol
   g. 2-nitro-2-cyclopenten-1-ol
   h. 2-amino-1-phenyl-1-ethanol
6. a. 2-butanone
   b. 3-cyclopenten-1-one
   c. E-3-hexen-5-yn-2-one
   d. N,N-diethyl-1-amino-3-pentanone
   e. 1-hydroxy-4,4-dimethyl-1-phenyl-3-pentanone

7. a. propanal
   b. butanal
   c. 1-cyclopentene-1-carbaldehyde
   d. E-2-butene-1,4-dial
   e. 2-oxo-1-cyclohexanecarbaldehyde
   f. 2-cyclohexyl-3-hydroxypropanal
8. a. 2-methylpropanoic acid
   b. 2-amino-3-methylbutanoic acid
   c. cyclopropanecarboxylic acid
   d. cyclobutanecarboxylic acid
   e. cyclopentanecarboxylic acid
   f. 2-methoxy-2-phenylacetic acid
   g. Z-3-chloro-5-hydroxy-3-pentenoic acid
   h. 2-formyl-4-oxocyclopentanecarboxylic acid
9. a. ethyl 2-methylpropanoate
   b. n-propyl 2-chloropropanoate
   c. isobutyl E-2-butenoate
   d. cyclohexyl
   e. phenyl cyclohexanecarboxylate
   f. t-butyl 2-phenylacetate
10. a. 4-nitrotoluene
    b. 2-methylphenol
    c. 3,5-dibromo-N-methylaniline
    d. 3-methoxybenzaldehyde
    e. 3-hydroxybenzoic acid
    f. 3-hydroxy-5-nitrobenzaldehyde
    g. phenyl 2-vinylbenzoate
    h. isobutyl 3-aminobenzoate
11. a. 5-hydroxy-2-cyclohexen-1-one
    b. 2-cyclopentyl-3-butynal
    c. ethyl 3,5-dihydroxybenzoate
    d. 4-methoxy-1,3-cyclopentanedione

634

Appendix 2: Summary of Acidity and Basicity

Imagine that you are at a family reunion where you can observe the competition for ice cream cones among your nieces and nephews. Pretty soon, you formulate a generalization: the older kids can hold onto their ice cream cones more strongly than the younger ones. Another way of saying it is that the older ones are less likely to give up their cones. You could even represent this information in a table showing a series of equilibria between the child with ice cream, and the free ice cream plus the hungry child. The differences in strength could also be quantitated: the larger the hunger factor, pK_H, the less likely the child will give up the ice cream.

Approximate pK_H Values of Children

pK_H			
12	12-year-old with ice cream	$\rightleftharpoons$	ice cream + hungry 12-year-old
10	10-year-old with ice cream	$\rightleftharpoons$	ice cream + hungry 10-year-old
8	8-year-old with ice cream	$\rightleftharpoons$	ice cream + hungry 8-year-old
6	6-year-old with ice cream	$\rightleftharpoons$	ice cream + hungry 6-year-old
4	4-year-old with ice cream	$\rightleftharpoons$	ice cream + hungry 4-year-old

What good is this table? It allows anyone to make predictions about what will happen when kids with ice cream are mixed with hungry kids.

$$\left.\begin{array}{c}\text{hungry 10-year-old}\\ +\\ \text{4-year-old with ice cream}\end{array}\right\} \overset{?}{\rightleftharpoons} \left\{\begin{array}{c}\text{10-year-old with ice cream}\\ +\\ \text{hungry 4-year-old}\end{array}\right.$$

If the hungry 10-year-old was left unattended in a room with the 4-year-old with ice cream, which side of this equilibrium would be favored when you came back in a few minutes: "reactants" or "products"? More likely than not, there would be an ice cream transfer in your absence; "products" would be favored. You could have predicted this from the table, and you could generalize: the hungry 10-year-old will be strong enough to rip the ice cream away from any kid lower on the table than the 10-year-old him/herself.

Let's predict the results of another equilibrium:

$$\left.\begin{array}{c}\text{hungry 6-year-old}\\ +\\ \text{12-year-old with ice cream}\end{array}\right\} \overset{?}{\rightleftharpoons} \left\{\begin{array}{c}\text{6-year-old with ice cream}\\ +\\ \text{hungry 12-year-old}\end{array}\right.$$

Is the hungry 6-year-old strong enough to pull the ice cream away from the 12-year-old? Not in most families. The table shows that the only chance for the hungry 6-year-old is to find a 4-year-old with ice cream. The 12-year-old with ice cream is pretty safe as long as the hunger table doesn't go to higher ages.

If you understand this analogy and can make predictions of ice cream transfer using the table, then you can understand how to predict the direction of equilibrium in acid-base reactions. Turn the page.

To set the stage ....

A few generalizations:

A) This Appendix deals with protic acids and bases, called Bronsted-Lowry acids. Similar statements can be made about Lewis acids but they are not the focus of this discussion.

B) Values of pK_a are measures of equilibrium constants, described further below. Values between 0 and 15.7 are measured by titration in water solution and are known accurately, to within 0.1 pK unit and sometimes better. Values outside of this range cannot be measured in water because of the leveling effect of water, and there is no universally accepted method for measuring these pK_a values. This lack of a single standard of measurement means that the values below 0 and above 15.7 should be considered relative, not absolute. If your instructor says the pK_a of methane is 46 and this book says it is 50, those should be considered the same value within experimental variation.

C) Acidity is a thermodynamic property, and the acid equilibrium constant, K_a, is a measure of the relative concentrations of species in the protonated and unprotonated form. As most organic acids are weak acids, meaning they are present mostly in the protonated form at equilibrium, the $K_a < 1$. Since the $pK_a = -\log K_a$, the pK_a values will be greater than 0, with the larger pK_a representing a weaker acid. If this is not clear, review text section 1-13.

D) Acids do not spontaneously spit out a proton! Despite our way of writing ionization equilibria as shown on the next page, acids do not give up a proton unless a base comes by to take the proton away. The reactions as drawn in the table should be considered half-reactions, just as the reactions in the electromotive series were half-reactions for balancing oxidation-reduction reactions in general chemistry.

I. Predicting equilibrium position

Look at the table on p. 637 and notice how it looks just like the "children-with-ice-cream" table. We can use this table to make predictions about equilibrium position in acid-base reactions just as we did for the children with ice cream.

1) *A base will deprotonate any acid stronger than its conjugate acid.* This is the most important principle of predicting acid-base reactions. On the table, this means that any base, hydroxide for example, can react with any acid more acidic than the conjugate acid of itself, water in our example. So hydroxide is a strong enough base to pull the proton from any of these: bicarbonate ion, a phenol, carbonic acid, a carboxylic acid, or a sulfonic acid. We can also predict that hydroxide is NOT a strong enough base to react with any acid above water on the table; for example, a mixture of hydroxide with an alkyne will favor the reactants at equilibrium, with only a small amount of products.

reactants favored

$$HO^- \ + \ RC\equiv C-H \ \xleftarrow{\hspace{1.5cm}}\xrightarrow{\hspace{1.5cm}} \ H_2O \ + \ RC\equiv C\colon^-$$

$$pK_a\ 25 \phantom{\xleftarrow{\hspace{1cm}}} pK_a\ 15.7$$

weaker	*weaker*	*stronger*	*stronger*
base	*acid*	*acid*	*base*

2) Another way of predicting the position of an equilibrium is to assign "stronger" and "weaker" to the acid and base on each side of the equation, using the table to determine which is stronger and which is weaker. *Equilibrium will always favor the weaker acid and base.* This method will always give the same answer as the principle in #1 above.

To lead into the next section, look again at the table on p. 637 and notice two things: a) with only a couple of exceptions, all the acidic protons are on either oxygen or carbon; and b) generalizations can be made about the acidity of functional groups. Learning to correlate acidity with functional group is important in predicting reactivity of the functional group.

Approximate pKa Values of Organic Compounds

		pKa			
weaker acid	alkane	≈ 50	$R-\underset{\mid}{\overset{\mid}{C}}-H \;\rightleftharpoons\; H^+ +$	$R-\underset{\mid}{\overset{\mid}{\ddot{C}}}{:}^{\,-}$	*stronger base*
	alkene	≈ 45	$=C\overset{/}{\underset{H}{\diagdown}} \;\rightleftharpoons\; H^+ +$	$=\overset{/}{\underset{\cdot\cdot}{C}}{}^{\,-}$	
	amine	35-40	$-\overset{\cdot\cdot}{\underset{\mid}{N}}-H \;\rightleftharpoons\; H^+ +$	$-\overset{\cdot\cdot}{\underset{\mid}{\ddot{N}}}{}^{\,-}$	
		≈ 35	$H-H \;\rightleftharpoons\; H^+ +$	$H{:}^{\,-}$	
	alkyne	≈ 25	$RC\equiv C-H \;\rightleftharpoons\; H^+ +$	$RC\equiv C{:}^{\,-}$	
	ketone and ester	20-25	$-\overset{\overset{O}{\parallel}}{C}-\underset{\mid}{\overset{\mid}{C}}-H \;\rightleftharpoons\; H^+ +$	$-\overset{\overset{O}{\parallel}}{C}-\underset{\mid}{\overset{\mid}{\ddot{C}}}{:}^{\,-}$	
	alcohol ≈ 18	≈ 18	$R-\underset{R}{\overset{R}{C}}-O-H \;\rightleftharpoons\; H^+ +$	$R-\underset{R}{\overset{R}{C}}-\ddot{\ddot{O}}{:}^{\,-}$	
		≈ 17	$R-\underset{R}{\overset{H}{C}}-O-H \;\rightleftharpoons\; H^+ +$	$R-\underset{R}{\overset{H}{C}}-\ddot{\ddot{O}}{:}^{\,-}$	
↑ cannot be measured in water solution		≈ 16	$R-\underset{H}{\overset{H}{C}}-O-H \;\rightleftharpoons\; H^+ +$	$R-\underset{H}{\overset{H}{C}}-\ddot{\ddot{O}}{:}^{\,-}$	

- -

↓ measured in water solution		15.7	$H_2O \;\rightleftharpoons\; H^+ +$	HO^-	
		10.3	$HCO_3^- \;\rightleftharpoons\; H^+ +$	CO_3^{2-}	
	phenol	≈ 10	$Ar-O-H \;\rightleftharpoons\; H^+ +$	$Ar-\ddot{\ddot{O}}{:}^{\,-}$	
		6.4	$H_2CO_3 \;\rightleftharpoons\; H^+ +$	HCO_3^-	
	carboxylic acid	4-5	$R-\overset{\overset{O}{\parallel}}{C}-O-H \;\rightleftharpoons\; H^+ +$	$R-\overset{\overset{O}{\parallel}}{C}-\ddot{\ddot{O}}{:}^{\,-}$	
	sulfonic acid	<0	$RSO_2-O-H \;\rightleftharpoons\; H^+ +$	$RSO_2-\ddot{\ddot{O}}{:}^{\,-}$	
stronger acid					*weaker base*

II. Correlation of Acidity with Functional Group

A. Oxygen Acids

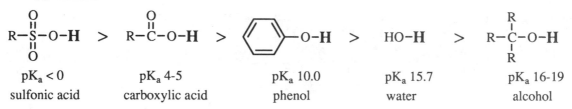

| pK$_a$ < 0 | pK$_a$ 4-5 | pK$_a$ 10.0 | pK$_a$ 15.7 | pK$_a$ 16-19 |
| sulfonic acid | carboxylic acid | phenol | water | alcohol |

Sulfonic acids are the strongest of the oxygen acids but are not common in organic chemistry. Carboxylic acids, however, are everywhere and are considered the strongest of the common organic oxygen acids. (Note that "strong" and "weak" are relative terms: acetic acid, pK$_a$ 4.74, was a "weak" acid in general chemistry in comparison to sulfuric and hydrochloric acids, but acetic acid is a "strong" acid in organic chemistry relative to the other oxygen acids.) Phenols having OH groups on benzene or other aromatic rings are still stronger acids than water.

Why are phenols, carboxylic acids, and sulfonic acids stronger acids than water? Because their anions are stabilized by resonance. (Refer to text sections 1-13, 10-6, and 20-4, especially Figure 20-1.) Let's look at that statement in more detail.

Remember that acidity is a thermodynamic property; that is, acidity equilibrium depends on the difference in energy between the reactants and products. The more the anion is stabilized by resonance, the lower in energy it is, and the less positive the ΔG, as shown on the reaction energy diagram:

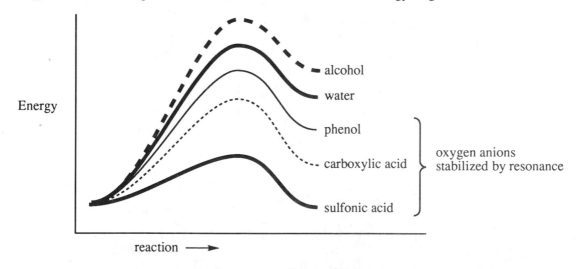

Why are alcohols weaker acids than water? There are two effects that contribute, both of which are consistent with the trend that 1° alcohols (pK$_a$ 16) are slightly stronger than 2° alcohols (pK$_a$ 17) which are slightly stronger than 3° alcohols (pK$_a$ 18). Alkyl groups are mildly electron-donating in their inductive effect (more about this later) and destabilize the anion, as shown in the energy diagram above. Second, the more crowded the anion is, the less it can be stabilized by hydrogen bonding with the solvent.

Appendix 2 continued, Summary of Acidity and Basicity

B. Carbon Acids

When we think of "acids", we do not usually think of protons on carbon, yet carbon acids and the carbanions that come from them are of tremendous importance in organic chemistry.

"Unstabilized" carbon acids are those that do not have any substituent to stabilize the anion. Alkanes, with only sp^3 carbons are the weakest acids with pK_a around 50. The vinyl carbon in a carbon-carbon double bond is sp^2 hybridized with the electrons of the anion slightly closer to the positive nucleus, leading to some stabilization of the anion. This type of stabilization is particularly important in alkynes with sp hybridized carbons.

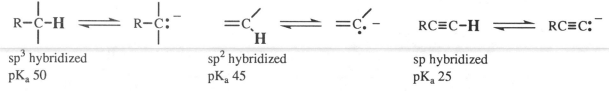

sp^3 hybridized
pK_a 50

sp^2 hybridized
pK_a 45

sp hybridized
pK_a 25

C. Carbon Acids Alpha to Carbonyl (This topic is described in detail in text section 22-2B.)

Look at these huge differences in pK_a when the acidic group is next to a carbonyl.

O–H $\Rightarrow$ RCH_2O-H pK_a 16-18

$$R-\overset{\overset{O}{\|}}{C}-O-H \qquad pK_a \ 4\text{-}5$$

N–H $\Rightarrow$ RCH_2NH-H pK_a 35-40

$$R-\overset{\overset{O}{\|}}{C}-\underset{H}{N}-H \qquad pK_a \ 16$$

C–H $\Rightarrow$ RCH_2CH_2-H pK_a 50

$$R-\overset{\overset{O}{\|}}{C}-\underset{H_2}{C}-H \qquad pK_a \ 20$$

Hydrogens alpha to carbonyl are unusually acidic because of resonance stabilization of the anionic conjugate base.

D. Acidities of Acyl Functional Groups

In addition to the significant variation in the acidity of alpha hydrogens depending on which atom the H is bonded to, what is on the other side of the carbonyl also has a dramatic influence. In this case, the stabilization is more important on the starting material, not on the conjugate base. See the energy diagram on p. 640.

aldehyde	ketone	ester	amide
$H_2\overset{\overset{H}{\|}}{C}-\overset{\overset{O}{\|}}{C}-H$	$H_2\overset{\overset{H}{\|}}{C}-\overset{\overset{O}{\|}}{C}-CH_3$	$H_2\overset{\overset{H}{\|}}{C}-\overset{\overset{O}{\|}}{C}-\overset{..}{\underset{..}{O}}CH_3$	$H_2\overset{\overset{H}{\|}}{C}-\overset{\overset{O}{\|}}{C}-\overset{..}{N}(CH_3)_2$
pK_a 17	pK_a 20	pK_a 25	pK_a 30
no stabilization of starting material	mild stabilization of starting material by weak electron donation from CH_3		

$$H_2\overset{\overset{H}{\|}}{C}-\overset{\overset{O^-}{\|}}{C}\overset{+}{=}\overset{..}{\underset{..}{O}}CH_3$$

significant resonance stabilization of starting material

$$H_2\overset{\overset{H}{\|}}{C}-\overset{\overset{O^-}{\|}}{C}\overset{+}{=}N(CH_3)_2$$

strongest resonance stabilization of starting material

Appendix 2 continued, Summary of Acidity and Basicity

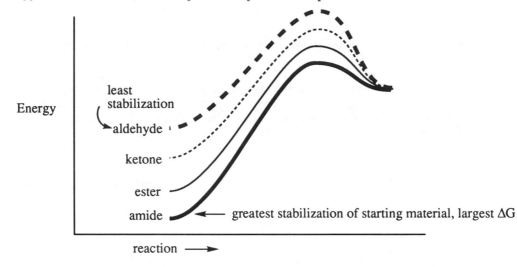

E. Carbon Acids Between Two Carbonyls (This topic is described in detail in text section 22-15.)

While a hydrogen alpha to one carbonyl moves into the pK_a 20-25 range for ketones and esters respectively, a hydrogen between two carbonyls (cyano and nitro are similar to carbonyl electronically) is more acidic than water. The increased resonance stabilization of the conjugate base is largely responsible, but there are subtle variations depending on the type of functional group as noted at the bottom of p. 639. Look at the enormous influence of the nitro group.

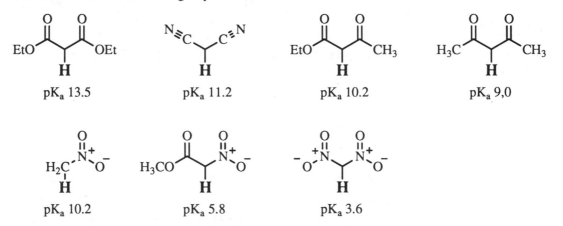

III. Correlation of Basicity with Functional Group

The bulk of this Appendix is on acidity because many more functional groups are acidic than are basic. Basically (oooh, sorry), only one functional group is basic: amines. There is variation among aliphatic, aromatic, and heteroaromatic amines; these are covered thoroughly in text sections 19-5 and 19-6. One point in the text, just before Table 19-3, deserves emphasis: for any conjugate acid-base pair:

$$pK_a + pK_b = 14$$

Appendix 2 continued, Summary of Acidity and Basicity

This simple algebraic relationship is very useful.

Sample problem. Is triethylamine (pK$_b$ 3.24) a strong enough base to deprotonate phenol (pK$_a$ 10.0)?

We need to calculate either the pK$_a$ of the conjugate acid of triethylamine or the pK$_b$ of the conjugate base of phenol to see which is stronger and weaker. Then we can say with certainty which side of the equilibrium will be favored.

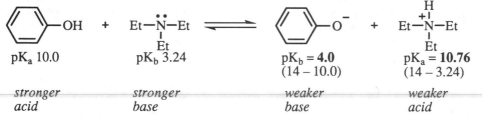

pK$_a$ 10.0	pK$_b$ 3.24	pK$_b$ = **4.0** (14 – 10.0)	pK$_a$ = **10.76** (14 – 3.24)
stronger acid	*stronger base*	*weaker base*	*weaker acid*

Aha! Products are favored at equilibrium, so the correct answer to the question is "Yes, triethylamine is a strong enough base to deprotonate phenol."

Try this for fun: How weak must a base be before it does NOT deprotonate phenol? What algebraic rule can you formulate to predict whether any combination of acid and base will favor products or reactants?

IV. Substituent Effects on Acidity

So far, we have focused on acidities of different functional groups. Let's turn to more minor, more subtle, structural changes to see what effect substituents will have on the acidity of a group. Primarily, we imply *electronic* effects as opposed to *steric* effects, but this Appendix will conclude with a discussion of how steric and electronic effects can work together.

A. Classification of Substituents—Induction and Resonance

Substituent groups can exert an electronic effect on an acidic functional group in two different ways: through sigma bonds, where this is called an *inductive effect*, or through p orbitals and pi bonds which is called a *resonance effect*. Groups can also be electron-donating or electron-withdrawing by either of the mechanisms, so there are four possible categories for groups. Note that a group can appear in more than one category, even in conflicting groups!

a) Electron-donating by induction: only alkyl groups (abbreviated R) have electrons to share by induction;

b) Electron-withdrawing by induction: every group that has a more electronegative atom than carbon is in this category; some examples: F, Cl, Br, I, OH, OR, NH$_2$, NHR, NR$_2$, NO$_2$, C=O, CN, SO$_3$H, CX$_3$ where X is halogen;

c) Electron-donating by resonance: groups that have electron pairs to share: F, Cl, Br, I, OH, OR, NH$_2$, NHR, NR$_2$;

d) Electron-withdrawing by resonance: NO$_2$, C=O, CN, SO$_3$H.

Appendix 2 continued, Summary of Acidity and Basicity

B. Generalizations on Electronic Effects on Acidity (refer to text section 20-4B)

Electric charge is the key to understanding substituent effects. An acid is always more positive than its conjugate base; in other words, the conjugate base is always more negative than the acid. Electron-donating and electron-withdrawing groups will have opposite effects on the acid-base conjugate pair.

Electron-donating groups stabilize the more positive acid form and destabilize the more negative conjugate base. From the diagram, it is apparent that electron-donating groups widen the energy gap between reactants and products, making ΔG more positive, favoring reactants more than products. In essence, this weakens the acid strength.

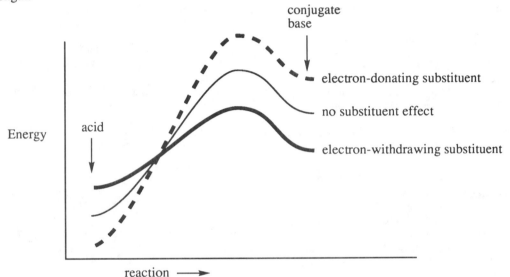

Electron-withdrawing groups destabilize the more positive acid form and stabilize the more negative conjugate base, narrowing the energy gap between reactants and products, making ΔG less positive. Products are increased in concentration at equilibrium which we define as a stronger acid.

Electron-withdrawing groups increase acid strength; electron-donating groups decrease acid strength.

C. Inductive Effects on Acidity

Text section 20-4B gives a thorough explanation of the inductive effect of electron-withdrawing groups on simple carboxylic acids, from which three generalizations arise:

A) Acidity increases with stronger electron-withdrawing groups. (See problem 20-31.)
B) Acidity increases with greater number of electron-withdrawing groups.
C) Acidity increases with closer proximity of the electron-withdrawing group to the acidic group.

We don't usually look to aromatic systems for examples of inductive effects, because the pi system of electrons is ripe for resonance effects. However, in analyzing the resonance forms of phenoxide on the next page, it becomes apparent that the negative charge is never distributed on the meta carbons. Meta substituents cannot exert any resonance stabilization or destabilization; at the meta position, substituents can exert only an inductive effect. The series of phenols demonstrates this phenomenon, consistent with aliphatic carboxylic acids.

Appendix 2 continued, Summary of Acidity and Basicity

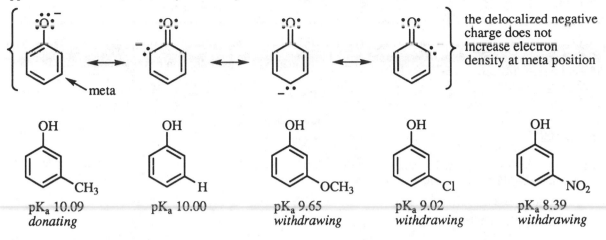

the delocalized negative charge does not increase electron density at meta position

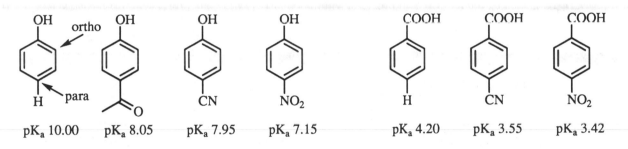

| pK$_a$ 10.09 | pK$_a$ 10.00 | pK$_a$ 9.65 | pK$_a$ 9.02 | pK$_a$ 8.39 |
| donating | | withdrawing | withdrawing | withdrawing |

D. Resonance Effects on Acidity—benzoic acids and phenols (review the solution to problem 20-41)

Resonance effects can be expressed with placement of substituents at ortho or para positions, but ortho has the complication of steric effects, so just para substitution is shown here.

Electron-withdrawing substituents increase the acidity of benzoic acids and phenols:

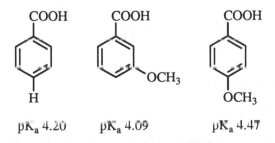

pK$_a$ 10.00   pK$_a$ 8.05   pK$_a$ 7.95   pK$_a$ 7.15   pK$_a$ 4.20   pK$_a$ 3.55   pK$_a$ 3.42

Electron-donating by resonance but electron-withdrawing by induction:

There is a group of substituents that donate by resonance but withdraw by induction: alkoxy groups and halogens are the most notable examples, and the acidity data provide an insight into which effect is stronger.

pK$_a$ 4.20   pK$_a$ 4.09   pK$_a$ 4.47

meta-Methoxybenzoic acid is stronger than benzoic acid, consistent with electron-withdrawing by induction which is expressed at the meta position. But the para isomer is *weaker* than benzoic acid; electron donation by induction has not only compensated for the inductive effect (which is still operative at the para position) but has decreased the acidity even further. Thus, the donating effect by resonance must be stronger than the withdrawing effect by induction for the methoxy group.

See another example on the next page.

Appendix 2 continued, Summary of Acidity and Basicity

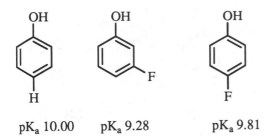

pK$_a$ 10.00   pK$_a$ 9.28   pK$_a$ 9.81

meta-Fluorophenol is stronger than phenol, consistent with electron-withdrawing by induction which is expressed at the meta position. The para isomer is still stronger than phenol; electron donation by induction has not compensated for the inductive effect (which is still operative at the para position). Thus, the donating effect by resonance must be weaker than the withdrawing effect by induction for the fluoro group.

Studying substituent effects on acidity is the standard method of determining whether a group is donating or withdrawing by induction and resonance.

E. Proximity Effects of Substituents

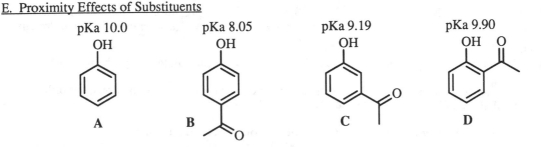

pKa 10.0   pKa 8.05   pKa 9.19   pKa 9.90

A   B   C   D

Three effects influence the pK$_a$ values of these substituted phenols. In **C**, the acetyl group at the meta position is electron-withdrawing by induction only. In **B**, the acetyl group at the para position exerts both resonance and inductive effects, both of which are electron-withdrawing, making the acid stronger. In theory, substituents at the ortho position should be like para, exerting both resonance and inductive effects; in fact, the inductive effect should be stronger because of closer proximity to the acidic group. So we would predict **D** to be a stronger acid than **B**, yet it is not. What other effect is operating?

Structure **E** shows that because of the proximity of the acetyl group to the OH, *intramolecular hydrogen bonding* is possible. Hydrogen bonding stabilizes the starting material, lowering the energy of the starting material and making ΔG more positive. Intuitively, it should be apparent that a hydrogen held between two oxygens will be more difficult to remove by a base. Also, after the proton has left as shown in structure **F**, the negative charge on the phenolic oxygen is close to the partial negative charge on the oxygen of the carbonyl, destabilizing product **F**, raising its energy, also making ΔG more positive. The proximity of the acetyl group influences both sides of the equation to make the acid weaker.

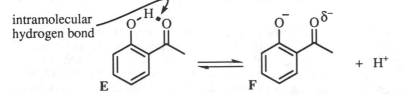

intramolecular hydrogen bond

E   F   + H$^+$

Here are two more examples where intramolecular hydrogen-bonding influences acidity.

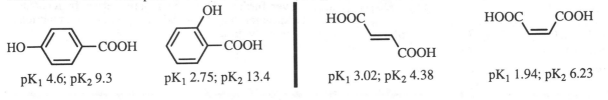

pK$_1$ 4.6; pK$_2$ 9.3   pK$_1$ 2.75; pK$_2$ 13.4   pK$_1$ 3.02; pK$_2$ 4.38   pK$_1$ 1.94; pK$_2$ 6.23

Appendix 2 continued, Summary of Acidity and Basicity

F. Steric Inhibition of Resonance

Another type of proximity effect arises when the placement of a substituent interferes with the orbital overlap required for resonance stabilization. This can be seen clearly in the acidity of substituted benzoic acids and in the basicity of substituted anilines.

Let's analyze this series of carboxylic acids.

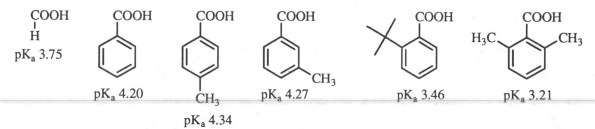

COOH
H
pK$_a$ 3.75

COOH
pK$_a$ 4.20

COOH
CH$_3$
pK$_a$ 4.34

COOH
CH$_3$
pK$_a$ 4.27

COOH
pK$_a$ 3.46

COOH
H$_3$C   CH$_3$
pK$_a$ 3.21

Formic acid, pK$_a$ 3.75, serves as the reference carboxylic acid. Benzoic acid is weaker because the phenyl group is electron-donating by resonance, stabilizing the protonated form. Methyl substituents are known to be electron-donating by induction, strengthening the electron-donating effect, making the meta- and para-substituted acids weaker than benzoic acid.

Then come the anomalies. Alkyl groups are electron-donating by induction and should weaken the acids, but the ortho-t-butyl and the 2,6-dimethylbenzoic acids are not only stronger than benzoic acid, they are stronger than formic acid! Something has happened to turn the phenyl group into an electron-withdrawing group.

Phenyl is electron-donating by resonance but electron-withdrawing by induction, so what has happened is that the ortho substituents have forced the COOH out of the plane of the benzene ring so that there is no resonance overlap between the benzene ring and the COOH orbitals. The COOH "feels" the benzene ring as simply an inductive substituent. Resonance has been "inhibited" because of the steric effect of the substituent.

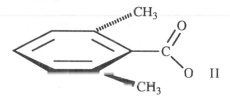

COOH group is not parallel with
the plane of the benzene ring—
no resonance interaction.

this three-dimensional view down the C-C bond
between the COOH and the benzene ring shows
that COOH is twisted out of the benzene plane

The same phenomenon is observed in substituted anilines. Anilines are usually much weaker bases than aliphatic amines because of resonance overlap of the nitrogen's lone pair of electrons with the pi system of benzene. When that resonance is disrupted, the aniline becomes closer in basicity to an aliphatic amine. (See problem 19-52(c).)

More examples on the next page.

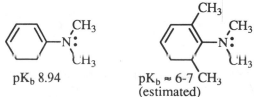

pK$_b$ 8.94

pK$_b$ ≈ 6-7
(estimated)

Appendix 2 continued, Summary of Acidity and Basicity

Examples of steric inhibition of resonance:

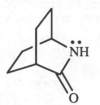

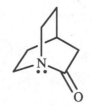

amide—not basic because of
resonance sharing of N lone
pair with carbonyl

strong base similar to aliphatic amine;
geometry of bridged ring prevents
overlap of N lone pair with carbonyl

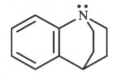

pK$_b$ 8.9

3° aromatic amine

pK$_b$ 3.4

3° aliphatic amine

pK$_b$ 6.2

3° amine, and the N is bonded
to a benzene, but the bridged
ring system prevents overlap
of N lone pair with benzene

$\ddot{N}(CH_3)_2$

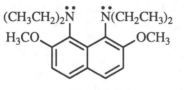

$(CH_3CH_2)_2\ddot{N}$ $\ddot{N}(CH_2CH_3)_2$

H$_3$CO ⟍   ⟋ OCH$_3$

pK$_b$ 8.9

pK$_b$ − 2.3 (yes, negative!)

Not only is steric inhibition of
resonance important in this example,
but so is intramolecular hydrogen-
bonding in the protonated form. Draw a
picture.